Chemical Modifications of Graphene-Like Materials

Chemical Modifications of Graphene-Like Materials

Nguyen Thanh Tien
Can Tho University, Vietnam

Thi Dieu Hien Nguyen
Vo Khuong Dien
Wen-Dung Hsu
Shih-Yang Lin
Yu-Ming Wang
Ming-Fa Lin
National Cheng Kung University, Taiwan

World Scientific

NEW JERSEY • LONDON • SINGAPORE • BEIJING • SHANGHAI • HONG KONG • TAIPEI • CHENNAI • TOKYO

Published by

World Scientific Publishing Co. Pte. Ltd.
5 Toh Tuck Link, Singapore 596224
USA office: 27 Warren Street, Suite 401-402, Hackensack, NJ 07601
UK office: 57 Shelton Street, Covent Garden, London WC2H 9HE

British Library Cataloguing-in-Publication Data
A catalogue record for this book is available from the British Library.

CHEMICAL MODIFICATIONS OF GRAPHENE-LIKE MATERIALS

ISBN 978-981-12-6793-2 (hardcover)
ISBN 978-981-12-6794-9 (ebook for institutions)
ISBN 978-981-12-6795-6 (ebook for individuals)

For any available supplementary material, please visit
https://www.worldscientific.com/worldscibooks/10.1142/13188#t=suppl

Desk Editor: Rhaimie Wahap

Typeset by Stallion Press
Email: enquiries@stallionpress.com

PREFACE

Graphene-like materials, with their remarkable properties and immense potential across multiple scientific disciplines, have attracted the attention of researchers in the fields of condensed matter physics, chemistry, and materials science. These materials have shown promise for a wide range of applications, ranging from energy storage to nanoelectronic, optoelectronic, and photonic devices.

Chemical modifications of graphene-like materials can further tune its physicochemical properties, as well as enhance its stability and functionality. Various types of chemical modifications have been explored, such as hydrogenation, fluorination, oxidation, and doping. These modifications can change the band gap, the Poisson's ratio, the Young's modulus, electric conductivity and the thermal conductivity of materials. They can also introduce new functionalities, such as magnetism, catalysis, and biocompatibility. Chemical modifications of materials can be used for various purposes, such as enhancing the performance, functionality, or compatibility of the materials for specific applications. However, chemical modifications can also have some drawbacks, such as affecting the stability, toxicity, or environmental impact of the materials. Therefore, chemical modifications of materials should be carefully designed and controlled to achieve the desired outcomes.

This book, "**Chemical Modifications of Graphene-Like Materials**", delves into the interesting world of graphene-like materials, offering a comprehensive exploration of their intriguing properties and the possible applications they hold. Through the lens of first-principles calculations and the grand quasiparticle framework, the diverse phenomena that take place when these materials undergo chemical transformations were thoroughly explored. Our aim is to unravel the critical mechanisms, orbital hybridizations, and spin configurations that arise from chemical adsorptions, intercalations, substitutions, decorations, and heterojunctions. The book also provides a comprehensive overview of the recent advances in the chemical modifications of graphene-like materials and its derivatives. It covers the synthesis, characterization, and applications of chemically modified graphene-like materials in various fields, such as energy, catalysis, sensors, biomedicine, nanoelectronics, and nanocomposites. The book also discusses the challenges and opportunities for the future development of graphene-like materials and devices.

This book is organized into 21 chapters, each meticulously written to provide a comprehensive understanding of the geometric, electronic, magnetic, and optical properties of 3D, 2D, 1D, and 0D graphene-like materials with various chemical modifications. It also explores recently discovered pentagonal materials, which exhibit asymmetric five-ring structures and offer novel physical effects. Throughout these chapters, we do not provide a thorough discussion of the basic concepts of physics and materials science. Instead, our aim is to provide clear and comprehensive guidance on how to assess and analyze first-principles results. By doing so, we ensure that readers can easily grasp the theoretical framework we present. The book is intended for researchers, students, and engineers who are interested in the chemistry and physics of graphene-like materials and its applications. It is also a valuable reference for anyone who wants to learn more about this fascinating nanomaterial and its potential impact on science and society.

This book is the result of collaborative efforts between research groups in Taiwan (Headed by Prof. Ming-Fa Lin) and Vietnam (Headed by Assoc. Prof. Nguyen Thanh Tien), representing a diverse range of contributors, including distinguished professors, post-doctoral researchers, and talented graduate students. It is our sincere hope that "Chemical Modifications of Graphene-Like Materials" will serve as a valuable resource and inspire future explorations, innovations, and discoveries in the captivating world of graphene-like materials.

We extend our gratitude to colleagues who reviewed and provided feedback on this book. This work was supported by the Hierarchical Green-Energy Materials (Hi-GEM) Research Center, from the Featured Areas Research Center Program within the framework of the Higher Education Sprout Project by the Ministry of Education (MOE) and the Ministry of Science and Technology (MOST 111-2112-M-006-020) in Taiwan. This work is also supported by the National Foundation for Science and Technology Development (NAFOSTED) in Vietnam through project code 103.01-2020.16 and the Ministry of Education and Training of Vietnam through project code B2020-TCT-14. The last but not least, we express our thanks to Lakshmi Narayanan, Rhaimie Wahap, and other staff of World Scientific Publishing for the pleasant collaboration and guidance.

Ming-Fa Lin
National Cheng Kung University, Taiwan

Nguyen Thanh Tien
Can Tho University, Vietnam

CONTENTS

CHAPTER 1

INTRODUCTION

Nguyen Thanh Tien*, Vo Khuong Dien†, Nguyen Thi Han†,
Nguyen Thi Dieu Hien† and Ming-Fa Lin‡

**Department of Physics, College of Natural Sciences, Can Tho University, Can Tho, Vietnam*
†Department of Physics, National Cheng Kung University, Tainan, Taiwan
‡Hierarchical Green-Energy Materials (Hi-GEM) Research Center, National Cheng Kung University, Tainan, Taiwan

The periodic table,[1] which clearly presents the unusual atomic configurations of all pure elements, can provide a concise overview of all the materials on Earth. These elements exhibit close relationships among the rich phenomena of physics,[2] chemistry,[3–5] and material sciences,[5] owing mainly to their diverse multi-orbital hybridizations[6–8] and spin configurations.[9] From Group I to Group VIII, their active orbitals in condensed-matter materials only come from (s, p_x, p_y, p_z) ones, such as (2s, $2p_x$, $2p_y$, $2p_z$),[10] (3s, $3p_x$, $3p_y$, $3p_z$),[11] (4s, $4p_x$, $4p_y$, $4p_z$),[12] (5s, $5p_x$, $5p_y$, $5p_z$),[13] and (6s, $6p_x$, $6p_y$, $6p_z$)[14] for the Group-IV C, Si, Ge, Sn, and Pb atoms, respectively. Group I and Group VII, respectively, display the largest charge transfers and the strongest electron affinities. However, the Group-VIII inert gases only show the very chemical bondings, that is, it is very difficult to generate many mainstream compounds from such atoms. The transition-metal atoms have the outermost orbitals of 3d, 4d, 5d, and 6d, in which each d-orbital wave function is characterized by d_{z^2}, $d_{x^2-y^2}$, d_{yz}, d_{zx}, d_{xy}.[15] Moreover, the rare-earth metal atoms present the active 4f and 5f orbitals,[Refs; Refs] with the featured wave functions of (f_{x^3}, f_{xz^2}, f_{yz^2}, f_{xyz}, $f_{z(x^2-y^2)}$, $f_{x(x^2-y^2)}$, $f_{x(x^2-3y^2)}$).[16] Most of the d and f orbitals are able to generate the spin-up and/or spin-down density arrangements and thus create the ferromagnetic or antiferromagnetic configuration. The above-mentioned sp^3, d^5, and f^7 orbitals are quite adequate for producing a super-giant superposition of various chemical bondings and spin configurations, that is, they are capable of creating many materials with greatly diversified charge[3] and spin-density distributions.[3,4] The orbital hybridizations could be divided into van der Walls interactions, covalent bondings, metallic bondings, and ionic forms, where the first, second, and third configurations might coexist in the specific regions of a condensed-matter system (e.g., stage-n graphite alkali compounds).[17,18] More interestingly, the spatial charge density distribution of each isolated atom reveals a uniformly spherical form. After the formation of a crystal structure, the position-dependent chemical bonds in a primitive unit cell will determine the critical multi-orbital hybridizations and spin configurations. Concise pictures can be achieved through delicate first-principles calculations and high-resolution experimental examinations. This is very useful for developing a grand quantum quasi-particle framework. For example, first-principle simulations have successfully identified the complete orbital hybridizations in graphite metal compounds, covering the intralayer σ-sp^2 bonding in a planar honeycomb crystal,[Refs] the interlayer van der Waals interactions between two neighboring graphitic sheets (the interlayer $2p_z$–$2p_z$ orbital hybridizations),[19] and the interlayer host–guest metallic bondings.[17,20] Furthermore, the metallic charge density distributions, respectively, belong to (I) 2s–/3s–/4s–/5s–/6s–$2p_z$ for Li–/Na–/K–/Rb–/Cs–C bonds, (II) (d_{z^2}, $d_{x^2-y^2}$, d_{yz}, d_{zx}, d_{xy}) for transition-metal–carbon orbital hybridizations, and (II) (f_{x^3}, f_{xz^2}, f_{yz^2}, f_{xyz}, $f_{z(x^2-y^2)}$, $f_{x(x^2-y^2)}$, $f_{x(x^2-3y^2)}$)–$2p_z$ for rare-earth–carbon bondings. These critical data are very useful for fully understanding the essential quasi-particle properties and promoting the applied sciences. Their collection together can build an international database of quantum particles and will be very helpful for all scientific endeavors.

A number of pure-element crystals[21] and binary,[22] ternary,[6,7,9] quaternary,[23] and pentanary compounds[24] have been successfully synthesized using various methods of chemical,[25] physical,[26] and material engineering.[27] Many related materials have become very suitable for thoroughly exploring the diversified quasi-particle phenomena, which constitute an outstanding research strategy. For example, carbon-based mainstream systems clearly present dimensionality-,[10] hexagon-,[28] stacking-,[29,30] layer-number-,[31,32] quantum-confinement-,[33,34] open-edge-,[35,36] hollow-cylinder-,[37] torus-,[38] onion-,[39] and ring-enriched[40] crystal symmetries, with strong relationships among them through position-dependent carbon–carbon interactions. These featured materials cover three-dimensional (3D) diamond (Fig. 1.1), AA-,[41] AB-,[42] or ABC-[43]

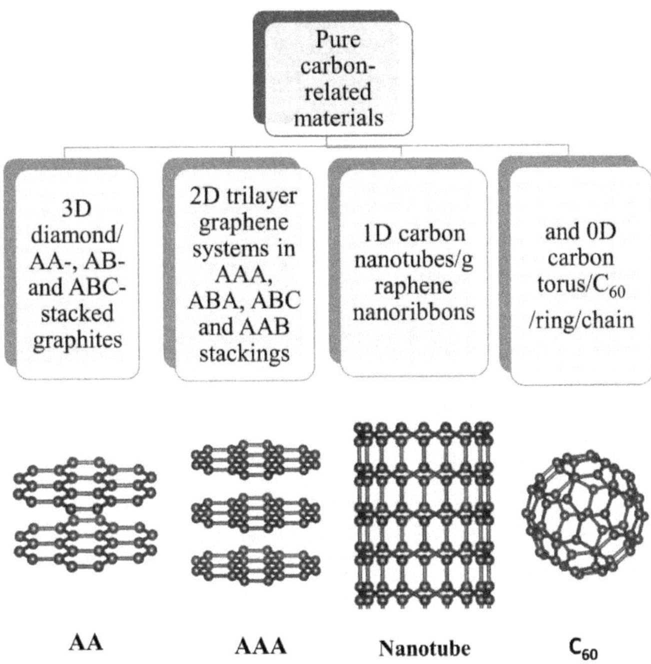

Fig. 1.1. Pure carbon-related materials: 3D diamond; AA-, AB-, and ABC-stacked graphite; 2D trilayer graphene systems with AAA, ABA, ABC, and AAB stackings; 1D carbon nanotubes/graphene nanoribbons; and 0D carbon torus/C60/ring/chain.

stacked graphite systems, carbon nanotube bundles,[44] cubic carbon fullerene crystals,[45] two-dimensional (2D) layered graphene with the distinct AAA,[46] ABA,[47] ABC,[48] or AAB[49] stackings, one-dimensional (1D) graphene nanoribbons with achiral and chiral edges,[50] cylindrical single-/multi-walled/coaxial carbon nanotubes,[44, 51, 52] zero-dimensional (0D) carbon tori,[53] carbon onions,[54] C60-related fullerenes,[55] carbon rings,[20] carbon lines,[56] and carbon disks.[57] More interestingly, each cylindrical carbon nanotube could be regarded as a rolled-up graphene along a specific direction, and a finite-width graphene nanoribbon is cut along the longitudinal direction of graphene. The close relations of geometric structures are successfully realized through experimental syntheses.[58, 59] The unusual crystal symmetries are directly reflected in the diverse quasi-particle phenomena (details in Section 2.2) in terms of electronic/magneto-electronic properties, Coulomb-field excitations/magneto-electronic excitations/static screening behaviors, optical transitions/magneto-optical selection rules, quantum Hall transports, and quantum phonons. The pure element crystals are able to further generate more compounds through strong chemical modifications (details in Section 2.1):

(1) chemisorption of guest adatoms or functional molecules on surfaces,[3, 4]
(2) intercalations between graphitic sheets/carbon nanotubes,[60, 61]
(3) substitutions of C-host atoms by suitable guest ones (e.g. B and N guest atoms),[62, 63]
(4) decorations on open edges (planar/curved/folded/scrolled graphene nanoribbons),[4, 35] and
(5) heterojunctions (few-layer graphene systems on the different substrates).[64]

Several popular commercial products, including semiconductor compounds/semiconductors (Fig. 1.2) and green-energy materials (Fig. 1.3), play critical roles in today's industries[65] and economics.[66] Furthermore, they are outstanding candidates for basic[67] and applied sciences.[68] For example, 3D silicon,[69] germanium,[70] binary III–VI,[71]

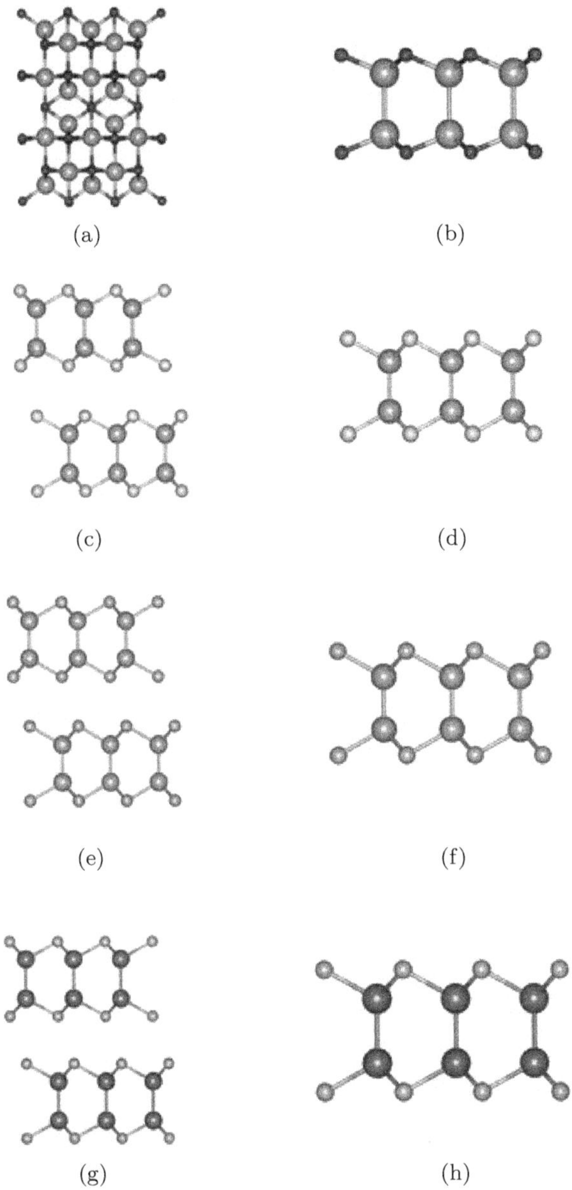

(a) (b) (c) (d) (e) (f) (g) (h)

Fig. 1.2. The 2D/3D III–VI semiconductor compounds: (a)/(b) gallium oxygen, (c)/(d) gallium sulfur, (e)/(f) gallium selenium, and (g)/(h) indium selenium.

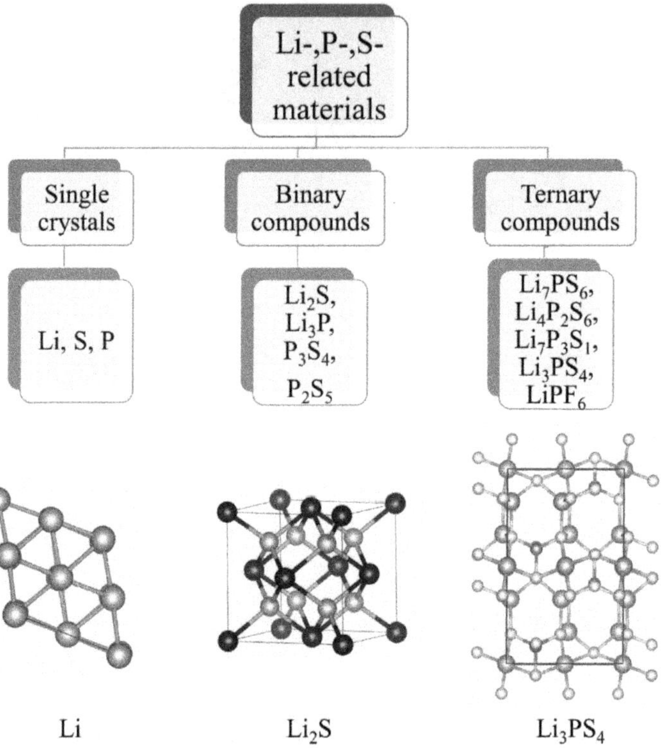

Fig. 1.3. The lithium-, phosphorus-, and sulfur-based materials: single-element crystals, binary compounds, and ternary ones.

III–V,[72] II–VI,[71] II–V, and IV–IV[73] compounds clearly exhibit semiconducting behavior. In general, the significant chemical bonds that belong to pure or impure sp^3–sp^3 multi-orbital hybridizations of covalent bondings are very sensitive to crystal symmetries. 3D binary compounds clearly reveal diamond- and/or graphite-like structures or other 3D cubic systems. The 3D layered crystals possess graphene-like vertical interlayer interactions but have buckled honeycomb lattices (the four-sublattice buckling; Fig. 1.1). The stacking configuration and the buckling degree are responsible for the unique dimensionality crossover behavior. A number of few-layer semiconductor compounds with distinct stackings have been successfully predicted using first-principles calculations.[74] The geometric mechanisms, the hexagonal or cubic symmetry, dimensionality, buckling, layer number, stacking, and the ratio among distinct atoms can determine the rich orbital hybridizations and thus create diverse quasi-particle phenomena. These are being currently investigated and are expected to be very useful in greatly expanding the quasi-particle framework.[3,4,32] On the green-energy side, the ternary lithium sulfide/oxide compounds[7,75] frequently serve as cathode, electrolyte, or anode materials in Li^+-ion-based batteries, owing mainly to their low cost, stability, and high-performance and wide-voltage factors. These 3D systems display semiconducting behavior almost without any free carriers at low temperatures. Their close partners mainly come from atom-built crystal structures. There are a number of lithium-, phosphorus-, and sulfur-enriched 3D condensed-matter systems (Fig. 1.3) formed by changing the atomic position, relative concentrations, and types of atoms, such as the BCC lithium crystal (Fig. 1.3), black and blue phosphorus crystals, most- and meta-stable sulfur crystals, binary lithium sulfide, binary lithium phosphorus materials, binary phosphorus sulfur of $P4S_x$ ($x < 10$), ternary lithium phosphorus sulfur compounds of $Li_2P_2S_6/Li_4P_2S_6/Li_7P_3S_{11}/Li_3PS_4$, and quaternary lithium phosphorus sulfide chloride. Such diverse crystals are also found in Li-, Fe-/Co-/Ni-O-related materials. These may generate interesting Moire superlattices with many atoms/active orbitals in a unit cell and further lead to the featured quantum quasi-particles in electronic,[3] optical, and phonon properties.[2,4] The unusual features are very helpful for highly potential applications in electronic/optical devices (details in Section 1.8).

Chemical modifications (details in Section 2.1; Fig. 1.4) are very powerful for producing a plenty of binary,[76] ternary,[77] quaternary,[78] or pentanary[28] compounds with specific functionalities. For example, monolayer graphene, silicene, germanene, tinene, and plumbene, with the planar or buckled honeycomb crystals,[12,79–81] are very suitable for alkalization,[3,4] hydrogenations,[82] oxidizations,[3] and halogenations.[83] The total changes in the chemical bondings and spin configurations are responsible for the dramatic transformations of the quantum quasi-particles. According to first-principles predictions, there are high-density free conduction electrons and

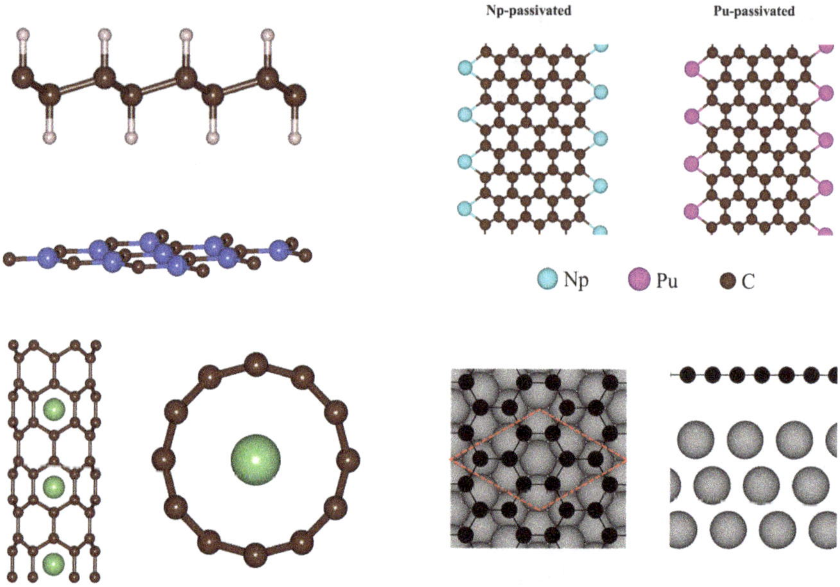

Fig. 1.4. Chemical modifications due to absorption, substitution, intercalation, decoration, and heterojunction.

valence holes, respectively, in alkali-[84] and halogen-adsorbed[85] graphene systems. The hydrogenated graphene compounds exhibit buckled crystal structures and insulating behavior because of the very prominent s–sp^3 covalent bondings between hydrogen and carbon atoms.[82] The 3D layered graphitic systems, which display well-behaved AA,[86] AB,[87] and ABC[88] stackings, are outstanding candidates for chemical intercalations/de-intercalations. They are used as very cheap anode materials in lithium-ion-based batteries,[5] where their high-performance operations arise from stable honeycomb crystals and the available inter-plane spacings. During the charging processes, electrons and lithium ions reach graphite simultaneously, and the latter will present the almost neutral configurations after the prominent charge screenings of the former. The time-dependent intermediate crystal structures, which will be achieved from the simulation of molecular dynamics, are expected to clearly illustrate Moire superlattices[89] and greatly diversified quasi-particle behaviors. $B_xC_yN_z$ compounds, which are built from the relative substitutions of boron, carbon, and nitrogen atoms,[90] can form single-element, binary, and ternary compounds. Their crystal phases include 3D/2D/1D tubes and nanoribbons or 0D crystal structures. Most of them belong to zero-, narrow-, middle-, and wide-gap semiconductors[90,91] and only a few exhibit metallic behavior.[91] Such semiconducting materials are considered the third generation from the viewpoint of the semiconductor industry.[92] The basic and applied sciences should be thoroughly explored through experimental observations,[93,94] first-principles simulations,[95,96] and phenomenological models.[97,98] Chemical decorations are frequently utilized at open edges with a number of very active dangling bonds. The 1D planar/curved/folded/scrolled graphene nanoribbons[4,33,35] have successfully become outstanding materials for examining their diverse quantum quasi-particles. Their hydrogen-atom passivation is frequently done before the experimental measurements of the essential properties since the bond-length fluctuation is greatly reduced for the stationary quasi-particles. The various guest adatoms are predicted to show three kinds of dressed edge structures (details in Section 2.1.4), in which the multi-orbital hybridizations and spin-dependent interactions of host–guest and host–host bonds are thoroughly identified as the critical mechanisms. Such interesting results could be generalized for the buckled open edges of silicene, germanene, tinene, and plumbene nanoribbons. Finally, heterojunctions, which are the composite boundaries associated with two or three subsystems,[99] can display rich and unique quantum quasi-particles. The semiconductor growths, which are efficiently produced by molecular beam epitaxy (MBE),[100] are successfully synthesized on various substrates (details in Section 2.1.5). The battery operations are realized through the charging and discharging processes of cathode,[101] electrolyte,[102] or anode[103] materials. The significant heterojunctions clearly illustrate the time-dependent exits and entrances of lithium atoms, so that these high-performance materials could exhibit a number of intermediate crystal phases.[63] Any composite condensed-matter system clearly shows the left-hand subsystem, middle heterojunction, and right-hand subsystem, with highly non-uniform chemical/physical environments.[63,104] Through thorough first-principles examinations and analyses, these emergent materials display slow modulation phenomena along the longitudinal and transverse directions in terms of charge and spin-density distributions, e.g., the greatly diversified orbital hybridizations and spin configurations of monolayer/bilayer silicene on Ag (111). In this book, the chemical modifications are mainly focused on the strong partnerships between the carbon-based systems and the transition-metal and rare-earth guest adatoms (3d/4d/5d/6d and 4f/5f orbitals). Such a strategy is very helpful for developing a grand quasi-particle framework.

Physical perturbations (Fig. 1.4; details in Section 2.2), which are fully assisted by the environmental symmetries of any condensed-matter system, can establish strong relations with intrinsic interactions due to charges and spins. More rich quantum quasi-particles will be very useful in expanding their framework.[2,3] The current successful external fields include stationary electric/magnetic fields,[32,105,106] high-energy electron beams,[105] electromagnetic waves,[86] mechanical stresses,[107] and thermal excitations.[21] A uniform perpendicular magnetic field is able to create rich and unique magnetic quantization in terms of the featured Landau levels,[32] diverse inter-LL electron–hole excitations and magneto–plasmon modes,[105] different magneto-optical selection rules,[108] and unusual quantum Hall effects.[62] The vector-potential-based viewpoints fully illustrate concise physics pictures. These magneto-quasiparticles are greatly diversified through further superposition with gate voltages (the split magneto-electronic states). It should be noted that the spatially modulated magnetic and electric fields/composite fields can also create drastic changes in the essential properties. The external field superposition is consistent with the linear response ability of charges/spins. The high-energy electron beams provide the time-dependent Coulomb-field perturbations. The screening charge fluctuations, which are created by all the energy-dependent orbitals, will determine the induced and total fields. The bare response ability is well characterized by the longitudinal dielectric function. The strong momentum and frequency dependences clearly indicate electron–electron Coulomb scatterings. This quasi-particle viewpoint is very useful in exploring and understanding the diverse (momentum, frequency)-excitation phase diagrams with the coexistent electron–hole pair regions and plasmon modes. The diversified electron–electron interactions are achieved through magnetic/electric fields, n- or p-type dopings, stacking configurations, layer numbers, and dimensionalities. An incident electromagnetic wave is able to induce significant couplings with all charges. The charge dynamics are well described by the Helmholtz equation, with a linear relationship between the perturbation and effective fields. The transverse dielectric function is well defined by the relative ratio of the electric field before and after charge screenings. With all charge screenings, the screened response function (the energy loss function) presents strong plasmon structures under the Landau dampings at long-wavelength limits. The single-particle and collective excitations are directly reflected in the reflectance spectra and absorption coefficients. Whether the many-body excitons could survive during the vertical optical transitions is mainly determined by the semiconducting behavior, but not the metallic/semi-metallic one. Under the perturbations of an external field, e.g., temperature, the ion cores move around their equivalence center. The vibrations of ions satisfy the Hamilton equation of motion and Hooke's law. The energy dispersions and phonon wave functions could be achieved by diagonalizing the dynamical matrix, which mainly involves the determination of the force constant tensors. The phonon properties of materials are directly related to the thermal properties of materials, such as heat capacity and thermal conductivity. Furthermore, information about the phonon is also very important to understand the coupling of electron and phonon in the normal and supercoiling phenomena. Under the perturbation of external mechanical stress, the geometric and thus the electronic and optical properties of low-dimensional materials are gradually changed. Homogeneous and non-homogeneous mechanical stresses are the two main types of mechanical modification, which usually appear in theoretical models or during experimental processes. Until now, a number of studies based on mechanical stress are one of the mainstream candidates and are expected to exhibit diverse quasi-particle phenomena. Such kinds of physical perturbations could be considered in the current development of a quasi-particle framework. Many studies based on the first-principles method have been considered for the strain effect on systems. For example, the effects of mechanically deforming graphene/graphite remain to be explored in detail and have been published only in a few previous works.[107,109]

The simultaneous progress among first-principles simulations,[110] phenomenological models,[111] and experimental measurements[112] are is the main focus of this book. The Vienna Ab-initio Simulation Package (VASP) calculations for all the orbitals of s, p_x, p_y, p_z, d^5, and f^7 are successfully described in Chapters 3–17. The multi-/single-orbital hybridizations and spin configurations are thoroughly explored through delicate calculations and analyses. The calculated results are consistent with one another within a unified quasi-particle framework:

(1) highly position-dependent chemical bonds in a unit cell,
(2) featured band structures and wave functions with the strong atom and spin dominances within the specific energy ranges,
(3) spatial charge density distributions in the covalent, metallic, van der Waals, and ionic forms,
(4) atom-, orbital-, and spin-decomposed magnetic moments,
(5) ferromagnetic or antiferromagnetic spin configurations,
(6) atom-, orbital-, and spin-projected density of states,
(7) optical transitions with/without excitonic effects, and
(8) phonon spectra phonon. The drastic deformations of orbital-dependent probability distributions and the merged van Hove singularities, respectively, qualitatively and quantitatively, provide critical information about the featured orbital hybridizations and spin configurations (detailed in a series of previously published books).[2–4]

More interestingly, the fundamental properties are greatly diversified into chemical adsorptions (details in Chapters 3 and 4), intercalations (Chapters 5 and 6), substitutions (Chapters 7 and 8), decorations (Chapters 9 and 10), and heterojunctions (Chapters 11 and 12), owing mainly to the dramatic transformations of orbital-bonding environments. The strong charge and spin relationships, which are clearly revealed in host–host, host–guest, and guest–guest bonds in the presence/absence of Moire superlattice, are built from the quasi-particle viewpoints. For example, those associated with 10-d and 14-f orbitals can be thoroughly explored under the different reactions, respectively, for 3d/4d/5d/6d transition metal atoms and 4f/5f rare-earth metal ones. The low-lying valence and conductions of VASP simulations are very useful in establishing the reliable parameters of the tight-binding model or the generalized tight-binding model,[2,113] including the orbital-dependent hopping integrals of the neighboring atoms,[32] on-site ionization energies,[32,105] and same-site Coulomb interactions from the spin-up and spin-down configurations (the Hubbard-like form). The occupied states are also available for conducting the high-resolution measurements of angle-resolved photoemission spectroscopy (ARPES)[114] especially for a number of valance energy subbands of Moire superlattices.[89,115,116]

In addition to the advanced viewpoints from the VASP simulations, the phenomenological models have been successfully explored in many books[2,32] and can clearly illustrate more concise pictures of the physical, chemical, and material sciences of high symmetries. For example, the generalized tight-binding model, modified random-phase approximations, and corrected self-energy method, which are conducted on the graphene-related layered honeycomb crystals (3D graphite/2D few-layer graphene/1D graphene nanoribbons/coaxial carbon nanotubes are thoroughly developed for magnetic quantization, (momentum, frequency)-Coulomb-excitation phase diagrams, and the unique wave vector/energy dependences of Coulomb decay rates. The complicated couplings mainly arise from the external magnetic\electric fields, intralayer/interlayer orbital hybridizations, intralayer/interlayer electron–electron Coulomb interactions, n- or p-type dopings, and thermal excitations. Such critical factors can be taken into consideration very efficiently since the Bloch wave functions the layer- and sublattice-induced tight-binding functions. All the following theories present close relationships even with the direct combinations of static and dynamic Kubo formulas (details in Chap. 2.2, 3), e.g., the featured Landau levels, the (momentum, frequency)-enriched magneto-electron–hole pairs and magneto-plasmon modes, the unique quantum Hall transports, and the diversified magneto-optical selection rules.[108,117] How to establish the reliable parameters of the tight-binding/generalized tight-binding models is currently being investigated. A greatly expanded quasi-particle framework could be achieved by the hydrogenations, alkalizations, oxidizations, and halogenations,[3] as well as other chemical modifications of graphene-related systems (Sections 2.1.1–2.1.5). Of course, the analytic models also possess several drawbacks, mainly due to Moire superlattices and complex orbital hybridizations and spin configurations. The effective tight-binding models of the ternary lithium oxides in ion-based batteries have almost never appeared in the published scientific works, while the VASP simulations (details in books) clearly illustrate a number of valence and conduction energy subbands close to the middle or large bandgaps. Similar issues might arise in other mainstream materials (details in Chapter 20). They need to be solved through the solid wav of step by step, e.g., directly linking with the high-performance AI techniques. The high-precision experimental measurements can observe, examine, and identify the various quasi-particle behaviors. X-ray diffraction,[118] reflection low-energy electron diffraction (RLEED),[119] tunneling electron microscopy (TEM),[120] and scanning tunneling microscopy (STM)[121] are very powerful for measuring crystal symmetries. The first and second techniques are suitable for 3D bulk condensed-matter systems, e.g., the unique chemical bonds of ternary lithium sulfides/oxides in ion-based batteries,[122,123] while the third and fourth methods can, respectively, verify the top and side views of the low-dimensional surface morphologies (e.g., 2D graphene/1D graphene nanoribbons and carbon nanotubes/0D carbon disks). The direct combinations of STM and STS clearly illustrate the close relationships between the geometric and electronic properties, where the latter is able to test the main features of the energy-dependent density of states. Their accurate measurements and analyses are very successful in the geometry-enriched bandgaps, such as, the metallic, narrow- and middle-gap carbon nanotubes, respectively, corresponding to $m = n$ (armchair),[4] $2m + n = 3I$, and others (each (m, n) carbon nanotube).[124] The angle-resolved photoemission spectra are available for the energy-spectrum measurements of occupied electronic states under normal physics conditions. Their powerful abilities are directly reflected in very thin/few-layer samples because of the negligible non-conservation of the perpendicular momentum (the minor surface effect), such as in the very successful ARPES measurements on layered graphene[125] and planar graphene nanoribbons.[126] Moreover, the inverse of the measured ARPES width corresponds to the quasi-particle lifetime, as done for alkali-doped graphene systems. On the other hand, the measured ARPES spectra almost disappear in 3D battery materials and 1D carbon nanotubes up to now since the former and the latter, respectively, possess Moire superlattices (too many valence subbands with strong energy dispersions) and non-straight geometries (without the well-defined momentum). STS,[127] an extension of STM,[127] is capable of examining the band property near the Fermi level and the energy-dependent van Hove singularities.[128] The high-precision measurements of quantum tunneling currents should only be suitable for few-layer materials, e.g., the well-behaved relations between bandgap and width for 1D graphene nanoribbons. The spin-polarized STS is developed to observe the ferromagnetic and antiferromagnetic spin-density distributions. The reflection/transmission electron energy loss spectroscopy can observe the (momentum, frequency)-dependent Coulomb-excitation spectra, The screened response functions will present the prominent plasmon peaks due to the coherent charge oscillations and the coexistent Landau dampings associated with the single-particle electron–hole pairs. The 3D graphite systems are verified to show the low-, middle-, and high-energy plasmon modes, respectively, from free carriers, π electrons. and $\pi + \sigma$ electrons. Optical reflection, absorption, transmission, and photoluminescence spectroscopies can identify the prominent absorption transitions in the presence or absence of excitonic effects for middle- and wide-gap semiconductors or arrow-gap ones and metals. Highly successful measurements have been made for 3D Bernal graphite, 2D few-layer graphene systems with electric and magnetic fields, 1D carbon nanotubes, graphene nanoribbons, and 0D C60-related fullerenes.

However, the optical measurements on the lithium oxide/sulfide compounds seem to be too rough, so that their critical excitation mechanisms have not been found until now. The experimental analyses require strong cooperation with the delicate VASP simulations (details in previously published books). How to establish close partnerships among the experimental measurements, VASP calculations, and phenomenological models is one of the focuses of the study.

The current book is organized as follows. Chapter 2 covers the chemical modifications and physical perturbations:

(1) absorptions, substitutions, intercalations, decorations, and heterojunctions; and
(2) the stationary magnetic and electric fields under the uniform/non-uniform superposition, high-energy electron beams, electromagnetic waves, mechanical stresses, and thermal excitations.

Such strong modulations are able to generate a number of semiconducting and metallic compounds and very efficiently drive the diverse quasi-particle phenomena. The VASP simulations can reach the final conclusions on the multi-/single-orbital hybridizations and spin configurations under a unified quasi-particle framework. The double-/single-side chemical adsorptions, which are conducted on monolayer graphene, are realized through the 3d-transition- and 4f-rare-earth-metal adatoms (Fe/Co and La/Gd), respectively, in Chapters 3 and 4. Five/seven kinds of 3d/4f orbitals and spin-up and spin-down configurations are simultaneously responsible for the van Hove singularities, the spatial charge/spin-density distributions, and the net magnetic moment. Whether metallic or covalent bondings dominate the low-energy quasi-particle behavior is thoroughly discussed. Chapters 5 and 6 clearly illustrate 4d-transition- and 5f-rare-earth-metal graphite intercalation compounds (GICs; Fe/Co and Ac/Th). The critical roles of intercalant-carbon bonds are achieved from all calculated results, and their strong effects on n-type doping are fully explored by modulating the guest-atom concentrations and arrangements. Certain important differences with alkali- and magnesium-related compounds can be obtained in detail. The close relations between the first-principles method and the tight-binding model are established by the carbon–carbon, carbon–intercalant, and intercalant–intercalant bonds. he 5d-transition- and 4f-rare-earth-meta-substituted graphene systems are, respectively, investigated in Chapters 7 and 8 (Fe/Co and Ac/Th). The pure sp^2 bondings become rather complicated impure sp^3–d^5/sp^3–f^7 ones (nine/eleven active orbitals in host/guest bonds).[Refs] How to accurately calculate and analyze the spatial charge/spin-density distributions is a very interesting question. These sensitive mechanisms are responsible for the unusual quasi-particle behaviors. Chapters 9 and 10, respectively, focus on the chemical decorations of graphene nanoribbon edges through the 6d-transition- and 4f-rare-earth-meta adatoms (Fe/Co and Ac/Th). The passivated carbon–adatom bonds are responsible for dramatic changes in the open-edge structure and thus the unusual quantum quasi-particles. Chapters 11 and 12, respectively, explore the unique heterojunctions of mono-/bi-layer graphene on transition- and rare-earth-metal substrates. Obtaining strong partnerships among three specific regions with different orbital hybridizations and spin configurations is an urgent requirement under a unified quasi-particle framework. The commensurate degree will determine the calculation efficiency and the spatial charge/spin-modulation behaviors. Low-dimensional materials present outstanding contributions to the basic and applied sciences. Pentagon quantum dots,[129] graphene quantum dots (GQDs),[130,131] and silicene–carbon-related systems are exceptional candidates for those groups due to their interesting applications in optoelectronic devices, sensors, and energy storage. The hybridization properties among σ, π, sp^2, and sp^3 play substantial roles in the comprehensive research of the featured properties of condensed-matter systems. Quantum dot materials possess discrete energy along with a wave-vector-dependent energy range, leading to restrictions in their applications. Therefore, chemical modifications offer innovative techniques to efficiently modify the geometric and electronic characteristics of core materials. To illustrate, hydrogenated pentagon-graphene quantum dots (PGQDs) have been thoroughly examined in diverse structural stabilities, including 12 H, 21 H, and 36 H compounds, which likely involve the opening of the bandgap up to 4.42, 3.94, and 3.60 eV,[129] respectively. The strong dominant chemical bondings of s–sp^3 and sp^3–sp^3 lead to buckling structures in the pentagon lattice. Such important properties are suitable for broadening the practical applications, particularly in nanoelectronics, engineering devices, and the energy industries. Besides the 0D quantum dots, Group-IV materials and related compounds have recently been the focus of extensive research. The formation of carbon and silicon can produce diverse layered nanomaterials, which can be denoted as Si_xC_y (where x and y represent the contributions of Si and C in the compounds). SiC and SiC_2 are two renowned chemical materials that possess planar structures with rich and unique properties. The pentagonal SiC_2 nanoribbons belong to interesting emerging 1D compounds that exhibit metallic behavior,[132,133] while SiC is predicted to have 2D graphene-like honeycomb structures in this research.[134,135] The strong competition between sp^2 from C and sp^3 preferred by Si might display rich physical, chemical, and material properties. Consequently, low-dimensional compounds are worth investigating due to their exceptional characteristics and significant applications. The highly potential application fields described in Chapter 19 include electronics, optics, green energy, chips, 3D printing, biology, medicine, and agriculture. Chapter 20 thoroughly explores open issues and near-future focuses: emergent materials, the time-dependent LDA of Coulomb-field excitations, semiconductor compounds, 3D diamond/silicon/germanium/tin/lead on binary/ternary oxides, inter-metallic compounds, ion transport, solar cells, hydrogen energy, and biochemistry. Such mainstream materials are expected to be rather outstanding candidates for developing a grander quasi-particle framework. Finally, the concluding remarks, which are focused on the close partnership between the theoretical predictions and experimental examinations, provide a very useful platform for expanding the grand quantum quasi-particle framework.

Since its discovery in 2004, graphene has always been viewed as an attractive material for various fields such as industry, agriculture, and environmental protection. Besides its attractive features, the fact that graphene is a semiconductor with zero bandgap is a huge obstacle that scientists need to consider. Many studies have been done on developing methods to change the bandgap of graphene, such as decoration and substitution, among which the adsorption method is considered one of the best methods. The choice of atom used for adsorption is one of the first things to consider when using an adsorption method. Previous research results have shown that transition metal atoms are a potential candidate because of their orbital hybridization with graphene. In Chapter 3, we have focused on studying the interesting electronic and magnetic properties of graphene after the adsorption of manganese and chromium atoms at different concentrations. We compared the binding energies, geometric structures, band structure, dominance of adatoms, the density of states, charge density, charge density difference, spin distribution, and magnetic moments of pristine graphene and adsorption samples. All results are computed using first principles based on the density functional theory (DFT).

The La and Ce atom-adsorbed graphene systems are investigated for a thorough understanding of the strong coupling effects associated with the very complicated quasi-particle 4f-orbital charges and spin configurations, such as the significant covalent bonds of C–sp^3 and La–$5d^5$/Ce–$4f^7$ orbitals and their non-magnetic/ferromagnetic spin configurations. Their quasi-particle behaviors are expected to present a sharp contrast with those of halogenations, alkalizations, oxidations, halogenations, and halogenations. How to commence the simultaneous progress among the VASP simulations, phenomenological models, and experimental observations is also one of the focuses of studies, as clearly illustrated in the previously published works.[6–8, 75, 136–138] Apparently, the geometric, electronic, and optical properties of the La and Ce adsorbed on monolayer graphene are determined using first-principles simulations based on the VASP software. The concentration and arrangement dependents are explored. The orbital hybridizations in the chemical bonding are achieved through a delicate analysis of the atom-dominated energy band structure, the spatial charge densities and charge density distribution, and the projected van Hove singularities in the density of states. The magnetic properties of the host and guest adatoms can be comprehensively studied through the spin-split/spin-degeneracy energy band structure, the partial/net magnetic moments, the spatial spin-density distribution, and the polarized density of states. Especially, the orbital character of the band-edge states was used to identify the optical excitations, e.g., the threshold frequency and the special prominence peaks in the optical adsorption spectrum; furthermore, the collective optical excitations could also be understood from the dielectric functions and electron energy loss functions. We can thus connect the electronic and optical excitations. All results in this chapter are required to be tested using high-resolution experimental techniques.[139,140]

GICs, which are inserted by foreign atoms or molecules into hexagonal graphene sheets of graphite, have attracted more attention because of their potential applications.[141–147] The electrons accepted or donated by the guest metallic intercalants could modify the fundamental properties of graphite. The defects induce magnetism in graphite due to the quasi-localized electronic states near the Fermi level, while pristine graphite belongs to a non-magnetic configuration. Furthermore, transition-metal atoms intercalated between graphene sheets of graphite could induce the s–p, s–d, and p–d hybridizations.[148,149] It indicates that a large magnetic moment might be present in GICs. Therefore, GICs are significant from the point of view of magnetic systems. The stable and tunable magnetism in GICs may lead to many technical applications, for example, temperature semiconductors and magnetic sensors, provided that the metal atom[150,151] intercalants exhibit structural and electronic modifications of graphite for wide applications. Inspired by these features, 4d transition-metal atoms (Zr, Nb) have become potential candidates for GICs due to their s and d outermost electronic configurations that could remarkably contribute to the hybridization in the system. Moreover, knowledge about the concept of transition-metal intercalation would be clarified in the system of 4d transition-metal intercalated graphite. The induced magnetism is expected to be found in these systems. Besides, the adjusted electronic and optical properties of the compounds are predicted to reveal Zr/Nb–C, which is suitable for energy storage or photoelectronic applications.

1D graphene nanoribbons, which consist of pure carbon atoms in the finite-width and open two-edge structurers, respectively, along the transverse and longitudinal directions (discussed later using Fig. 10.1)[152] can exhibit diverse quasi-particle phenomena.[153] More interestingly, such mainstream materials have successfully garnered simultaneous progress in first-principles simulations,[154] phenomenological models, and experimental observations. The unusual geometric symmetries mainly come from the highly position-dependent bond lengths, leading to small gaps in armchair and zigzag systems. Their cooperation with the zigzag-induced ferromagnetic/antiferromagnetic configuration on the same side/across the nanoribbon center leads to the splitting/a finite bandgap of the partially flat valence and conduction energy subbands near the zone boundary.[19] All 1D graphene nanoribbons, with any number of chiral edges,[155] belong to the middle- or small-gap semiconductors, where the magnitudes decrease with an increase in width. When the boundary dangling bonds are passivated by the guest adatoms, the essential quasi-particle properties are expected to display dramatic changes. The rare-earth metal atoms Pd and Pu with 5f orbitals are outstanding candidates for fully exploring the significant decoration effects, owing mainly to the seven types of probability functions (the coordinate-dependent cubic powers; details in Chapter 10). The very complicated orbital hybridizations and spin configurations closely related to C–Pd/C–Pu, C–C, and Pd–Pd/Pu–Pu bonds are fully examined under a unified framework. That is, the featured edge passivation,[156] the unique energy spectra and wave functions,[157] the unusual charge/spin-density distributions,[158] the atom-, orbital-, and spin-decomposed van Hove singularities, and the dielectric functions/energy loss spectra with stable/quasi-stable excitons. Their consistence is available for determining the critical mechanisms. This work is very useful in developing a grand quasi-particle framework.

Substitution is one of the popular methods used to enhance the essential properties of host systems. Chapter 7 describes 5d transition metals substituted on 2D graphene. The rich feature properties of this system focus on four elements, namely tungsten (W), osmium (Os), iridium (Ir), and platinum (Pt). The optimal stabilities, chemical electronic phenomena, magnetic configurations, and optical features exposed to unique physical and material properties dramatically modulate the potential applications. First-principles calculations have been applied to investigate the crucial chemical bonds of C–C and M–C (M is W, Os, Ir, or Pt) based on the critical single- or multi-orbital hybridizations. The chemical modifications which use substitution methods show diverse semiconducting and semi-metallic behavior. Moreover, the spin-split properties have been found in W and Os impure systems which have ferromagnetic configurations, while the Ir and Pt substitutions have shown non-magnetic characteristics. Remarkably, the complex dielectric functions are important physical quantities, which describe the charge screening abilities when responding to an electromagnetic field. The dielectric functions significantly depend on the wave-vector band structures and the joint densities of states from the initial to final states. The theoretical prediction of optical characteristics is confirmed using optical spectroscopy in experimental studies.

There are many interesting problems which are revealed from the periodic table, typically concerning atomic configurations that present many interesting physical and chemical properties of each atom in the periodic table. This has important implications for the basic and applied sciences. Its main properties are very useful in considering the electronic properties and many characteristics of any condensed-matter system. Specifically, the various geometric structures of many materials closely relate to the active orbitals of the single atom or pure elements. Based on theoretical and experimental research, the orbitals of atoms can be classified into three types according to the features of chemical bonding: (1) (s, p_x, p_y, p_z) orbitals in Group-I/-II/-III/-IV/-V/-VI/-VII/VIII systems, (2) (d_{z^2}, $d_{x^2-y^2}$, d_{xy}, d_{yz}, d_{xz}) orbitals in transition-metal atoms, (e.g., Ti,[159] Fe,[160] W,[161] Co,[162] Ni,[163] Zn,[164] Cu,[165] Cd,[166] Ag,[167] or Au[168] with 3d or 4d orbitals and (f_{x^3}, f_{xz^2}, f_{yz^2}, f_{xyz}, $f_{z(x^2-y^2)}$, $f_{x(x^2-3y^2)}$, $f_{y(3x^2-y^2)}$) orbitals of rare-earth metal atoms (such as La,[169] Ce,[170] Pr,[171] Nd,[172] Pm,[173] Sm,[174] Eu,[175] Gd,[176] or Lu with 4f orbital).

The diversified materials are constructed from these charge distributions through weak, strong, or dramatic single- or multi-orbital hybridizations, e.g. compounds with transition and rare-earth metals and lithium compounds; metal–graphene systems (transition/rare-earth metal substitution in[177] or absorption on a graphene system[178]); graphene-related systems (graphitic systems,[179] graphene nanoribbons,[50] carbon nanotubes,[124] layered graphene,[180] and carbon tori/fullerene;[181] binary semiconductor compounds (III–VI,[71] III–V,[182] II–VI,[183] and IV–IV[184]); and lithium oxygen- and sulfur-based mainstream materials. Generally, these condensed-matter systems exhibit highly anisotropic and non-uniform environments of chemical, physical, and material engineering, owing mainly to their very prominent covalent bonds. Both chemical modifications and physical perturbations can greatly diversify the composite quantum quasi-particles, in which the former and the latter, respectively, include intercalations, adsorptions, heterojunctions, substitutions, decorations, or composites and a uniform perpendicular magnetic field, electric field, thermal excitations, mechanical stresses and their spatially modulated forms, and the superposition of uniform and non-uniform fields, electromagnetic waves, or time- and position-dependent Coulomb fields. Chapter 8 considers rare-earth element substitution in graphene systems for a thorough understanding of the strong coupling effects associated with the very complicated quasi-particle 4f-orbital charges and spin configurations, such as the significant covalent bindings of C-sp^2/sp^3 and La-/Gd-4f orbitals and ferromagnetic/non-magnetic spin configurations of these systems. The quasi-particle behaviors of these material systems are expected to present a sharp contrast with those of alkalization, oxygenations, and halogenations. The calculations are done using first principles based on the DFT, with simultaneous progress in the VASP simulations. Besides, phenomenological models and experimental observations are also among the focuses of the study, as clearly illustrated in the previously published books.[32]

Graphene is a famous truly 2D material, possessing a cone-like energy structure near the Fermi level and treated as a gapless semiconductor. Its unique properties have motivated researchers to find applications for it. The gapless feature restricts the development of graphene nanoelectronics. Making 1D strips of graphene nanoribbons (GNRs) could be one of the promising routes to modulating the electronic and optical properties of graphene. The electronic and optical properties are highly sensitive to the edge and width. The tunability of electronic and optical properties further implies the possibilities of GNR application. However, the dangling bonds at ribbon edges remain an open question in GNR systems. Various passivations at the ribbon edge might change the physical properties. In this work, 5d transition-metal elements are considered the guest atoms at the edges. The geometric structure, energy bands, densities of states, charge distribution, and optical transitions are discussed.

1D graphene nanoribbons, which consist of pure carbon atoms in the finite-width and open two edge structurers, respectively, along the transverse and longitudinal directions, can exhibit the diverse quasi-particle phenomena. Very interesting, such mainstream materials have successfully provide the simultaneously progresses from the first-principles simulations [185, 186] the phenomenological models, and the experimental observations detail in books.[2, 105] The unusual geometric symmetries mainly come from the highly position-dependent bond lengths, leading to small gaps in armchair and zigzag systems. Their cooperation with the zigzag-induced ferromagnetic/antiferromagnetic configuration on the same side/across the nanoribbon center leads to the splitting/a finite bandgap of the partially flat valence and conduction energy sub-bands near the zone boundary. All 1D graphene nanoribbons, with any chiral edges [33, 35], belong to middle- or small-gap semiconductors, where the magnitudes decline in the increment of width. When the boundary dangling bonds are passivated by the guest adatoms, the essential quasi-particle properties are expected to display the dramatic changes. The rare-earth-metal atoms. Pd and Pu with 5f orbitals, are outstanding candidates in fully exploring the significant decoration effects, mainly owing to the seven types of probability functions [the coordinate-dependent cubic powers; details in Chap. 10]. The very complicated orbital hybridizations and spin configurations closely related to C-Pd/C-Pu, C-C and Pd-Pd/Pu-Pu bonds are fully examined under a unified framework. That is to say, the featured edge passivation [187, 188], the unique energy spectra and wave function [35, 83, 189, 190], the unusual charge/spin-density distributions [17, 190], the atom-, orbital- and spin-decomposed van Hove singularities, and the dielectric functions/energy loss spectra with stable/quasi-stable excitons. Their consistence is available in determining the critical mechanisms. This work is very useful in developing a grand quasi-particle framework.

The specific heterojunctions, which frequently arise in composite compounds,[191] are outstanding candidates for providing diversified quasi-particle phenomena and thus developing a grander framework. For example, layered Group-IV systems,[10, 12–14, 192] and the binary and ternary semiconductor compounds of III–VI,[193] III–V,[194] II–VI,[71, 195] II–V,[195] and IV–IV[196] have been successfully grown on various substrates. Furthermore, the composite cathode/electrolyte/anode materials clearly illustrate stationary ion transport through the real boundaries during the charging and discharging processes of ion-based batteries. Such mainstream materials present, at least, three different orbital-hybridization/spin-modulation regions[63, 99] because of the creation of significant host–guest chemical bonds within the complicated boundary. Generally, there exists a Moire superlattice with a number of active atoms, even for a commensurate crystal environment. This will lead to a giant barrier to the first-principles calculations and subsequent delicate analyses. All calculated results, which are consistent with one another within an expanded framework (detailed in previous books[3, 4]), are available for achieving the critical mechanisms associated with charge- and spin-dependent interactions. The concise chemical bondings and lattice-site-dependent spin-density distributions are determined from the optimal crystal symmetries, the featured band structure and wave functions with strong atom and spin dependences, the atom- and orbital-projected magnetic moments, the spatial charge/spin-density modulation distribution along the top and side views, the merged structures of the atom-/orbital-/spin-decomposed van Hove singularities, and the orbital- and spin-enriched optical excitations in the presence/absence of excitonic effects. In Chapter 11, monolayer and bilayer graphene systems of Pt[111] are chosen as a model study. Their fundamental properties are expected to reveal the dramatic transformations under the strong multi-orbital hybridizations and many-body interactions, e.g., the low-lying valence and conduction energy subbands due to Moire unit cells. It should be noted that too thick substrates would become meaningless in theoretical simulations since their electronic states cannot escape from their surfaces, even using high-resolution ARPES measurements.[114,197]

Cerium oxide has recently become a popular material for scientific studies due to its wide application as catalysts for the elimination of toxic gases,[198–201] the detection of oxygen gases,[200–203] and the use of fuel cells material.[204–208] The storage and transport properties of oxygen in CeO_2 are strongly related to these applications. The magnitude of electronic conductivity of CeO_2 could be largely enhanced when it becomes a nanocrystalline material.[209–212] The 5d and 4f orbitals of cerium are the critical factors in investigating the multi-hybridization of orbitals between them and carbon atoms. On the other hand, graphene exhibits an ultrathin thickness and a honeycomb lattice, on which a number of theoretical and experimental studies have been performed.[213–222] Its feature-rich properties cover tunable electronic densities, ultrafast carrier mobilities, and Dirac cone-shaped electronic structures.[223–226] The electronic properties can be tuned by the substrate effect, which is the most effective way of dramatically changing the energy gaps or inducing the metal–semiconductor transitions. Through detailed analyses and simulations, the critical orbital hybridizations are introduced to comprehend the fundamental properties.

Recently, 2D penta-graphene (PG) and its 1D ribbons (PGNRs) have received considerable attention because they are semiconductors that can be optimally combined with graphene and 2D-like materials.[5,6] Their potential applications have been recently explored, as have the possibilities of functionalization, adsorption, and atomic substitution to modify their properties[7] Some of these modifications may change the semiconducting character of PGs and

PGNRs[8–11] Another penta-graphene derivative is the PGQDs, in which the quantum confinement and edge morphology can offer additional functionality to penta-graphene.[12] Moreover, the surface or edges of PGQDs can be chemically functionalized to increase their interactivity, tune the energy gap, enhance physicochemical capabilities, and exhibit new physical properties. These structures have a valuable asymmetry property.[13] Therefore, the study of the structural stability and the electronic and optical properties of PGQDs is essential. Theoretical calculations reveal that the electronic and optical properties of the PGQD structures depend strongly on their size, morphology, type of edge functional atom, and type of doping atom. The cause of these effects can be traced back to the variety of atomic orbital hybridizations.

GQDs, a 0D material with a size of less than 10 nm, is a newly developed member of the graphene family. GQDs have become highly promising materials in various applications, including catalysis, bio-imaging, energy conversion, and storage, owing to their unique physicochemical properties, such as excellent photoluminescence and biocompatibility, low toxicity, easy fabrication, and edge effects. Additionally, oxygen-containing functional groups, such as -COOH, -COC-, -OH, -CHO, and -OCH_3, play a crucial role in the structure of GQDs. Oxygen is located at different edges and on the basal plane of GQDs. The unique architecture of oxygen inside the structure of GQDs and the outstanding properties and performance of GQDs provide a powerful impetus to use GQDs as a promising material with applicability in various fields. Herein, we describe the possible structure, application, and critical role of the oxygen-containing functional group on the essential properties and application of GQDs. Finally, we provide a brief outlook to point out the issues that need to be resolved for further development.

A series of DFT calculations were performed to understand the bonding and interaction of hydrogen adsorption on 2D silicon carbide. The converged energy results pointed out that the H atom can sufficiently bond to 2D SiC at the top sites (atop Si and C), of which the most stable adsorption site is T_{Si}. The vibrational properties, along with the zero-point energy, were incorporated into the energy calculations to further understand the phonon effect of the adsorbed H. Most of the 2D SiC structure deformations caused by the H atoms were found at the adsorbent atom along the vertical axis. For the first time, five SiC defect formations, including the quadrilateral-octagon linear defect (8-4), the silicon interstitial defect, the divacancy (4-10-4) defect, the divacancy (8-4-4-8) defect, and the divacancy (4-8-8-4) defect, were investigated. The linear defect (8-4) has the lowest formation energy and is most likely to form. This study could be an interesting source for future applications of hydrogen energy.

Two-dimensional monolayer pentagonal silicon dicarbide (pSiC_2) and its nanoribbon 1D derivative (pSiC_2NR) possess novel electronic properties, possibly leading to many potential applications. In this chapter, we have systematically investigated the structural, electronic, and transport properties of pSiC2NRs using chemical functions and strain engineering. The energy gaps of the pSiC_2NRs (ZZ-ribbon, ZA-ribbon, AA-ribbon, and SS-ribbon) are created mainly owing to the competition in the edge structures, finite-size confinements, and asymmetry of chemical bonds in the tetrahedral lattice or chemical modification. By applying uniaxial tensile or compressive strain, it is possible to modulate the physical properties of pSiC_2NRs, which subsequently changes the transport behavior of the carriers. Interestingly, the pentagon network of SS-pSiC_2NRs is still maintained, but the bond length along the strained direction undergoes a large change. The electronic band structure and bandgap are strongly affected by the uniaxial compressive strain. The evolution of bandgap versus strain is linear. The I–V characteristic of SS-pSiC_2NR seems to be more sensitive to compressive strain than stretch strain. The ultrahigh and strain-modulated carrier mobility in monolayer penta-SiC_2 may lead to many novel applications in high-performance electronic devices. Furthermore, the unusual properties of the pSiC_2 nanoribbons have great potential for applications in optoelectronic devices, especially in photovoltaics.

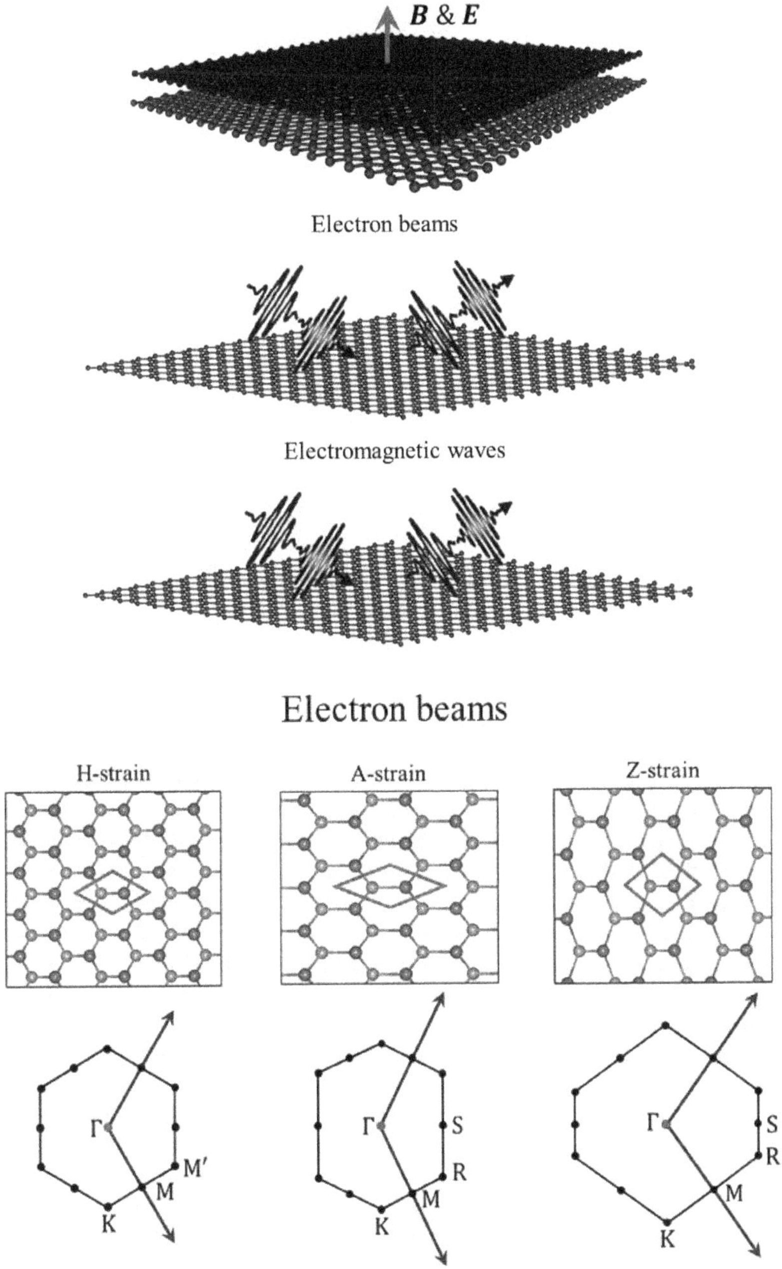

Fig. 1.5. Physical perturbations arising from the stationary electric/magnetic fields, high-energy electron beams, electromagnetic waves, mechanical stresses, and thermal excitations.

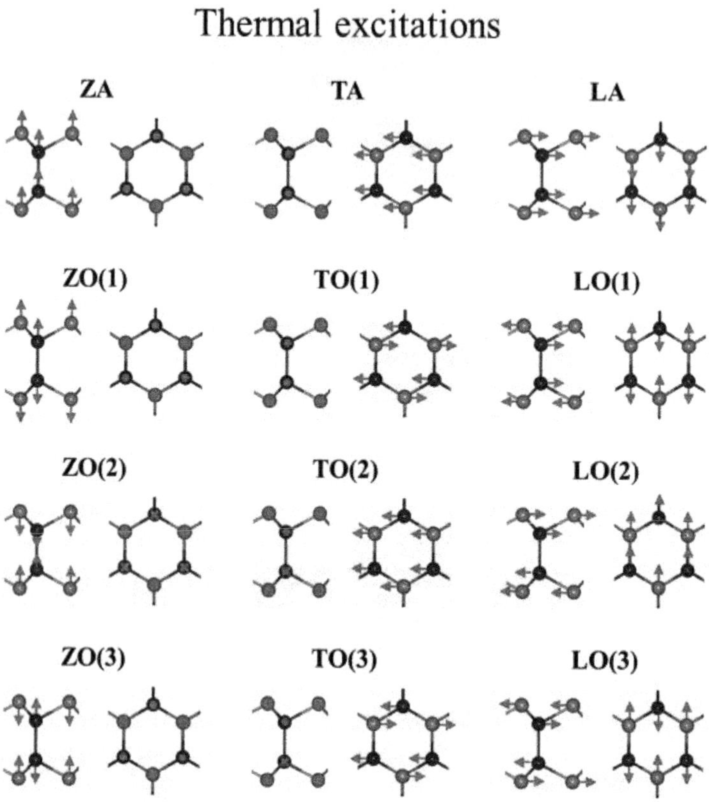

Fig. 1.5. (*Continued*)

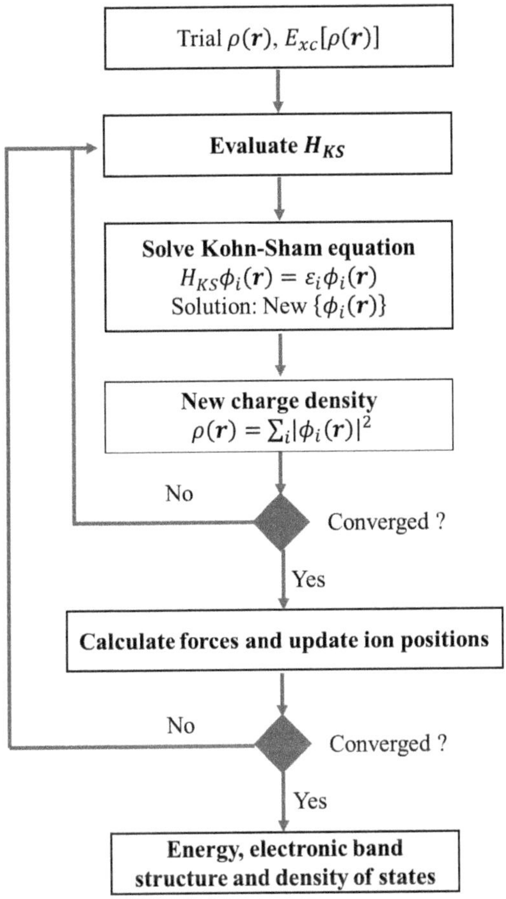

Fig. 1.6. Flowchart of first-principles simulation.

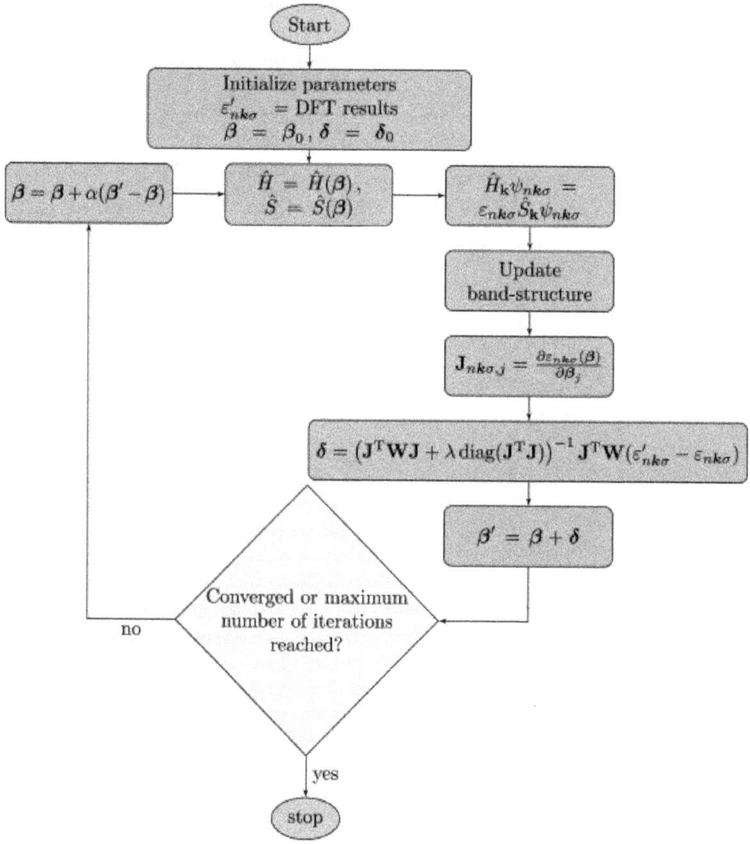

Fig. 1.7. Flowchart of phenomenological model.

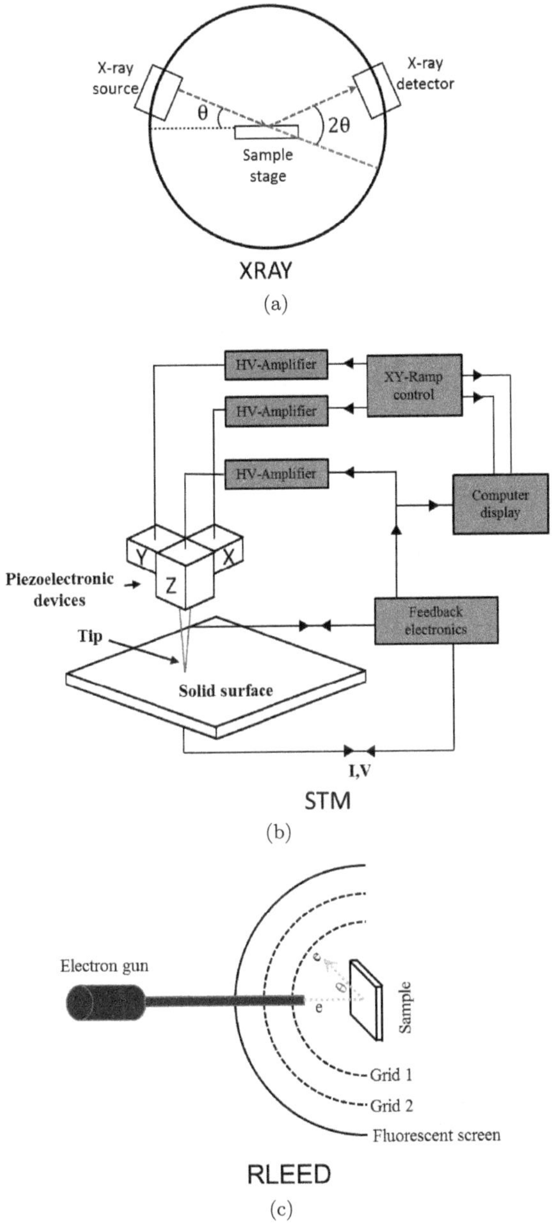

Fig. 1.8. High-resolution experimental measurements: (a) X-ray diffractions, (b) scanning-Tunneling spectroscopy (STM), (c) low-energy electron diffraction (LEED), (d) transmission electron microscopy (TEM), (e) scanning tunneling spectroscopy (STS), and (f) angle-resolved photoemission spectroscopy (ARPES).

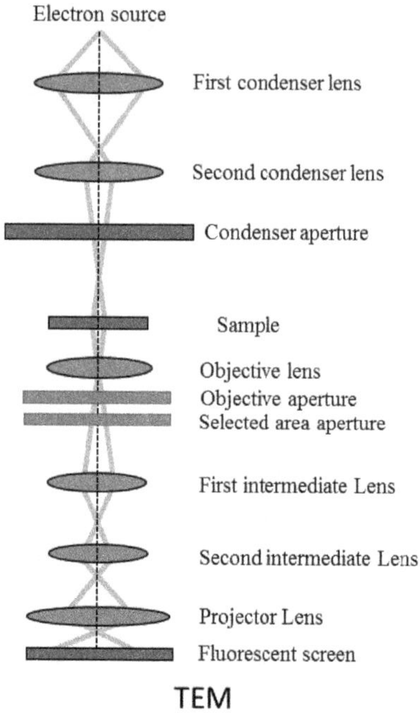

(d)

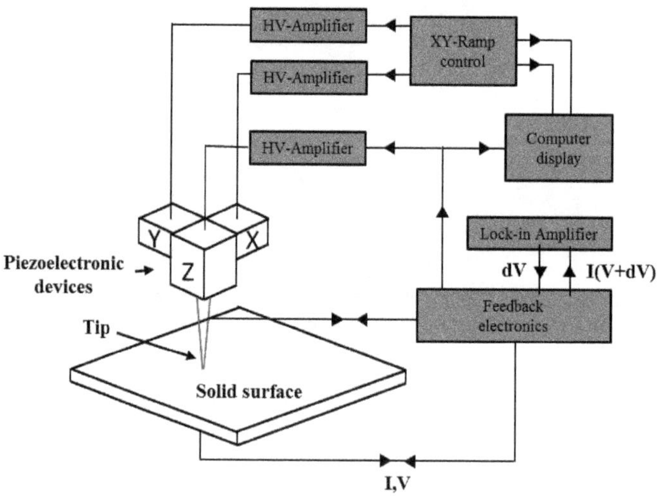

(e)

Fig. 1.8. (*Continued*)

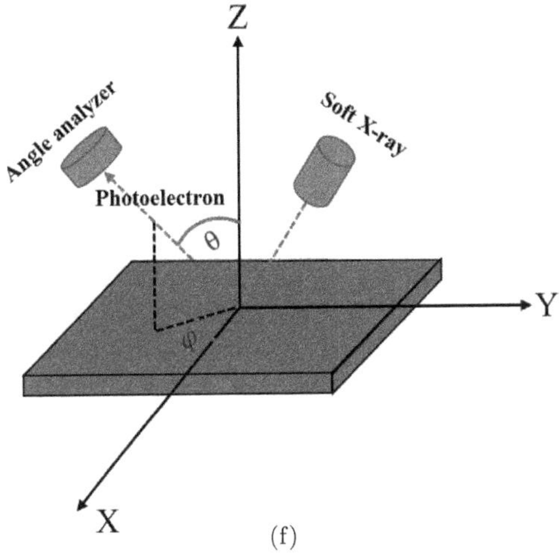

(f)

Fig. 1.8. (*Continued*)

References

[1] Levi, P.; Rosenthal, R. *The Periodic Table.* New York: Schocken Books, 1984.

[2] Lin, C.-Y.; Do, T.-N.; Huang, Y.-K.; Lin, M.-F. *Optical Properties of Graphene in Magnetic and Electric Fields*, 2017.

[3] Tran, N. T. T.; Lin, S.-Y.; Lin, C.-Y.; Lin, M.-F. *Geometric and Electronic Properties of Graphene-related Systems: Chemical Bonding Schemes.* CRC Press, 2017.

[4] Lin, S.-Y.; Tran, N. T. T.; Chang, S.-L.; Su, W.-P.; Lin, M.-F. *Structure-and Adatom-enriched Essential Properties of Graphene Nanoribbons.* CRC press, 2018.

[5] Tran, N. T. T.; Hsu, W.-D.; Huang, J.-L.; Lin, M.-F. *Lithium-Related Batteries: Advances and Challenges*, 2022.

[6] Han, N. T.; Dien, V. K.; Lin, M.-F. Excitonic effects in the optical spectra of Li2SiO3 compound. *Scientific Reports* **2021**, *11*, 1–10.

[7] Dien, V. K.; Pham, H. D.; Tran, N. T. T.; Han, N. T.; Huynh, T. M. D.; Nguyen, T. D. H.; *et al.* Orbital-hybridization-created optical excitations in Li2GeO3. *Scientific Reports* **2021**, *11*, 1–10.

[8] Khuong Dien, V.; Thi Han, N.; Nguyen, T. D. H.; Huynh, T. M. D.; Pham, H. D.; Lin, M.-F. Geometric and electronic properties of Li 2 GeO 3," *Frontiers in Materials* **2020**, *7*, 288.

[9] Dien, V. K.; Han, N. T.; Su, W.-P.; Lin, M.-F. Spin-dependent optical excitations in LiFeO2. *ACS Omega* **2021**, *6*, 25664–25671.

[10] Geim, A. K.; Novoselov, K. S. The rise of graphene. In *Nanoscience and Technology: A Collection of Reviews from Nature Journals.* World Scientific, 2010, pp. 11–19.

[11] Cahangirov, S.; Audiffred, M.; Tang, P.; Iacomino, A.; Duan, W.; Merino, G.; *et al.* Electronic structure of silicene on Ag (111): Strong hybridization effects. *Physical Review B* **2013**, *88*, 035432.

[12] Acun, A.; Zhang, L.; Bampoulis, P.; Farmanbar, M. V.; van Houselt, A.; Rudenko, A.; *et al.* Germanene: The germanium analogue of graphene. *Journal of Physics: Condensed Matter* **2015**, *27*, 443002.

[13] Zhu, F.-F.; Chen, W.-J.; Xu, Y.; Gao, C.-l.; Guan, D.-D.; Liu, C.-H.; *et al.* Epitaxial growth of two-dimensional stanene. *Nature Materials* **2015**, *14*, 1020–1025.

[14] Yuhara, J.; He, B.; Matsunami, N.; Nakatake, M.; Le Lay, G.; Graphene's latest cousin: Plumbene epitaxial growth on a "nano WaterCube. *Advanced Materials* **2019**, *31*, 1901017.

[15] Scerri, E. R. Transition metal configurations and limitations of the orbital approximation. *Journal of Chemical Education* **1989**, *66*, 481.

[16] Koehler, W. C. Magnetic properties of rare-earth metals and alloys. *Journal of Applied Physics* **1965**, *36*, 1078–1087.

[17] Li, W.-B.; Lin, S.-Y.; Tran, N. T. T.; Lin, M.-F.; Lin, K.-I. Essential geometric and electronic properties in stage-n graphite alkali-metal-intercalation compounds. *RSC Advances* **2020**, *10*, 23573–23581.

[18] Li, W.-B.; Lin, S.-Y.; Nguyen, T. D. H.; Chung, H.-C.; Tran, N. T. T.; Han, N. T.; *et al.* Graphite alkali-atom-/ion-intercalated compounds: Li+-related-based battery anodes. *First-Principles Calculations for Cathode* **2021**, 10–1.

[19] Shyu, F.-L.; Lin, M.-F. Plasmons and optical properties of semimetal graphite. *Journal of the Physical Society of Japan* **2000**, *69*, 3781–3784.

[20] Hunter, J.; Fye, J.; Jarrold, M. F. Carbon rings. *The Journal of Physical Chemistry* **1993**, *97*, 3460–3462.

[21] Peng, B.; Zhang, H.; Shao, H.; Xu, Y.; Ni, G.; Zhang, R.; *et al.* Phonon transport properties of two-dimensional group-IV materials from ab initio calculations. *Physical Review B* **2016**, *94*, 245420.

[22] Zaima, S.; Nakatsuka, O.; Taoka, N.; Kurosawa, M.; Takeuchi, W.; Sakashita, M. Growth and applications of GeSn-related group-IV semiconductor materials. *Science and Technology of Advanced Materials* **2015**, *16*, 043502.

[23] Sevik, C.; Çağın, T. Ab initio study of thermoelectric transport properties of pure and doped quaternary compounds. *Physical Review B* **2010**, *82*, 045202.
[24] Kumar, J.; Ingole, S. Optical phonons in pentanary compound (AgxCu1− x) 2ZnSnS4 semiconductor: A Raman study. *Journal of Alloys and Compounds* **2021**, *865*, 158113.
[25] Murota, J.; Sakuraba, M.; Tillack, B. Atomically controlled processing for group IV semiconductors by chemical vapor deposition. *Japanese Journal of Applied Physics* **2006**, *45*, 6767.
[26] Eletskii, A. V. Iskandarova, I. M.; Knizhnik, A. A.; Krasikov, D. N. Graphene: Fabrication methods and thermophysical properties. *Physics-Uspekhi* **2011**, *54*, 227.
[27] Vassilakos, A.; Giannatsis, J.; Dedoussis, V. Fabrication of parts with heterogeneous structure using material extrusion additive manufacturing. *Virtual and Physical Prototyping* **2021**, *16*, 267–290.
[28] Liu, J.; Lu, H. Azugraphene: A new graphene-like hexagonal carbon allotrope with Dirac cones. *RSC Advances* **2019**, *9*, 34481–34485.
[29] Lin, C.-Y.; Wu, J.-Y.; Chiu, Y.-H.; Chang, C.-P.; Lin, M.-F. Stacking-dependent magnetoelectronic properties in multilayer graphene *Physical Review B* **2014**, *90*, 205434.
[30] Do, T.-N.; Chang, C.-P.; Shih, P.-H.; Wu, J.-Y.; Lin, M.-F. Stacking-enriched magneto-transport properties of few-layer graphenes. *Physical Chemistry Chemical Physics* **2017**, *19*, 29525–29533.
[31] Lin, C.-Y.; Wu, J.-Y.; Ou, Y.-J.; Chiu, Y.-H.; Lin, M.-F. Magneto-electronic properties of multilayer graphenes. *Physical Chemistry Chemical Physics* **2015**, *17*, 26008–26035.
[32] Lin, C.-Y.; Ho, C.-H.; Wu, J.-Y.; Do, T.-N.; Shih, P.-H.; Lin, S.-Y.; *et al. Diverse Quantization Phenomena in Layered Materials.* CRC Press, 2019.
[33] Chang, S.-L.; Wu, B.-R.; Wong, J.-H.; Lin, M.-F. Configuration-dependent geometric and electronic properties of bilayer graphene nanoribbons. *Carbon* **2014**, *77*, 1031–1039.
[34] Chung, H.-C.; Lee, M.-H.; Chang, C.-P.; Huang, Y.-C.; Lin, M.-F. Effects of transverse electric fields on quasi-landau levels in zigzag graphene nanoribbons. *Journal of the Physical Society of Japan* **2011**, *80*, 044602.
[35] Chang, S.-L.; Lin, S.-Y.; Lin, S.-K.; Lee, C.-H.; Lin, M.-F. Geometric and electronic properties of edge-decorated graphene nanoribbons. *Scientific Reports* **2014**, *4*, 1–8.
[36] Chang, S.-L.; Wu, B.-R.; Yang, P.-H.; Lin, M.-F. Geometric, magnetic and electronic properties of folded graphene nanoribbons. *RSC Advances* **2016**, *6*, 64852–64860.

[37] Mahan, G. Oscillations of a thin hollow cylinder: Carbon nanotubes *Physical Review B* **2002**, *65*, 235402.

[38] Latgé, A.; Rocha, C.; Wanderley, L.; Pacheco, M.; Orellana, P.; Barticevic, Z. Defects and external field effects on the electronic properties of a carbon nanotube torus. *Physical Review B* **2003**, *67*, 155413.

[39] Zeiger, M.; Jäckel, N.; Mochalin, V. N.; Presser, V. carbon onions for electrochemical energy storage," *Journal of Materials Chemistry A* **2016**, *4*, 3172–3196.

[40] Treydte, K. S.; Frank, D. C.; Saurer, M.; Helle, G.; Schleser, G. H.; Esper, J. Impact of climate and CO2 on a millennium-long tree-ring carbon isotope record. *Geochimica et Cosmochimica Acta* **2009**, *73*, 4635–4647.

[41] Lee, J.-K.; Lee, S.-C.; Ahn, J.-P.; Kim, S.-C.; Wilson, J. I.; John, P. The growth of AA graphite on (111) diamond. *The Journal of Chemical Physics* **2008**, *129*, 234709.

[42] Habib, K. M.; Sylvia, S. S.; Ge, S.; Neupane, M.; Lake, R. K. The coherent interlayer resistance of a single, rotated interface between two stacks of AB graphite. *Applied Physics Letters* **2013**, *103*, 243114.

[43] Wong, J.-H.; Wu, B.-R.; Lin, M.-F. Electronic properties of rhombohedral graphite. *Computer Physics Communications* **2011**, *182*, 77–80.

[44] Lin, M. F.; Shung, K. W. K. Optical and magneto-optical properties of carbon nanotube bundles. *Journal of the Physical Society of Japan* **1997**, *66*, 3294–3302.

[45] Quo, Y.; Karasawa, N.; Goddard, W. A. Prediction of fullerene packing in C60 and C70 crystals. *Nature* **1991**, *351*, 464–467.

[46] Mohammadi, Y.; Moradian, R.; Tabar, F. S. Effects of doping and bias voltage on the screening in AAA-stacked trilayer graphene. *Solid State Communications* **2014**, *193*, 1–5.

[47] Cong, C.; Yu, T.; Sato, K.; Shang, J.; Saito, R.; Dresselhaus, G. F.; *et al.* Raman characterization of ABA-and ABC-stacked trilayer graphene. *ACS Nano* **2011**, *5*, 8760–8768.

[48] Koshino, M. Interlayer screening effect in graphene multilayers with A B A and A B C stacking. *Physical Review B* **2010**, *81*, 125304.

[49] Lin, C.-Y.; Huang, B.-L.; Ho, C.-H.; Gumbs, G.; Lin, M.-F. Geometry-diversified Coulomb excitations in trilayer AAB stacking graphene. *Physical Review B* **2018**, *98*, 195442.

[50] Cai, J.; Pignedoli, C. A.; Talirz, L.; Ruffieux, P.; Söde, H.; Liang, L.; *et al.* Graphene nanoribbon heterojunctions. *Nature Nanotechnology* **2014**, *9*, 896–900.

[51] Li, T.-S.; Lin, M.-F. Electronic properties of carbon nanotubes under external fields. *Physical Review B* **2006**, *73*, 075432.

[52] Lin, M.-F. Magnetic properties of toroidal carbon nanotubes. *Journal of the Physical Society of Japan* **1998**, *67*, 1094–1097.

[53] Haddon, R. Electronic properties of carbon toroids. *Nature* **1997**, *388*, 31–32.

[54] Tomita, S.; Sakurai, T.; Ohta, H.; Fujii, M.; Hayashi, S. Structure and electronic properties of carbon onions. *The Journal of Chemical Physics* **2001**, *114*, 7477–7482.

[55] Dresselhaus, M.; Dresselhaus, G.; Eklund, P. Fullerenes. *Journal of Materials Research* **1993**, *8*, 2054–2097.

[56] Samsonidze, G. G.; Saito, A. R.; Jorio, D. A.; Pimenta, E.; Souza, F.; Grüneis, F. A.; *et al.* The concept of cutting lines in carbon nanotube science. *Journal of Nanoscience and Nanotechnology* **2003**, *3*, 431–458.

[57] Feldman, B. J.; Osterloh, J. D.; Hata, B. H.; D'Alessandro, A. Determination of lead in blood by square wave anodic stripping voltammetry at a carbon disk ultramicroelectrode. *Analytical Chemistry* **1994**, *66*, 1983–1987.

[58] Shams, S. S.; Zhang, R.; Zhu, J. Graphene synthesis: A Review. *Materials Science - Poland* **2015**, *33*, 566–578.

[59] Bhuyan, M.; Alam, S.; Uddin, M.; Islam, M.; Bipasha, F. A.; Hossain, S. S. Synthesis of graphene. *International Nano Letters* **2016**, *6*, 65–83.

[60] Nobuhara, K.; Nakayama, H.; Nose, M.; Nakanishi, S.; Iba, H. First-principles study of alkali metal-graphite intercalation compounds. *Journal of Power Sources* **2013**, *243*, 585–587.

[61] Charlier, J.-C.; Gonze, X.; Michenaud, J.-P. First-principles study of graphite monofluoride (CF) n. *Physical Review B* **1993**, *47*, 16162.

[62] Shih, P.-H.; Do, T.-N.; Gumbs, G.; Huang, D.; Pham, T. P.; Lin, M.-F. Magneto-transport properties of B-, Si-and N-doped graphene. *Carbon* **2020**, *160*, 211–218.

[63] Lin, S.-Y.; Liu, H.-Y.; Nguyen, D. K.; Tran, N. T. T.; Pham, H. D.; Chang, S.-L.; *et al.* *Silicene-Based Layered Materials*. IOP Publishing Ltd., 2020.

[64] Guo, Y.-N.; Lu, X.; Weng, J.; Leng, Y. Density functional theory study of the interaction of arginine-glycine-aspartic acid with graphene, defective graphene, and graphene oxide. *The Journal of Physical Chemistry C* **2013**, *117*, 5708–5717.

[65] Malinauskas, M.; Žukauskas, A.; Hasegawa, S.; Hayasaki, Y.; Mizeikis, V.; Buividas, R.; *et al.* Ultrafast laser processing of materials: From science to industry. *Light: Science & Applications* **2016**, *5*, e16133–e16133.

[66] Ali, I.; Alharbi, O. M.; Tkachev, A.; Galunin, E.; Burakov, A.; Grachev, V. A. Water treatment by new-generation graphene materials: Hope for bright future. *Environmental Science and Pollution Research* **2018**, *25*, 7315–7329.

[67] Banwaskar, M.; Dachawar, S. Graphene basics and applications. In *Advanced Materials Research*, 2013, pp. 259–262.

[68] Shahil, K. M.; Balandin, A. A. Thermal properties of graphene and multilayer graphene: Applications in thermal interface materials. *Solid State Communications* **2012**, *152*, 1331–1340.

[69] Bahadur, M.; Norris, A. W.; Zarisfi, A.; Alger, J. S.; Windiate, C. C. Silicone materials for LED packaging. In *Sixth International Conference on Solid State Lighting*, 2006, pp. 56–62.

[70] Claeys, C.; Simoen, E. *Germanium-based Technologies: From Materials to Devices*. Elsevier, 2011.

[71] Afzaal M.; O'Brien, P. Recent developments in II–VI and III–VI semiconductors and their applications in solar cells. *Journal of Materials Chemistry* **2006**, *16*, 1597–1602.

[72] Rideout, V. A review of the theory and technology for ohmic contacts to group III–V compound semiconductors. *Solid-State Electronics* **1975**, *18*, 541–550.

[73] Mukhopadhyay, S.; Acharyya, A.; Mitra, M. Relative study on MM-wave performance of group IV-IV and group III-V materials based IMPATT sources. *IETE Journal of Research* **2020**, 1–11.

[74] Tomić, S.; Montanari, B.; Harrison, N. The group III–V's semiconductor energy gaps predicted using the B3LYP hybrid functional. *Physica E: Low-Dimensional Systems and Nanostructures* **2008**, *40*, 2125–2127.

[75] Han, N. T.; Dien, V. K.; Tran, N. T. T.; Nguyen, D. K.; Su, W.-P.; Lin, M.-F. First-principles studies of electronic properties in lithium metasilicate (Li 2 SiO 3). *RSC Advances* **2020**, *10*, 24721–24729.

[76] Ansara, I.; Chatillon, C.; Lukas, H.; Nishizawa, T.; Ohtani, H.; Ishida, K.; *et al.* A binary database for III–V compound semiconductor systems. *Calphad* **1994**, *18*, 177–222.

[77] Soref, R. A. Optical band gap of the ternary semiconductor Si1− x−y Ge x C y. *Journal of Applied Physics* **1991**, *70*, 2470–2472.
[78] Li, H.; Malliakas, C. D.; Liu, Z.; Peters, J. A.; Jin, H.; Morris, C. D.; *et al.* CsHgInS3: A new quaternary semiconductor for γ-ray detection. *Chemistry of Materials* **2012**, *24*, 4434–4441.
[79] Seabra, A. B.; Paula, A. J.; de Lima, R.; Alves, O. L.; Durán, N. Nanotoxicity of graphene and graphene oxide. *Chemical Research in Toxicology* **2014**, *27*, 159–168.
[80] Molle, A.; Grazianetti, C.; Tao, L.; Taneja, D.; Alam, M. H.; Akinwande, D. Silicene, silicene derivatives, and their device applications. *Chemical Society Reviews* **2018**, *47*, 6370–6387.
[81] Ezawa, M. Monolayer topological insulators: Silicene, germanene, and stanene. *Journal of the Physical Society of Japan* **2015**, *84*, 121003.
[82] Huang, H.-C.; Lin, S.-Y.; Wu, C.-L.; Lin, M.-F. Configuration- and concentration-dependent electronic properties of hydrogenated graphene. *Carbon* **2016**, *103*, 84–93.
[83] Tran, N. T. T.; Nguyen, D. K.; Glukhova, O. E.; Lin, M.-F. Coverage-dependent essential properties of halogenated graphene: A DFT study. *Scientific Reports* **2017**, *7*, 1–13.
[84] Jin, K.-H.; Choi, S.-M.; Jhi, S.-H. Crossover in the adsorption properties of alkali metals on graphene. *Physical Review B* **2010**, *82*, 033414.
[85] Xu, C.; Brown, P. A.; Lu, J.; Shuford, K. L. Electronic properties of halogen-adsorbed graphene. *The Journal of Physical Chemistry C* **2015**, *119*, 17271–17277.
[86] Ho, Y.-H.; Wu, J.-Y.; Chen, R.-B.; Chiu, Y.-H.; Lin, M.-F. Optical transitions between Landau levels: AA-stacked bilayer graphene *Applied Physics Letters* **2010**, *97*, 101905.
[87] Chuang, Y.-C.; Wu, J.-Y.; Lin, M.-F. Electric field dependence of excitation spectra in AB-stacked bilayer graphene. *Scientific Reports* **2013**, *3*, 1–5.
[88] Ho, C.-H.; Chang, C.-P.; Lin, M.-F. Evolution and dimensional crossover from the bulk subbands in ABC-stacked graphene to a three-dimensional Dirac cone structure in rhombohedral graphite. *Physical Review B* **2016**, *93*, 075437.
[89] McGilly, L. J.; Kerelsky, A.; Finney, N. R.; Shapovalov, K.; Shih, E.-M.; Ghiotto, A.; *et al.* Visualization of moiré superlattices. *Nature Nanotechnology* **2020**, *15*, 580–584.
[90] Raidongia, K.; Nag, A.; Hembram, K.; Waghmare, U. V.; Datta, R.; Rao, C. BCN: A graphene analogue with remarkable adsorptive properties. *Chemistry–A European Journal* **2010**, *16*, 149–157.

[91] Azevedo, S.; De Paiva, R. Structural stability and electronic properties of carbon-boron nitride compounds. *EPL (Europhysics Letters)* **2006**, *75*, 126.

[92] Beard, M. C.; Luther, J. M.; Semonin, O. E.; Nozik, A. J. Third generation photovoltaics based on multiple exciton generation in quantum confined semiconductors. *Accounts of Chemical Research* **2013**, *46*, 1252–1260.

[93] Weissenbacher, R.; Haubner, R.; Aigner, K.; Lux, B. Interactions of Ta-filaments during hot-filament CVD of BCN-layers. *Diamond and Related Materials* **2002**, *11*, 191–197.

[94] Bi, Y.-S.; Liu, B.; Liu, X.-Y.; Qin, Y.; Zou, B.-X. A h-BCN for electrochemical sensor of dopamine and uric acid. *Journal of Nanomaterials* **2020**, *2020*.

[95] Kim, S. Y.; Park, J.; Choi, H. C.; Ahn, J. P.; Hou, J. Q.; Kang, H. S. X-ray photoelectron spectroscopy and first principles calculation of BCN nanotubes. *Journal of the American Chemical Society* **2007**, *129*, 1705–1716.

[96] Bafekry, A.; Naseri, M.; Fadlallah, M. M.; Abdolhosseini Sarsari, I.; Faraji, M.; Bagheri Khatibani, A.; *et al.* A novel two-dimensional boron–carbon–nitride (BCN) monolayer: A first-principles insight. *Journal of Applied Physics* **2021**, *130*, 114301.

[97] Ebrahimi, M.; Horri, A.; Sanaeepur, M.; Tavakoli, M. B. Tight-binding description of graphene–BCN–graphene layered semiconductors *Journal of Computational Electronics* **2020**, *19*, 62–69.

[98] Yoshioka, T.; Suzuura, H.; Ando, T. Electronic states of BCN alloy nanotubes in a simple tight-binding model. *Journal of the Physical Society of Japan* **2003**, *72*, 2656–2664.

[99] Wang, W.; Chen, S.; Yang, P.-X.; Duan, C.-G.; Wang, L.-W. Si: WO 3 heterostructure for Z-scheme water splitting: An AB initio study. *Journal of Materials Chemistry A* **2013**, *1*, 1078–1085.

[100] Joyce, B. Molecular beam epitaxy. *Reports on Progress in Physics* **1985**, *48*, 1637.

[101] Fergus, J. W. Recent developments in cathode materials for lithium ion batteries. *Journal of Power Sources* **2010**, *195*, 939–954.

[102] Zhong, C.; Deng, Y.; Hu, W.; Qiao, J.; Zhang, L.; Zhang, J. A review of electrolyte materials and compositions for electrochemical supercapacitors. *Chemical Society Reviews* **2015**, *44*, 7484–7539.

[103] Wu, Y.; Rahm, E.; Holze, R. Carbon anode materials for lithium ion batteries. *Journal of Power Sources* **2003**, *114*, 228–236.

[104] Lin, S.-Y.; Liu, H.-Y.; Nguyen, D. K.; Tran, N. T. T.; Pham, H. D.; Chang, S.-L.; *et al.* Silicene on Ag (111).

[105] Lin, C.-Y.; Wu, J.-Y.; Chiu, C.-W.; Lin, M.-F. *Coulomb Excitations and Decays in Graphene-related Systems.* CRC Press, 2019.

[106] Lin, C.-Y.; Chen, R.-B.; Ho, Y.-H.; Lin, M.-F. *Electronic and Optical Properties of Graphite-related Systems.* CRC Press, 2017.

[107] Wong, J.-H.; Wu, B.-R.; Lin, M.-F. Strain effect on the electronic properties of single layer and bilayer graphene. *The Journal of Physical Chemistry C* **2012**, *116*, 8271–8277.

[108] Ho, Y.-H.; Chiu, Y.-H.; Lin, D.-H.; Chang, C.-P.; Lin, M.-F. Magneto-optical selection rules in bilayer Bernal graphene. *ACS Nano* **2010**, *4*, 1465–1472.

[109] Sing Li, T.; Hung Ho, Y.; Fa Lin, M. Influences of mechanical deformation on Bilayer Graphene's magneto-optical absorption properties. *Journal of the Physical Society of Japan* **2013**, *82*, 034701.

[110] Van de Walle, C. G.; Neugebauer, J. First-principles calculations for defects and impurities: Applications to III-nitrides. *Journal of Applied Physics* **2004**, *95*, 3851–3879.

[111] Foulkes, W. M. C.; Haydock, R. Tight-binding models and density-functional theory. *Physical Review B* **1989**, *39*, 12520.

[112] Zhang, X.; Chen, A.; Chen, L.; Zhou, Z. 2D materials bridging experiments and computations for electro/photocatalysis. *Advanced Energy Materials* **2022**, *12*, 2003841.

[113] Shyu, F.-L.; Lin, M. F.; Chang, C.; Chen, R.-B.; Shyu, J.; Wang, Y.; *et al.* Tight-binding band structures of nanographite multiribbons," *Journal of the Physical Society of Japan* **2001**, *70*, 3348–3355.

[114] Lv, B.; Qian, T.; Ding, H. Angle-resolved photoemission spectroscopy and its application to topological materials. *Nature Reviews Physics* **2019**, *1*, 609–626.

[115] Liu, Y.; Zeng, C.; Yu, J.; Zhong, J.; Li, B.; Zhang, Z.; *et al.*, Moiré superlattices and related moiré excitons in twisted van der Waals heterostructures. *Chemical Society Reviews* **2021**, *50*, 6401–6422.

[116] Huang, D.; Choi, J.; Shih, C.-K.; Li, X. Excitons in semiconductor moiré superlattices. *Nature Nanotechnology* **2022**, *17*, 227–238.

[117] Lin, C.-Y.; Wu, J.-Y.; Chang, C.-P.; Lin, M.-F. Magneto-optical selection rules of curved graphene nanoribbons and carbon nanotubes. *Carbon* **2014**, *69*, 151–161.

[118] Warren, B. E. *X-ray Diffraction.* Courier Corporation, 1990.

[119] Pendry, J. The application of pseudopotentials to low-energy electron diffraction II: Calculation of the reflected intensities. *Journal of Physics C: Solid State Physics* **1969**, *2*, 2273.

[120] Garnaes, J.; Kragh, F.; Mo/rch, K.; Thölén, A. Transmission electron microscopy of scanning tunneling tips. *Journal of Vacuum Science & Technology A: Vacuum, Surfaces, and Films* **1990**, *8*, 441–444.

[121] Hansma, P. K.; Tersoff, J. Scanning tunneling microscopy. *Journal of Applied Physics* **1987**, *61*, R1–R24.

[122] Son, Y.; Lee, J. S.; Son, Y.; Jang, J. H.; Cho, J. Recent advances in lithium sulfide cathode materials and their use in lithium sulfur batteries. *Advanced Energy Materials* **2015**, *5*, 1500110.

[123] Xia, C.; Kwok, C.; Nazar, L. A high-energy-density lithium-oxygen battery based on a reversible four-electron conversion to lithium oxide. *Science* **2018**, *361*, 777–781.

[124] Dresselhaus, M. S.; Dresselhaus, G.; Eklund, P.; Rao, A. Carbon nanotubes. In *The Physics of Fullerene-based and Fullerene-related Materials*. Springer, 2000, pp. 331–379.

[125] Hwang, E.; Sarma, S. D. Quasiparticle spectral function in doped graphene: Electron-electron interaction effects in ARPES. *Physical Review B* **2008**, *77*, 081412.

[126] Celis, A.; Nair, M. N.; Taleb-Ibrahimi, A.; Conrad, E.; Berger, C.; De Heer, W.; *et al.* Graphene nanoribbons: Fabrication, properties and devices. *Journal of Physics D: Applied Physics* **2016**, *49*, 143001.

[127] Kano, S.; Tada, T.; Majima, Y. Nanoparticle characterization based on STM and STS. *Chemical Society Reviews* **2015**, *44*, 970–987.

[128] Li, G.; Luican, A.; Lopes dos Santos, J.; Castro Neto, A.; Reina, A.; Kong, J.; *et al.* Observation of Van Hove singularities in twisted graphene layers. *Nature Physics* **2010**, *6*, 109–113.

[129] Dang, M. T.; Bich Thao, P. T.; Ngoc Thao, T. T.; Tien, N. T. First-principles study of electronic and optical properties of small edge-functionalized penta-graphene quantum dots. *AIP Advances* **2022**, *12*, 065008.

[130] Sun, H.; Wu, L.; Wei, W.; Qu, X. Recent advances in graphene quantum dots for sensing. *Materials Today* **2013**, *16*, 433–442.

[131] Yan, Y.; Gong, J.; Chen, J.; Zeng, Z.; Huang, W.; Pu, K.; *et al.* Recent advances on graphene quantum dots: From chemistry and physics to applications. *Advanced Materials* **2019**, *31*, 1808283.

[132] Mi, T. Y.; Khanh, N. D.; Ahuja, R.; Tien, N. T. Diverse structural and electronic properties of pentagonal SiC2 nanoribbons: A first-principles study. *Materials Today Communications* **2021**, *26*, 102047.

[133] Li, Y.; Li, F.; Zhou, Z.; Chen, Z. SiC2 silagraphene and its one-dimensional derivatives: Where planar tetracoordinate silicon happens. *Journal of the American Chemical Society* **2011**, *133*, 900–908.

[134] Shi, Z.; Zhang, Z.; Kutana, A.; Yakobson, B. I. Predicting two-dimensional silicon carbide monolayers. *ACS Nano* **2015**, *9*, 9802–9809.

[135] Lin, X.; Lin, S.; Xu, Y.; Hakro, A. A.; Hasan, T.; Zhang, B.; *et al.* Ab initio study of electronic and optical behavior of two-dimensional silicon carbide. *Journal of Materials Chemistry C* **2013**, *1*, 2131–2135.

[136] Han, N. T.; Dien, V. K.; Lin, S.-Y.; Chung, H.-C.; Li, W.-B.; Tran, N. T. T.; *et al.* Comprehensive understanding of electronic and optical properties of Li2SiO3 compound.

[137] Han, N. T.; Tran, N. T. T.; Dien, V. K.; Nguyen, D. K.; Lin, M.-F. Diversified properties in 3D ternary oxide compound. In *Lithium-Ion Batteries and Solar Cells: Physical, Chemical, and Materials Properties.* 2021.

[138] Chung, N. T. T. T.; Han, N. T.; Liu, H.-Y.; Pham, H. D.; Lin, M.-F. Experimental measurements.

[139] Hanke, W.; Sham, L. Many-particle effects in the optical excitations of a semiconductor. *Physical Review Letters* **1979**, *43*, 387.

[140] Dodd, R.; Edwards, M.; Clark, C. W.; Burnett, K. Collective excitations of Bose-Einstein-condensed gases at finite temperatures. *Physical Review A* **1998**, *57*, R32.

[141] Luo, X. C.; Chugh, R.; Biller, B. C.; Hoi, Y. M.; Chung, D. D. L. Electronic applications of flexible graphite. *Journal of Electronic Materials* **2002**, *31*, 535–544.

[142] Wissler, M. Graphite and carbon powders for electrochemical applications. *Journal of Power Sources* **2006**, *156*, 142–150.

[143] Bonal, J. P.; Gosmain, L. Graphites for nuclear applications. *Actualite Chimique* **2006**, 23–27.

[144] Tong, X.; Wei, Q. L.; Zhan, X. X.; Zhang, G. X.; Sun, S. H. The new graphene family materials: Synthesis and applications in oxygen reduction reaction. *Catalysts* **2017**, *7*.

[145] Zhang, H.; Yang, Y.; Ren, D. S.; Wang, L.; He, X. M. Graphite as anode materials: Fundamental mechanism, recent progress and advances. *Energy Storage Materials* **Apr 2021**, *36*, 147–170.

[146] Shuvo, M. A. I.; Khan, M. A. R.; Karim, H.; Morton, P.; Wilson, T.; Lin, Y. R. Investigation of modified graphene for energy storage applications. *ACS Applied Materials & Interfaces* **2013**, *5*, 7881–7885.

[147] Zhang, M.; Song, X. H.; Ou, X. W.; Tang, Y. B. Rechargeable batteries based on anion intercalation graphite cathodes. *Energy Storage Materials* **2019**, *16*, 65–84.

[148] Grechnev, G. E.; Lyogenkaya, A. A.; Kolesnichenko, Y. A.; Prylutskyy, Y. I.; Hayn, R. Electronic structure and magnetic properties of graphite intercalated with 3d-metals. *Low Temperature Physics* **2014**, *40*, 450–453.

[149] Singh, R.; Prakash, S. First-principles investigation into structural and magnetic properties of binary graphite 3d-transition metal intercalated compounds (XC6; X = Cr, Mn, Fe). *Carbon* **2010**, *48*, 1341–1344.

[150] Assouik, J.; Hajji, L.; Boukir, A.; Chaouqi, M.; Lagrange, P. Heavy alkali metal-arsenic alloy-based graphite intercalation compounds: Investigation of their synthesis and of their physical properties. *Comptes Rendus Chimie* **2017**, *20*, 116–124.

[151] Speyer, L.; Fontana, S.; Cahen, S.; Herold, C. Intercalation of sodium and heavy alkali metals into graphenic foams. *Microporous and Mesoporous Materials* **2020**, *306*.

[152] Eshkalak, K. E.; Sadeghzadeh, S.; Jalaly, M. Studying the effects of longitudinal and transverse defects on the failure of hybrid graphene-boron nitride sheets: A molecular dynamics simulation. *Physica E: Low-dimensional Systems and Nanostructures* **2018**, *104*, 71–81.

[153] Gusynin, V.; Sharapov, S.; Carbotte, J. Unusual microwave response of Dirac quasiparticles in graphene. *Physical Review Letters* **2006**, *96*, 256802.

[154] Segall, M.; Lindan, P. J.; Probert, M. A.; Pickard, C. J.; Hasnip, P. J.; Clark, S.; *et al.* First-principles simulation: Ideas, illustrations and the CASTEP code. *Journal of Physics: Condensed Matter* **2002**, *14*, 2717.

[155] Lin, Y.-M.; Perebeinos, V.; Chen, Z.; Avouris, P. Electrical observation of subband formation in graphene nanoribbons. *Physical Review B* **2008**, *78*, 161409.

[156] Koskinen, P.; Malola, S.; Häkkinen, H. Self-passivating edge reconstructions of graphene. *Physical Review Letters* **2008**, *101*, 115502.

[157] Subramaniam, D.; Libisch, F.; Li, Y.; Pauly, C.; Geringer, V.; Reiter, R.; *et al.* Wave-function mapping of graphene quantum dots with soft confinement. *Physical Review Letters* **2012**, *108*, 046801.

[158] Wang, W. L.; Meng, S.; Kaxiras, E. Graphene nanoflakes with large spin. *Nano Letters* **2008**, *8*, 241–245.

[159] Narath, A. Nuclear spin-lattice relaxation in hexagonal transition metals: Titanium. *Physical Review* **1967**, *162*, 320.

[160] Bacon, A. D.; Zerner, M. C. An intermediate neglect of differential overlap theory for transition metal complexes: Fe, Co and Cu chlorides. *Theoretica Chimica Acta* **1979**, *53*, 21–54.

[161] Khazaei, M.; Ranjbar, A.; Arai, M.; Yunoki, S. Topological insulators in the ordered double transition metals $M'_2M''C_2$ MXenes (M' = Mo, W; M'' = Ti, Zr, Hf). *Physical Review B* **2016**, *94*, 125152.

[162] Kim, C.; Kim, H.-S.; Park, J.-G. Spin-orbital entangled state and realization of Kitaev physics in 3d cobalt compounds: A progress report. *Journal of Physics: Condensed Matter* **2021**, *34*, 023001.
[163] Mott, N. F. The basis of the electron theory of metals, with special reference to the transition metals. *Proceedings of the Physical Society. Section A* **1949**, *62*, 416.
[164] Butler, M. J.; Crimmin, M. R. Magnesium, zinc, aluminium and gallium hydride complexes of the transition metals. *Chemical Communications* **2017**, *53*, 1348–1365.
[165] Burdett, J. K.; Eisenstein, O. From three-to four-coordination in copper (I) and silver (I). *Inorganic Chemistry* **1992**, *31*, 1758–1762.
[166] Streltsov, S. V. Orbital-selective behavior in Y5Mo2O12 and (Cd, Zn) V2O4. *Journal of Magnetism and Magnetic Materials* **2015**, *383*, 27–29.
[167] Fan, Y.; Zhang, J.; Qiu, Y.; Zhu, J.; Zhang, Y.; Hu, G. A DFT study of transition metal (Fe, Co, Ni, Cu, Ag, Au, Rh, Pd, Pt and Ir)-embedded monolayer MoS2 for gas adsorption. *Computational Materials Science* **2017**, *138*, 255–266.
[168] Qi, B.; Perez, I.; Ansari, P.; Lu, F.; Croft, M. L 2 and L 3 measurements of transition-metal 5 d orbital occupancy, spin-orbit effects, and chemical bonding. *Physical Review B* **1987**, *36*, 2972.
[169] Pati, A.; Kundu, T.; Pal, S. DFT studies on interaction between lanthanum and hydroxyamide. In *IOP Conference Series: Materials Science and Engineering* **2018**, 012025.
[170] Dolg, M.; Stoll, H.; Savin, A.; Preuss, H. Energy-adjusted pseudopotentials for the rare earth elements. *Theoretica Chimica Acta* **1989**, *75*, 173–194.
[171] Xie, K.; Jia, Q.; Wang, Y.; Zhang, W.; Xu, J. The electronic structure and optical properties of anatase TiO2 with rare earth metal dopants from first-principles calculations. *Materials* **2018**, *11*, 179.
[172] Moeller, T.; Martin, D. F.; Thompson, L. C.; Ferrús, R.; Feistel, G. R.; Randall, W. J. The coordination chemistry of yttrium and the rare earth metal ions. *Chemical Reviews* **1965**, *65*, 1–50.
[173] El Hachimi, A.; Zaari, H.; Benyoussef, A.; El Yadari, M.; El Kenz, A. First-principles prediction of the magnetism of 4f rare-earth-metal-doped wurtzite zinc oxide. *Journal of Rare Earths* **2014**, *32*, 715–721.
[174] Das, B.; Choudhary, R.; Skomski, R.; Balasubramanian, B.; Pathak, A. K.; Paudyal, D.; *et al.* Anisotropy and orbital moment in Sm-Co permanent magnets. *Physical Review B* **2019**, *100*, 024419.

[175] Turek, I.; Rusz, J.; Diviš, M. Electronic structure and volume magnetostriction of rare-earth metals and compounds. *Journal of Magnetism and Magnetic Materials* **2005**, *290*, 357–363.

[176] Myers, C. E.; Norman, L. J.; Loew, L. M. A molecular orbital study of rare earth metal trihalide molecules. *Inorganic Chemistry* **1978**, *17*, 1581–1584.

[177] Lee, S. J.; Lim, Y. R.; Ji, S.; Kim, S. K.; Yoon, Y.; Song, W.; *et al.* Long-term air-stable Au doping of graphene by layer-by-layer assembly with graphene oxide for flexible transparent electrodes. *Carbon* **2018**, *126*, 241–246.

[178] Zhou, Y.; Zhang, W.; Pan, Z.; Zhao, B. Graphene-doped polyaniline nanocomposites as electromagnetic wave absorbing materials. *Journal of Materials Science: Materials in Electronics* **2017**, *28*, 10921–10928.

[179] Botello-Mendez, A. R.; Dubois, S. M.-M.; Lherbier, A.; Charlier, J.-C. Achievements of DFT for the investigation of graphene-related nanostructures. *Accounts of Chemical Research* **2014**, *47*, 3292–3300.

[180] He, X.; Kitipornchai, S.; Liew, K. Resonance analysis of multi-layered graphene sheets used as nanoscale resonators. *Nanotechnology* **2005**, *16*, 2086.

[181] Han, J. Energetics and structures of fullerene crop circles. *Chemical Physics Letters* **1998**, *282*, 187–191.

[182] Adachi, S. *Physical Properties of III-V Semiconductor Compounds.* John Wiley & Sons, 1992.

[183] Jain, M. *II-VI Semiconductor Compounds.* World Scientific, 1993.

[184] Causa, M.; Dovesi, R.; Roetti, C. Pseudopotential Hartree-Fock study of seventeen III-V and IV-IV semiconductors. *Physical Review B* **1991**, *43*, 11937.

[185] Matsunaga, K; Tanaka, T.; Yamamoto, T.; Ikuhara, Y. First-principles calculations of intrinsic defects in Al2O3. *Physical Review B* **2003**, *68*, 085110.

[186] Togo, A.; Tanaka, I. First principles phonon calculations in materials science. *Scripta Materialia* **2015**, *108*, 1–5.

[187] Pak, Y.; Kim, S.-M.; Jeong, H.; Kang, C. G.; Park, J. S.; Song, H.; *et al.* Palladium-decorated hydrogen-gas sensors using periodically aligned graphene nanoribbons. *ACS Applied Materials & Interfaces* **2014**, *6*, 13293–13298.

[188] Zheng, N.; Yang, S.; Xu, H.; Lan, Z.; Wang, Z.; Gu, H.; A DFT study of the enhanced hydrogen storage performance of the Li-decorated graphene nanoribbons. *Vacuum* **2020**, *171*, 109011.

[189] Tran, N. T. T.; Gumbs, G.; Nguyen, D. K.; Lin, M.-F. Fundamental properties of metal-adsorbed silicene: A DFT Study. *ACS Omega* **2020**, 5, 13760–13769.

[190] Nguyen, D. K.; Tran, N. T. T.; Chiu, Y.-H.; Lin, M.-F. Concentration-diversified magnetic and electronic properties of halogen-adsorbed silicene. *Scientific Reports* **2019**, *9*, 1–15.

[191] Sharma, B.; Purohit, R. *Semiconductor Heterojunctions,* Vol. 5. Elsevier, 2015.

[192] Arafune, R.; Lin, C.-L.; Kawahara, K.; Tsukahara, N.; Minamitani, E.; Kim, Y.; *et al.* Structural transition of silicene on Ag (111). *Surface Science* **2013**, *608*, 297–300.

[193] Bahuguna, B. P.; Saini, L.; Sharma, R. O.; Tiwari, B. Hybrid functional calculations of electronic and thermoelectric properties of GaS, GaSe, and GaTe monolayers. *Physical Chemistry Chemical Physics* **2018**, *20*, 28575–28582.

[194] Mohammad, S. N.; Morkoç, H. Progress and prospects of group-III nitride semiconductors. *Progress in Quantum Electronics* **1996**, *20*, 361–525.

[195] Lazarev, V.; Shevchenko, V. I.; Grinberg, I. K.; Sobolev, V. Group II-V semiconductor compounds. *Moscow Izdatel Nauka*, **1978**.

[196] Hu, Z.; Ding, Y.; Hu, X.; Zhou, W.; Yu, X.; Zhang, S. Recent progress in 2D group IV–IV monochalcogenides: Synthesis, properties and applications. *Nanotechnology* **2019**, *30*, 252001.

[197] Jin, W.; Yeh, P.-C.; Zaki, N.; Zhang, D.; Sadowski, J. T.; Al-Mahboob, A.; *et al.* Direct measurement of the thickness-dependent electronic band structure of MoS 2 using angle-resolved photoemission spectroscopy. *Physical Review Letters* 2013, vol. 111, p. 106801.

[198] Ohno, H.; Ikhlayel, M.; Tamura, M.; Nakao, K.; Suzuki, K.; Morita, K.; *et al.* Direct dimethyl carbonate synthesis from CO 2 and methanol catalyzed by CeO 2 and assisted by 2-cyanopyridine: A cradle-to-gate greenhouse gas emission study. *Green Chemistry* **2021**, *23*, 457–469.

[199] Renganathan, B.; Rao, S. K.; Ganesan, A.; Deepak, A. High proficient sensing response in clad modified ceria doped tin oxide fiber optic toxic gas sensor application. *Sensors and Actuators A: Physical* **2021**, *332*, 113114.

[200] Motaung, D. E.; Mhlongo, G. H.; Makgwane, P. R.; Dhonge, B. P.; Cummings, F. R.; Swart, H. C.; *et al.* Ultra-high sensitive and selective H2 gas sensor manifested by interface of n–n heterostructure of CeO2-SnO2 nanoparticles. *Sensors and Actuators B: Chemical* **2018**, *254*, 984–995.

[201] Naganaboina, V. R.; Singh, S. G. Graphene-CeO2 based flexible gas sensor: Monitoring of low ppm CO gas with high selectivity at room temperature. *Applied Surface Science* **2021**, *563*, 150272.
[202] Ramshanker, N.; Ganapathi, K. L.; Bhat, M.; Mohan, S. RF sputtered CeO 2 thin films-based oxygen sensors. *IEEE Sensors Journal* **2019**, *19*, 10821–10828.
[203] Beie, H.-J.; Gnörich, A. Oxygen gas sensors based on CeO2 thick and thin films. *Sensors and Actuators B: Chemical* **1991**, *4*, 393–399.
[204] Wang, B.; Zhu, B.; Yun, S.; Zhang, W.; Xia, C.; Afzal, M.; *et al.* Fast ionic conduction in semiconductor CeO2-δ electrolyte fuel cells. *NPG Asia Materials* **2019**, *11*, 1–12.
[205] Akbar, M.; Tu, Z.; Jin, B.; Mushtaq, N.; He, Z.; Dong, W.; *et al.* Demonstrating the dual functionalities of CeO2–CuO composites in solid oxide fuel cells. *International Journal of Hydrogen Energy* **2021**, *46*, 9938–9947.
[206] Li, L.; Zhu, B.; Zhang, J.; Yan, C.; Wu, Y. Electrical properties of nanocube CeO2 in advanced solid oxide fuel cells. *International Journal of Hydrogen Energy* **2018**, *43*, 12909–12916.
[207] Li, J.; Xie, J.; Li, D.; Yu, L.; Xu, C.; Yan, S.; et al. An Interface Heterostructure of NiO and CeO2 for using electrolytes of low-temperature solid oxide fuel cells. *Nanomaterials* **2021**, *11*, 2004.
[208] Xing, Y.; Akbar, M.; Yousaf, M.; Shah, M. Y.; Xia, C.; Gao, J.; *et al.* CeO2 coated NaFeO2 proton-conducting electrolyte for solid oxide fuel cell. *International Journal of Hydrogen Energy* **2021**, *46*, 9855–9860.
[209] Melchionna, M.; Bevilacqua, M.; Fornasiero, P. The electrifying effects of carbon-CeO2 interfaces in (electro) catalysis. *Materials Today Advances* **2020**, *6*, 100050.
[210] Lucid, A. K.; Keating, P. R.; Allen, J. P.; Watson, G. W. Structure and reducibility of CeO2 doped with trivalent cations. *The Journal of Physical Chemistry C* **2016**, *120*, 23430–23440.
[211] Kogut, I.; Wollbrink, A.; Steiner, C.; Wulfmeier, H.; El Azzouzi, F.-E.; Moos, R.; *et al.* Linking the electrical conductivity and non-stoichiometry of thin film Ce1$-$ xZrxO2$-\delta$ by a resonant nanobalance approach. *Materials* **2021**, *14*, 748.
[212] Sun, L.; Zhou, L.; Yang, C.; Yuan, Y. CeO2 nanoparticle-decorated reduced graphene oxide as an efficient bifunctional electrocatalyst for oxygen reduction and evolution reactions. *International Journal of Hydrogen Energy* **2017**, *42*, 15140–15148.
[213] Lai, L.; Lu, J.; Wang, L.; Luo, G.; Zhou, J.; Qin, R.; *et al.* Magnetism in carbon nanoscrolls: Quasi-half-metals and half-metals in pristine hydrocarbons. *Nano Research* **2009**, *2*, 844–850.

[214] Nishitani, S.; Sekiya, R.; Haino, T. Chirality-embedded nanographenes. *Angewandte Chemie* **2020**, *132*, 679–683.

[215] Liu, N.; Bo, G.; Liu, Y.; Xu, X.; Du, Y.; Dou, S. X. Recent progress on germanene and functionalized germanene: Preparation, characterizations, applications, and challenges. *Small* **2019**, *15*, 1805147.

[216] Goodenough, J. B.; Park, K.-S. The Li-ion rechargeable battery: A perspective. *Journal of the American Chemical Society* **2013**, *135*, 1167–1176.

[217] Nguyen, C. V.; Ngoc Hieu, N.; Duque, C.; Quoc Khoa, D.; Van Hieu, N.; Van Tung, L.; *et al.* Linear and nonlinear magneto-optical properties of monolayer phosphorene. *Journal of Applied Physics* **2017**, *121*, 045107.

[218] Wu, M.; Xu, B.; Chen, L.; Ouyang, C. Geometry and fast diffusion of AlCl4 cluster intercalated in graphite. *Electrochimica Acta* **2016**, *195*, 158–165.

[219] Siebentritt, S.; Pues, R.; Rieder, K.-H.; Shikin, A. M. Surface phonon dispersion in graphite and in a lanthanum graphite intercalation compound. *Physical Review B* **1997**, *55*, 7927.

[220] Lee, C.; Chen, S.; Yang, C.; Su, W.-S.; Lin, M. Low-energy electronic structures of nanotube–nanoribbon hybrid systems. *Computer Physics Communications* **2011**, *182*, 68–70.

[221] Okamoto, Y. Density functional theory calculations of alkali metal (Li, Na, and K) graphite intercalation compounds. *The Journal of Physical Chemistry C* **2014**, *118*, 16–19.

[222] Thess, A.; Lee, R.; Nikolaev, P.; Dai, H.; Petit, P.; Robert, J.; *et al.* Crystalline ropes of metallic carbon nanotubes. *Science* **1996**, 273, 483–487.

[223] Wang, Y.; Ding, Y. Mechanical and electronic properties of stoichiometric silicene and germanene oxides from first-principles. *Physica Status Solidi (RRL)-Rapid Research Letters* **2013**, *7*, 410–413.

[224] Pennington, G.; Goldsman, N. Semiclassical transport and phonon scattering of electrons in semiconducting carbon nanotubes. *Physical Review B* **2003**, *68*, 045426.

[225] Chu, S. W.; Baek, S. J.; Kim, D. C.; Seo, S.; Kim, J. S.; Park, Y. W. Charge transport in graphene doped with diatomic halogen molecules (I2, Br2) near Dirac point. *Synthetic Metals* **2012**, *162*, 1689–1693.

[226] Zhang, X.; Lu, G. The spin-orbit coupling induced spin flip and its role in the enhancement of the photocatalytic hydrogen evolution over iodinated graphene oxide. *Carbon* **2016**, *108*, 215–224.

CHAPTER 2

CHEMICAL AND PHYSICAL ENVIRONMENTS

Nguyen Thanh Tien*, Nguyen Thi Dieu Hien†, Vo Khuong Dien†, Shih-Yang Lin†, Wen-Dung Su‡, Wang Yu-Ming† and Ming-Fa Lin†

**Department of Physics, College of Natural Sciences, Can Tho University, Can Tho, Vietnam*
†Department of Physics, National Cheng Kung University, Tainan, Taiwan
‡Department of Materials Science and Engineering, National Cheng Kung University, Tainan, Taiwan

Abstract

This chapter delves into the diverse chemical and physical environments encountered in materials science. It investigates the phenomena of chemical absorptions, intercalations, substitutions, decorations, hetero-junctions, and physical perturbations. By examining the effects of these factors on materials, it uncovers insights into the modification and optimization of material properties for targeted applications. This comprehensive exploration equips researchers and engineers with the knowledge necessary to manipulate and enhance material behavior in specific chemical and physical environments.

Keywords: Chemical and Physical modifications.

The periodic table,[1,2] which clearly shows the unusual atomic configurations of pure elements, is response for all condensed-matter systems. Each isolated atom has multi-spherical charge distributions according to the binding-dependent orbitals. After chemical reactions, the neighboring atoms are able to combine together through the drastic changes in the spatial charge density distribution. The various 3D, 2D, 1D, and 0D materials have been successfully synthesized by the various methods through the variation of chemical and physical environments, such as graphite/diamond (Fig. 2.1), graphene,[3–5] carbon nanotubes/graphene nanoribbons,[6–8] and carbon fullerenes/disks/rings/chains/tori/onions. Any condensed-matter system can provide a very active chemical environment through the dangling bonds at its open surfaces and edges, the defects, impurities, and vacancies inside a sample, and the significant deformations. As a result of dramatic changes in orbital hybridizations[9–11] and spin configurations,[12] their quasi-particle behaviors will be greatly diversified by chemical modifications, such as: (1) chemical adsorptions (discussed later in Section 2.1.1), (2) intercalations (Section 2.1.2), (3) substitutions (Section 2.1.3), (4) decorations (Section 2.1.4), and (5) heterojunctions (Section 2.1.5). Moreover, the physical conditions, which correspond to the crystal symmetries, are easily modulated by external perturbations, such as: (1) a uniform perpendicular magnetic field/a spatially modulated magnetic or electric field/a combined field of the first and second (five types of external fields); (2) position- and time-dependent Coulomb fields (the incident electron beam), electromagnetic waves, mechanical stresses, and thermal excitations. Efficient research strategies can be initiated in outstanding mainstream materials, leading to an expanded quasi-particle framework, e.g., graphene-related systems,[13,14] binary semiconductor compounds of III–VI, III–V, II–VI, II–V, or IV–IV,[15–24] and cathode/electrolyte/anode of ion-based batteries. The simultaneous progress among the first-principles simulations, the phenomenological models, and the experimental observations is also one of the focuses of the study.

2.1. Chemical Modifications

Chemical modifications are very powerful in creating various crystal phases, leading to great developments in the basic and applied scientific research under a unified quasi-particle framework through the simultaneous use of numerical simulations, phenomenological models, and experimental observations.

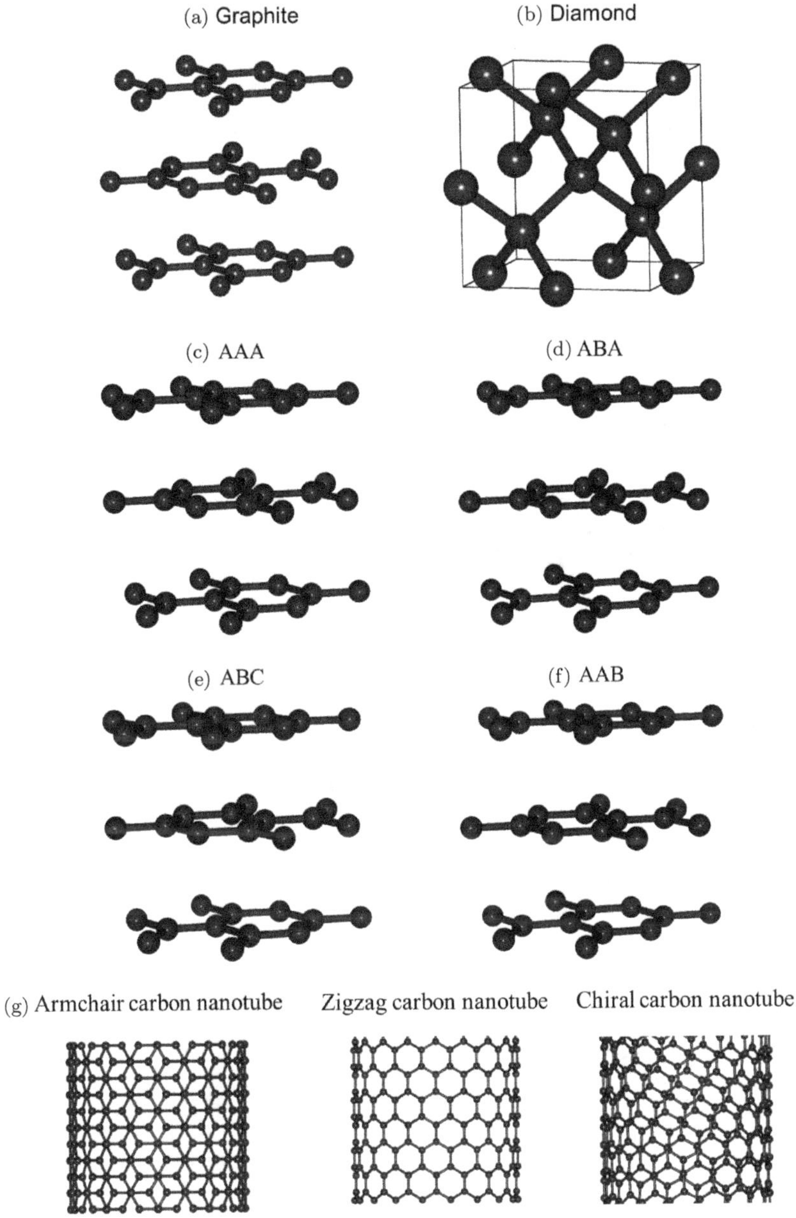

Fig. 2.1. The optimal crystal structures for (a) graphite, (b) diamond, graphene with (c) AAA, (d) ABA, (e) ABC, and (f) AAB stackings, (g) carbon nanotubes, (h) graphene nanoribbons, (i) carbon fullerenes, (j) disks, (k) rings, (l) chains, (m) tori, and (n) onions.

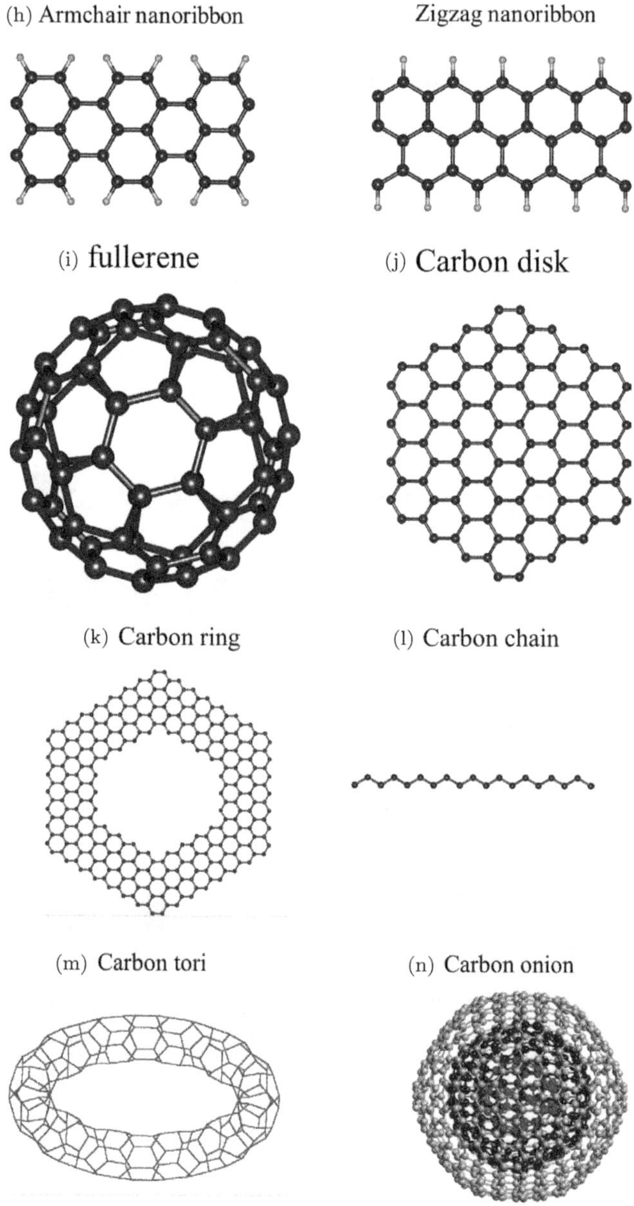

Fig. 2.1. (*Continued*)

2.1.1. *Chemical absorptions*

Any planar, curved, folded, scrolled, buckled, cylindrical, or spherical surface, which frequently appears in honeycomb networks, is an outstanding candidate for providing a platform for chemical modifications. A number of dangling bonds are able to generate very strong attractive interactions between host atoms and guest atoms/molecules/functional groups. The diversified single-/multi-orbital hybridizations comprise the van der Waals,[25,26] metallic, covalent, and ionic bondings, in which their characteristics are mainly determined by the various atomic configurations of distinct chemical bonds. Moreover, the ferromagnetic, antiferromagnetic, and non-magnetic spin configurations emerge under specific physical environments. For example, most of the transition- and rare-earth-metal atoms can exhibit ferromagnetic spin-density distributions, and the two open edges of 1D zigzag graphene nanoribbons clearly exhibit antiferromagnetic/ferromagnetic magnetism across the ribbon center/on the same boundary. The drastic changes, even the dramatic transformations, of the spatial charge/spin-density distributions are responsible for the composite quasi-particles, which are very suitable for developing a grander framework (a great development described in a series of books[13,14,27–34]).

The low-dimensional systems, which possess the high-symmetry honeycomb structures, are very suitable for implementing hydrogenations,[35] alkalizations,[36,37] oxidizations, and halogenations,[38,39] such as the chemical absorptions of (s, p_x, p_y, p_z)-guest atoms on monolayer graphene (Figs. 2.2(a) and 2.2(b)), silicene (Figs. 2.2(c) and 2.2(d)), germanene (Figs. 2.2(e) and 2.2(f)), tinene, (Figs. 2.2(g) and 2.2(h)), plumbene (Figs. 2.2(i) and 2.2(j)), phosphorene, antimonyene, and bismuthene, as well as binary GaO/GaS/GaSe/GaTe/InSe/InTe/GaAs compounds. More interestingly, the hydrogenated graphene systems under the double- or single-side full reactions clearly indicate the thorough destruction of carbon-honeycomb lattices and thus cause a total change in the essential quasi-particle behavior. A pristine carbon-sp^2 bonding hexagonal network has a planar structure, where the σ- and π-electronic bondings are perpendicular to each other, characterizing

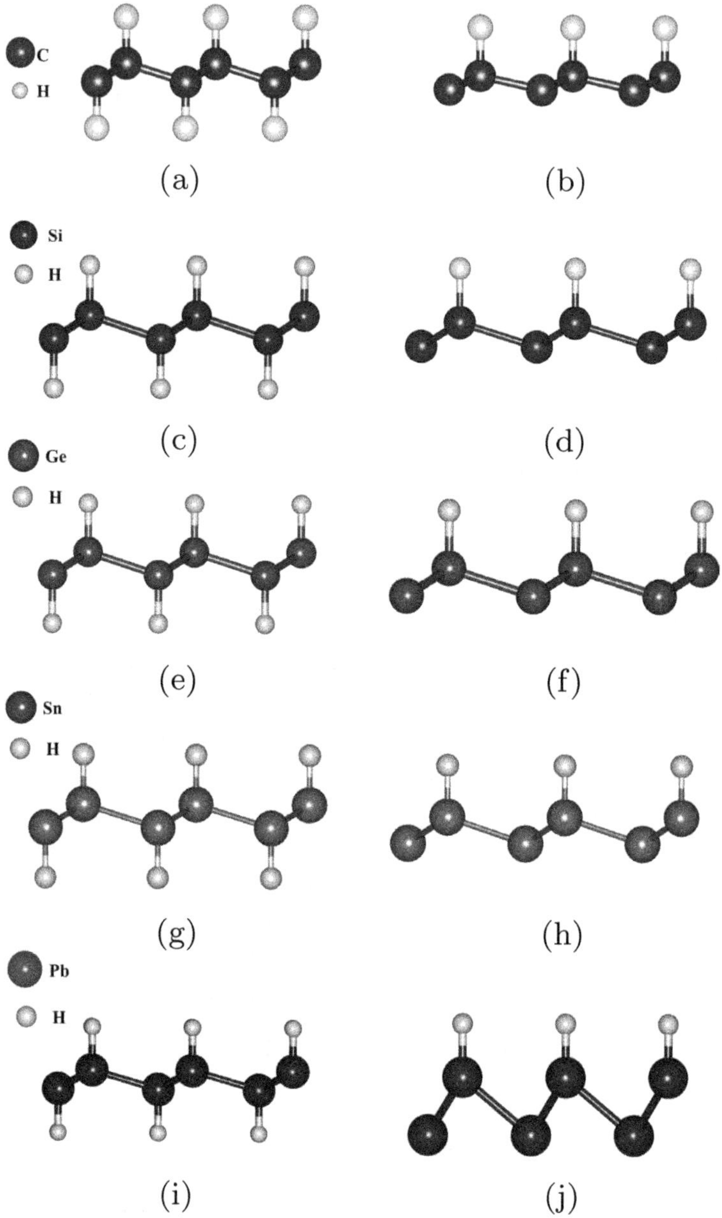

Fig. 2.2. Double- and single-side hydrogenations of (a)/(b) monolayer graphene, (c)/(d) silicene, (e)/(f) germanene, (g)/(h) tinene, and (i)/(j) plumbene systems.

their non-anticrossing subband behaviors. However, the very strong attractions between the hydrogen and carbon atoms are realized by the top-side adatom positions[35,40,41] and the buckled honeycomb structure. The sp^2 and π bondings are totally changed into the more complicated ssp^3 ones through the delicate VASP simulations,[42–45] so that the essential composite quasi-particles display more unusual phenomena, e.g., a zero- to wide-gap semiconductor transition, the absence of the well-defined π- and σ-electronic states, the presence of excitonic effects, a great enhancement in the optical threshold frequency, more complex optical excitation channels associated with the distinct orbitals of the initial and final states, and the irregular plasmon modes. Obviously, similar chemical adsorptions could be generalized to a number of 2D layered systems and serve as the modulation strategy for observing the diversified quasi-particle (as detailed in several previous books[13,14,27,29,31–33]).

1D zigzag graphene nanoribbons are predicted to exhibit the highest charge transfer of free conduction electrons and the dramatic transitions of spin configurations after active alkalizations (as described in a previous book[30]). A pristine system behaves as a small-gap semiconductor at the Γ point with the splitting of two partially flat valence and conduction subbands because of the antiferromagnetic spin arrangement across the Fermi level. This direct gap will decrease with an increase in the nanoribbon width. The net magnetic moment vanishes since the spin configuration is ferromagnetic along the same edge. The spatial spin-density distribution is very sensitive to the position-dependent metallic bond between alkali and carbon atoms, as clearly illustrated in Fig. 2.3, showing the lithium-adsorbed zigzag graphene nanoribbon at different positions (a)/(c)/(e)/(g) near the open edge to understand (b)/(d)/(f)/(h) the distinct magnetisms. The unusual magnetic configurations cannot survive in armchair graphene nanoribbons. The Hubbard-like spin interactions and the on-site electron–electron Coulomb ones strongly cooperate or compete with orbital-dependent kinetic energy, the ionization Coulomb potential, and the exchange and correlation energies. This is worthy of systematic investigations for 1D silicene, germanene, tinene, and plumbene nanoribbons.[8,46–48] It is very interesting to evaluate the magnitude of free carrier density from the

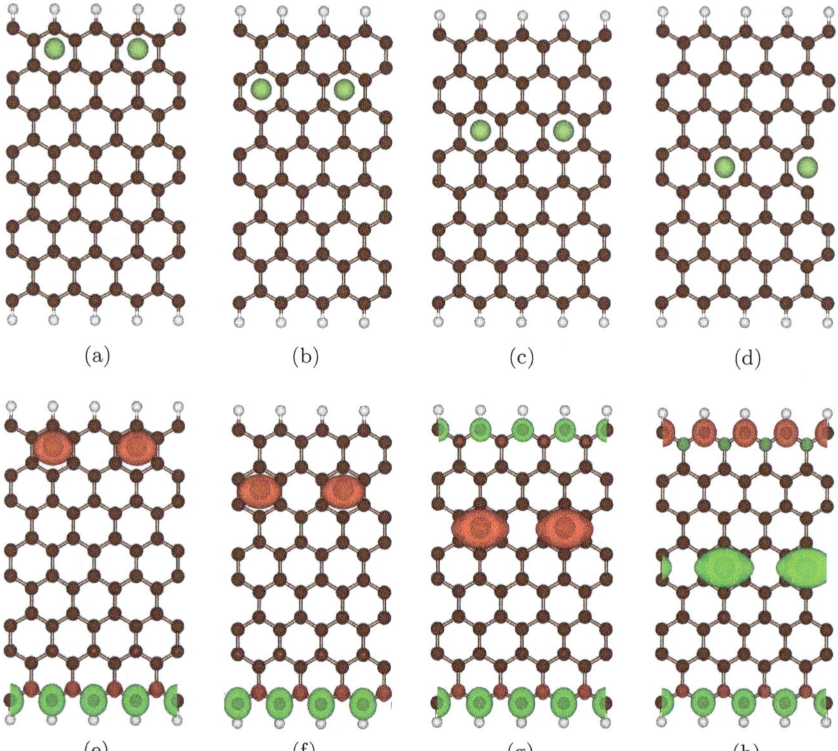

Fig. 2.3. Lithium-adsorbed zigzag graphene nanoribbon under the distinct positions (a)/(b)/(c)/(d) near the open edge clearly illustrating (e)/(f)/(g)/(h) the different magnetisms (the spatial spin-density distributions).

metallic bondings. This quantity is proportional to the magnitude of the Fermi momentum, and its value is identical to the adatom concentration, being independent of position. The alkalized graphene nanoribbons could be outstanding 1D conductors and have great potential applications in electronic nanodevices,[49] as do the 1D armchair carbon nanotubes.

2.1.2. *Intercalations*

The layered crystal structures, which are very suitable for the chemical intercalations/de-intercalations,[50] can greatly diversify the quantum quasi-particles. For example, pure carbon atoms can clearly form 3D bulk graphite (Fig. 2.1(a)) and diamond (Fig. 2.1(b)) in nature. The former and the latter, respectively, possess the sp^2 and sp^3 chemical bondings with three and four nearest neighbors. There exists the dimensionality crossover with a decrease in the number of graphite sheets. The interlayer van der Waals interactions and the interlayer C-2p_z orbital hybridizations are very weak, but there are significant attractive interactions between two neighboring graphitic sheets. Such a mechanism is strongly dependent on the stacking configurations of honeycomb lattices, e.g., the existence of the AB-, ABC-, and AA-stacked graphite systems. Most/A few of the natural materials belong to the first/second type (~99%/1%), while the third one is almost vanishing. The small overlaps of valence and conduction subbands occur near the K/H valleys (the hexagonal corners of the first Brillouin zone) and thus exhibit the temperature-dependent semi-metallic behaviors, such as the diverse (momentum, frequency)-excitation phase diagrams (the electron–hole excitations and plasmon modes) and the unusual optical/magneto-optical properties. More interestingly, the graphite-like crystal structures also survive in binary semiconductor compounds. For example, GaSe/GaTe/InSe/InTe (Figs. 2.4(a)–2.4(h)) have been successfully synthesized using the experimental technique of molecular beam epitaxy (MBE).[51–53] These mainstream layered materials possess a hexagonal honeycomb symmetry on the (x, y) plane and prominent four-sublattice bucklings on the $(x, z)/(y, z)$ planes. The attractive interactions mainly arise from the weak vertical orbital hybridizations, as thoroughly examined in a previous book.[28] They depend on covalent bondings to create the middle-gap characteristics. Chemical interactions and de-intercalations can be outstanding modulation strategies.

Graphite intercalation compounds, which include the donor[54] and acceptor types,[55] have been extensively studied experimentally and theoretically for more than 100 years. After chemical intercalations, carbon-honeycomb structures remain in the planar form because of the very strong $sp^2\,\sigma$ bonding, and the intercalant atoms/molecules can generate a plane crystal. This well-characterized symmetry structure is the so-called stage-n crystal, where two periodic intercalant layers possess n graphitic sheets. Figure 2.5 clearly shows the layered graphite potassium compounds,

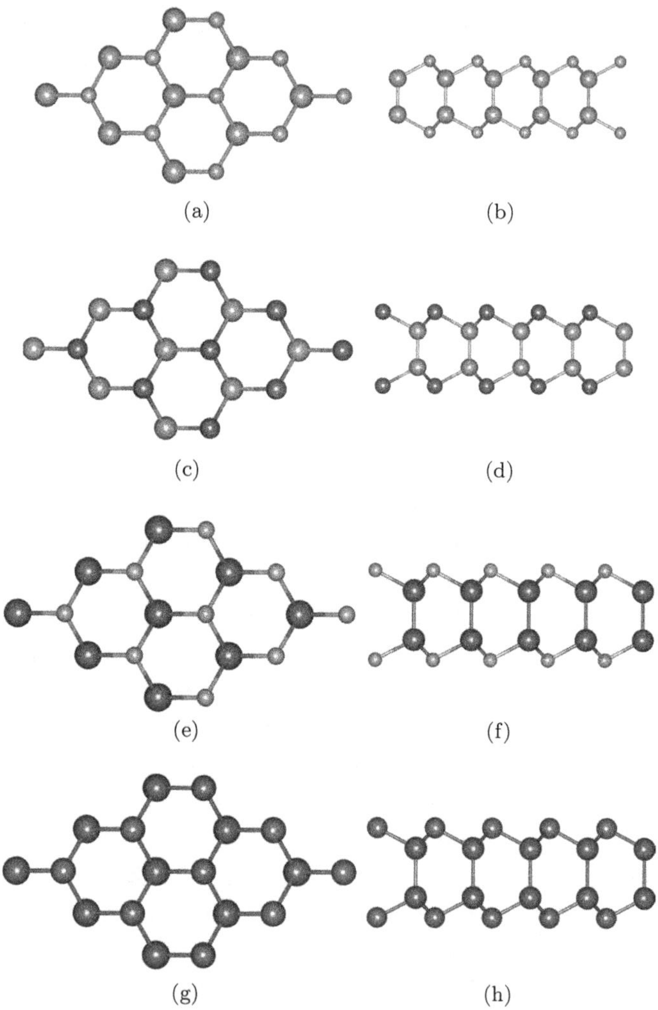

Fig. 2.4. The graphite-like binary semiconductor compounds: (a)/(b) GaSe, (c)/(d) GaTe, (e)/(f) InSe, and (g)/(h) InTe under the top/side views.

namely (1) stage-1 C_8K, (2) stage-2 $C_{16}K$, (3) stage-3 $C_{24}K$, and (4) stage-4 $C_{32}K$. The identical structures correspond to Na, Rb, Cs, and Mg, while the lithium intercalations have the LiC_{6n} form with the highest intercalant concentrations. Through delicate first-principles simulations and analyses, the semi-metal–metal transitions can be attributed to the creation of the metallic bondings between

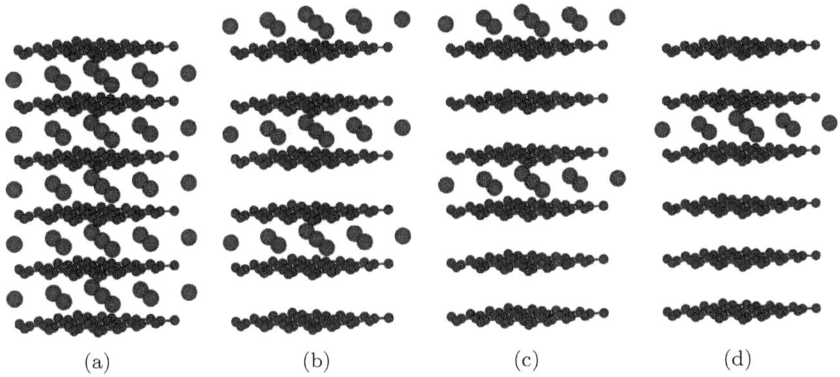

Fig. 2.5. Layered graphite potassium compounds: (a) stage-1 C_8K, (b) stage-2 $C_{16}K$, (c) stage-3 $C_{24}K$, and (d) stage-4 $C_{32}K$.

the carbon–intercalant bonds, that is, more C-2p$_z$ orbitals exhibit wider charge density distributions along the perpendicular direction. Concerning the electron/hole doping cases, electron charges are greatly transferred from the host/guest atoms to the guest/host ones through the vertical orbital hybridizations, e.g., the graphite alkali/$AlCl_4^-$ compound. Furthermore, there exists the blue-/redshift of the Fermi level, which is responsible for drastic changes in other quasi-particle behaviors. Its magnitude is proportional to the intercalant concentration and could be estimated from the featured van Hove singularities and threshold absorption spectra. Pristine graphite has the minimum density of states (DOS) at the Fermi level, and then, its energy splitting occurs after electron or hole doping. Such a semi-metallic system only shows a vanishing threshold frequency due to the gapless interband optical transitions. However, the vertical excitations in graphite intercalation compounds are effective only under the specific condition of $\omega > 2E_F$. In addition to the middle-/deep-energy π/σ carriers, the low-energy conduction electrons or valence holes generate unusual quasi-particle phenomena, such as copper-like electrical conductivities, greatly enhanced superconductivity temperatures, diverse (momentum, frequency)-excitation phase diagrams, stacking-enriched optical properties,[56–60] and rich magneto-optical spectra.[57,61]

Graphite intercalation compounds[50,54,62–66] are outstanding candidates for study in the basic and applied sciences and for commercial batteries. The current experimental studies clearly illustrate their critical roles as effective cathode/anode materials through the high-performance stationary transports in positive-/negative-ion-based batteries, e.g., the Li^+-, Na^+-, and Mg^{2+}-/$AlCl_4^-$ and $Al_2Cl_7^-$-based batteries (the mainstream operation systems). Their significant features, which have enabled them to become commercial products, have been well characterized in previous books.[34] During the charging/discharging reaction processes in lithium-based batteries,[67,68] the I^+ ions and free electrons first escape from the ternary or quaternary lithium oxides (e.g., LiFeO/LiCoO/LiNiO or LIFePO)[69–74] and then enter the electrolyte and outside the lead line, respectively. Both of them will reach graphite within a very short time, where their very strong attractions come to exist because of their natural charge screening ability. Consequently, the guest intercalants should recover to an almost neutral status, as clearly illustrated in the current work using the stage-n crystal structures.[65] The very sensitive time dependence, which mainly arises from the quick chemical reactions, obviously leads to unusual creations of the time-dependent intermediate configurations. Such irregular crystal phases are expected to possess giant Moire superlattices, in which their complicated crystal symmetries could be simulated by the VASP calculations. How to fully comprehend the prominent partnerships among the temporary/quasi-stable crystal phases in anode/cathode/electrolyte materials[75–78] is a very interesting research topic, e.g., the very sensitive relation of the vertical carbon–intercalant orbital hybridizations with guest-adatom density and arrangement. The chemical reactions, which determine the intercalant concentration and distribution, can be simulated using molecular dynamics.[79,80] A strong cooperation between two numerical methods will be useful for the further development of a grand quasi-particle framework. It should be noted that the layered crystals of binary semiconductor compounds of III–VI/III–V/II–VI/II–V/IV–IV are worthy of systematic investigations of chemical interactions.[15,18–21,23,24,81]

2.1.3. *Substitutions*

Chemical substitutions can generate a number of crystal phases by changing the relative concentrations and positions of the distinct atom-based materials, such as the 2D binary semiconductor compounds of III–VI/III–V/II–VI/II–V/IV–IV (Fig. 2.6), the planar/tubular $B_xC_yN_z$,[82–84] the 3D Li-, P-, and S-related compounds (details in book), and the ternary or quaternary lithium oxides in ion batteries.[72–74, 85–88] These outstanding condensed-matter systems are very suitable for the development of a grand quasi-particle framework since they exhibit diversified phenomena in terms of geometric,[89, 90] electronic, magnetic,[38, 89, 91] optical,[92, 93] Coulomb-excitation,[13, 94–96] transport,[97, 98] mechanical,[99] and thermal properties. It is well known that monolayer graphene is a carbon-honeycomb network with pure

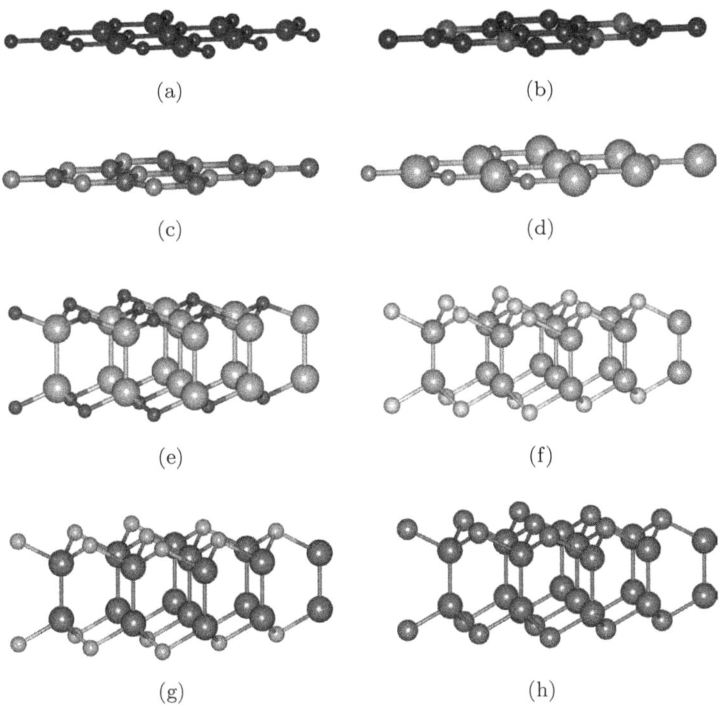

Fig. 2.6. Two-dimensional binary compounds: (a) SiC, (b) BC_3 (c) BN, (d) GaN, (e) GaO, (f) GaS, (g) InSe, and (f) InTe.

sp^2 and π bondings, and its hexagonal structure remains unchanged under full silicon substitution, whereas the 2D binary silicon carbide has successfully become the third-generation semiconductor compound in the industry because of its specific functionalities. The former and the latter, respectively, belong to the zero- and middle-gap semiconductors. The dramatic transformation of SiC mainly arise from the impure (2s, $2p_x$, $2p_y$)–(3s, $3p_x$, $3p_y$) and $2p_z$–$3p_z$ orbital hybridizations. Their hopping integrals[100,101] and on-site ionization potentials,[100,101] which appear in the tight-binding model,[102–106] will be response for the diverse quasi-particle quasi-particles. The binary Si_xC_{1-x} compounds will dramatically change from graphene to silicene with a gradual increase in the silicon component (x varies from 0 to 1). The coexistence of C–C, Si–C, and Si–Si bonds can be investigated through systematic studies. That is, the superposition of pure and impure covalent bondings is very useful in understanding all essential properties and serves as a modulation mechanism for observing diversified phenomena.

The lithium-, phosphorus-, and sulfur-based materials clearly present a number of crystal phases (details in book), such as lithium BCC crystal, black and blue phosphorus crystals, sulfur crystals, binary lithium sulfide, lithium phosphorus materials,[107–109] phosphorus sulfur compounds of $P4S_x$ ($x \leq 10$; Figs. 2.7(a) and 2.7(b)), ternary lithium phosphorus sulfur compounds of $Li_2P_2S_6/Li_4P_2S_6/Li_7P_3S_{11}/Li_3PS_4$ (Figs. 2.7(c)–2.7(e)), and quaternary lithium phosphorus sulfide chloride (Fig. 2.7(f)). Using the delicate VASP simulations, all systems, except for metallic lithium crystals, reveal highly non-uniform covalent bondings, owing mainly to the active (s, p_x, p_y, p_z) orbitals. For example, the ternary compounds possess the impure mixings of 2s–($2s, 2p_x, 2p_y, 2p_z$) and ($2s, 2p_x, 2p_y, 2p_z$)–($2s, 2p_x, 2p_y, 2p_z$) orbital hybridizations, respectively, in the Li–S and P–S bonds, in which they survive in Moire superlattices. The unusual charge density distributions are strongly dependent on the positions and relative concentrations of the lithium, phosphorus, and sulfur atoms. How to build very close relationships between the impure sp^3-orbital bondings and all essential properties could be a focus of future studies, e.g., the concise

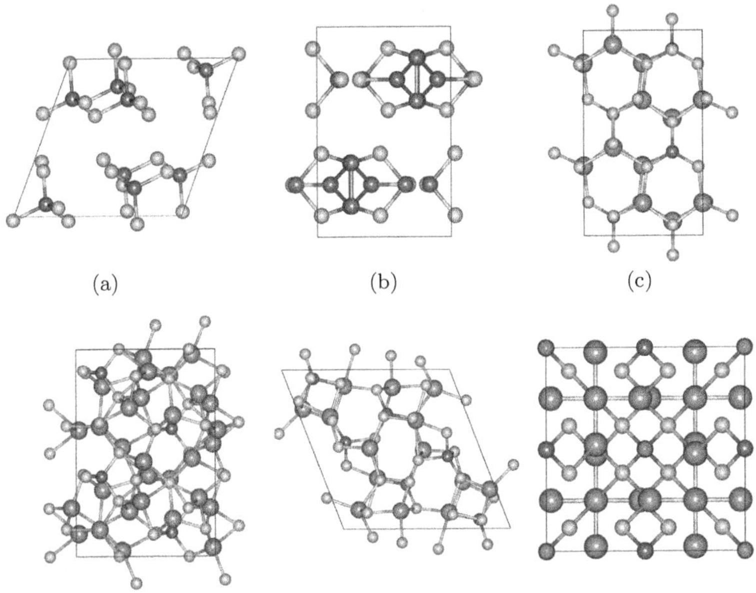

(a) (b) (c)

(d) (e) (f)

Fig. 2.7. Lithium-, phosphorus-, and sulfur-based materials: (a) binary P_4S_{10}, (b) P_4S_3, (c) ternary Li_3PS_4, (d) Li_7PS_6, (e) $Li_7P_3S_{11}$ and (f) quaternary lithium phosphorus sulfide chloride.

orbital-hybridization pictures for determining the featured quasi-particle properties. Interestingly, lithium sulfide batteries might become a popular product with extensive use in industries owing to their high operating performance (details in book). Both basic and applied scientific studies need to achieve strong cooperation in the current and near-future investigations.

The 3D ternary and quaternary lithium oxides have been very successful in serving as cathode/electrolyte/anode materials in lithium-ion-based batteries, which is clearly demonstrated in a series of published books.[34] Such operation systems, which are part of everyday electrical equipment and tools, have presented the highest performance up to now. Specifically, they play critical roles in achieving environmental targets associated with the car industry. How to build a generalized model to fully understand the critical mechanisms of stationary ion transports should be an open issue for the near-future studies. Generally, the electrical conductivities of condensed-matter systems mainly arise from the electron fermions, and the static Kubo formula is sufficient for exploring the temperature scattering behavior.[110,111] However, the ion transports are closely related to the heavy particles with integer spins, so whether the Bose–Einstein statistics need to be taken into account is an interesting question. During the charging process of a lithium-ion-based battery, very few Li atoms and free electrons, respectively, escape from the cathode and enter the lead. Lithium oxide will present vacancies at a very low density, e.g., a low-symmetry LiFeO (Fig. 2.8(a)) or LiFePO (Fig. 2.8(b)) with plenty of atoms in a primitive unit cell. A rather small variation of atom positions, which represents the classical chemical substitutions, will initiate the effective Coulomb field in driving the ion transports. The close relationships among the vacancy concentrations, the greatly reduced crystal symmetries, and the sensitive electric fields should be the focus of studies.

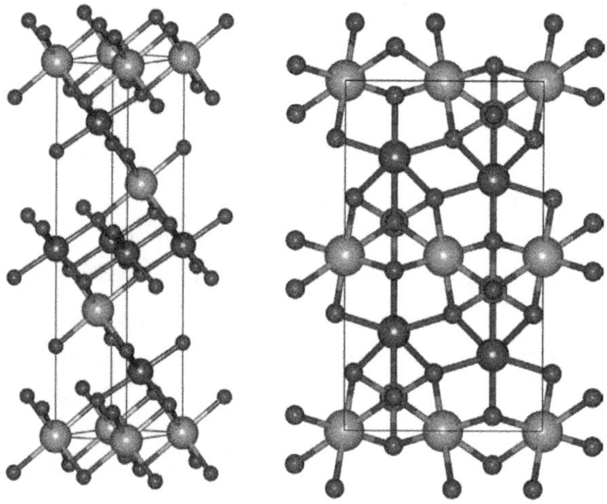

(a) (b)

Fig. 2.8. The ternary and quaternary lithium iron oxides: (a) LiFeO and (b) LiFePO.

2.1.4. *Decorations*

Chemical decorations, which are very suitable for open-edge structures, can greatly diversify the composite quasi-particles and thus expand the grand framework established in previously published books.[14,30] Such chemical modifications are clearly revealed in open boundaries with a number of dangling bonds under the quantum-confinement effects.[91,112] For example, drastic changes occur in the essential properties of 1D graphene/silicene/germanene/tinene/plumbene nanoribbons[46–48] and 0D finite carbon nanotubes. Furthermore, the current experimental efforts have successfully synthesized planar, curved, folded, scrolled, and stacked carbon-honeycomb crystals with various boundary conditions. The edge dangling bonds can create observable bond-length fluctuations, produce non-uniform hopping integrals of the neighboring orbitals, and thus greatly diversify the geometric properties after chemical decorations. The strongly modified crystal symmetries are responsible for all quasi-particle properties. The electronic states of pristine graphene nanoribbons could be regarded as those sampled from the Dirac cone structure in monolayer graphene under a suitable open boundary condition. An m-dimer-/zigzag-line graphene nanoribbon, corresponding to an armchair/zigzag system, is predicted to exhibit a well-behaved bandgap through hydrogen passivation at two open edges. The unusual relations with ribbon widths (w_Ds) are characterized as follows: (1) small gap widths are inversely proportional to the square of w_D for $m = 3I + 2$ armchair systems (where I is an integer);[113] (2) the middle gaps of $E_g s$\proport $1/w_D$ for $m = 3I + 1$ and $3I + 2$; and (3) an obvious decline in bandgap in the widened zigzag nanoribbons.[114] The absence of a uniform C–C network is responsible for the narrow-gap semiconductors and the splitting of extremely localized valence and conduction states near the zone boundary (the split-edge states). In short, the longitudinal edge structures, transverse quantum confinement, chemical passivations, and non-uniform honeycomb crystals cooperate with one another in determining all the fundamental quasi-particle behaviors. Also, the

1D buckled silicene/germanene/tinene/plumbene nanoribbons[46–48] display more complicated chemical modifications. A 1D graphene nanoribbon is available for examination through the very strong STS cutting along the longitudinal direction of 2D monolayer graphene or the rather prominent chemical reaction in destroying the uniform C–C network of 1D carbon nanotubes. More interestingly, there exist two open edges in a curved/folded/scrolled/planar honeycomb crystal. The curved graphene nanoribbons (Fig. 2.9(a)) are outstanding candidates for observing the unusual phenomena. Through delicate VASP simulations and analytic calculations, three types of stable crystal structures have been identified: (I) the closed carbon nanotubes for B/C/N decorations with very strong covalent bondings, (II) the attractive arc structures through the classical dipole–dipole interactions between Be/Mg/Al guest adatoms, and (III) the planar crystals under the repulsive interactions H/Li/Na/K/Ca/Si/P/O/S/F/Cl/I/Cr/Mn/Fe/Co/Ni/Cu/Ti/Sc/Zn. Furthermore, the main features of charge transfer, atom–carbon distance, atom–atom distance, magnetism of carbon/atom, and AFM/FM are explored in detail. Such 1D materials are predicted to exhibit the featured metallic and semiconducting band structures, respectively, for type (I)/(II) and (III). The simultaneous measurements of high-resolution scanning tunneling microscopy (STM) for geometric structures[115,116] and scanning tunneling spectroscopy (STS) for van Hove singularities[117–119] are very useful in identifying the

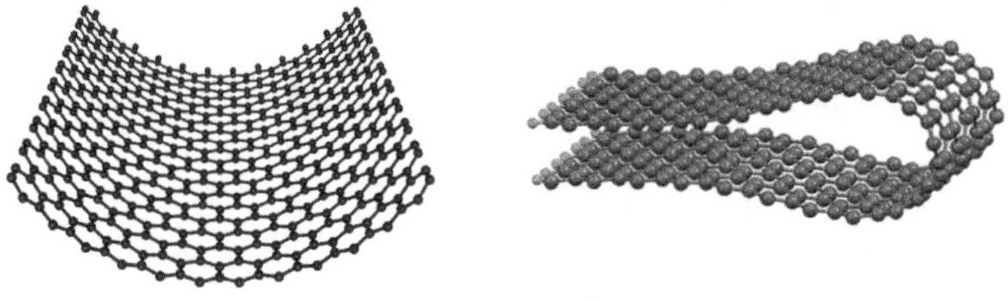

(a) (b)

Fig. 2.9. The (a) curved and (b) folded graphene nanoribbons.

geometry-enriched electronic properties, as done for the single-walled carbon nanotubes. Folded graphene nanoribbons (Fig. 2.9(b)), which have been successfully synthesized in laboratories, present unusual crystal phases and thus rich quasi-particle properties.[8] Their optimal geometries are similar to irregular book frameworks and are characterized by two open edges,[120, 121] bilayer overlap (different stacking configurations), and their combination through the curved region. The achiral hexagon arrangement might consist of passivated/non-passivated boundaries. Most of the folded systems should have well-behaved zigzag or armchair edges, of which the former have magnetic spin arrangements. The upper and lower parts overlap each other and display the well-known vertical attractions of van der Waals interactions.[25, 26, 122] The interlayer carbon-$2p_z$ orbital hybridizations play a critical role in a stable crystal and its energy spectrum. The curvature effect,[123, 124] as clearly illustrated in a cylindrical carbon nanotube surface, is able to create a misorientation of the carbon-$2p_z$ orbitals (or even the prominent sp^3 bondings in a small radius system). The fundamental quasi-particle properties, the rich crystal symmetries, featured band structures, orbital mixings (covalent or metallic bondings), van Hove singularities, spatial spin-density distributions (the Hubbard-like electron–electron Coulomb interactions),[125] and net magnetic moments clearly present the diverse phenomena, which are very suitable for forming a search modulation strategy. For example, the metal–semiconductor transition occurs under the variation of stacking configuration and edge structure. Furthermore, the significant width dependence of the bandgap might be quite different from that of pristine graphene nanoribbons. Of course, further chemical modifications can greatly diversify quasi-particle behaviors because of more complicated guest-adatom environments due to charge and spin distributions. More interestingly, both geometric structures and electronic properties, which correspond to passivated armchair graphene nanoribbons,[126] are greatly enriched by strong chemical bondings between the edge carbons and various adatoms through investigations based on first-principles calculations. Adatom edge arrangements, bond length fluctuations, charge density distributions, and energy spectra are dramatically changed by a specific edge passivation. Elements with an atomic number of less than 20 could be classified into the following three types of well-defined crystal symmetries, depending on the optimal geometric structures: (I) planar and (II) & (III) buckled structures, the latter of which are associated with specific arrangements and stacked configurations of guest adatoms. H-, N-, Be-, and B-edge decorations (I) only present a planar honeycomb, respectively, corresponding to two saturated open edges, heptagon-pentagon closed structures (the most stable among the various crystal phases), triangle-pentagon structures, and non-uniform hexagons. O-, S-F, and Cl-C reveal non-planar passivation edges (II) on the double side with unusual forms. Type III, consisting of Li/Na/K/Ca/Mg/Al/Si/P guest adatoms, presents an arrangement where they symmetrically stack up at the edges. Distorted pentagons are clearly revealed through the adatom bonding to the nanoribbon, while the carbon hexagons have no distortion even near the edges. Apparently, the unique orbital hybridizations are responsible for the edge geometries and the featured band structures. For example, the H–C, O–C, Ca–C, and C-C bonds, respectively, have multi-orbital mixings: 1s–(2p$_x$, 2p$_y$, 2p$_z$), (2s, 2p$_x$, 2p$_y$, 2p$_z$)–(2s, 2p$_x$, 2p$_y$, 2p$_z$), (3s, 3p$_x$, 3p$_y$, 3p$_z$)–(2s, 2p$_x$, 2p$_y$, 2p$_z$), (2s, 2p$_x$, 2p$_y$)–(2s, 2p$_x$, 2p$_y$) and 2p$_z$–2p$_z$. The low-lying band structures are drastically varied, displaying non-monotonous energy dispersions and adatom-dominated bands. A concise relationship between energy gap and nanoribbon width, as clearly illustrated in pristine systems, no longer exists; furthermore, certain adatoms can generate semiconductor–metal transitions. All the main characteristics are directly reflected in van Hove singularities, revealing dip structures, plateaus, symmetric peaks, and square-root divergent asymmetric peaks. More interesting results will be achieved if the transition- and rare-earth-metal adatoms are available in the edge decorations of graphene/silicene/germanene/tinene/plumbene nanoribbons. Complicated orbital bondings and rich spin configurations are fully supported by each other and can display greatly diversified quasi-particles.

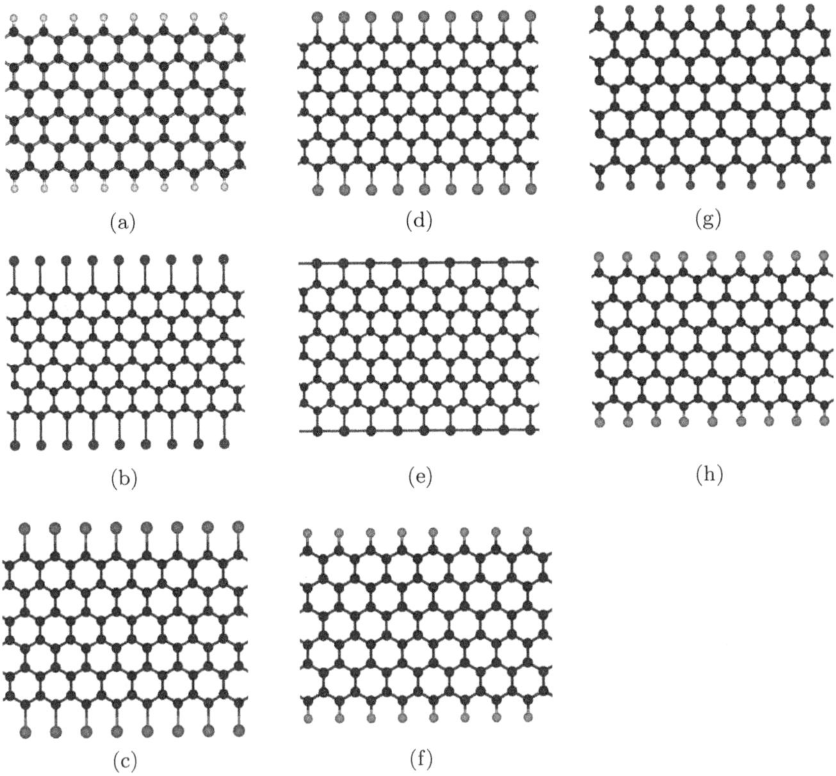

Fig. 2.10. Dangling-bond passivation of planar zigzag graphene nanoribbon at open edges through the guest adatoms: (a) hydrogen, (b) potassium, (c) magnesium, (d) aluminum, (e) silicon, (f) nitrogen, (g) oxygen, and (h) fluorine.

2.1.5. *Heterojunctions*

A heterojunction, which consists of, at least, two different subsystems,[127–129] exhibits rich and unique quasi-particle behaviors and is very suitable for developing its framework. Interestingly, it can form rather strong chemical bondings and spin-dependent configurations at a specific boundary to directly link the left- and right-hand-side subsystems. The dramatic transformation is revealed in the unusual arrangement of host–guest bonds. Their modulation phenomena along the transverse and longitudinal directions frequently induce a giant primitive unit cell under the well-behaved commensurate case. A giant Moire superlattice has a number of host and guest atoms (many active orbitals). For example, the mainstream materials have been successfully grown on metallic or semiconducting substrates, such as few-layer graphene/silicene/germanene/tinene/plumbene/phosphorene/bithmuthene,[130–135] binary/ternary semiconductor compounds of III–VI/III–V/II–VI/II–V/IV–IV,[15,18–21,23,24,136] and lithium oxide/sulfide compounds. The real operations of ion-based batteries require cathode–electrolyte and electrolyte–anode composites. Quantum transports in low-dimensional materials must cover their contacts with outside leads. Moreover, the nanoscale probes in STM/STS clearly illustrate an unusual heterojunction with a normal surface morphology. How to characterize and identify the main features/roles of heterojunction should be a study focus. In general, high-resolution angle-resolved photoemission spectroscopy (ARPES),[137–140] STS,[117–119,141] and optical spectroscopies[142–147] are available for observing the essential quasi-particle properties associated with the heterojunction effects. The Group-IV 2D materials, which include C, Si, Ge, Sn, and Pb atoms, have been successfully synthesized in laboratories since the first discovery of few-layer graphene systems through mechanical exfoliation from the Bernal 3D graphite in 2004.[5] Only the 2D carbon-related materials are produced using the various growth methods (details in books), while the others are generated through molecular beam epitaxial growth (MBE).[51–53] The former are thoroughly examined to find that they possess the AAA-/ABA-/ABC-/AAB-stacked configurations,[94,95,148–158] with different layer numbers, based on the high-resolution TEM measurements.[159] Using direct combinations of STM with LEED/RHEED,[160–163] the graphene-like honeycomb crystals are clearly shown by the 2D C-, Si-, Ge-, Sn, and Pb-adlayers, with the hexagonal symmetries, respectively, on SiO_2, Ag(111), Ir(111), and ZrB_2(0001),[164–166] Pt(111)/Al(111/Au(111),[167–169] Bi_2Te_3(111),[170] and Au(111)/AlO & Si(111)[171,172] Also, note that the surface reconstructions frequently come to exist at the adlayer and substrate (details in a previous book[32]). Among them, only graphene systems exhibit the planar structure in the absence of two-sublattice buckling, owing mainly to the dominating sp^2 bonding and its orthogonality with significant π bondings. The other Group-IV materials are fully identified to present buckled structures associated with the height difference between A and B sublattices. The currently available data show that the buckling of each layer is, respectively, ~0.24, 0.26, 0.32, and 0.35 Å for silicene,[32] germanene,[32] tinene,[32] and plumbene.[32] The bucking degree grows with an increase in the atomic number, clearly illustrating the great enhancement of the sp^3 bondings. As for the experimental observations on the band properties near the Fermi level, the semiconducting and semi-metallic behaviors in layered graphenes are confirmed by the STS and ARPES measurements (detailed in previous books[27,33]). The latter are also available for testing the existence or absence of specific Dirac cone-like energy bands in other Group-IV systems. This result indicates the critical roles of the adlayer–substrate chemical bondings. Obviously, the semiconducting or metallic adlayers have become one of the mainstream 2D materials, in which a plenty of unusual quasi-particles are worthy of systematic investigations. In addition, few-layer graphene systems can exhibit rich and unique physical, chemical, and material phenomena, such as the best mechanical responses, stacking-\layer-number-diversified electronic properties,[150,173–175] rich magnetic quantization phenomena,[31,176] distinct optical/magneto-optical selection rules,[58] special quantum Hall conductivities,[177] and various electron–hole pair regions and plasmon/magneto-plasmon modes.[178,179]

To thoroughly elucidate the substrate effects,[180,181] the silicon-adlayer/Ag(111)[32] and the bilayer AB-bottom-top silicene/Ag(111)[32] are outstanding candidates for a model study (Fig. 2.11). Both silicon and silver atoms, with atomic numbers 14 and 47, respectively, have unusual orbital configurations, making them very suitable for generating active chemical bonds. The Moire superlattices, which consist of distinct subsystems and a sufficiently thick substrate, should be taken into account simultaneously through the VASP simulations,[42–45] but not the very complicated calculations of the generalized tight-binding model.[182] The extremely non-uniform environments within the enlarged unit cells are thoroughly explored through the various bond lengths, the different bucklings

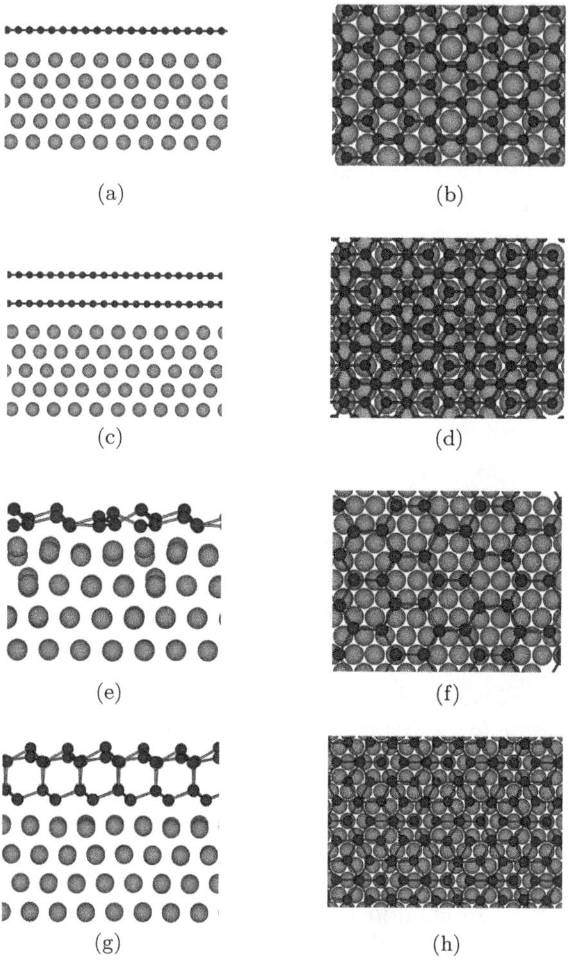

Fig. 2.11. Active orbital hybridizations near heterojunctions through the chemical cases under the top/side views: (a)/(b) & (c)/(d) monolayer & bilayer graphene on Au(111) substrate and (e)/(f) & (g)/(h) monolayer & bilayer silicene on Ag(111) substrate.

of the height differences of A and B sublattices, and the pnservablet fluctuations of the interlayer distances. Then, the optimal composite structures are available for thoroughly examining the atom- and spin-dominated electronic energy spectra, the spatial charge density distributions, the spin arrangements, the atom- and orbital-decomposed magnetic moments, and the adlayer-, substrate-, orbital- and spin-projected van Hove singularities. The detailed examinations are conducted on the main features of valence and conduction subbands, the substrate-enhanced asymmetry of hole and electron bands about the Fermi level, more band-edge states in the energy–wave-vector space, the oscillatory energy dispersions or modified Dirac cone structures near E_F, the undefined/defined π- and σ-electronic states, the silicon and silver dominances corresponding to the specific energy ranges, and the spin-degenerate or spin-split states across E_F (the vanishing or finite magnetic moments). The spin-density distributions, which might survive on all the Si and Ag layers, are very useful in understanding the creation of magnetic configurations due to the former and/or the latter. There are certain important differences among adlayer/bilayer silicene, substrates, and their composites. The above-mentioned complex results could be delicately analyzed to achieve the concise pictures of chemistry and physics. The significant multi-orbital hybridizations of Si–Si, Si–Ag, and Ag–Ag bonds are identified to be dominated by (2s, $2p_x, 2p_y, 2p_z$)–(2s, $2p_x, 2p_y, 2p_z$), (2s, $2p_x$, $2p_y, 2p_z$)–($4d_{z^2}, 4d_{x^2-y^2}, 4d_{yz}, 4d_{zx}, 4d_{xy}$), and ($4d_{z^2}, 4d_{x^2-y^2}, 4d_{yz}$, $4d_{zx}, 4d_{xy}$)–($4d_{z^2}, 4d_{x^2-y^2}, 4d_{yz}, 4d_{zx}, 4d_{xy}$). Furthermore, the ferromagnetic spin configurations are initiated from the transition-metal Ag atoms. Similar studies could be generalized to rare-earth metals with more complex 3f or 5f orbitals (as detailed in Chapter 12). This is helpful for developing a grand quasi-particle framework.

2.2. Physical Perturbations

The static and time-dependent field perturbations are able to dramatically change the environmental symmetries and can thus greatly diversify the composite quasi-particles. Such modulation strategies are clearly illustrated in the following five sections, which are consistent with a grand quasi-particle framework established in previously published books.

2.2.1. *Stationary fields: Uniform and non-uniform electric and magnetic fields*

The stationary magnetic and electric fields can dramatically change the crystal symmetries and thus greatly diversify the essential quasi-particle properties, as clearly shown by a number of theoretical predictions about electronic/magneto-electronic and optical/magneto-optical properties and quantum Hall effects. Such fields include: (I) a perpendicular uniform magnetic field (Fig. 2.12(a)), (II)/(III) a spatially modulated magnetic/electric field (Fig. 2.12(b)), (IV)/(V) their composite fields through a direct superposition, (VI) a transverse electric field (a gate voltage), and (VII) the Lorentz fields (Fig. 2.13 showing a single-walled carbon nanotube). A primitive unit cell is mainly determined by the commensurate period of a pristine crystal and a perturbation one. Furthermore, the magnetic

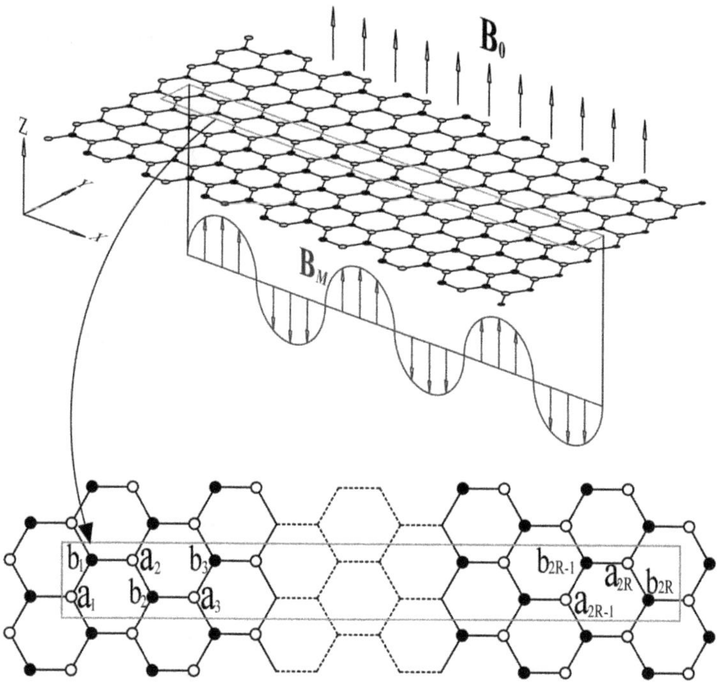

Fig. 2.12. The (a) uniform and (b) spatially modulated magnetic fields on monolayer graphene.

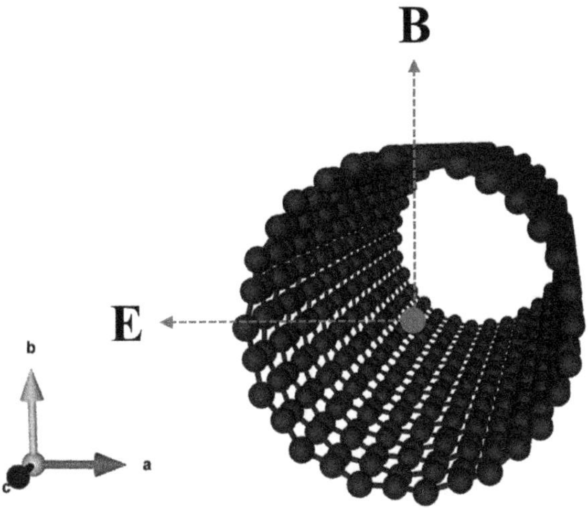

Fig. 2.13. The longitudinal magnetic field and the transverse electric field through a cylindrical carbon nanotube.

and electric fields, respectively, dominate the hopping integrals of the neighboring orbitals and the on-site ionization energies. In general, a giant Moire superlattices frequently come to exist because of magnetic or spatial modulation effects. This will lead to the greatly diversified quasi-particle phenomena. The exact diagonalization method is available for solving a very large Hamiltonian matrix by using the featured matrix elements of the specific neighboring interactions. A band-like Hermitian matrix is produced when the tight-binding functions are rearranged using a specific ordering.[31] Such a Hamiltonian is built from the generalized tight-binding model,[183] but not from numerical calculations. The magnetic quantization phenomena have been successfully explored in a number of mainstream materials, including 3D AA-, AB-, and ABC-stacked bulk graphites,[96,148,153] 2D few-layer graphene materials with AAA, ABA, ABC, and AAB stackings,[56,59,94,95,149,151,152,154–157,184] bilayer sliding rotational graphene systems,[185–188] 1D graphene nanoribbons,[8,189–192] coaxial/single-walled carbon nanotubes, 0D carbon tori,[193–195] monolayer silicene/germanene/tinene/plumbene/phosphorene/bismuthine,[131,135,196–200] and layered MoS_2-related materials.[181,201,202]

A perpendicular magnetic field ($Bz\hat{z}$ in Fig. 2.12(a)) can flock the neighboring 2D (k_x, k_y) states into the dispersionless Landau levels (LLs)[203] in which their degeneracy is proportional to the magnetic field strength. The stationary vector potential of $\boldsymbol{A}(r)$, which characterizes $Bz\hat{z}$, mainly determines the Peierls phases in the modified hopping integrals through its path integration between two lattice points (detailed in previously published books[31]). Very complicated calculations have been done for the significant sp^3 multi-orbital hybridizations and spin–orbital couplings, e.g., the magnetic Hamiltonian of monolayer tinene.[204] The band-like Hermitian is achieved through a specific ordering of the sub-lattice-dependent tight-binding functions.[31] It could be solved very efficiently by using the main features of the spatial probability distributions. In general, the magnetic wave functions exhibit the localized and oscillatory features, as observed in a harmonic oscillator, that is, the superposition of the exponential decay function and the n-order Hermit polynomials (the zero-point number of n). The generalized tight-binding model is capable of revealing the square of amplitude at a neighboring lattice site. Since the magnetic length is much larger than the atom–atom bond length, its spatial behavior could be regarded as continuous Landau wave functions with a well-behaved localization and an oscillatory zero point. The localization centers are dominated by the crystal symmetry, e.g., the four-fold degenerate LLs at 1/6, 2/6, 4/6, and 5/6 of the magnetic Moire superlattice.

The diverse magneto-quasiparticle are clearly revealed in dimensionality-dependent graphene-related systems, as published in a series of books.[13,14,27–29,31,32] The featured LL energy spectra and magneto-wave functions cover their sensitive dependences on the dominating quantum numbers, the magnetic-field strengths,[205,206] the perpendicular gate voltages,[184] the layer numbers,[173] the stacking configurations about AAA, ABA, ABC, and AAB, the sliding/rotational bilayer structures, the interlayer van der Waals of 3D graphite crystals (simply hexagonal, Bernal, and rhombohedral),[60,93,207–210] the width-induced quantum confinement and open two edges in 1D graphene nanoribbons,[91,211] and the cylindrical surfaces and periodic boundary conditions in coaxial carbon nanotubes. The magneto-electronic states of layered graphene systems are quantized from the low-lying valence and conduction energy subbands, in which they are closely related to the pristine or strongly modified Dirac cone structures because of very important interlayer atomic hybridizations. The main features of LLs mainly come from the energy dispersions of the linear, parabolic, partially flat, and sombrero-like subbands with/without the unusual anticrossing behaviors. Monolayer graphene exhibits a unique square-root dependence on B_z and n, which is also displayed in AA-stacked systems with the vertical Dirac cone pairs. An AB-stacked graphene has linear (B_z, n) relationships,[212] as observed in 2D electron gas.[213] Furthermore, the trilayer ABA stacking could be regarded as a superposition of the monolayer and bilayer ones. More interestingly, both ABC and AAB trilayer stackings can show the prominent LL-anticrossing phenomena frequently through the B_z-dependent energy spectra. Most importantly, all LLs are further divided into three kinds according to the featured quantum modes of the sliding bilayer graphene systems: (I) well-defined, (II) perturbed, and (III) undefined LLs, respectively, with specific oscillations, dominating and minor quantum numbers, and infinite modes (drastically changing with B_z). Such LLs have greatly diversified the magnetic quantization phenomena in terms of magneto-electronic properties, magneto-Coulomb excitations (electron–hole pairs and plasmon modes)[214] optical selection rules, and quantum Hall transports. Such investigations have been very successful in developing a grand quasi-particle framework (details in previously published books[13,14,27–34]).

In addition to the dispersionless LLs with the B_z-dependent (k_x, k_y) degeneracy, the 1D Landau subbands, which, respectively, present the strong k_z and k_x dispersion relationships, frequently come to exist in 3D graphite systems and 1D graphene nanoribbons.[114] The former possesses prominent 3D band structures with non-negligible (k_x, k_y, k_z)-dependences, owing mainly to the weak but significant van der Walls $2p_z$-orbital hybridizations and the well-behaved stacking configurations. The k_z-dependent LS energy spectra are achieved for the AA-, AB-, and ABC-stacked graphite systems, in which the classical quantization of the Onsager method is utilized in the rhombohedral graphite. The rich features of LSs have resulted in unusual magneto-optical absorption spectra[56, 60, 208] and quantum Hall effects. On the other hand, the magnetic flocking and quantum confinement clearly show the strong partnerships in 1D graphene nanoribbons and carbon nanotubes. For example, each graphene nanoribbon has a finite width and thus exhibits a number of 1D π-electronic valence and conduction energy subbands, which are sensitive to the open-edge structures. The VASP simulations agree with the tight-binding model calculations. After magnetic quantization, the Landau subbands appear at the lower energies close to the Fermi level. Such LSs are partially flat about the $k_x = 0$ center, where the dispersionless range declines with an increase in the quantum number. This clearly illustrates the prominent competition between magnetic quantization and quantum confinement. The featured magneto-electronic spectra of LSs in graphene nanoribbons are expected to greatly diversify the composite quasi-particles, e.g., the coexistence of edge- and B_z-dependent magneto-optical selection rules. However, LLs and LSs almost disappear in 1D carbon nanotubes, except for a giant magnetic field under the perpendicular case. A magnetic field, being parallel to the tubular axis, is able to create the periodic Aharnov–Bohm effects[215–217] for any essential properties. That the traverse angular momentum clearly presents a dramatic variation of $\boldsymbol{J}$ to $J + \frac{\phi}{\phi_0}$ (where ϕ is the magnetic flux through a cross-section) is the main reason. Obviously, the quasi-particle behaviors strongly depend on the nanotube radii/chiralities[195, 218, 219] and the magnitude and direction of the magnetic field, e.g., energy gaps, semiconductor–metal transitions, magneto-optical selection rules, magneto-plasmon modes, and ballistic magneto-transport behaviors. Similar magnetic study results could be generalized to 2D buckled silicene/germanene/tinene/plumbene systems and their 1D partners.

The magnetic quantization phenomena, which simultaneously cover the intralayer and interlayer multi-orbital hybridizations and the spin-orbital couplings[220–223] the coupling effects, are thoroughly explored using the generalized tight-binding model.

However, the effective-mass approximation[224] is suitable only for the magneto-electronic properties of condensed-matter systems with simple energy dispersions, as done for the analytic LL energy spectra of the AA-stacked graphene systems. In this method, we need to first complete the Taylor expansion about the high symmetry and then do the magnetic quantization as for an electron gas. The super-high barriers come to exist under the multi-fold degeneracy of constant-energy loops. The low-energy perturbations would become meaningless, and they cannot be utilized to get the magnetically quantized LLs, e.g., those of ABA-, ABC-, and AAB-stacked few-layer graphene systems. To develop a grand quasi-particle framework, the strong relations among the generalized tight-binding model,[182] the modified random-phase approximation (RPA),[225,226] the self-energy method,[227] and the static/dynamic Kubo formulas,[110,111] are directly established in the absence of perturbations. For example, the rich and unique (momentum, frequency)-magneto-Coulomb excitations are clearly revealed in monolayer, bilayer AA- and AB-stacked graphenes through the simultaneous considerations of the intralayer and interlayer orbital hybridizations and Coulomb interactions, as well as the magnetic quantization. More interestingly, the main features of LL wave functions are available for greatly reducing the numerical evaluation time. The magneto-electron–hole pairs display strong delta-function-like peaks in the imaginary-part dielectric functions because of the inter-LL transitions, Corresponding to such single-particle excitations, the screened response function further presents the unusual magneto-plasmon modes. Their momentum dependence only shows irregular behavior, with the positive, vanishing, and negative group velocities due to the very strong competition between the longitudinal Coulomb forces and the transverse cyclotron ones. It should be noted that a number of neighboring discrete magneto-plasmons would merge together and behave as 2D plasmons of an electron gas under the high carrier doping.

The direct combination of the generalized tight-binding model and the dynamic/static Kubo formula is very useful in fully exploring the rich magneto-optical properties/the quantum Hall transports. An electromagnetic wave can induce the electric-dipole perturbations between the initial and final LLs, leading to the specific magneto-optical selection rules. The scattering probabilities are evaluated within the gradient approximation, as successfully done for graphite,[228,229] graphite intercalation compounds, carbon nanotubes, and graphene nanoribbons. Their finite magnitudes arise under a suitable selection rule. The prominent absorption peaks have been thoroughly examined for the quantum modes of the initial and final magneto-electronic states on the same layer (the zero point of the oscillatory LL wave functions). Basically, monolayer graphene exhibits the magneto-optical selection rule of $\Delta_n^c - \Delta_n^v = \pm 1$, in which the available conditions are also suitable for the same layer sublattice. The extra magneto-optical selection rules arise mainly from the significant minor modes during the anticrossing phenomena. Specifically, such rules become vanishing for the magneto-optical transitions of the undefined LLs. The diverse magneto-optical quasi-particles are clearly revealed in the stacking-enriched graphene systems. For example, the trilayer AAA, ABA, ABC, and AAB stackings are able to present five, five, nine, and nine magneto-absorption categories because of the band-structure symmetries. Furthermore, there are a number of merged peak structures, strongly depending on the stacking configurations. On the quantum transport side, the scatterings of any two LLs do not need to satisfy the conservation of energy. The generalized tight-binding model is capable of evaluating the real contributions of the occupied LL states under the suitable magneto-selection rules. Its investigations on quantum Hall effects of few-layer graphene/silicene systems with any stacking configurations have been successful up to now, compared with those by the effective-mass approximation.[224] Similar studies should be extended to layered silicene/germanene/tinene/plumbene/phosphorene/bismuthene, where their quantum pictures are very useful in establishing the prominent partnerships among the sp^3 orbital hybridizations, observable spin–orbital couplings, and magneto-effects. This is under current investigation.

2.2.2. *Electron beams*

When a high-energy electron beam penetrates a condensed-matter system, the Coulomb-field perturbation is able to induce time- and position-dependent charge density fluctuations. All charges, which are characterized by different intrinsic vibration frequencies, present their prominent screening abilities. The propagating waves of charge densities (Figs. 2.14(a) and 2.14(b)) directly reflect the main features of orbital hybridizations in chemical bonds. For example, the graphite alkali compounds[65] have the low-lying conduction electrons, the middle-energy π electrons, and the high-energy $\pi + \sigma$ electrons. Three kinds of charge carriers will dominate the electron–electron interactions within the distinct frequency ranges (as detailed in previous books[13,33]). According to the classical Maxwell equations, the Poisson formula, assisted by the boundary conditions, might be available for exploring the significant charge couplings. In general, the screened charge density mainly arises from the induced dipole moments, and it is proportional to the induced/total Coulomb potential under a linear response.[110,111] How to include the full quantum

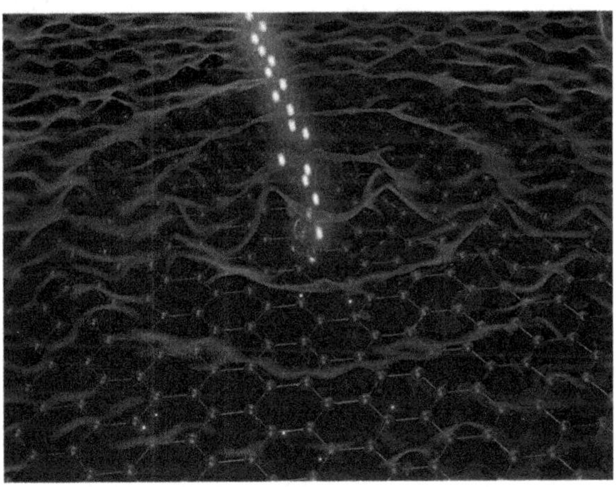

Fig. 2.14. An incident electron beam with moment and frequency of $(\boldsymbol{q}, \omega)$, leading to the position- and time-dependent charge density waves and electron–hole pair excitations.

picture can be realized from a quasi-particle viewpoint. An electron beam has momentum and energy, and it can provide sufficient resources for drastic changes in condensed-matter charges through electron–electron Coulomb interactions. Such inelastic scatterings need to satisfy the conservation of the transferred momentum and energy, as clearly indicated in the classical mechanisms. The current quantum quasi-particle framework is only generalized based on the classical field theory of the Lagrange and Hamiltonian equations. Their quantized relationships consist of many-particle quantum interactions. This theoretical strategy is very useful in fully understanding the frequent scatterings in many materials, including the electron–electron, electron–phonon, and electron–impurity/electron–defect interactions.[230] The longitudinal Coulomb dielectric function[231,232] and energy loss function,[233] which are characterized in the $(\boldsymbol{q}, \omega)$-space, can, respectively, elucidate the electron–hole pair excitations and plasmon modes. The former is defined as the ratio of electric fields, Coulomb potentials, or charge densities before and after charge screenings, as done for the classical Coulomb fields. In general, the band-structure effects on the Coulomb matrix elements of the initial and final quantum states and the conservation of $(\boldsymbol{q}, \omega)$ are fully included in the calculations of the bare response function through the RPA. The linear relationships among the screening charge density, the induced Coulomb potential, and their total are achieved using the self-consistent method and the many-body Feynman diagrams. The imaginary-part dielectric function possesses the complicated $(\boldsymbol{q}, \omega)$-regions, in which they are surrounded by the Fermi momenta or the interband boundaries due to multi-energy dispersions. Its prominent structures mainly arise from the joint van Hove singularities. As for the screened response function, it should cover all charge screenings so that the perturbation field is changed from the electric displacement to the effective field. Through delicate calculations of the Fermi golden rule, the energy loss spectrum, $\mathrm{Im}(-1/\varepsilon)$, is able to illustrate and explain the high-precision measurements of reflection/transmission electron energy loss spectroscopy (REELS/TEELS).[147,234–236] More interestingly, the $(\boldsymbol{q}, \omega)$-Coulomb excitation phase diagrams are clearly illustrated in the previous books using one million EELS calculation data. Such delicate figures are very helpful for all readers in understanding the concise pictures. The layered graphene systems/the coaxial carbon nanotubes[61,154,155,237] are outstanding candidates for exploring the diverse, many-particle phenomena since the intralayer/interlayer orbital hybridizations[238] and Coulomb interactions,[100,149,239–243] the spin–orbital couplings, the uniform perpendicular magnetic field, and the gate voltage should be taken into account simultaneously. Also, note that the 2D transferred momentum in graphene planes $\boldsymbol{q} = (q_x, q_y)$ is only changed into the longitudinal momentum and transverse angular momentum of (L, q_y) in carbon nanotubes. Such cylindrical systems can display the L-decoupled modes of the single- and many-particle Coulomb excitations, clearly indicating the 1D characteristics. Both theoretical predictions and experimental examinations of single-walled carbon nanotube bundles have successfully identified the discrete plasmon modes of the inter-π-band transitions. The screening charge densities on any honeycomb crystals strongly affect one another by establishing the layer-enriched Coulomb potentials. The time-dependent electron–hole pair productions could arise at two different layers. More complicated screening phenomena are covered within the modified RPA, which is expressed as the Dyson equation.[230] The $N \times N$ dielectric function tensor (where N is the layer number) appears under a linear response, and the bare polarization function $P(\boldsymbol{q}, \omega)$ is determined using the layer subindex. The dimensional energy loss spectrum is expressed by this tensor function through the Born scattering approximation. Researchers need to determine the layer decompositions for all valence and conduction energy subbands. This viewpoint is very useful in dealing with the inelastic Coulomb decay of the quasi-particle self-energy. The quasi-particle Coulomb decay rates in layered graphenes[13,244] and coaxial carbon nanotubes[245] are predicted to exhibit unusual dependences on wave vectors, doping densities, stacking configurations, and layer numbers. These lifetime behaviors have never appeared in 3D, 2D, and 1D electron gases or in graphite, graphite intercalation compounds, and monolayer graphene/silicene/germanene. More interestingly, few-layer graphene systems obviously show diverse Coulomb excitation phenomena, owing mainly to the stacking- and layer-number-enriched band structure across the Fermi level. For example, monolayer, bilayer AA/AB stackings, and trilayer AAA/ABA/ABC & AAB stackings, respectively, behave as a zero-gap semiconductor, a semi-metal, and a semi-metal & a narrow-gap semiconductor. These band properties and the low-lying energy dispersions of valence and conduction energy subbands are responsible for all the featured Coulomb-excitation phase diagrams, as clearly illustrated in Figs. 2.15(a)–2.15(d) for trilayer graphene with AAA/ABA/ABC/AAB stackings under the pristine case. In general, there are six independent polarization functions for trilayer systems. Their imaginary parts can well characterize the existence of electron–hole pair excitations, in which the prominent responses arise from the joint van Hove singularities. The Landau-damping regions are fully dominated by the band-edge states and the Fermi moments. The intensities of collective excitations are evaluated and examined using the prominent energy loss spectra; furthermore, their momentum dispersion relationships are very useful for observing plasmon group velocities. The plasmon modes will experience serious Landau dampings, so that a critical

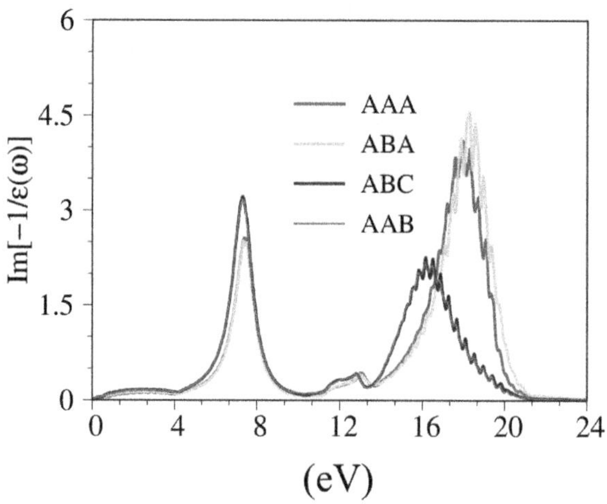

Fig. 2.15. The energy loss spectra of trilayer graphene systems under the pristine case for (a) AAA, (b) ABA, (c) ABC, and (d) AAB stackings.

momentum arises with an increase in the transferred momentum. Whether a quasi-particle plasmon belongs to an optical or acoustic mode[246] is associated with its finite or vanishing vibration frequency within the long-wavelength limit. Of course, the blue-/redshift of the Fermi level dramatically changes the Coulomb excitation spectra because of the strong asymmetry of hole and electron bands due to the interlayer van der Waals interactions. The above-mentioned studies have successfully provided sufficient information about the many-body electron–electron Coulomb interactions. 1D coaxial carbon nanotubes can present angular-momentum-decoupled modes, according to the electron–electron Coulomb interactions on a cylindrical surface. This unusual quasi-particle behavior is clearly illustrated by an 1D electron gas without the band-structure effects. The analytic formulas for the bare and screened response functions are achieved within the RPA.[225,226] The total strength of the single-particle and collective excitations is identified as being equal to one for each discrete (q, L) mode. When carbon-honeycomb crystals are included in the delicate calculations, their energy spectra, and wave functions can greatly enrich the Coulomb-field excitation spectra. For example, the $L = 0$ plasmon mode at zero temperature, with the main features of an 1D electron gas, only survives in type-I metallic armchair carbon nanotubes because of free carriers in linearly intersecting valence and conduction subbands. This plasmon frequency exhibits the $(\sqrt{q}\ln(\boldsymbol{qr}))$ dispersion relationship at small q values in the absence of Landau dampings. The 1D acoustic $L = 0$ plasmon will gradually disappear with an increase in the parallel/perpendicular magnetic field. Its strength declines quickly under the enlarged bandgap or stronger couplings of the L-discrete modes. Similar magnetic-field dependences are revealed in type-II and type-III carbon nanotubes (narrow- and middle-gap semiconductors).[48,218,240,247] The field-induced semiconductor–metal transitions, which occur at certain magnetic fluxes, can lead to the creation of the longitudinal plasma wave without a zero point. The charged impurity screenings are totally different between the metallic and semiconducting carbon nanotubes/few-layer graphene systems.[149,156,241,249–251] For example, boron and nitrogen guest adatoms could serve as the negative and positive impurities, respectively. The effective long-range Coulomb potential is thoroughly screened by the free carriers in a metallic system, while it maintains the $1/r$ behavior for insufficient screening of valence electrons in a semiconducting system. Furthermore, the Fermi surfaces are able to generate the well-known Friedel oscillations at long distances, e.g., $\sin(2k_{\mathrm{F}}r)/r$ and $\sin(2k_{\mathrm{F}}r)/r^2$, respectively, for the metallic 1D carbon nanotubes[252,253] and layered graphene materials. More interestingly, the beating phenomenon frequently arises in metallic coaxial armchair carbon nanotubes since such materials possess multi-Fermi surfaces. The intertube couplings, which are achieved by Coulomb interactions and orbital hybridizations, also lead to unusual quasi-particle lifetimes. The theoretical predictions could be verified by the measured widths of the angle-resolved photoemission spectra.[238,254–256]

2.2.3. *Electromagnetic waves*

An electromagnetic wave fully couples with all the energy-dependent electrons through the natural charge screenings and thus exhibits the material-enriched quasi-particle properties. Under the classical picture, the induced current density has a linear relationship with the total electric field, in which the former is created by the position- and time-dependent charge density within a linear response (Fig. 2.16). For the non-magnetic materials, the Maxwell equations can be combined using the Coulomb gauge (the zero divergence of the vector potential) to obtain the Helmholtz equation with a propagating wave form. The dielectric features of any condensed-matter system are responsible for the charge dynamics and the screened electromagnetic wave. The transverse dielectric function, which depends on momentum and frequency, is characterized by the variation of an electric field before and after electromagnetic wave perturbations. Its definition is similar to the longitudinal one discussed earlier. It is meaningful only at the long-wavelength limit since the photon momentum is much smaller than the electron one (or compared to their wavelengths). This function possesses

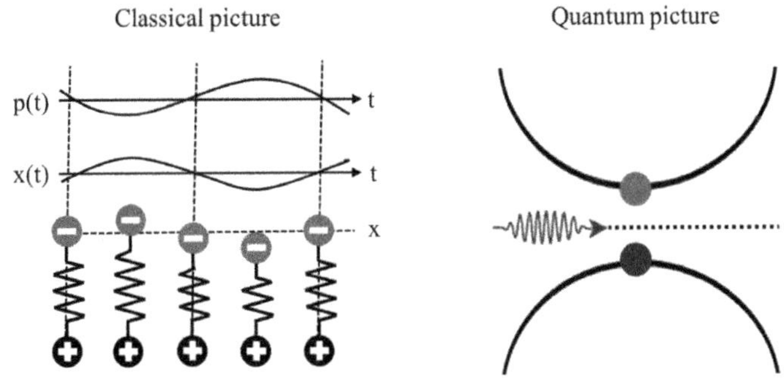

(a) (b)

Fig. 2.16. The significant couplings between an electromagnetic wave and any condensed-matter system under (a) a classical picture and (b) a quantum absorption process.

imaginary and real parts even when using the oscillator model for the different orbitals of atoms/molecules. A well-behaved propagating electromagnetic wave will gradually exhibit a widened deformation since the interband optical transitions can induce a prominent energy loss. Moreover, the group velocity of a wave packet, which is defined as the first derivative of frequency versus wave vector, is very sensitive to the charge screening ability. Of course, the linear relationship of $\omega = ck$ is strongly modified through the obvious dielectric effects. In addition, the frequency-related optical phenomena are reduced to the static carrier transport at zero frequency, that is, both optical and electrical conductivities are identical to each other under $\omega = 0$. An electromagnetic wave is further quantized into a number of photons with momenta and energies, according to the standard quantum field theory.[257]

These quasi-particles are absorbed by the occupied electronic states and then exhibit almost vertical transitions with the unoccupied ones. The imaginary-part dielectric function, which mainly comes from the single-particle electron–hole pair excitations, is determined by the joint DOS and the square of electric dipole moments. The special absorption structures are associated with the van Hove singularities of the band-edge states. The scattering amplitude accounts for the existence of optical/magneto-optical selection rules.[58,258] The gradient approximation of the initial and final states is evaluated using the layer- and sublattice-dependent tight-binding functions, which are consistent with wave functions under a uniform perpendicular magnetic field. This clearly indicates a very suitable combination of the Kubo formula and the generalized tight-binding model. The optical and magneto-optical properties have thoroughly illustrated the greatly diversified phenomena, such as the diverse quantum quasi-particles in graphite (as discussed later), layered graphene systems, graphene nanoribbons, and coaxial carbon nanotubes. The composite composites, being characterized by the multi-orbital hybridizations, are proposed for each prominent absorption peak and the frequency-dependent plasmon modes. Their developments in fully hydrogenated Group-IN materials make them outstanding candidates even in the presence of excitonic effects. In general, the many-body optical absorption processes only occur in middle- and wide-gap semiconductors, but not in narrow-gap ones and metallic conductors. Whether the Coulomb attractions of excited electron–hole pairs are strong enough for generating the stable/quasi-stable bound states strongly depends on the screening ability of all charges and the effective masses/band curvatures. 2D layered graphene systems and 3D graphites/graphite intercalation compounds have clearly displayed the greatly diversified phenomena of optical excitations, as indicted in previously reported simultaneous progress in high-resolution experimental observations and delicate theoretical predictions. The rich vertical transitions due to π electrons are easily modulated by the stacking configuration (the sliding and rotational symmetries)[187,188,259,260] layer number, magnetic field, gate voltage, and valence hole or conduction electron dopings (p- or n-type cases). The weak but significant van der Waals interactions, which mainly arise from the interlayer position-dependent $2p_z$-orbital hybridizations, play critical roles in determining the low-lying valence and conduction energy subbands, e.g., the unusual bandgaps and energy dispersions of monolayer, bilayer AA and AB stackings, and trilayer AAA, ABA, ABC, and AAB stackings. The Taylor expansions of the stable K/K′ valleys, which are conducted on well-behaved crystal structures, could be completed by the effective-mass approximation. The analytic energy spectra and wave functions are obtained from the exact diagonalization method, e.g., those of the trilayer AAA, ABA, and ABC stackings but not the ABA stacking. It should be noted that the low-symmetry AAB-stacked graphene presents tine hopping integrals for the well-fitting between the tight-binding model[102–104, 261] and VASP simulations.[42–45] Furthermore, this system has shown a narrow gap of ∼10 meV, strongly oscillatory/Sombrero-shape/parabolic energy dispersions, and a high asymmetry of hole and electron subbands about the Fermi level. In general, the concise relations among the subband-index-defined energy subbands cannot survive because of the complex contributions from the different layers and sublattices, except for the AA-stacked systems with the highest symmetry. As a result, it would be very difficult to observe and examine the optical selection rules in 2D and 3D carbon-related mainstream materials. There exist four kinds of 2D prominent absorption structures: (1) the broadening shoulders (the parabolic band-edge states), (2) the logarithmic divergences (the saddle points), (3) the delta-function-like symmetric peaks (the dispersionless states), and (4) the asymmetric peaks in the square-root forms (the 1D constant-energy loops during the subband anticrossings or the parabolic LSs). On the other hand, the 0D LLs and 1D LSs, which are created from a uniform perpendicular magnetic field (as detailed in Section 2.2.1), are able to show well-behaved magneto-optical selection rules ($\Delta n = \pm 1$) and the extra ones ($\Delta n = \pm 2, 3, 4, \ldots$ due to the LLs/LSs anticrossings). However, such rules are totally absent for the undefined LLs since the third-type magneto-electronic states display a lot of delta-function-like peaks with lower intensities. The unique magneto-optical quasi-particles are clearly revealed in few-layer graphene systems[57–59] and AA-/AB-/ABC-stacked graphites.[59, 96, 148, 151, 153–155, 157, 262, 263] For example, the trilayer ABA stacking is capable of displaying nine categories of magneto-optical excitations with many merged absorption structures, owing mainly to the highly asymmetric LL spectra. Such concise absorption pictures clearly indicate the successful development of a grand quasi-particle framework, which is helpful in unifying a series of published books.

1D planar graphene nanoribbons, which possess the longitudinal open edges[120,264] and the transverse width quantum confinement,[91] are able to clearly reveal the coexistence of geometry- and magnetic-field-enriched selection rules. Such materials have a lot of π-electronic valence and conduction energy subbands because of the presence of many carbon atoms in a transverse primitive unit cell (equal to the dimer-/zigzag-line number.[30]). Most of their energy dispersions belong to the parabolic ones, as verified from the high-resolution measurements of angle-resolved photoemission spectra. Any graphene nanoribbons are small- and middle-gap semiconductors, owing mainly to a strong cooperation among the quantum confinement, the significant bond-length fluctuations, and the ferromagnetic spin configuration on the same zigzag side/the anti-ferromagnetic one across the nanoribbon center. The van Hove singularities of parabolic energy dispersions and partially flat Landau subbands, respectively, display the asymmetric peaks in the inverse of square-root form and the delta-function-like symmetric peaks. Through delicate calculations and analyses, the strong absorption structures in armchair and zigzag systems are found to be only present under the specific partnerships between the valence and conduction subbands.[33] Even the subband indices cannot serve as a dominating quantum number with a physical meaning, which is totally different from the decoupled angular momentum modes in a cylindrical carbon nanotube.[252] However, according to the analytic calculations, the subband-index-dependent wave functions have close and strong relations between two valence/conduction/valence–conduction subbands through hexagonal honeycomb symmetries. As a result, the geometry-induced prominent absorption peaks satisfy the different optical selections for the armchair and zigzag graphene nanoribbons, where they are expressed as the number difference of subband indices about the occupied and unoccupied states. When the external magnetic field gradually grows in 1D systems, the lower energy will start displaying the partially flat Landau subbands with specific zero points. The magnetic wave functions, which frequently appear at π-electronic states near $k_x = 0$, are similar to those of a well-defined oscillator. On the other hand, the quantum-confinement-enriched dispersive energy subbands can present the original wave functions. These two kinds of wave functions can show a strong competition between the magnetic and quantum-confinement effects. The low-frequency and other frequency ranges are, respectively, dominated by the symmetric and asymmetric absorption peaks. The coexistent quasi-particle phenomena are worthy of experimental examinations.

1D coaxial carbon nanotubes have cylindrical symmetries and can create unusual optical/magneto-optical selection rules. The main features of crystal structures, radii, chiralities, and stacking configurations present strong relationships with the other essential properties. The electric polarization of an electromagnetic wave is parallel or perpendicular to the tubular axis. Through detailed calculations,[33] only the initial and final states, with the same angular momentum, are available for clearly revealing the vertical transitions through the 1D asymmetric absorption peaks for the first case. Furthermore, their differences in angular momenta should be ± 1 in the second case. That is, $\Delta l = l^c . l^v = 0$ and ± 1, respectively, for the parallel and perpendicular cases. Such optical selection rules remain the same when carbon nanotubes are threaded by a parallel magnetic flux. The decoupled angular momentum exhibits an obvious variation from $l^{c,v}$ to $l^{c,v} + \varphi/\varphi_0$ (flux quantum of $\varphi_0 hc/e$) and thus displays the periodic Aharnov–Bohm effect.[215–217] It is a good quantum number, accounting for the unchanged selection rules. The theoretical predictions are thoroughly consistent with the high-precision magneto-optical measurements. With the application of non-parallel magnetic fields, the extra magneto-optical selection rules are predicted to appear frequently in absorption spectra, e.g., the coexistence of $\Delta l = 0, \pm 1, \pm 2$, and ± 2. It is relatively easy to observe more prominent absorption peaks using the perpendicular magnetic field. Similar behaviors are induced by the transverse gate voltages because of the strong coupling of the decoupled angular momentum modes. More interestingly, the excited valence holes and conduction electrons simultaneously occur during the photon excitations. These excited quasi-particles will further decay the inelastic scatterings, e.g., the electron–electron Coulomb interactions (as detailed in Section 2.2.2) and the temperature-dependent electron–phonon scatterings. According to delicate and thorough calculations, their Coulomb lifetimes strongly depend on the band property close to the Fermi level. There are certain important differences among the metallic (type-I), narrow-gap (type-II), and middle-gap (type-III) carbon nanotubes, such as the sensitive dependences on the angular momenta and wave vectors. The high-resolution experimental measurements, which cover the time-resolved absorption spectra at the specific frequencies, the photo-electron emission spectra about the time-dependent Fermi–Dirac distributions, and the femtosecond fluorescence spectra.

2.2.4. *Mechanical stresses*

Mechanical stress is a measure of the internal resistance exhibited by a body or material when an external force is applied to it. This phenomenon is expressed by the equation $\sigma = \frac{R}{A}$, where R is the resistive force and A is the unit area. Its value at any point can be determined by considering that A goes to zero. Typically, we have two kinds of mechanical stress: uniaxial normal and shear stress, in which the former corresponds to homogenous deformations and thus does not modify the material's shape. The opposite is true for shear stress.

Until now, a number of studies have considered mechanical stress as one of the mainstream options and expected it to exhibit diverse quasi-particle phenomena. This kind of physical perturbation could be considered in the current development of the quasi-particle framework. Many studies based on the first-principles method have been considered for the strain effect on systems. For example, the effects of mechanically deforming graphene/graphite remain to be explored in detail and are published only in a few previous works.

How to examine the relationship among the featured electronic and optical properties is worthy of systematic investigation. Similarly, reliable verifications of different physical properties require high-precision experimental observations. For example, the X-ray diffraction (XRD) peaks[34,35] and the measurement of ARPES[36,37] of the occupied energy spectrum can examine the crystal symmetry and the energy gap in a band structure, respectively. The STS can be used to identify the van Hove singularities in the DOSs. The reflectance, absorption, transmittance, and photoluminescence spectra could be examined using optical spectroscopies.[37,38]

2.2.5. *Thermal excitations*

2.2.5.1. *Phonon energy dispersions*

Under the perturbations of an external field or temperature, the ion cores will vibrate about their equivalence center. These vibrations will gradually expand into the whole lattice due to the mutual interactions between the atoms, and thus, the local excitations will rapidly transfer into the collective excitation of all ions in the system. From a quantum mechanics point of view, the lattice displacement could be regarded as new quasi-particles and the associated quanta as elementary excitations called phonons. The information about these new quasi-particles is very useful to understand the dynamical and electrical transport behaviors, e.g., the investigations of electron–phonon interactions play a critical role in normal and superconducting transport properties.[39]

The crystal potential $U(r)$ of the lattice could be expressed as a function of the displacement r. Since most of the phonon excitations/atomic vibrations around the equivalence center are much less than the lattice constant, the problem of phonon energy dispersion could be easier to treat by adopting only the second-order term of the Taylor expansion (Fig. 2.17). As has been discussed in many textbooks, the phonon dispersions and their wave functions could be obtained by diagonalizing the dynamical matrices:

$$D^{i,j}(\boldsymbol{k}) = \frac{1}{\sqrt{M_i M_j}} \sum_j C_{ij} e^{i\boldsymbol{k}\Delta \boldsymbol{R}_{ij}},$$

Analogous to that of the electrons, the first term C_{ij} in the expression is related to the atomic interactions between the neighboring atoms, while the second part $e^{i\boldsymbol{k}\Delta \boldsymbol{R}_{ij}}$ corresponds to the phase difference factors. To avoid self-interactions, the summation here is

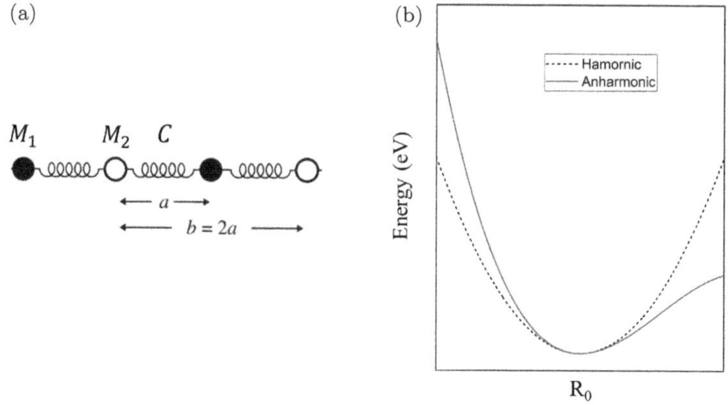

Fig. 2.17. (a) A diatomic crystal structure with masses M_1 and M_2 connected by force constant C between adjacent planes. (b) The vibration of the atoms around the equilibrium position is really an anharmonic oscillator but can be approximated as a harmonic oscillator at low energies or small displacements.

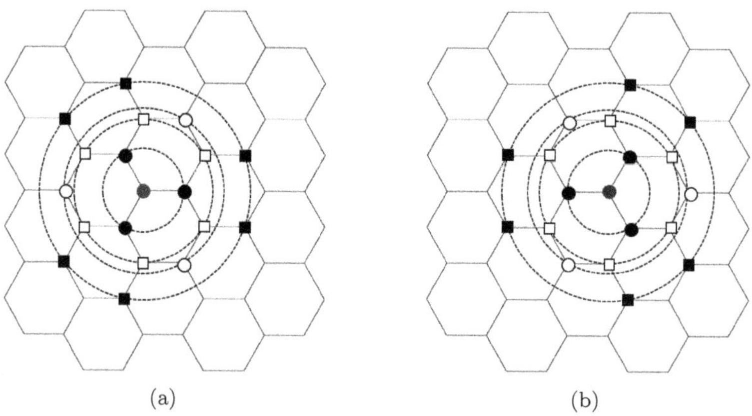

Fig. 2.18. Neighboring atoms of a graphene plane up to the fourth nearest neighbors for (a) an A atom and (b) a B atom at the center, denoted by the red and blue circles, respectively. The first, second, third, and fourth neighboring atoms are, respectively, denoted by the solid black circles, open squares, open circles, and solid black squares.

usually taken over a few neighbors' interactions respective to the ith side. For example, the inclusion of at least four nearest atomic interactions can reproduce the experimental phonon band structure of 2D graphene (Fig. 2.18).

The dynamic behaviors and other thermal properties are thoroughly studied through the energy dispersion spectrum of the phonon

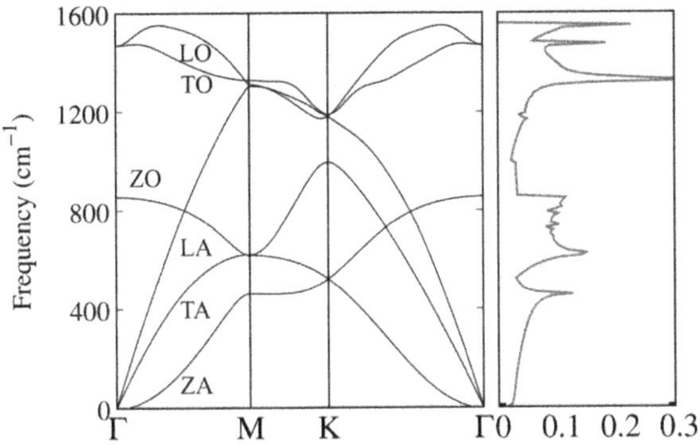

Fig. 2.19. Graphene phonon band structure and corresponding phonon density of states.

and the atom-decomposed phonon density of states. Moreover, such information is also useful for calculating the electron–phonon properties.[40,41] The essential properties of the phonon energy dispersions could be further verified by the inelastic scattering of neutrons with the emission or absorption of a phonon,[42,43] e.g., the k^2 energy-dependence of the out-of-plane ZA mode and the highest frequency of the degenerate LO and TO modes at the Γ point of 2D graphene (Fig. 2.19) have been determined experimentally.[44,45] The harmonic approximation has given very accurate phonon results for a variety of materials, ranging from metals to semiconductors to complex oxides.

2.2.5.2. *Phonon–phonon scatterings and thermal conductivity*

The harmonic approximations have been discussed earlier as a very efficient way to predict the phonon energy dispersions; However, neglecting the anharmonic terms leads to some important consequences, as follows: (i) The thermal expansion cannot be considered since the back and forth vibrations are almost the same; (ii) the volumetric heat capacity is maintained at a constant value at high temperatures; and (iii) two phonons do not interact with each other since the creation and destruction operators are applied to the same phonon states.

These consequences are not possible when we take higher terms in the expansion of the potential. The effects of the non-harmonic contributions could be easy to understand in a three-phonon scattering process, in which two phonons states with frequencies ω_1 and ω_2 scatter to produce a new phonon with frequency ω_3. The thermal conductivity κ is one of the properties directly related to phonon–phonon couplings. In short, the general form of κ could be expressed as

$$\kappa = \frac{1}{V}\sum_{\lambda} C_\lambda v_\lambda^2 \tau_\lambda,$$

where C_λ is the specific heat capacity, v_λ is the velocity of phonon, and τ_λ is the phonon lifetime. The thermal conductivity κ is limited by the intrinsic three-phonon scattering process and by extrinsic

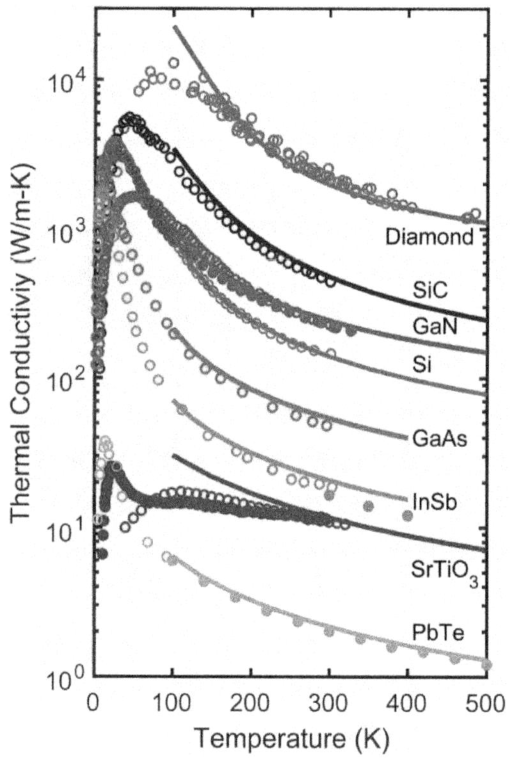

Fig. 2.20. Thermal conductivity data from experimental measurements (markers) and first-principles calculations for diamond, SiC, GaN, Si, GaAs, InSb, $SrTiO_3$, and PbTe.

point defect scattering owing to isotopic impurities. For high-quality semiconductors, the intrinsic anharmonic scattering rates $1/\tau_{\lambda}^{anh}$ are determined from the scattering processes involving three phonons that satisfy the conservation of energy and momentum: $\omega_{\lambda} \pm \omega_{\lambda'} = \omega_{\lambda''}$ and $\vec{q} \pm \overrightarrow{q'} = \overrightarrow{q''} + \vec{G}$. The $\pm$ signs correspond to the two types of possible three-phonon processes, and $\vec{G}$ is a reciprocal lattice vector, which is zero for normal (N) processes and non-zero for umklapp (U) processes. Under the theoretical framework of the perturbation theory and Boltzmann transport equation (BTE), the anharmonic scattering rates $1/\tau_{\lambda}^{anh}$ and, therefore, phonon thermal conductivity of a wide range of materials could be determined (Fig. 2.20).

References

[1] Hargittai, B.; Hargittai, I. *Year of the Periodic Table: Mendeleev and the Others*. Springer, 2019, pp. 1–7.

[2] Pyykkö, P. The physics behind chemistry and the periodic table. *Chemical Reviews* **2012**, *112* (1), 371–384.

[3] Geim, A. K.; Novoselov, K. S. The rise of graphene. *Nature Materials* **2007**, *6* (3), 183–191.

[4] Castro Neto, A. H.; *et al.* The electronic properties of graphene. *Reviews of Modern Physics* **2009**, *81* (1), 109–162.

[5] Novoselov, K. S. Electric field effect in atomically thin carbon films. *Science* **2004**, *306* (5696), 666–669.

[6] Saito, R.; *et al.* Electronic structure of chiral graphene tubules. *Applied Physics Letters* **1992**, *60* (18), 2204–2206.

[7] Dresselhaus, M.; Dresselhaus, G.; Saito, R. Physics of carbon nanotubes. *Carbon* **1995**, *33* (7), 883–891.

[8] Terrones, M.; *et al.* Graphene and graphite nanoribbons: Morphology, properties, synthesis, defects and applications. *Nano Today* **2010**, *5* (4), 351–372.

[9] Pauling, L. The nature of the chemical bond. Application of results obtained from the quantum mechanics and from a theory of paramagnetic susceptibility to the structure of molecules. *Journal of the American Chemical Society* **1931**, *53* (4), 1367–1400.

[10] Sreeprasad, T. S.; Berry, V. How Do the Electrical Properties of Graphene Change with its Functionalization? *Small* **2013**, *9* (3), 341–350.

[11] Alabugin, I. V.; Bresch, S.; dos Passos Gomes, G. Orbital hybridization: A key electronic factor in control of structure and reactivity. *Journal of Physical Organic Chemistry* **2015**, *28* (2), 147–162.

[12] Lyons, D.; Kaplan, T. Method for determining ground-state spin configurations. *Physical Review* **1960**, *120* (5), 1580.

[13] Lin, C.-Y.; *et al. Coulomb Excitations and Decays in Graphene-related Systems*. CRC Press, 2019.

[14] Tran, N. T. T.; *et al. Geometric and Electronic Properties of Graphene-related Systems: Chemical Bonding Schemes*. CRC Press, 2017.

[15] Zhao, J.; Zeng, H.; Yao, G. Computational design of a polymorph for 2D III–V orthorhombic monolayers by first principles calculations: Excellent anisotropic, electronic and optical properties. *Physical Chemistry Chemical Physics* **2021**, *23* (6), 3771–3778.

[16] Denis, A.; Goglio, G.; Demazeau, G. Gallium nitride bulk crystal growth processes: A review. *Materials Science and Engineering: R: Reports* **2006**, *50* (6), 167–194.

[17] Baliga, B. J. Gallium nitride devices for power electronic applications. *Semiconductor Science and Technology* **2013**, *28* (7), 074011.

[18] Tamargo, M. C. *II-VI Semiconductor Materials and Their Applications*, Vol. 12. CRC Press, 2002.

[19] Turner, W. J.; Fischler, A. S.; Reese, W. E. Physical properties of several II-V semiconductors. *Physical Review* **1961**, *121* (3), 759–767.

[20] Mohammad, S. N.; Morkoç, H. Progress and prospects of group-III nitride semiconductors. *Progress in Quantum Electronics* **1996**, *20* (5–6), 361–525.

[21] Afzaal, M.; O'Brien, P. Recent developments in II–VI and III–VI semiconductors and their applications in solar cells. *Journal of Materials Chemistry* **2006**, *16* (17), 1597–1602.

[22] She, X.; *et al.* Review of silicon carbide power devices and their applications. *IEEE Transactions on Industrial Electronics* **2017**, *64* (10), 8193–8205.

[23] Izhevskyi, V.; *et al.* Silicon carbide. Structure, properties and processing. *Cerâmica* **2000**, *46* (297), 4–13.

[24] Cai, H.; *et al.* Synthesis and emerging properties of 2D layered III–VI metal chalcogenides. *Applied Physics Reviews* **2019**, *6* (4), 041312.

[25] Holstein, B. R. The van der Waals interaction. *American Journal of Physics* **2001**, *69* (4), 441–449.

[26] Charlier, J.-C.; Gonze, X.; Michenaud, J.-P. Graphite interplanar bonding: Electronic delocalization and van der Waals interaction. *EPL (Europhysics Letters)* **1994**, *28* (6), 403.

[27] Chen, S.-C.; *et al. Theory of Magnetoelectric Properties of 2d Systems.* IOP Publishing, 2017.
[28] Lin, C.-Y.; *et al. Electronic and Optical Properties of Graphite-related Systems.* CRC Press, 2017.
[29] Lin, C.-Y.; *et al. Optical Properties of Graphene in Magnetic and Electric Fields.* IOP Publishing, 2017.
[30] Lin, S.-Y.; *et al. Structure-and Adatom-enriched Essential Properties of Graphene Nanoribbons.* CRC Press, 2018.
[31] Lin, C.-Y.; *et al. Diverse Quantization Phenomena in Layered Materials.* CRC Press, 2019.
[32] Lin, S.-Y.; *et al. Silicene-Based Layered Materials.* IOP Publishing Ltd., 2020.
[33] Lin, C.-Y.; *et al. Rich Quasiparticle Properties of Low Dimensional Systems.* IOP Publishing, 2021.
[34] Nguyen, T. D. H.; *et al.* First-principles calculations for cathode, electrolyte and anode battery materials. *First-Principles Calculations for Cathode*, 2021.
[35] Huang, H.-C.; *et al.* Configuration- and concentration-dependent electronic properties of hydrogenated graphene. *Carbon* **2016**, *103*, 84–93.
[36] Sahin, H.; Peeters, F. M. Adsorption of alkali, alkaline-earth, and 3d transition metal atoms on silicene. *Physical Review B* **2013**, *87* (8).
[37] Jin, K.-H.; Choi, S.-M.; Jhi, S.-H. Crossover in the adsorption properties of alkali metals on graphene. *Physical Review B* **2010**, *82* (3), 033414.
[38] Nguyen, D. K.; *et al.* Concentration-diversified magnetic and electronic properties of halogen-adsorbed silicene. *Scientific Reports* **2019**, *9* (1), 1–15.
[39] Karlicky, F.; *et al.* Halogenated graphenes: Rapidly growing family of graphene derivatives. *ACS Nano* **2013**, *7* (8), 6434–6464.
[40] Pumera, M.; Wong, C. H. A. Graphane and hydrogenated graphene. *Chemical Society Reviews* **2013**, *42* (14), 5987.
[41] Peng, Q.; *et al.* A theoretical analysis of the effect of the hydrogenation of graphene to graphane on its mechanical properties. *Physical Chemistry Chemical Physics* **2013**, *15* (6), 2003–2011.
[42] Hafner, J.; Kresse, G. The Vienna ab-initio simulation program VASP: An efficient and versatile tool for studying the structural, dynamic, and electronic properties of materials. In *Properties of Complex Inorganic Solids.* Springer, 1997, pp. 69–82.
[43] Hafner, J. Materials simulations using VASP — A quantum perspective to materials science. *Computer Physics Communications* **2007**, *177* (1–2), 6–13.

[44] Hacene, M.; *et al.* Accelerating VASP electronic structure calculations using graphic processing units. *Journal of Computational Chemistry* **2012**, *33* (32), 2581–2589.

[45] Hafner, J. Ab-initio simulations of materials using VASP: Density-functional theory and beyond. *Journal of Computational Chemistry* **2008**, *29* (13), 2044–2078.

[46] Xiong, W.; *et al.* Spin–orbit coupling effects on electronic structures in stanene nanoribbons. *Physical Chemistry Chemical Physics* **2016**, *18*(9), 6534–6540.

[47] Wang, Y.; *et al.* Half-metallic silicene and germanene nanoribbons: Towards high-performance spintronics device. *Nano* **2012**, *7* (05), 1250037.

[48] De Padova, P.; *et al.* Evidence of graphene-like electronic signature in silicene nanoribbons. *Applied Physics Letters* **2010**, *96* (26), 261905.

[49] Martel, R.; *et al.* Single-and multi-wall carbon nanotube field-effect transistors. *Applied Physics Letters* **1998**, *73* (17), 2447–2449.

[50] Dresselhaus, M.; Dresselhaus, G. Intercalation compounds of graphite. *Advances in Physics* **1981**, *30* (2), 139–326.

[51] Cho, A. Recent developments in molecular beam epitaxy (MBE). *Journal of Vacuum Science and Technology* **1979**, *16* (2), 275–284.

[52] Cho, A. Y.; Arthur, J. R. Molecular beam epitaxy. *Progress in Solid State Chemistry* **1975**, *10*, 157–191.

[53] McCray, W. P. MBE deserves a place in the history books. *Nature Nanotechnology* **2007**, *2* (5), 259–261.

[54] Guerard, D.; *et al.* Electronic structure of donor-type graphite intercalation compounds. *Il Nuovo Cimento B (1971–1996)* **1977**, *38* (2), 410–417.

[55] Pietronero, L.; *et al.* Charge distribution in c direction in lamellar graphite acceptor intercalation compounds. *Physical Review Letters* **1978**, *41* (11), 763.

[56] Do, T.-N.; *et al.* Rich magneto-absorption spectra of AAB-stacked trilayer graphene. *Physical Chemistry Chemical Physics* **2016**, *18* (26), 17597–17605.

[57] Abergel, D.; Fal'ko, V. I. Optical and magneto-optical far-infrared properties of bilayer graphene. *Physical Review B* **2007**, *75* (15), 155430.

[58] Ho, Y.-H.; *et al.* Magneto-optical selection rules in bilayer Bernal graphene. *ACS Nano* **2010**, *4* (3), 1465–1472.

[59] Lin, Y.-P.; *et al.* Magneto-optical properties of ABC-stacked trilayer graphene. *Physical Chemistry Chemical Physics* **2015**, *17* (24), 15921–15927.

[60] Ho, Y.-H.; *et al.* Magneto-absorption spectra of Bernal graphite. *Applied Physics Letters* **2011**, *99* (1), 011914.
[61] Koshino, M.; Ando, T. Magneto-optical properties of multilayer graphene. *Physical Review B* **2008**, *77* (11), 115313.
[62] Moissette, A.; *et al.* Sulfate graphite intercalation compounds: New electrochemical data and spontaneous intercalation. *Carbon* **1995**, *33* (2), 123–128.
[63] Lin, M.; Shung, K. W.-K. Self-energy of electrons in graphite intercalation compounds. *Physical Review B* **1996**, *53* (3), 1109.
[64] Lin, M.; Huang, C.; Chuu, D. Plasmons in graphite and stage-1 graphite intercalation compounds. *Physical Review B* **1997**, *55* (20), 13961.
[65] Li, W.-B.; *et al.* Essential geometric and electronic properties in stage-n graphite alkali-metal-intercalation compounds. *RSC Advances* **2020**, *10* (40), 23573–23581.
[66] Sorokina, N.; *et al.* Acceptor-type graphite intercalation compounds and new carbon materials based on them. *Russian Chemical Bulletin* **2005**, *54* (8), 1749–1767.
[67] Du, Z.; Wood III, D. L.; Belharouak, I. Enabling fast charging of high energy density Li-ion cells with high lithium ion transport electrolytes. *Electrochemistry Communications* **2019**, *103*, 109–113.
[68] Kang, B.; Ceder, G. Battery materials for ultrafast charging and discharging. *Nature* **2009**, *458* (7235), 190–193.
[69] Kanno, R.; *et al.* Synthesis, structure, and electrochemical properties of a new lithium iron oxide, LiFeO2, with a corrugated layer structure. *Journal of the Electrochemical Society* **1996**, *143* (8), 2435.
[70] Molenda, J. Structural, electrical and electrochemical properties of LiNiO2. *Solid State Ionics* **2002**, *146* (1–2), 73–79.
[71] Scott, I. D.; *et al.* Ultrathin coatings on nano-LiCoO2for Li-ion vehicular applications. *Nano Letters* **2011**, *11* (2), 414–418.
[72] Ong, S. P.; *et al.* Li- Fe- P- O2 phase diagram from first principles calculations. *Chemistry of Materials* **2008**, *20* (5), 1798–1807.
[73] Fongy, C.; *et al.* Ionic vs electronic power limitations and analysis of the fraction of wired grains in LiFePO4 composite electrodes. *Journal of the Electrochemical Society* **2010**, *157* (7), A885.
[74] Nakayama, M.; *et al.* Density functional studies of olivine-type LiFePO4 and NaFePO4 as positive electrode materials for rechargeable lithium and sodium ion batteries. *Solid State Ionics* **2016**, *286*, 40–44.

[75] Huie, M. M.; *et al.* Cathode materials for magnesium and magnesium-ion based batteries. *Coordination Chemistry Reviews* **2015**, *287*, 15–27.

[76] Liang, J.; *et al.* In situ Li3PS4 solid-state electrolyte protection layers for superior long-life and high-rate lithium-metal anodes. *Advanced Materials* **2018**, *30* (45), 1804684.

[77] Liu, K.; *et al.* Stretchable lithium metal anode with improved mechanical and electrochemical cycling stability. *Joule* **2018**, *2* (9), 1857–1865.

[78] Wang, A.; *et al.* Review on modeling of the anode solid electrolyte interphase (SEI) for lithium-ion batteries. *npj Computational Materials* **2018**, *4* (1), 1–26.

[79] Alder, B. J.; Wainwright, T. E. Studies in molecular dynamics. I. General method. *The Journal of Chemical Physics* **1959**, *31* (2), 459–466.

[80] Rapaport, D. C.; Rapaport, D. C. R. *The Art of Molecular Dynamics Simulation.* Cambridge University Press, 2004.

[81] Harris, G. L. *Properties of Silicon Carbide.* London: INSPEC, the Institution of Electrical Engineers, 1995.

[82] Blase, X.; *et al.* Structural and electronic properties of composite BxCyNz nanotubes and heterojunctions. *Applied Physics A* **1999**, *68* (3), 293–300.

[83] Kumar, N.; *et al.* Borocarbonitrides, BxCyNz. *Journal of Materials Chemistry A* **2013**, *1* (19), 5806–5821.

[84] Terrones, M.; *et al.* Pyrolytically grown BxCyNz nanomaterials: Nanofibres and nanotubes. *Chemical Physics Letters* **1996**, *257* (5–6), 576–582.

[85] Yi, T.-F.; *et al.* Recent development and application of Li4Ti5O12 as anode material of lithium ion battery. *Journal of Physics and Chemistry of Solids* **2010**, *71* (9), 1236–1242.

[86] Musa, N.; *et al.* Optimization of Li 2 SnO 3 synthesis for anode material application in Li-ion batteries. *Materials Today: Proceedings* **2017**, *4* (4), 5169–5177.

[87] Dien, V. K.; *et al.* Orbital-hybridization-created optical excitations in Li2GeO3. *Scientific Reports* **2021**, *11* (1), 1–10.

[88] Han, N. T.; Dien, V. K.; Lin, M.-F. Excitonic effects in the optical spectra of Li2SiO3 compound. *Scientific Reports* **2021**, *11* (1).

[89] Chang, S.-L.; *et al.* Geometric, magnetic and electronic properties of folded graphene nanoribbons. *RSC Advances* **2016**, *6* (69), 64852–64860.

[90] Ho, Y., Lin, H.-P.; Lin, M.-F. Effects of geometric structures and temperature on electronic properties of double-walled armchair carbon nanotubes. *Diamond and Related Materials* **2008**, *17* (7–10), 1550–1553.
[91] Huang, Y., Chang, C.; Lin, M.-F. Magnetic and quantum confinement effects on electronic and optical properties of graphene ribbons. *Nanotechnology* **2007**, *18* (49), 495401.
[92] Chang, C.; *et al.* Optical properties of simple hexagonal and rhombohedral few-layer graphenes in an electric field. *Journal of Applied Physics* **2008**, *103* (10), 103109.
[93] Chiu, C.-W.; *et al.* Low-frequency electronic and optical properties of rhombohedral graphite. *Physical Chemistry Chemical Physics* **2011**, *13* (13), 6036.
[94] Lin, C.-Y.; *et al.* Geometry-diversified Coulomb excitations in trilayer AAB stacking graphene. *Physical Review B* **2018**, *98* (19), 195442.
[95] Lin, C.-Y., Lee, M.-H.; Lin, M.-F. Coulomb excitations in ABC-stacked trilayer graphene. *Physical Review B* **2018**, *98* (4), 041408.
[96] Ho, J. H.; *et al.* Coulomb excitations in AA- and AB-stacked bilayer graphites. *Physical Review B* **2006**, *74* (8).
[97] Ezawa, M. Quantum Hall effects in silicene. *Journal of the Physical Society of Japan* **2012**, *81* (6), 064705.
[98] Sheng, D. N.; Sheng, L.; Weng, Z. Y. Quantum Hall effect in graphene: Disorder effect and phase diagram. *Physical Review B* **2006**, *73* (23).
[99] Wong, J.-H.; Wu, B.-R.; Lin, M.-F. Strain effect on the electronic properties of single layer and bilayer graphene. *The Journal of Physical Chemistry C* **2012**, *116* (14), 8271–8277.
[100] Schüler, M.; *et al.* Optimal Hubbard models for materials with nonlocal coulomb interactions: Graphene, silicene, and benzene. *Physical Review Letters* **2013**, *111* (3).
[101] Hubbard, J. Electron correlations in narrow energy bands. *Proceedings of the Royal Society of London. Series A. Mathematical and Physical Sciences* **1963**, *276* (1365), 238–257.
[102] Slater, J. C.; Koster, G. F. Simplified LCAO method for the periodic potential problem. *Physical Review* **1954**, *94* (6), 1498–1524.
[103] Goringe, C. M.; Bowler, D. R.; Hernández, E. Tight-binding modelling of materials. *Reports on Progress in Physics* **1997**, *60* (12), 1447–1512.
[104] Reich, S.; *et al.* Tight-binding description of graphene. *Physical Review B* **2002**, *66* (3).

[105] Sapra, S.; Shanthi, N.; Sarma, D. Realistic tight-binding model for the electronic structure of II-VI semiconductors. *Physical Review B* **2002**, *66* (20), 205202.
[106] Schulz, S.; Czycholl, G. Tight-binding model for semiconductor nanostructures. *Physical Review B* **2005**, *72*(16), 165317.
[107] Zhang, L.; *et al.* Revealing the electrochemical charging mechanism of nanosized Li2S by in situ and operando X-ray absorption spectroscopy. *Nano Letters* **2017**, *17* (8), 5084–5091.
[108] Zhao, Z.; *et al.* Pressure-induced stable Li5P for high-performance lithium-ion batteries. *The Journal of Physical Chemistry C* **2017**, *121* (39), 21199–21205.
[109] Hayashi, A.; *et al.* Preparation of Li2S–P2S5 amorphous solid electrolytes by mechanical milling. *Journal of the American Ceramic Society* **2001**, *84* (2), 477–479.
[110] Kubo, R.; Yokota, M.; Nakajima, S. Statistical-mechanical theory of irreversible processes. II. Response to thermal disturbance. *Journal of the Physical Society of Japan* **1957**, *12* (11), 1203–1211.
[111] Kubo, R. Statistical-mechanical theory of irreversible processes. I. General theory and simple applications to magnetic and conduction problems. *Journal of the Physical Society of Japan* **1957**, *12* (6), 570–586.
[112] Dolui, K.; Quek, S. Y. Quantum-confinement and structural anisotropy result in electrically-tunable dirac cone in few-layer black phosphorous. *Scientific Reports* **2015**, *5* (1), 1–12.
[113] Li, H.; *et al.* Carbon nanomaterials for next-generation interconnects and passives: Physics, status, and prospects. *IEEE Transactions on Electron Devices* **2009**, *56* (9), 1799–1821.
[114] Huang, Y.-C.; *et al.* Low-energy Landau levels of Bernal zigzag graphene ribbons. *Journal of Applied Physics* **2009**, *106* (1), 013711.
[115] Binnig, G.; *et al.* Surface studies by scanning tunneling microscopy. *Physical Review Letters* **1982**, *49* (1), 57.
[116] Binnig, G.; Rohrer, H. Scanning tunneling microscopy — From birth to adolescence. *Reviews of Modern Physics* **1987**, *59* (3), 615–625.
[117] Chen, C. J. Theory of scanning tunneling spectroscopy. *Journal of Vacuum Science & Technology A: Vacuum, Surfaces, and Films* **1988**, *6* (2), 319–322.
[118] Feenstra, R. M. Scanning tunneling spectroscopy. *Surface Science* **1994**, *299–300*, 965–979.
[119] LeRoy, B.; *et al.* Scanning tunneling spectroscopy of suspended single-wall carbon nanotubes. *Applied Physics Letters* **2004**, *84* (21), 4280–4282.

[120] Ritter, K. A.; Lyding, J. W. The influence of edge structure on the electronic properties of graphene quantum dots and nanoribbons. *Nature Materials* **2009**, *8* (3), 235–242.

[121] Liu, Z.; *et al.* Open and closed edges of graphene layers. *Physical Review Letters* **2009**, *102* (1), 015501.

[122] Berland, K.; *et al.* van der Waals forces in density functional theory: A review of the vdW-DF method. *Reports on Progress in Physics* **2015**, *78* (6), 066501.

[123] Gülseren, O.; Yildirim, T.; Ciraci, S. Systematic ab initio study of curvature effects in carbon nanotubes. *Physical Review B* **2002**, *65* (15), 153405.

[124] Popov, V. N. Curvature effects on the structural, electronic and optical properties of isolated single-walled carbon nanotubes within a symmetry-adapted non-orthogonal tight-binding model. *New Journal of Physics* **2004**, *6* (1), 17.

[125] Bonča, J.; Pruschke, T. Van Hove singularities in the paramagnetic phase of the Hubbard model: DMFT study. *Physical Review B* **2009**, *80* (24), 245112.

[126] Andersen, D. R.; Raza, H. Plasmon dispersion in semimetallic armchair graphene nanoribbons. *Physical Review B* **2012**, *85* (7), 075425.

[127] Yang, Z.; Hao, J. Recent progress in 2D layered III–VI semiconductors and their heterostructures for optoelectronic device applications. *Advanced Materials Technologies* **2019**, *4* (8), 1900108.

[128] Bimberg, D.; Grundmann, M.; Ledentsov, N. N. *Quantum Dot Heterostructures*. John Wiley & Sons, 1999.

[129] Hong, Y. J.; Fukui, T. Controlled van der Waals heteroepitaxy of InAs nanowires on carbon honeycomb lattices. *ACS Nano* **2011**, *5* (9), 7576–7584.

[130] Lalmi, B.; *et al.* Epitaxial growth of a silicene sheet. *Applied Physics Letters* **2010**, *97* (22), 223109.

[131] Liu, H.; *et al.* Phosphorene: An unexplored 2D semiconductor with a high hole mobility. *ACS Nano* **2014**, *8* (4), 4033–4041.

[132] Acun, A.; *et al.* Germanene: The germanium analogue of graphene. *Journal of Physics: Condensed Matter* **2015**, *27* (44), 443002.

[133] Zhu, F.-F.; *et al.* Epitaxial growth of two-dimensional stanene. *Nature Materials* **2015**, *14* (10), 1020–1025.

[134] Zhang, L.; *et al.* Structural and electronic properties of germanene on MoS2. *Physical Review Letters* **2016**, *116* (25), 256804.

[135] Yuhara, J.; *et al.* Graphene's Latest Cousin: Plumbene Epitaxial Growth on a "Nano WaterCube". *Advanced Materials* **2019**, *31* (27), 1901017.

[136] Casady, J.; Johnson, R. W. Status of silicon carbide (SiC) as a wide-bandgap semiconductor for high-temperature applications: A review. *Solid-State Electronics* **1996**, *39* (10), 1409–1422.

[137] Damascelli, A. Probing the electronic structure of complex systems by ARPES. *Physica Scripta* **2004**, *T109*, 61.

[138] Kordyuk, A. A. ARPES experiment in fermiology of quasi-2D metals (review article). *Low Temperature Physics* **2014**, *40* (4), 286–296.

[139] Sobota, J. A.; He, Y.; Shen, Z.-X. Angle-resolved photoemission studies of quantum materials. *Reviews of Modern Physics* **2021**, *93* (2).

[140] Lv, B.; Qian, T.; Ding, H. Angle-resolved photoemission spectroscopy and its application to topological materials. *Nature Reviews Physics* **2019**, *1* (10), 609–626.

[141] Kano, S.; Tada, T.; Majima, Y. Nanoparticle characterization based on STM and STS. *Chemical Society Reviews* **2015**, *44* (4), 970–987.

[142] Weightman, P.; *et al.* Reflection anisotropy spectroscopy. *Reports on Progress in Physics* **2005**, *68* (6), 1251.

[143] Karrai, K.; Warburton, R. J. Optical transmission and reflection spectroscopy of single quantum dots. *Superlattices and Microstructures* **2003**, *33* (5–6), 311–337.

[144] Wooten, F. *Optical Properties of Solids.* Academic Press, 2013.

[145] John, R.; Merlin, B. Optical properties of graphene, silicene, germanene, and stanene from IR to far UV–A first principles study. *Journal of Physics and Chemistry of Solids* **2017**, *110*, 307–315.

[146] Gangopadhyay, P.; *et al.* Optical absorption and photoluminescence spectroscopy of the growth of silver nanoparticles. *Physical Review Letters* **2005**, *94* (4), 047403.

[147] Egerton, R. F. *Electron Energy-loss Spectroscopy in the Electron Microscope.* Springer Science & Business Media, 2011.

[148] L. Lu, C.; *et al.* Electronic properties of AA-and ABC-stacked few-layer graphites. *Journal of the Physical Society of Japan* **2007**, *76* (2), 024701.

[149] Koshino, M. Interlayer screening effect in graphene multilayers with A B A and A B C stacking. *Physical Review B* **2010**, *81* (12), 125304.

[150] Mak, K. F.; Shan, J.; Heinz, T. F. Electronic structure of few-layer graphene: Experimental demonstration of strong dependence on stacking sequence. *Physical Review Letters* **2010**, *104*(17).

[151] Chiu, C.-W.; *et al.* Excitation spectra of ABC-stacked graphene superlattice. *Applied Physics Letters* **2011**, *98* (26), 261920.
[152] Yuan, S.; Roldán, R.; Katsnelson, M. I. Landau level spectrum of ABA-and ABC-stacked trilayer graphene. *Physical Review B* **2011**, *84* (12), 125455.
[153] Chiu, C.-W.; *et al.* Anisotropy of π-Plasmon dispersion relation of AA-stacked graphite. *Journal of the Physical Society of Japan* **2012**, *81* (10), 104703.
[154] Lin, M.-F.; Chuang, Y.-C.; Wu, J.-Y. Electrically tunable plasma excitations in AA-stacked multilayer graphene. *Physical Review B* **2012**, *86* (12), 125434.
[155] Chiu, C.-W.; *et al.* Critical optical properties of AA-stacked multilayer graphenes. *Applied Physics Letters* **2013**, *103* (4), 041907.
[156] Roldán, R.; Brey, L. Dielectric screening and plasmons in AA-stacked bilayer graphene. *Physical Review B* **2013**, *88* (11), 115420.
[157] Lin, Y.-P.; *et al.* Energy spectra of ABC-stacked trilayer graphene in magnetic and electric fields. *RSC Advances* **2014**, *4* (100), 56552–56560.
[158] Do, T.-N.; *et al.* Configuration-enriched magneto-electronic spectra of AAB-stacked trilayer graphene. *Carbon* **2015**, *94*, 619–632.
[159] Muller, D. A. Structure and bonding at the atomic scale by scanning transmission electron microscopy. *Nature Materials* **2009**, *8* (4), 263–270.
[160] Jona, F.; Strozier Jr, J.; Yang, W. Low-energy electron diffraction for surface structure analysis. *Reports on Progress in Physics* **1982**, *45* (5), 527.
[161] Lagally, M. G. The present status of low-energy electron diffraction. *Applications of Surface Science* **1982**, *13* (1–2), 260–281.
[162] Lagally, M. G.; Martin, J. A. Instrumentation for low-energy electron diffraction. *Review of Scientific Instruments* **1983**, *54* (10), 1273–1288.
[163] Sojka, F.; *et al.* Determination and correction of distortions and systematic errors in low-energy electron diffraction. *Review of Scientific Instruments* **2013**, *84* (1), 015111.
[164] Kiraly, B.; *et al.* Solid-source growth and atomic-scale characterization of graphene on Ag (111). *Nature Communications* **2013**, *4* (1), 1–7.
[165] Coraux, J.; *et al.* Structural coherency of graphene on Ir (111). *Nano Letters* **2008**, *8* (2), 565–570.

[166] Walker, L. S.; *et al.* Toughening in graphene ceramic composites. *ACS Nano* **2011**, *5* (4), 3182–3190.
[167] Li, L.; *et al.* Buckled germanene formation on Pt (111). *Advanced Materials* **2014**, *26* (28), 4820–4824.
[168] Derivaz, M.; *et al.* Continuous germanene layer on Al (111). *Nano Letters* **2015**, *15* (4), 2510–2516.
[169] Zhuang, J.; *et al.* Cooperative electron–phonon coupling and buckled structure in germanene on Au (111). *ACS Nano* **2017**, *11* (4), 3553–3559.
[170] Xu, Y.; Tang, P.; Zhang, S.-C. Large-gap quantum spin Hall states in decorated stanene grown on a substrate. *Physical Review B* **2015**, *92* (8), 081112.
[171] Ganz, E.; *et al.* Growth and morphology of Pb on Si (111). *Surface Science* **1991**, *257* (1–3), 259–273.
[172] Yakes, M.; Tringides, M. Probing the buried Pb/Si (111) interface with SPA LEED and STM on Si (111)-Pb$\alpha\sqrt{3} \times \sqrt{3}$. *The Journal of Physical Chemistry A* **2011**, *115* (25), 7096–7104.
[173] Hao, Y.; *et al.* Probing layer number and stacking order of few-layer graphene by Raman spectroscopy. *Small* **2010**, *6* (2), 195–200.
[174] Lui, C. H.; *et al.* Imaging stacking order in few-layer graphene. *Nano Letters* **2011**, *11* (1), 164–169.
[175] Bao, C.; *et al.* Stacking-dependent electronic structure of trilayer graphene resolved by nanospot angle-resolved photoemission spectroscopy. *Nano Letters* **2017**, *17* (3), 1564–1568.
[176] Huang, Y.-K.; *et al.* Feature-rich magnetic quantization in sliding bilayer graphenes. *Scientific Reports* **2014**, *4* (1), 1–10.
[177] Kühne, P.; *et al.* Polarization selection rules for inter-landau-level transitions in epitaxial graphene revealed by the infrared optical Hall effect. *Physical Review Letters* **2013**, *111* (7), 077402.
[178] Yan, H.; *et al.* Infrared spectroscopy of tunable Dirac terahertz magneto-plasmons in graphene. *Nano Letters* **2012**, *12* (7), 3766–3771.
[179] Politano, A.; Chiarello, G. Plasmon modes in graphene: Status and prospect. *Nanoscale* **2014**, *6* (19), 10927–10940.
[180] Wang, Y. Y.; *et al.* Raman studies of monolayer graphene: The substrate effect. *The Journal of Physical Chemistry C* **2008**, *112* (29), 10637–10640.
[181] Liu, K.-K.; *et al.* Growth of large-area and highly crystalline MoS2 thin layers on insulating substrates. *Nano Letters* **2012**, *12* (3), 1538–1544.

[182] Korshunov, M.; *et al.* Hybrid LDA and generalized tight-binding method for electronic structure calculations of strongly correlated electron systems. *Physical Review B* **2005**, *72* (16), 165104.

[183] Hancock, Y.; *et al.* Generalized tight-binding transport model for graphene nanoribbon-based systems. *Physical Review B* **2010**, *81* (24).

[184] Tsai, S.-J.; *et al.* Gate-voltage-dependent Landau levels in AA-stacked bilayer graphene. *Chemical Physics Letters* **2012**, *550*, 104–110.

[185] Li, G.; *et al.* Observation of Van Hove singularities in twisted graphene layers. *Nature Physics* **2010**, *6* (2), 109–113.

[186] Dodaro, J. F.; *et al.* Phases of a phenomenological model of twisted bilayer graphene. *Physical Review B* **2018**, *98* (7), 075154.

[187] Stauber, T.; Kohler, H. Quasi-flat plasmonic bands in twisted bilayer graphene. *Nano Letters* **2016**, *16* (11), 6844–6849.

[188] Brihuega, I.; *et al.* Unraveling the intrinsic and robust nature of van hove singularities in twisted bilayer graphene by scanning tunneling microscopy and theoretical analysis. *Physical Review Letters* **2012**, *109* (19).

[189] Savoskin, M. V.; *et al.* Carbon nanoscrolls produced from acceptor-type graphite intercalation compounds. *Carbon* **2007**, *45* (14), 2797–2800.

[190] Zhang, J.; *et al.* Free folding of suspended graphene sheets by random mechanical stimulation. *Physical Review Letters* **2010**, *104* (16), 166805.

[191] Kumar, P.; Panchakarla, L.; Rao, C. Laser-induced unzipping of carbon nanotubes to yield graphene nanoribbons. *Nanoscale* **2011**, *3* (5), 2127–2129.

[192] Kosynkin, D. V.; *et al.* Longitudinal unzipping of carbon nanotubes to form graphene nanoribbons. *Nature* **2009**, *458* (7240), 872–876.

[193] Meunier, V.; Lambin, P.; Lucas, A. A. Atomic and electronic structures of large and small carbon tori. *Physical Review B* **1998**, *57* (23), 14886–14890.

[194] L. Shyu, F.; *et al.* Electronic properties of carbon tori in external fields. *Journal of the Physical Society of Japan* **2006**, *75* (10), 104710.

[195] Shyu, F.-L.; *et al.* Magnetoelectronic properties of chiral carbon nanotubes and tori. *Journal of Physics: Condensed Matter* **2006**, *18* (35), 8313.

[196] Chen, S.-C.; Wu, J.-Y.; Lin, M.-F. Feature-rich magneto-electronic properties of bismuthene. *New Journal of Physics* **2018**, *20* (6), 062001.

[197] Huang, S.; Kang, W.; Yang, L. Electronic structure and quasiparticle bandgap of silicene structures. *Applied Physics Letters* **2013**, *102* (13), 133106.

[198] Ezawa, M. Monolayer topological insulators: Silicene, germanene, and stanene. *Journal of the Physical Society of Japan*, **2015**, *84* (12), 121003.

[199] Cai, Y.; *et al.* Stability and electronic properties of two-dimensional silicene and germanene on graphene. *Physical Review B* **2013**, *88* (24).

[200] Mahdavifar, S.; Shayesteh, S. F.; Tagani, M. B.; Electronic and mechanical properties of Plumbene monolayer: A first-principle study. *Physica E: Low-dimensional Systems and Nanostructures* **2021**, *134*, 114837.

[201] Splendiani, A.; *et al.* Emerging photoluminescence in monolayer MoS2. *Nano Letters* **2010**, *10* (4), 1271–1275.

[202] Ramakrishna Matte, H.; *et al.* MoS2 and WS2 analogues of graphene. *Angewandte Chemie International Edition* **2010**, *49* (24), 4059–4062.

[203] Kallin, C.; Halperin, B. Excitations from a filled Landau level in the two-dimensional electron gas. *Physical Review B* **1984**, *30* (10), 5655.

[204] Chen, S.-C.; *et al.* Magnetic quantization of s p 3 bonding in monolayer gray tin. *Physical Review B* **2016**, *94* (4), 045410.

[205] Gopinadhan, K.; *et al.* Giant magnetoresistance in single-layer graphene flakes with a gate-voltage-tunable weak antilocalization. *Physical Review B* **2013**, *88*(19), 195429.

[206] Wray, L. A.; *et al.* A topological insulator surface under strong Coulomb, magnetic and disorder perturbations. *Nature Physics* **2011**, *7* (1), 32–37.

[207] García, N.; *et al.* Evidence for semiconducting behavior with a narrow band gap of Bernal graphite. *New Journal of Physics* **2012**, *14* (5), 053015.

[208] Lu, C.-L.; *et al.* Absorption spectra of trilayer rhombohedral graphite. *Applied Physics Letters* **2006**, *89* (22), 221910.

[209] McClure, J. Electron energy band structure and electronic properties of rhombohedral graphite. *Carbon* **1969**, *7* (4), 425–432.

[210] Wong, J.-H.; Wu, B.-R.; Lin, M.-F. Electronic properties of rhombohedral graphite. *Computer Physics Communications* **2011**, *182* (1), 77–80.
[211] Sako, R.; Hosokawa, H.; Tsuchiya, H. Computational study of edge configuration and quantum confinement effects on graphene nanoribbon transport. *IEEE Electron Device Letters* **2010**, *32* (1), 6–8.
[212] Ho, J.; *et al.* Landau level spectrum of bilayer Bernal graphene. *Diamond and Related Materials* **2009**, *18* (2–3), 374–379.
[213] Schlesinger, Z.; Hwang, J.; Allen Jr, S. Subband-Landau-level coupling in a two-dimensional electron gas. *Physical Review Letters* **1983**, *50* (26), 2098.
[214] Wu, J.-Y.; Gumbs, G.; Lin, M.-F. Combined effect of stacking and magnetic field on plasmon excitations in bilayer graphene. *Physical Review B* **2014**, *89* (16), 165407.
[215] Peshkin, M.; Tonomura, A. *The Aharonov-Bohm Effect*, Vol. 340. Springer, 1989.
[216] Ajiki, H.; Ando, T. Aharonov-Bohm effect in carbon nanotubes. *Physica B: Condensed Matter* **1994**, *201*, 349–352.
[217] Zaric, S.; *et al.* Optical signatures of the Aharonov-Bohm phase in single-walled carbon nanotubes. *Science* **2004**, *304* (5674), 1129–1131.
[218] Ding, J.; Yan, X.; Cao, J. Analytical relation of band gaps to both chirality and diameter of single-wall carbon nanotubes. *Physical Review B* **2002**, *66* (7), 073401.
[219] Neyts, E. C.; Van Duin, A. C.; Bogaerts, A. Changing chirality during single-walled carbon nanotube growth: A reactive molecular dynamics/Monte Carlo study. *Journal of the American Chemical Society* **2011**, *133* (43), 17225–17231.
[220] Liu, C. C.; Jiang, H.; Yao, Y. G. Low-energy effective Hamiltonian involving spin-orbit coupling in silicene and two-dimensional germanium and tin. *Physical Review B* **2011**, *84* (19), 11.
[221] Kurpas, M.; *et al.* Spin-orbit coupling in elemental two-dimensional materials. *Physical Review B* **2019**, *100* (12).
[222] Winkler, R.; *et al. Spin-Orbit Coupling in Two-Dimensional Electron and Hole Systems*, Vol. 41. Springer, 2003.
[223] Konschuh, S.; Gmitra, M.; Fabian, J. Tight-binding theory of the spin-orbit coupling in graphene. *Physical Review B* **2010**, *82* (24), 245412.

[224] Koshino, M.; Ando, T. Orbital diamagnetism in multilayer graphenes: Systematic study with the effective mass approximation. *Physical Review B* **2007**, *76* (8), 085425.
[225] Ren, X.; *et al.* Random-phase approximation and its applications in computational chemistry and materials science. *Journal of Materials Science* **2012**, *47* (21), 7447–7471.
[226] Chen, G. P.; *et al.* Random-phase approximation methods. *Annual Review of Physical Chemistry* **2017**, *68*, 421–445.
[227] Quinn, J. J.; Ferrell, R. A. Electron self-energy approach to correlation in a degenerate electron gas. *Physical Review* **1958**, *112* (3), 812.
[228] Perdew, J. P.; Burke K.; Ernzerhof, M. Generalized gradient approximation made simple. *Physical Review Letters* **1996**, *77* (18), 3865.
[229] Perdew, J. P.; Burke, K.; Wang, Y. Generalized gradient approximation for the exchange-correlation hole of a many-electron system. *Physical Review B* **1996**, *54* (23), 16533.
[230] Mahan, G. D. *9. Many-Particle Systems.* Princeton University Press, 2008.
[231] Scholz, A.; Stauber, T.; Schliemann, J. Dielectric function, screening, and plasmons of graphene in the presence of spin-orbit interactions. *Physical Review B* **2012**, *86* (19).
[232] Penn, D. R. Wave-number-dependent dielectric function of semiconductors. *Physical Review* **1962**, *128* (5), 2093.
[233] Planes, D. J.; *et al.* Wavenumber dependence of the energy loss function of graphite and aluminium. *Journal of Electron Spectroscopy and Related Phenomena* **1996**, *82* (1–2), 23–29.
[234] Brydson, R. A brief review of quantitative aspects of electron energy loss spectroscopy and imaging. *Materials Science and Technology* **2000**, *16*(10), 1187–1198.
[235] Egerton, R. F. Electron energy-loss spectroscopy in the TEM. *Reports on Progress in Physics* **2008**, *72* (1), 016502.
[236] Ibach, H.; Mills, D. L. *Electron Energy Loss Spectroscopy and Surface Vibrations.* Academic Press, 2013.
[237] Yannouleas, C.; Bogachek, E. N.; Landman, U. Collective excitations of multishell carbon microstructures: Multishell fullerenes and coaxial nanotubes. *Physical Review B* **1996**, *53* (15), 10225.
[238] Ohta, T.; *et al.* Interlayer interaction and electronic screening in multilayer graphene investigated with angle-resolved photoemission spectroscopy. *Physical Review Letters* **2007**, *98* (20), 206802.
[239] Kane, C.; Balents, L.; Fisher, M. P. Coulomb interactions and mesoscopic effects in carbon nanotubes. *Physical Review Letters* **1997**, *79* (25), 5086.

[240] Chiu, C.-W.; *et al.* Coulomb scattering rates of excited carriers in moderate-gap carbon nanotubes. *Physical Review B* **2006**, *73* (23), 235407.
[241] Wang, X.-F.; Chakraborty, T. Coulomb screening and collective excitations in a graphene bilayer. *Physical Review B* **2007**, *75* (4), 041404.
[242] Zhu, J.-J., Badalyan, S.; Peeters, F. Plasmonic excitations in Coulomb-coupled N-layer graphene structures. *Physical Review B* **2013**, *87* (8), 085401.
[243] Wehling, T.; *et al.* Strength of effective coulomb interactions in graphene and graphite. *Physical Review Letters* **2011**, *106* (23), 236805.
[244] Ho, J.-H.; *et al.* Electron decay rates in a zero-gap graphite layer. *Physics Letters A* **2006**, *357* (4–5), 401–406.
[245] Chiu, C.-W.; *et al.* Electronic decay rates in semiconducting carbon nanotubes. *Physica E: Low-dimensional Systems and Nanostructures* **2006**, *34* (1–2), 658–661.
[246] Kittel, C.; McEuen, P.; McEuen, P. *Introduction to Solid State Physics*, Vol. 8. New York: Wiley, 1996.
[247] Zhou, C.; Kong, J.; Dai, H. Intrinsic electrical properties of individual single-walled carbon nanotubes with small band gaps. *Physical Review Letters* **2000**, *84* (24), 5604.
[248] W. Chiu, C.; *et al.* Electron lifetime in armchair carbon nanotubes. *Journal of the Physical Society of Japan* **2004**, *73* (11), 2936–2939.
[249] Wang, X.-F.; Chakraborty, T. Coulomb screening and collective excitations in biased bilayer graphene. *Physical Review B* **2010**, *81* (8), 081402.
[250] Pyatkovskiy, P. Dynamical polarization, screening, and plasmons in gapped graphene. *Journal of Physics: Condensed Matter* **2008**, *21* (2), 025506.
[251] Hwang, E.; Sarma, S. D. Screening, Kohn anomaly, Friedel oscillation, and RKKY interaction in bilayer graphene. *Physical Review Letters* **2008**, *101* (15), 156802.
[252] Lin, M.; Chuu, D.; Shung, K.-K. Low-frequency plasmons in metallic carbon nanotubes. *Physical Review B* **1997**, *56* (3), 1430.
[253] Chiu, C.; *et al.* Temperature-dependent carrier dynamics in metallic carbon nanotubes. *Physics Letters A* **2005**, *346* (5–6), 347–354.
[254] Ku, W.; Berlijn, T.; Lee, C.-C. Unfolding first-principles band structures. *Physical Review Letters* **2010**, *104* (21), 216401.
[255] Cattelan, M.; Fox, N. A perspective on the application of spatially resolved ARPES for 2D materials. *Nanomaterials* **2018**, *8* (5), 284.

[256] Sprinkle, M.; *et al.* First direct observation of a nearly ideal graphene band structure. *Physical Review Letters* **2009**, *103* (22).
[257] Peskin, M. E. *An Introduction to Quantum Field Theory*. CRC Press, 2018.
[258] Lin, C.-Y.; *et al.* Magneto-optical selection rules of curved graphene nanoribbons and carbon nanotubes. *Carbon* **2014**, *69*, 151–161.
[259] Li, G.; *et al.* Observation of Van Hove singularities in twisted graphene layers. *Nature Physics* **2010**, *6* (2), 109–113.
[260] Sato, K.; *et al.* Zone folding effect in RamanG-band intensity of twisted bilayer graphene. *Physical Review B* **2012**, *86* (12).
[261] Grüneis, A.; *et al.* Tight-binding description of the quasiparticle dispersion of graphite and few-layer graphene. *Physical Review B* **2008**, *78* (20).
[262] Ho, Y.-H.; *et al.* Optical transitions between Landau levels: AA-stacked bilayer graphene. *Applied Physics Letters* **2010**, *97* (10), 101905.
[263] Chang, C.-P.; *et al.* Magnetoelectronic properties of the AB-stacked graphite. *Carbon* **2005**, *43* (7), 1424–1431.
[264] Patlatiuk, T.; *et al.* Evolution of the quantum Hall bulk spectrum into chiral edge states. *Nature Communications* **2018**, *9*(1).

CHAPTER 3

3D TRANSITION METAL-ADSORBED GRAPHENE

Le Vo Phuong Thuan and Ming-Fa Lin

Department of Physics, National Cheng Kung University, Tainan, Taiwan

Abstract

The adsorption of manganese and chromium on graphene is investigated using first principles based on density functional theory. The binding energy, optimal geometric structures, band structure, density of states, charge density, and charge density difference, spin distribution, and magnetic moment of each adatom-adsorbed graphene system are calculated in detail. The calculated results show that the most stable adsorption positions of both Mn and Cr on graphene are hollow sites. The calculated adsorption of all systems illustrates strong hybridization between transition metal adatoms and graphene, as well as metallic behavior. Besides, the positive magnetic moment shows that the Mn- and Cr-adsorbed graphenes have ferromagnetic configurations. The electronic and magnetic properties of Mn- and Cr-adsorbed graphenes reveal that the transition metal-adsorbed graphene systems have potential for applications in the future.

Keywords: Graphene, Transition metals absorbance, electronic properties.

3.1. Introduction

Nearly two decades since its discovery in 2004, graphene has continued to attract the attention of many scientists thanks to its remarkable physical and chemical properties, such as high electron mobility and the half-integer quantum Hall effect.[1–5] Since its discovery, graphene has been considered a potential material for many products, including its use in manufacturing lithium batteries,[6] electronic devices, and gas sensors.[7,8] Besides its interesting properties, graphene is a zero-gap semiconductor, which leads to many limitations in its application.[9] In order to expand its potential applications, many chemical transformations have been studied, such as substitution,[10–12] intercalation,[13–15] heterojunction,[16–18] and adsorption.[19–21] Among the chemical modifications, the adsorption of graphene has attracted considerable attention from both experimental and theoretical studies.[22,23] Adsorbents play an important role in the reconstruction of graphene, as well as conferring remarkable properties on the new structure. Recently, graphene metal-doped structures have become the focus of adsorption studies. The transition metals, after being adsorbed onto graphene, have become one of the main causes of the magnetic properties of graphene. The adsorption samples thus show promise as potential materials for application in spintronic devices and batteries.[24,25] In this chapter, we focus on the essential electronic and magnetic properties of graphene upon the adsorption of 3d transition metals (X = Mn, Cr).

Transition metals, when adsorbed onto different substrates, offer various potential applications, such as hydrogen stores,[26–28] spintronics,[29,30] and transistors.[31,32] The materials used to make up the substrates need to be sufficiently inert to avoid strong interactions with the adsorbent but must also bind to it. As with previous research results, graphene is considered a good candidate because it has a sufficiently inert base plane. Transition metal atoms adsorb onto the graphene surface, resulting in interesting modifications to the honeycomb lattice structure of the carbon atoms without disrupting it completely. Moreover, the geometric structure, electronic properties, and magnetic properties strongly depend on the adsorption concentrations of different types of atoms as well as their distribution on the substrate surface. To obtain complete information on the variety of fundamental properties that the adsorbent exerts on the substrate, we investigated the binding energies, the most stable site, bond lengths, the height between transition metals and the graphene surface, the band structure, the orbital-projected density of states, the charge density, and the spin-density distributions of pristine graphene and Mn- and Cr-adsorbed ones with different concentrations. Based on first-principles calculations, the analyses of the detailed results can provide useful information about the multi-orbital hybridizations and the competitive relations in the honeycomb lattice of graphene before and after adsorption. The reliable results of our research can be very helpful in the design and development of potential future applications.

3.2. Computational Method

In this study, we investigated the electronic and magnetic properties of Mn- and Cr-adsorbed graphene based on density functional theory (DFT)[33–38] using the Vienna Ab-initio Simulation Package (VASP)[39–42] software. To fully explore the spin-dependent magnetic configurations, GGA+U is utilized to deal with the single- and many-particle intrinsic interactions.[43,44] This simulation method principally covers the frequent electron–ion scatterings due to the atom-dependent crystal potentials. The projector-augmented wave (PAW) method[45–48] was used to characterize the valence electron and ionic interactions, whereas the electron–electron Coulomb interactions belong to the many-particle exchange and correlation energies under the Perdew–Burke–Ernzerhof generalized gradient approximation method.[49] To exclude van der Waals interactions between the two nearest unit cells, the vacuum distance in the z-direction was set to 15 Å. The cutoff energy for planar wave base expansion was set to 600 eV, which is available in the calculations of Bloch wave functions. To optimize the structure, the central sampling technique was used to integrate the Brillouin region[50] with a $30 \times 30 \times 1$ unique k-point grid and then with $100 \times 100 \times 1$ k-points for calculation of the electronic and magnetic properties. Equivalent k-point meshes were built for other enlarged cells depending on their sizes. A ground-state convergence condition of 10^{-8} eV between two successive simulation steps was set. The atoms can fully expand during the geometric optimization until the Hellmann–Feynman force[51–53] acting on each particle is less than 0.01 eV.

3.3. Discussion and Results

3.3.1. *Geometric structure*

The essential properties of graphene after adsorption of transition metals are investigated for the distinct adatoms and the various concentrations. The optimal geometric structures of Mn- and Cr-adsorbed graphenes are displayed in Figs. 3.1 and 3.2, respectively. Based on the calculated results, the most stable adsorption site of transition metals on graphene is the hollow position, which has been reported in previous theoretical studies.[24,25,54–56] The geometric structure of X-adsorbed graphene has two kinds of bonds, namely the C–C and X–C bonds. The bond lengths and heights of the transition metal adatoms relative to the graphene surface are significantly dependent on the adatom structures, as shown in Table 3.1. The top and side view configurations of X-adsorbed graphene do not create a buckling structure at various concentrations. That means the planar lattice of pristine graphene remains flat, so that chemisorption of the transition metals hardly changes the sigma bond by the p orbitals

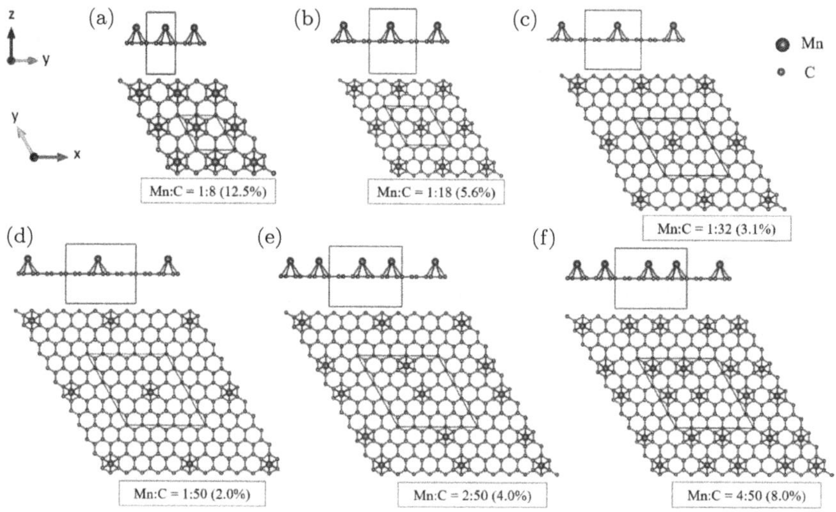

Fig. 3.1. The geometric structures of Mn-adsorbed graphene with top and side views under various concentrations: (a) Mn:C = 1:8 (12.5%), (b) Mn:C = 1:18 (5.6%), (c) Mn:C = 1:32 (3.1%), (d) Mn:C = 1:50 (2.0%), (e) Mn:C = 2:50 (4.0%), and (f) Mn:C = 4:50 (8%).

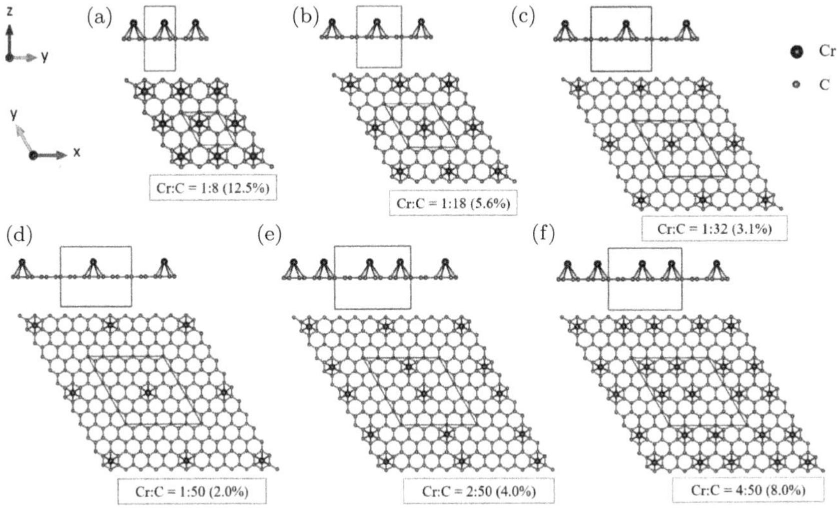

Fig. 3.2. The geometric structures of Cr-adsorbed graphene with top and side views under various concentrations: (a) Cr:C = 1:8 (12.5%), (b) Cr:C = 1:18 (5.6%), (c) Cr:C = 1:32 (3.1%), (d) Cr:C = 1:50 (2.0%), (e) Cr:C = 2:50 (4.0%), and (f) Cr:C = 4:50 (8%).

Table 3.1. The calculated C–C and C–X bond lengths, adatom heights (h), binding energies (E_b), energy gaps (E_g), and magnetic moments (M) per unit cell of (Mn, Cr)-adsorbed graphene systems. All the samples have the adsorbed adatoms arranged in a zigzag direction.

		Bond length (Å)					
Samples	**X**:C	**C–C**	**C–X**	h (Å)	M (μ_B)	E_g (eV)	E_b (eV)
Pristine graphene	—	1.42	—	—	—	—	—
Mn	1:8 (12.5%)	1.41–1.42	2.55–2.63	2.49	5.17	0	−5.07
	1:18 (5.6%)	1.41–1.44	2.52–2.58	2.48	5.23	0	−5.22
	1:32 (3.1%)	1.42–1.43	2.51–2.58	2.43	5.44	0	−5.22
	1:50 (2.0%)	1.42–1.43	2.55–2.52	2.41	5.47	0	−5.22
	2:50 (4.0%)	1.42–1.44	2.54–2.61	2.47	10.7	0	−10.43
	4:50 (8.0%)	1.42–1.43	2.43–2.54	2.40	20.9	0	−20.96
Cr	1:8 (12.5%)	1.40–1.44	2.50–2.59	2.48	5.50	0	−3.97
	1:18 (5.6%)	1.41–1.43	2.53–2.59	2.43	5.62	0	−5.52
	1:32 (3.1%)	1.41–1.44	2.49–2.57	2.41	5.64	0	−5.52
	1:50 (2.0%)	1.42–1.43	2.56–2.52	2.40	5.65	0	−5.62
	2:50 (4.0%)	1.42–1.43	2.54–2.62	2.48	11.2	0	−11.23
	4:50 (8.0%)	1.42–1.43	2.54–2.62	2.47	22.3	0	−22.36

of the carbon atom in graphene. The C–C bonds in graphene undergo a slight change in the range of 1.41–1.44 Å, which indicates that transition metals tend to induce small modifications in the sigma bonds. Besides, when concentration decreases, the X–C bond lengths and the adatom heights also decrease. Specifically, the Mn–C and Cr–C bond lengths lie in the ranges of 2.52–2.63 Å and 2.49–2.62 Å, respectively. The Mn atom height increase from 2.40 to 2.49 Å, corresponding to the increase in concentration. Similarly, the Cr atom height increases from 2.41– to 2.48 Å with the increase in concentration. When the transition metal concentration is increased, the X–C bond lengths and adatom heights of all adsorbed systems are enlarged. Moreover, we calculated the binding energy (E_b) to gauge the stabilities of the graphene after adsorption, which characterizes the lowered total ground-state energy. The binding energy is calculated according to the following formula:

$$E_b = (E_{\text{total}} - E_{\text{gra}} - nE_X)/n$$

where E_{total} is the total energy of the adsorbed graphene, and E_{gra} and E_X are the isolated energies of the graphene monolayer and the transition metal atoms, respectively. n is the number of transition metal atoms. The more negative the values of the binding energy, the more stable the structure. Specifically, at the same concentrations, Mn-adsorbed graphene structures are less stable than Cr-adsorbed ones. In addition, the stabilities of the adsorbed systems rise, which corresponds to the increasing concentrations.

3.3.2. *Diversified electronic and magnetic properties*

To gauge the major characteristics of the electronic properties of transition metal-adsorbed graphene, we investigated the band structures along the high-symmetry direction. The band structures of Mn- and Cr-adsorbed graphenes are depicted in Figs. 3.3 and 3.4, respectively. The pristine graphene structure (shown in Fig. 3.3(a)) has a linear valence band that intersects a conduction band at the K point. According to previous theoretical studies,[2,57,58] the

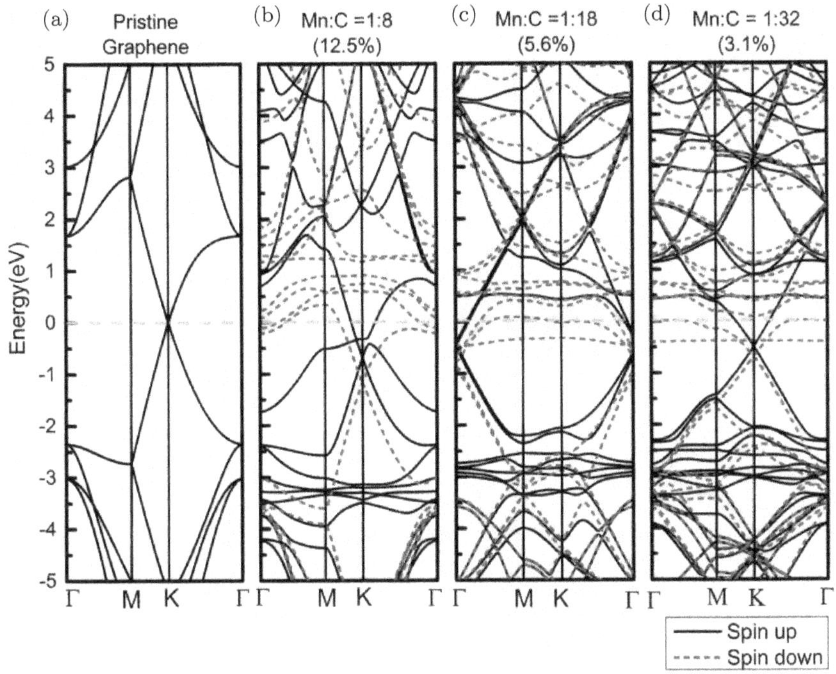

Fig. 3.3. The band structure of pristine graphene and Mn-adsorbed graphene with different concentrations: (a) pristine graphene, (b) Mn:C = 1:8 (12.5%), (c) Mn:C = 1:18 (5.6%), and (d) Mn:C = 1:32 (3.1%). The black and red dashed curves correspond to spin-up and spin-down states. The green dashed line indicates the location of the Fermi level.

Dirac point is the point of intersection, which is located at the Fermi level ($E = 0$ eV). The Dirac cone structures at the low-lying energy states created by the sp^2 orbitals of ($2s$, $2p_x$, $2p_y$) form strong covalent bonds between the nearest-neighbor carbon atoms, and their perpendicular π bonds come from the $2p_z$ orbitals. Specifically, the σ orbitals take shape in the valence band at the deeper-energy range of $E < -2.5$ eV. The π and π^* conduction bands contribute to the band structures at energy states less than 2.5 eV. Besides, the saddle points of the middle-energy bands are located at the M points. The low-lying Dirac cone structure of graphene slightly changes after the adsorption of a transition

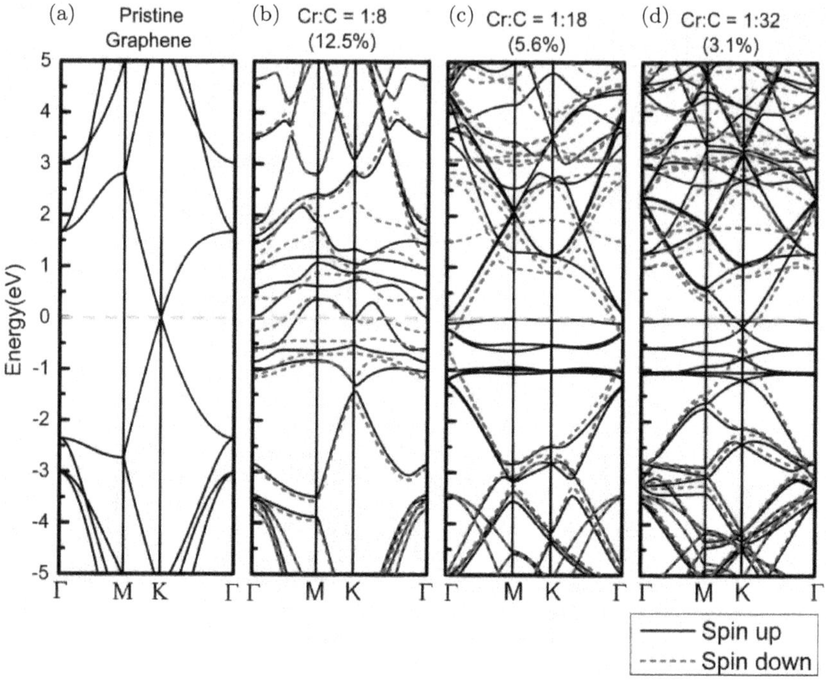

Fig. 3.4. The band structure of pristine graphene and Cr-adsorbed graphene with different concentrations: (a) pristine graphene, (b) Cr:C = 1:8 (12.5%), (c) Cr:C = 1:18 (5.6%), and (d) Cr:C = 1:32 (3.1%). The black and red dashed curves correspond to spin-up and spin-down states. The green dashed line indicates the location of the Fermi level.

metal. For the various concentrations, all adsorbed systems have metal structures, which can be seen from the site of the Dirac cone configuration compared with the Fermi level. For the 12.5% concentration (shown in Figs. 3.3(b) and 3.4(b) for Mn-adsorbed graphene) and the 3.1% concentration (Figs. 3.3(d) and 3.4(d) for Cr-adsorbed graphene), the Dirac point appears in the valence bands at the K point. Conversely, for the 5.6% concentration, this point appears at the Γ point, as illustrated in Figs. 3.3(c) and 3.4(c). The Dirac cone structure created by the $2p_z$ orbitals of C atoms in pristine graphene changes from isotropic to anisotropic in the presence of the adatom-dominated bands. Moreover, the transition metal-adsorbed graphene systems have spin–state separation. The new bands created in the energy range of 2 to −3 eV are mainly contributed by a transition metal, as marked by the pink and blue circles for Mn and Cr atoms, respectively. Specifically, for Mn-adsorbed graphene, the Mn atoms dominate the valence states in the spin-up band structure (Figs. 3.5(b), 3.6(b), and 3.7(b)). On the contrary, in the spin-down band structure, the Mn atoms make major contributions to the conduction bands (Figs. 3.5(c), 3.6(c), and 3.7(c)). The Cr-adsorbed configurations exhibit domination of the Cr atoms, similar to Mn-adsorbed systems, except for the case with 12.5% concentration (Figs. 3.8–3.10). For Cr-adsorbed

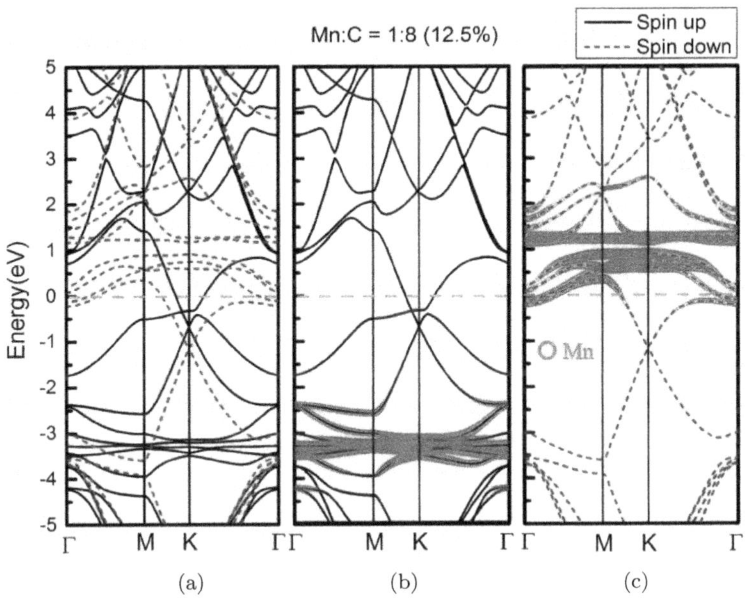

Fig. 3.5. Valence hole and conduction electron energy spectra for Mn:C = 1:8 (12.5%) along the high-symmetry points of the first Brillouin zone: (a) the total band structure with the black and red curves corresponding to the spin-up and spin-down states, (b) the band structure for the spin-up states, and (c) the band structure for the spin-down states. The pink circles represent the contributions of Mn atoms. The green dashed line indicates the location of the Fermi level.

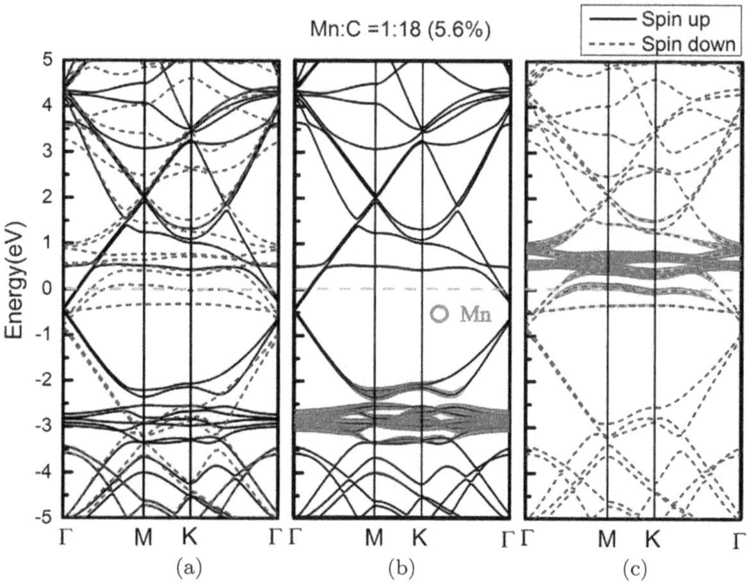

Fig. 3.6. Valence hole and conduction electron energy spectra for Mn:C = 1:18 (5.6%) along the high-symmetry points of the first Brillouin zone: (a) the total band structure with the black and red curves corresponding to the spin-up and spin-down states, (b) the band structure for the spin-up states, and (c) the band structure for the spin-down states. The pink circles represent the contributions of Mn atoms. The green dashed line indicates the location of the Fermi level.

graphene with 12.5% concentration, the Cr atom dominated both the conduction and valence bands near the Fermi level in the spin-up band structure, while only contributing to the conduction bands in the spin-down configuration. This indicates that transition metals make major contributions to the low-lying energy states near the Fermi level. Based on the previous studies, the Mn-/Cr-adsorbed graphene structures are classified as magnetic materials.[24,25,54] The calculated results show that Mn- and Cr-adsorbed structures have spin-up and spin-down energy splitting. Moreover, as noted in Table 3.1, they possess positive magnetic moment values, indicating that the adsorbed systems have ferromagnetic configurations (as shown in Figs. 3.3 and 3.4).

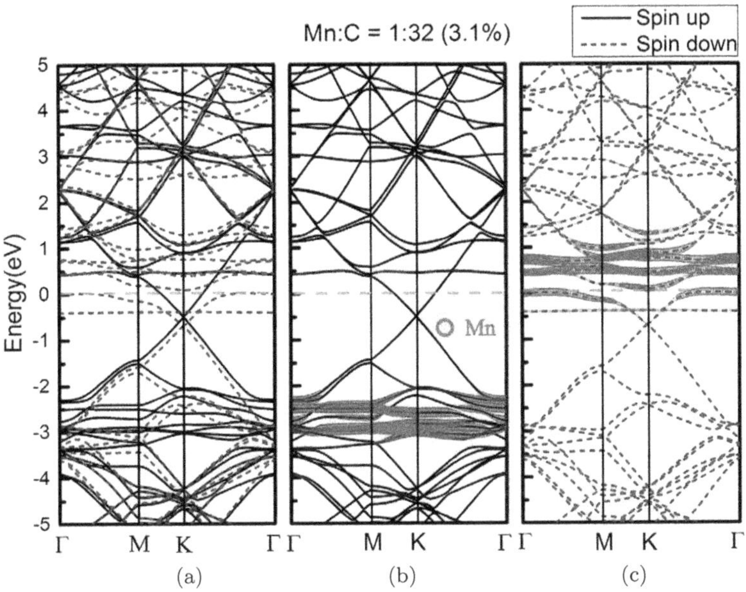

Fig. 3.7. Valence hole and conduction electron energy spectra for Mn:C = 1:32 (3.1%) along the high-symmetry points of the first Brillouin zone: (a) the total band structure with the black and red curves corresponding to the spin-up and spin-down states, (b) the band structure for the spin-up states, and (c) the band structure for the spin-down states. The pink circles represent the contributions of Mn atoms. The green dashed line indicates the location of the Fermi level.

To understand the orbital hybridization as well as the changes in the electronic properties, we calculated charge density (presented in Figs. 3.11(a)–3.11(d) and Figs. 3.12(a)–3.12(d)) for Mn- and Cr-adsorbed graphenes, respectively) and the charge density differences (shown in Figs. 3.11(e)–3.11(g) and Figs. 3.12(e)–3.12(g), respectively). The latter is defined as the total charge density of adsorbed graphene minus the charge densities of graphene and transition metal atoms. The red region between the two nearest-neighbor carbon atoms illustrates the exit of the strong σ bonds, as presented in Figs. 3.11(a)–3.11(d) and Figs. 3.12(a)–3.12(d). Besides, the charge density differences of Mn-/Cr-adsorbed graphene exhibit a slight modification compared to those of pristine graphene.

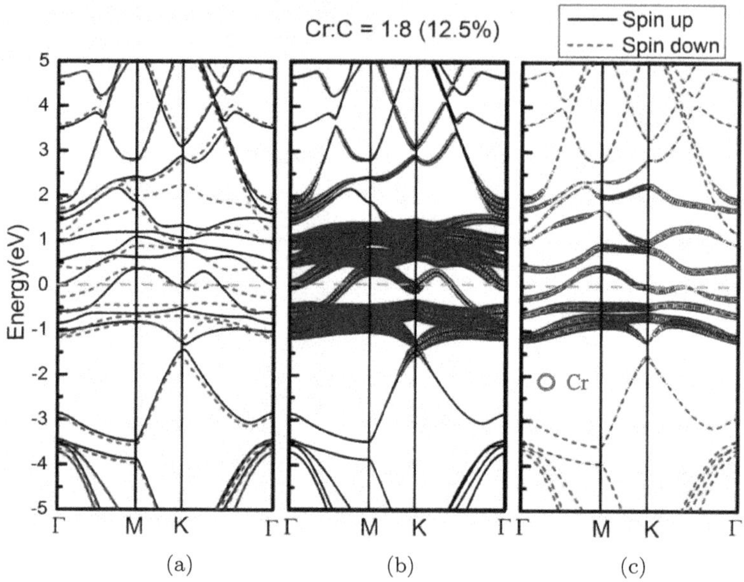

Fig. 3.8. Valence hole and conduction electron energy spectra for Cr:C = 1:8 (12.5%) along the high-symmetry points of the first Brillouin zone: (a) the total band structure with the black and red curves corresponding to the spin-up and spin-down states, (b) the band structure for the spin-up states, and (c) the band structure for the spin-down states. The blues circles represent the contributions of Cr atoms. The green dashed line indicates the location of the Fermi level.

The yellow regions in Figs. 3.11(e)–3.11(g) are described as having van der Waals interactions between the Mn and C atoms. A similar observation is made in the Cr- adsorbed samples. This implies the existence of multi-orbital hybridization between the (3d, 4s) orbitals of the adatom and the $2p_z$ orbitals of the C atoms. At high concentrations, the values of charge density differences are larger than their values at low concentrations.

The density of states (DOS) can provide full information about orbital hybridization as well as their contributions to chemical bonds. The band-edge states in the band structure are reflected by the form, number, and energy of the special configurations in the DOS. To observe the change in DOS after the adsorption of transition

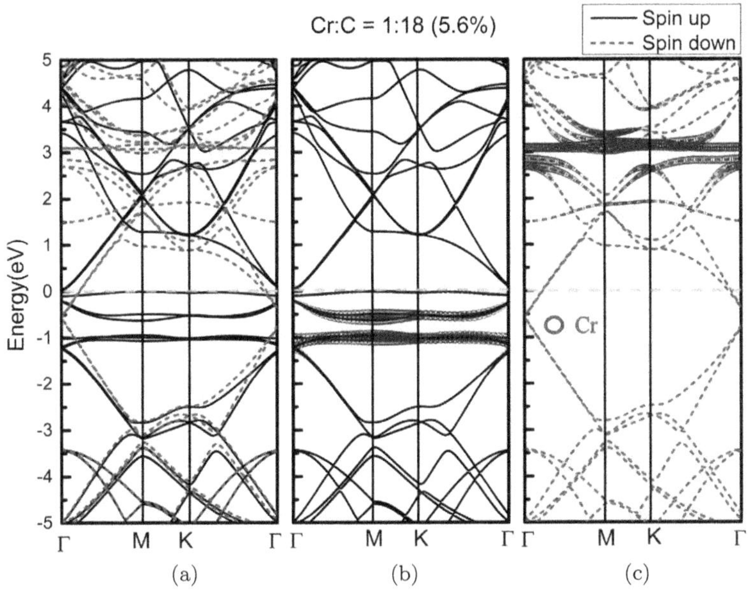

Fig. 3.9. Valence hole and conduction electron energy spectra for Cr:C = 1:18 (5.6%) along the high-symmetry points of the first Brillouin zone: (a) the total band structure with the black and red curves corresponding to the spin-up and spin-down states, (b) the band structure for the spin-up states, and (c) the band structure for the spin-down states. The blues circles represent the contributions of Cr atoms. The green dashed line indicates the location of the Fermi level.

metals, we calculated the DOS of the pristine graphene structure. Figure 3.11(a) shows that the $2p_z$–$2p_z$ orbital bonds of C atoms created a pair of significantly symmetric peaks near the Fermi level, which dominated the DOS in the energy range of −2.5–2.5 eV. Besides, the isotropic Dirac cone structure of pristine graphene causes the DOS value to be equal to zero at the Fermi level. After the adsorption of Mn and Cr atoms, the DOS of C atoms undergoes many changes. The complex van Hove singularities in the orbital-projected DOS show the drastic multi-orbital hybridization in X–C and C–C bonds. The ($3d_{xy}$, $3d_{yz}$, $3d_{xz}$, $3d_{z^2}$, $3d_{x^2-y^2}$, 4s)-orbitals of transition metals dominated some unusual van Hove singularities in the DOS of the adsorbed systems. For the Mn-adsorbed graphene structures,

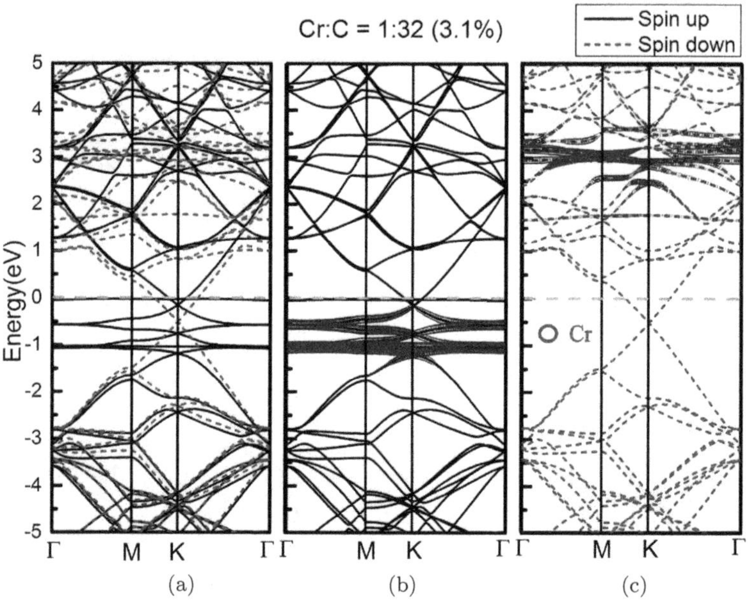

Fig. 3.10. Valence hole and conduction electron energy spectra for Cr:C = 1:32 (3.1%) along the high-symmetry points of the first Brillouin zone: (a) the total band structure with the black and red curves corresponding to the spin-up and spin-down states, (b) the band structure for the spin-up states, and (c) the band structure for the spin-down states. The blues circles represent the contributions of Cr atoms. The green dashed line indicates the location of the Fermi level.

at high concentrations, the orbital hybridizations of ($3d_{xz}$, $3d_{yz}$) and ($3d_{xy}$, $3d_{x^2-y^2}$), and their hybridization with $3d_{z^2}$ mainly dominated the prominent peaks near the Fermi level. On the other hand, at low concentrations, it is difficult to distinguish the contributions of the ($3d_{xy}$, $3d_{yz}$, $3d_{xz}$, $3d_{z^2}$, $3d_{x^2-y^2}$, 4s)orbitals of the Mn atoms and the $2p_z$ of the C atoms, indicating significant hybridization in the Mn–C bonds. The orbitals of the Mn atoms dominated the spin-up DOS in the range of −4 to −2 eV, while the spin-down states are contributed by them at energies ranging from 0 to 2 eV, as shown in Fig. 3.13. For the Cr-adsorbed graphene systems, at high concentrations, there exist the complex multi-orbital hybridizations of ($3d_{xy}$, $3d_{yz}$, $3d_{xz}$, $3d_{z^2}$, $3d_{x^2-y^2}$, 4s)-orbitals of Cr atoms and $2p_z$ orbitals of C atoms in both the spin-up and spin-down states at

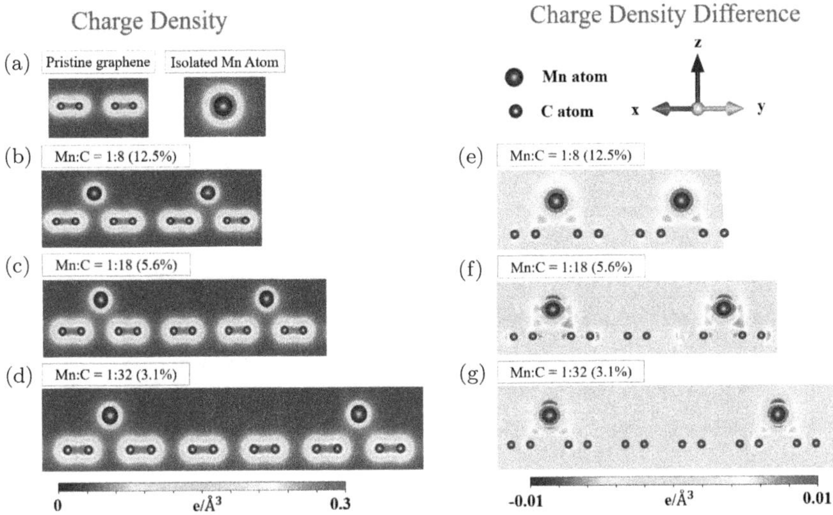

Fig. 3.11. The spatial charge density of Mn-adsorbed graphene: (a) pristine graphene and isolated Mn atom, (b) Mn:C = 1:8 (12.5%), (c) Mn:C = 1:18 (5.6%), and (d) Mn:C = 1:32 (3.1%). The corresponding charge density differences of Mn-adsorbed graphene are shown in (e)–(g), respectively.

the Fermi level. In addition, at low concentrations, there exist the hybridizations of ($3d_{xz}$, $3d_{z^2}$, $3d_{x^2-y^2}$)-orbitals in the spin-up states at energies ranging from -1.5 to 0 eV, while their hybridizations exist in the range of 2–4 eV in the spin-down states (as shown in Fig. 3.14). Moreover, the DOS corresponds to the results from the band structures and spatial charge densities that the sigma bonds between carbon atoms do not participate in the X–C bonds.

Besides diverse electronic properties, graphene after transition metal adsorption also exhibits interesting magnetic properties. To obtain more useful information about this property, we calculated the spin-density distributions, as is clearly illustrated in Fig. 3.15. All the samples have the spin-up states mainly dominating the spin-density arrangement, except at the high-concentration levels of Cr-adsorbed graphene, which corresponds to the behavior of spin split in the band structure (as shown in Figs. 3.3–3.10) and the DOS of transition metal orbitals (Figs. 3.13 and 3.14). The magnetic moment of the adsorbed systems is sensitive to the adatom concentrations

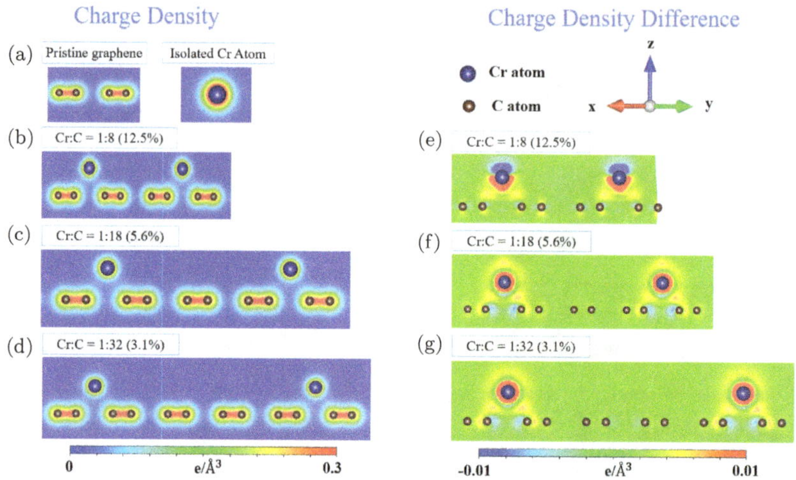

Fig. 3.12. The spatial charge density of Cr-adsorbed graphene: (a) pristine graphene and isolated Cr atom, (b) Cr:C = 1:8 (12.5%), (c) Cr:C = 1:18 (5.6%), and (d) Cr:C = 1:32 (3.1%). The corresponding charge density differences of Cr-adsorbed graphene are shown in (e)–(g), respectively.

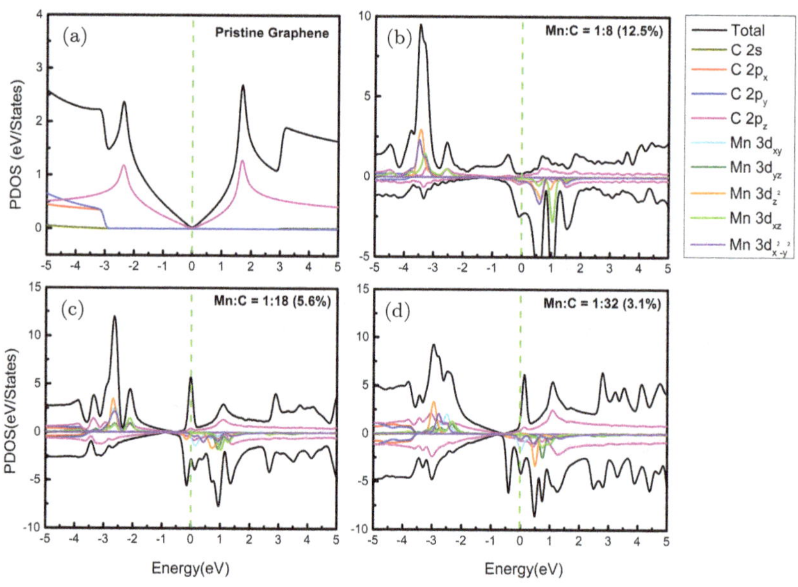

Fig. 3.13. The orbital-projected density of states of pristine graphene and Mn-adsorbed graphene with different concentrations: (a) pristine graphene, (b) Mn:C = 1:8 (12.5%), (c) Mn:C = 1:18 (5.6%), and (d) Mn:C = 1:32 (3.1%). The green dashed line indicates the location of the Fermi level.

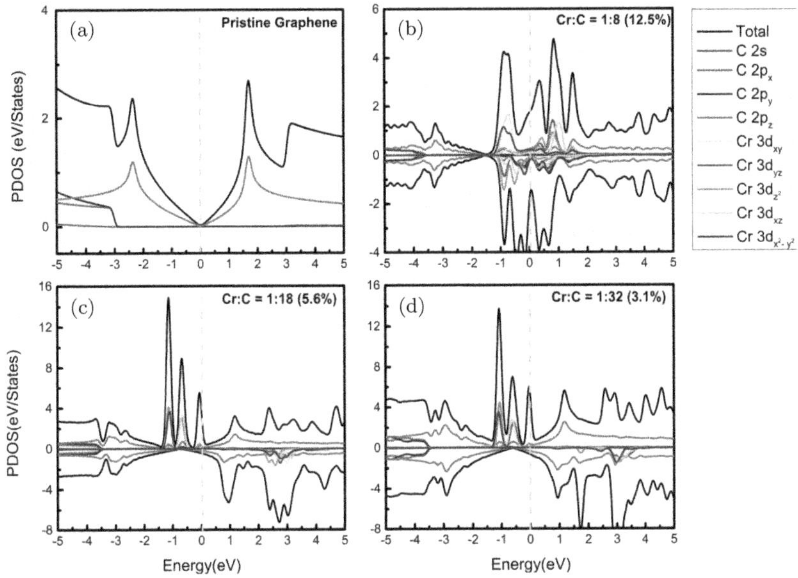

Fig. 3.14. The orbital-projected density of states of pristine graphene and Cr-adsorbed graphene with different concentrations: (a) pristine graphene, (b) Cr:C = 1:8 (12.5%), (c) Cr:C = 1:18 (5.6%), and (d) Cr:C = 1:32 (3.1%). The green dashed line indicates the location of the Fermi level.

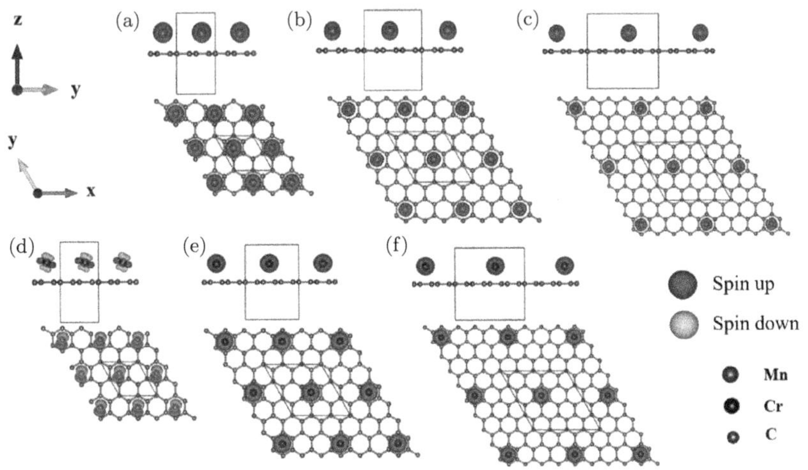

Fig. 3.15. The spin-density distributions of Mn- and Cr-adsorbed graphenes with top and side views under various concentrations: (a) Mn:C = 1:8 (12.5%), (b) Mn:C = 1:18 (5.6%), (c) Mn:C = 1:32 (3.1%), (d) Cr:C = 1:8 (12.5%), (e) Cr:C = 1:18 (5.6%), and (f) Cr:C =1:32 (3.1%). The red and green iso-surfaces display the charge density of spin-up and spin-down structures, respectively.

and distribution structures (Table 3.1). For the various concentrations, the magnetic moments are 5.12–20.9 μ_B and 5.5–22.3 μ_B for Mn- and Cr-adsorbed graphenes, respectively. Their positive values imply that both Mn- and Cr-adsorbed graphenes display ferromagnetic behavior. The ferromagnetic moments of Cr-adsorbed graphene are slightly higher than those of Mn-adsorbed systems. In addition, the magnetic moment values rapidly increase in the order of the increasing concentration, which can be due to the interaction between adatoms. The above results reveal that transition metals can create diverse magnetic properties on graphene surfaces.

3.4. Conclusions

The calculated results reveal that the chemical and physical properties of graphene after the adsorption of transition metals are enriched by the distribution and concentration of adatoms. The most stable site of the Mn/Cr atom adsorbed on graphene is the hollow position. The geometric, electronic, and magnetic properties of Mn-/Cr-adsorbed graphene with different concentrations have complicated orbital hybridizations between the ($3d_{xy}$, $3d_{yz}$, $3d_{xz}$, $3d_{z^2}$, $3d_{x^2-y^2}$, 4s) orbitals of the transition metal atoms and the $2p_z$ orbitals of the C atoms. Transition metals mainly dominated the low-lying energy states near the Fermi level in the band structures. After adsorption, graphene systems show an anisotropic and spin-split Dirac cone configuration. Besides, the Dirac point changes its position when compared to that in pristine graphene. In addition, the spin-density distribution reveals that all adsorbed graphene structures exhibit ferromagnetic metallic behavior.

References

[1] Abergel, D.; Apalkov, V.; Berashevich, J.; Ziegler, K.; Chakraborty, T. Properties of graphene: A theoretical perspective. *Advances in Physics* **2010**, *59* (4), 261–482.

[2] Neto, A. C.; Guinea, F.; Peres, N. M.; Novoselov, K. S.; Geim, A. K. The electronic properties of graphene. *Reviews of Modern Physics* **2009**, *81* (1), 109.

[3] Ovid'Ko, I. Mechanical properties of graphene. *Reviews on Advanced Materials Science* **2013**, *34* (1), 1–11.

[4] Pop, E.; Varshney, V.; Roy, A. K. Thermal properties of graphene: Fundamentals and applications. *MRS Bulletin* **2012**, 37 (12), 1273–1281.

[5] Falkovsky, L. A. Optical properties of graphene. *Journal of Physics: Conference Series* **2008**, *129* (1), 012004.

[6] Wang, C.; Li, D.; Too, C. O.; Wallace, G. G. Electrochemical properties of graphene paper electrodes used in lithium batteries. *Chemistry of Materials* **2009**, *21* (13), 2604–2606.

[7] Yoon, H. J.; Yang, J. H.; Zhou, Z.; Yang, S. S.; Cheng, M. M.-C. Carbon dioxide gas sensor using a graphene sheet. *Sensors and Actuators B: Chemical* **2011**, *157* (1), 310–313.

[8] Yuan W.; Shi, G. Graphene-based gas sensors. *Journal of Materials Chemistry A* **2013**, *1* (35), 10078–10091.

[9] Mohan, V. B.; Lau, K.-T.; Hui, D.; Bhattacharyya, D. Graphene-based materials and their composites: A review on production, applications and product limitations. *Composites Part B: Engineering* **2018**, *142*, 200–220.

[10] Kawai, S.;*et al.* Multiple heteroatom substitution to graphene nanoribbon. *Science Advances* **2018**, *4* (4), eaar7181.

[11] Chen, N.; Huang, X.; Qu, L. Heteroatom substituted and decorated graphene: Preparation and applications. *Physical Chemistry Chemical Physics* **2015**, *17* (48), 32077–32098.

[12] Sun, M.; Ren, Q.; Zhao, Y.; Chou, J.-P.; Yu, J.; Tang, W. Electronic and magnetic properties of 4d series transition metal substituted graphene: A first-principles study. *Carbon* **2017**, *120*, 265–273.

[13] Petrović, M.;*et al.* The mechanism of caesium intercalation of graphene. *Nature communications* **2013**, *4* (1), 1–8.

[14] Liu, Y.;*et al.* Mechanism of metal intercalation under graphene through small vacancy defects. *The Journal of Physical Chemistry C* **2021**, *125* (12), 6954–6962.

[15] Sugawara, K.; Kanetani, K.; Sato, T.; Takahashi, T. Fabrication of Li-intercalated bilayer graphene. *AIP Advances* **2011**, *1* (2), 022103.

[16] Behura, S. K.; Wang, C.; Wen, Y.; Berry, V. Graphene–semiconductor heterojunction sheds light on emerging photovoltaics. *Nature Photonics* **2019**, 13 (5), 312–318.

[17] Li, X., Shen, R., Ma, S., Chen, X.; Xie, J. Graphene-based heterojunction photocatalysts. *Applied Surface Science* **2018**, 430, 53–107.

[18] He, F.; Meng, A.; Cheng, B.; Ho, W.; Yu, J. Enhanced photocatalytic H2-production activity of WO3/TiO2 step-scheme heterojunction by

graphene modification. *Chinese Journal of Catalysis* **2020**, *41* (1), 9–20.

[19] Chan, K. T.; Neaton, J.; Cohen, M. L. First-principles study of metal adatom adsorption on graphene. *Physical Review B* **2008**, *77* (23), 235430.

[20] Moon, H. S.; Lee, J. H.; Kwon, S.; Kim, I. T.; Lee, S. G. Mechanisms of Na adsorption on graphene and graphene oxide: Density functional theory approach. *Carbon Letters* **2015**, *16* (2), 116–120.

[21] Dai, J.; Yuan, J.; Giannozzi, P. Gas adsorption on graphene doped with B, N, Al, and S: A theoretical study. *Applied Physics Letters* **2009**, 95 (23), 232105.

[22] Jauris, I.;*et al.* Adsorption of sodium diclofenac on graphene: A combined experimental and theoretical study. *Physical Chemistry Chemical Physics* **2016**, *18* (3), 1526–1536.

[23] Manadé, M.; Viñes, F.; Illas, F. Transition metal adatoms on graphene: A systematic density functional study. *Carbon* **2015**, *95*, 525–534.

[24] Sevinçli, H.; Topsakal, M.; Durgun, E.; Ciraci, S. Electronic and magnetic properties of 3 d transition-metal atom adsorbed graphene and graphene nanoribbons. *Physical Review B* **2008**, *77* (19), 195434.

[25] Hu, L.; Hu, X.; Wu, X.; Du, C.; Dai, Y.; Deng, J. Density functional calculation of transition metal adatom adsorption on graphene. *Physica B: Condensed Matter* **2010**, *405* (16), 3337–3341.

[26] Tozzini, V.; Pellegrini, V. Prospects for hydrogen storage in graphene. *Physical Chemistry Chemical Physics* **2013**, *15* (1), 80–89.

[27] Spyrou, K.; Gournis, D.; Rudolf, P. Hydrogen storage in graphene-based materials: Efforts towards enhanced hydrogen absorption. *ECS Journal of Solid State Science and Technology* **2013**, *2* (10), M3160.

[28] Huang, C.-C.; Pu, N.-W.; Wang, C.-A.; Huang, J.-C.; Sung, Y.; Ger, M.-D. Hydrogen storage in graphene decorated with Pd and Pt nanoparticles using an electroless deposition technique. *Separation and Purification Technology* **2011**, *82*, 210–215.

[29] Avsar, A.; Ochoa, H.; Guinea, F.; Özyilmaz, B.; Van Wees, B.; Vera-Marun, I. J. Colloquium: Spintronics in graphene and other two-dimensional materials. *Reviews of Modern Physics* **2020**, *92* (2), 021003.

[30] Han, W.; Kawakami, R. K.; Gmitra, M.; Fabian, J. Graphene spintronics. *Nature Nanotechnology* **2014**, *9*, (10), 794–807.

[31] Liu, Y.; Duan, X.; Huang, Y.; Duan, X. Two-dimensional transistors beyond graphene and TMDCs. *Chemical Society Reviews* **2018**, *47* (16), 6388–6409.

[32] Schwierz, F. Graphene transistors. *Nature Nanotechnology* **2010**, *5* (7), 487–496.
[33] Parr, R. G. Density functional theory. *Annual Review of Physical Chemistry* **1983**, *34* (1), 631–656.
[34] Orio, M.; Pantazis, D. A.; Neese, F. Density functional theory. *Photosynthesis Research* **2009**, *102* (2), 443–453.
[35] Gross E. K.; Dreizler, R. M. *Density Functional Theory*. Springer Science & Business Media, 2013.
[36] Engel, E.; Dreizler, R. M. Density functional theory. *Theoretical and Mathematical Physics* **2011**, 351–399.
[37] Kohn, W. Density functional theory. In *Introductory Quantum Mechanics with MATLAB: for Atoms, Molecules, Clusters, and Nanocrystals*, 2019.
[38] Argaman, N.; Makov, G. Density functional theory: An introduction. *American Journal of Physics* **2000**, *68* (1), 69–79.
[39] Sun, G.; Kürti, J.; Rajczy, P.; Kertesz, M.; Hafner, J.; Kresse, G. Performance of the Vienna ab initio simulation package (VASP) in chemical applications. *Journal of Molecular Structure: THEOCHEM 2003*, *624* (1–3), 37–45.
[40] Hafner, J. Ab-initio simulations of materials using VASP: Density-functional theory and beyond. *Journal of Computational Chemistry* **2008**, *29* (13), 2044–2078.
[41] Bartlett, R. J.; Lotrich, V. F.; Schweigert, I. V. Ab initio density functional theory: The best of both worlds? *The Journal of Chemical Physics* **2005**, *123* (6), 062205.
[42] Bartlett, R. J.; Grabowski, I.; Hirata, S.; Ivanov, S. The exchange-correlation potential in ab initio density functional theory. *The Journal of Chemical Physics* **2005**, *122* (3), 034104.
[43] Jafarova, V.; Orudzhev, G. Structural and electronic properties of ZnO: A first-principles density-functional theory study within LDA (GGA) and LDA (GGA)+ U methods. *Solid State Communications* **2021**, **325**, 114166.
[44] Bo, S.; Ping, Z. First-principles local density approximation (LDA)+ U and generalized gradient approximation (GGA)+ U studies of plutonium oxides. *Chinese Physics B* **2008**, *17* (4) 1364.
[45] Kresse, G.; Joubert, D. From ultrasoft pseudopotentials to the projector augmented-wave method. *Physical Review B* **1999**, *59* (3), 1758.
[46] Gajdoš, M.; Hummer, K.; Kresse, G.; Furthmüller, J.; Bechstedt, F. Linear optical properties in the projector-augmented wave methodology. *Physical Review B* **2006**, *73* (4), 045112.

[47] Mortensen, J. J.; Hansen, L. B.; Jacobsen, K. W. Real-space grid implementation of the projector augmented wave method. *Physical Review B* **2005**, *71* (3), 035109.

[48] Holzwarth, N.; Tackett, A.; Matthews, G. A Projector Augmented Wave (PAW) code for electronic structure calculations, Part I: Atompaw for generating atom-centered functions. *Computer Physics Communications* **2001**, *135* (3), 329–347.

[49] Peng, H.; Perdew, J. P. Rehabilitation of the Perdew-Burke-Ernzerhof generalized gradient approximation for layered materials. *Physical Review B* **2017**, *95* (8), 081105.

[50] Griffin, A. Brillouin light scattering from crystals in the hydrodynamic region. *Reviews of Modern Physics* **1968**, *40* (1), 167.

[51] Politzer, P.; Murray, J. S. The Hellmann-Feynman theorem: A perspective. *Journal of Molecular Modeling* **2018**, *24* (9), 1–7.

[52] Di Ventra, M.; Pantelides, S. T. Hellmann-Feynman theorem and the definition of forces in quantum time-dependent and transport problems. *Physical Review B* **2000**, *61* (23), 16207.

[53] Bakken, V.; Helgaker, T.; Klopper, W.; Ruud, K. The calculation of molecular geometrical properties in the Hellmann — Feynman approximation. *Molecular Physics* **1999**, *96* (4), 653–671.

[54] Mao, Y.; Yuan, J.; Zhong, J. Density functional calculation of transition metal adatom adsorption on graphene *Journal of Physics: Condensed Matter* **2008**, *20* (11), 115209.

[55] Zhou, M.; Lu, Y.-H.; Cai, Y.-Q.; Zhang, C.; Feng, Y.-P. Adsorption of gas molecules on transition metal embedded graphene: A search for high-performance graphene-based catalysts and gas sensors. *Nanotechnology* **2011**, *22* (38), 385502.

[56] Cao, C.; Wu, M.; Jiang, J.; Cheng, H.-P. Transition metal adatom and dimer adsorbed on graphene: Induced magnetization and electronic structures. *Physical Review B* **2010**, *81* (20), 205424.

[57] Thuy Tran, N. T.; Dahal, D.; Gumbs, G.; Lin, M.-F. Adatom doping-enriched geometric and electronic properties of pristine graphene: A method to modify the band gap. *Structural Chemistry* **2017**, *28* (5), 1311–1318.

[58] Novoselov, K.;*et al.* Electronic properties of graphene. *Physica Status Solidi (B)* **2007**, *244* (11), 4106–4111.

CHAPTER 4

4f RARE-EARTH ELEMENT-ADSORBED GRAPHENE

Nguyen Thi Han, Vo Khuong Dien
and Ming-Fa Lin

*Department of Physics, National Cheng Kung University,
701 Tainan, Taiwan*

Abstract

Lanthanum (La) and cerium (Ce) atom-adsorbed graphene systems are investigated for a thorough understanding of the strongly coupling effects associated with the very complicated quasi-particle 4f orbital charges and spin configurations, such as the significant covalent bonds of C-sp3 and La-5d5/Ce-4f7 orbitals and their non-magnetic/ferromagnetic spin configurations. Their quasi-particle behaviors are expected to present a sharp contrast with those observed in alkalizations, oxidations, and halogenations.

Keywords: Graphene, 4f rare-earth elements, absorbance, electronic and optical properties.

4.1. Introduction

To date, a periodic table[1,2] shows the dominating roles in the basic and applied sciences and can provide the unusual atomic configurations. Its main features are very useful in fully understanding the critical mechanisms and presenting a concise picture of any system, specifically the various crystal structures closely related to the active orbitals of pure elements.

So far, the experimental and theoretical studies have clearly indicated the following three kinds of atomic orbitals according to the real chemical bonds: (i) (s, p_x, p_y, p_z) orbitals of group I–VIII systems,[3–6] (ii) d_{yz}, d_{xy}, d_{z^2}, d_{xz}, and $d_{x^2-y^2}$ orbitals of transition metal atoms (e.g., Fe/Co/Ni/Cu/Ag/Au[7,8] with 3d or 4d orbitals), and (iii) $4f_{xyz}$, $4f_{yz^2}$, $4f_{z^3}$, $4f_{xz^2}$, $4f_{z(x^2-3y^2)}$, $4f_{x(x^2-3y^2)}$, $4f_{y(y^2-3x^2)}$ orbitals of rare-earth metal atoms (from Ce to Lr atoms with 4f and 5f orbitals). These charge distributions are capable of building weak, strong, or dramatic orbital hybridizations, leading to diversified materials, e.g., the graphene-related systems (graphite systems,[9,10]/layered graphene systems[11]/graphene nanoribbons[12–14]/carbon nanotubes[15–17]/fullerenes[18,19]). In general, these condensed-matter crystals expose the highly anisotropic and non-uniform environments[20–22] of chemical and physical engineering, mainly owing to the very prominent covalent bonds.

Both chemical modifications[23,24] and physical perturbations[25] expose the adsorptions[26–28]/substitutions[29]/intercalations[30]/decorations[31]/heterojunction[32] composites, which can greatly diversify the composite quantum quasi-particles, as well as a uniform perpendicular magnetic field, an electric field, their spatially modulated fields, the superposition of uniform and non-uniform fields, electromagnetic (EM) waves,[33] time- and position-dependent Coulomb fields,[34] mechanical stresses,[35,36] or thermal excitations.[37,38]

In this study, the Lanthanum (La) and Cerium (Ce) atom-adsorbed graphene systems are investigated for a thorough understanding of the strongly coupling effects associated with the very complicated quasi-particle 4f-orbital charges and spin configurations, such as the significant covalent bonds of C-sp^3 and La-$5d^5$/Ce-$4f^7$ orbitals and their non-magnetic/ferromagnetic spin configurations. Their quasi-particle behaviors are expected to present a sharp contrast with alkalizations, oxidations, and halogenations. How to commence the simultaneous progress among the Vienna Ab-initio Simulation Package (VASP) simulations, phenomenological models, and experimental observations is also one of the focuses of this study, as clearly illustrated in the previously published works[20,21,39–43] Apparently, the geometric, electronic, and optical properties of the La and Ce atoms adsorbed on monolayer graphene are determined using first-principles simulations based on the VASP software. The concentration and arrangement dependents are explored. The orbital hybridizations in the chemical bonding are achieved by the delicate analysis of the atom-dominated energy band structure, the spatial charge densities and charge density distribution, and the projected van Hove singularities in the density of states (DOS). The magnetic properties of the host and guest adatoms can be comprehensively determined through the spin-split/spin-degeneracy energy band structure, the partial/net magnetic moments, the spatial spin-density distribution, and the polarized DOS. In particular, the orbital character of the band-edge states was used to identify the optical excitations, e.g., the threshold frequency and the special prominence peaks in the optical adsorption spectrum; furthermore, the collective optical excitations could also be understood from the dielectric and electron energy loss functions. We can connect electronic and optical excitations. All results reported in this chapter are required to be tested with high-resolution experimental techniques.[44,45]

4.2. Computational Details

For the current analysis, we used the density functional theory (DFT) method[21,46,47] based on the VASP[48] software to perform the optimization of crystal structures and the calculations of the electronic properties. The Perdew–Burke–Ernzerhof (PBE)[49] generalized gradient approximation was used for the exchange–correlation functional. The interaction between the valence electrons and ion cores was evaluated using the projector-augmented wave (PAW) method.[50] The cutoff energy for the expansion of the plane wave basis is set at 400 eV for all calculations. The Brillouin zone was integrated with special k-point meshes of $25 \times 25 \times 1$ and $30 \times 30 \times 1$ in the Monkhorst–Pack sampling technique for geometric relaxation and electronic/optical calculation, respectively. The convergence condition of the ground state is set to be 10^{-6} eV between two consecutive simulation steps, and all atoms were allowed to fully relax during the geometric optimization until the Hellmann–Feynman force[51] acting on each atom was smaller than 0.01 eV/Å. Spin-polarized calculations were performed for the geometric optimization and calculation of the band structure.[52] In the k-point sample, the cutoff energy was checked for convergence of the calculations. Besides, we can consider the optical excitations[39–41] without electron–hole interactions. When the quasi-particle energies were obtained within the approximation for the self-energy, the cutoff energy for the response function was set to 400 eV, and a $20 \times 20 \times 20$ Γ-centered k-point sampling was used to represent reciprocal space. These can be described by the Kubo formula:[53]

$$\epsilon_2(\omega) = \frac{8\pi^2 e^2}{\omega^2} \sum_{v\mathbf{ck}} |e\langle v\mathbf{k}|\mathbf{v}|c\mathbf{k}\rangle|^2 \delta(\omega - (E_{c\mathbf{k}} - E_{v\mathbf{k}}))(f(E^v(k) - f(E^c(k)).$$

The $|e\langle v\mathbf{k}|\mathbf{v}|c\mathbf{k}\rangle|^2$ term is the square of the electric moment and is responsible for the strength of the excitation peaks. $\delta(\omega-(E_{c\mathbf{k}}-E_{v\mathbf{k}}))$ is the combined DOS and is useful to provide the available excitation transition channels, and $(f(E^v(k) - f(E^c(k))$ goes to 1. In addition, other optical properties, such as the energy loss function,[54] can be obtained from the dielectric function through the following relation:

$$L(\omega) = \frac{\varepsilon_2(\omega)}{\varepsilon_1^2(\omega) + \varepsilon_2^2(\omega)}.$$

4.3. Results and Discussions

La and Ce adatom adsorption monolayer graphene can form unusual geometric structures. The La-/Ce-related graphene network is modeled with La/Ce atoms in five cases, named as $\sqrt{3} \times \sqrt{3}$, $2\sqrt{3} \times 2\sqrt{3}$, $2\sqrt{3}\times2\sqrt{3}$, 6×6, and 6×6, in which the corresponding concentrations are 16.67%, 8.30%, 4.20%, 2.80%, and 1.40%, respectively. The most stable configurations are the hollow sites, regardless of the rare-earth type and concentration. This could be verified through STM measurements.[55,56]

Apparently, La-/Ce-adsorbed graphene systems are stable under single-side adsorption, with specific distributions and concentrations, as indicated in Fig. 4.1 and Table 4.1. It is very difficult to form high-concentration adsorption systems since the La/Ce radius is very big and the Coulombic repulsive interactions between them are significant within a short distance. Apparently, pristine graphene belongs

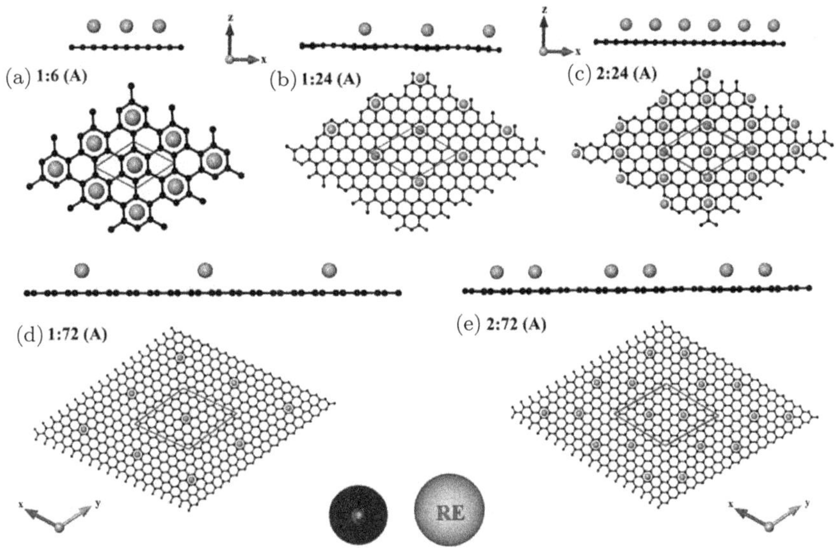

Fig. 4.1. Geometric structures of rare earth (RE) (La/Ce atoms)-adsorbed graphene (C) corresponding to five cases: (a) 1:6, (b) 1:24, (c) 2:24, (d) 1:72, and (e) 2:72 armchair stable configurations.

to hexagonal primitive unit cells with two atoms in the unit cells. The lattice constant and the C–C chemical bond length are 2.46 and 1.42 Å, respectively,[57,58] which agree well with previous theoretical and experimental results. As the highest La/Ce concentrations are reduced to the lowest ones, the C–C bonds nearest to the adatom and the RE-height are varied from 1.429 to 1.439 Å and 2.595 to 2.742 Å, respectively. These phenomena are strongly related to the orbital hybridizations in La–C/Ce–C bonds.

We analyzed pristine graphene with a 3×3 superlattice, which belongs to zero-gap semiconductors at Γ points (Fig. 4.2(a)). Apparently, the π bands near the Fermi level are contributed by $2p_z$–$2p_z$ in the C–C bonds. The π and π^* bands are highly asymmetrical at Γ points. Also, the σ-electronic energy bands with a local minimum of $E^v = -4.00$ eV at the Γ point arise mainly from $2p_x$ and $2p_y$ bonds. The electronic properties of the graphene monolayer are enriched by tuning the rare-earth concentration. Considering the cases of La-/Ce-adsorbed graphene, the energy bands are

Table 4.1. Calculated C–C and C–RE bond lengths, adatom heights, and total magnetic moment per unit cell of La- and Ce-adsorbed graphenes.

Adatom	**Unit cell**	**Concentration**	**Nearest C–C (Å)**	**RE–C**	**RE height (Å)**	**Bulkling (Å)**	M_{tot} (μ_B)
La	$\sqrt{3}\times\sqrt{3}$	1: 6 (16.67%)	1.429	2.742	2.34	0	0
	$2\sqrt{3}\times 2\sqrt{3}$	2:24 (8.3%)	1.434	2.669	2.26	0	0
	$2\sqrt{3}\times 2\sqrt{3}$	1:24 (4.2%)	1.439	2.595	2.16	0	0
	6×6	2:72 (2.8%)	1.440	2.610	2.18	0	0
	6×6	1:72 (1.4%)	1.436	2.685	2.27	0	0
Ce	$\sqrt{3}\times\sqrt{3}$	1: 6 (16.67%)	1.426	2.721	2.32	0	
	$2\sqrt{3}\times 2\sqrt{3}$	2:24 (8.3%)				0	
	$2\sqrt{3}\times 2\sqrt{3}$	1:24 (4.2%)				0	0.732
	6×6	1:72 (1.4%)				0	
	6×6	2:72 (2.8%)				0	

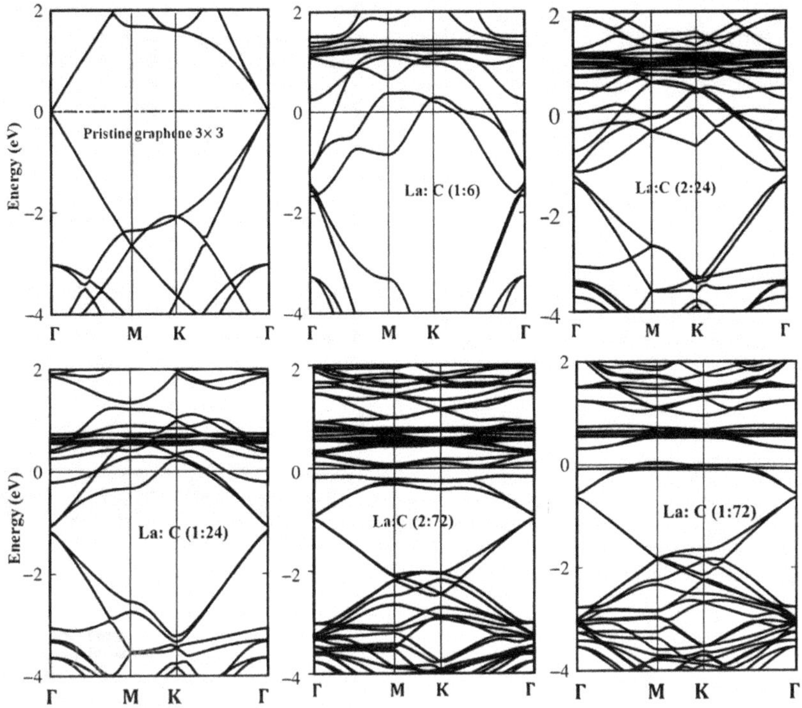

Fig. 4.2. Quasi-band structure of La-adsorbed graphene, corresponding to stable configurations: 1:6, 2: 24, 1:24, 2:72, and 1:72.

dramatically shifted by the La/Ce adsorptions at low concentrations, and the modified Dirac cone structure could be observed (Figs. 4.3 and 4.4). It means that these systems are negative type or blueshifted with respect to the Femi level. Moreover, there exist spin-up and spin-down energy bands splitting near the K and M points of Ce-adsorbed graphene. By contrast, it is true for the case of La doping. In particular, all the La-/Ce-adsorbed graphene types exhibit metallic or semi-metallic behavior. The La/Ce concentrations are contributed to focus at unoccupied states and flat subbands near the Fermi level at occupied states. In broad terms, there exist dramatically modified low-lying energy dispersions owing mainly to the highly non-uniform environment of an enlarged unit cell and more significant or complicated interlayer orbital hybridizations.

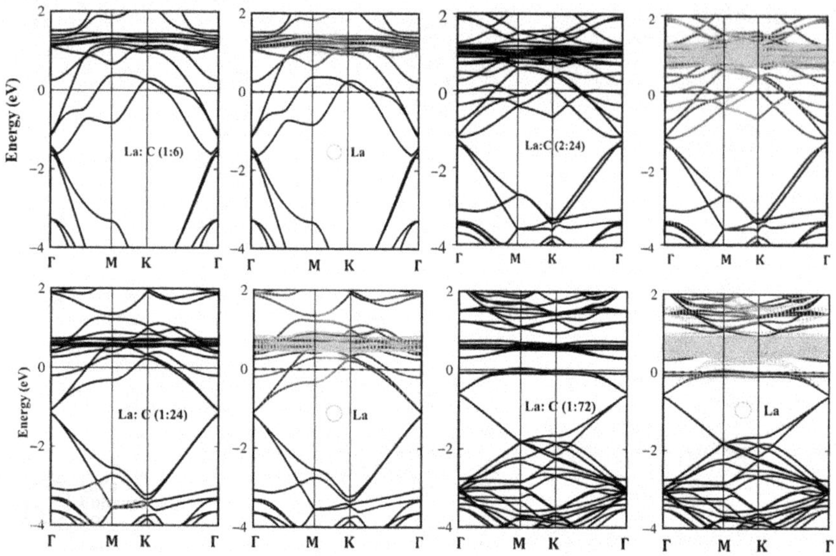

Fig. 4.3. Project-quasi band structure of La atom-adsorbed graphene (green circles).

The spatial charge density distributions/charge density differences (Fig. 4.5) are the first step in understanding the orbital hybridizations in all C–C, La–C, and Ce–C bonds of La-/Ce-adsorbed graphene. Indeed, there are three kinds of chemical bonds in the honeycomb lattices that could survive simultaneously. The σ electronic orbital hybridizations, with a high charge density in the C–C bonds, are slightly affected by the La or Ce adsorptions, corresponding to a slight variation in charge density or a minor charge density difference. This is responsible for a rigid redshift of the σ energy bands and the absence of co-domination of the La/Ce and C atoms in the valence band. In significantly contract, the slight modifications of the π bondings are absorbed through variations in the charges between C and La-/Ce- atoms. These features suggest the existence of C-(2p$_z$) –La-(5d$_{yz}$, 5d$_{xy}$, 5d$_{z^2}$, 5d$_{xz}$, 5d$_{x^2-y^2}$)/ C-(2p$_z$)– Ce-(4f$_{xyz}$, 4f$_{yz^2}$, 4f$_{z^3}$, 4f$_{xz^2}$, 4f$_{z(x^2-3y^2)}$, 4f$_{x(x^2-3y^2)}$, 4f$_{y(y^2-3x^2)}$), accounting for the (C, La), (C, Ce) co-domination of conduction bands. In addition, the La–La/Ce–Ce bonds are very weak

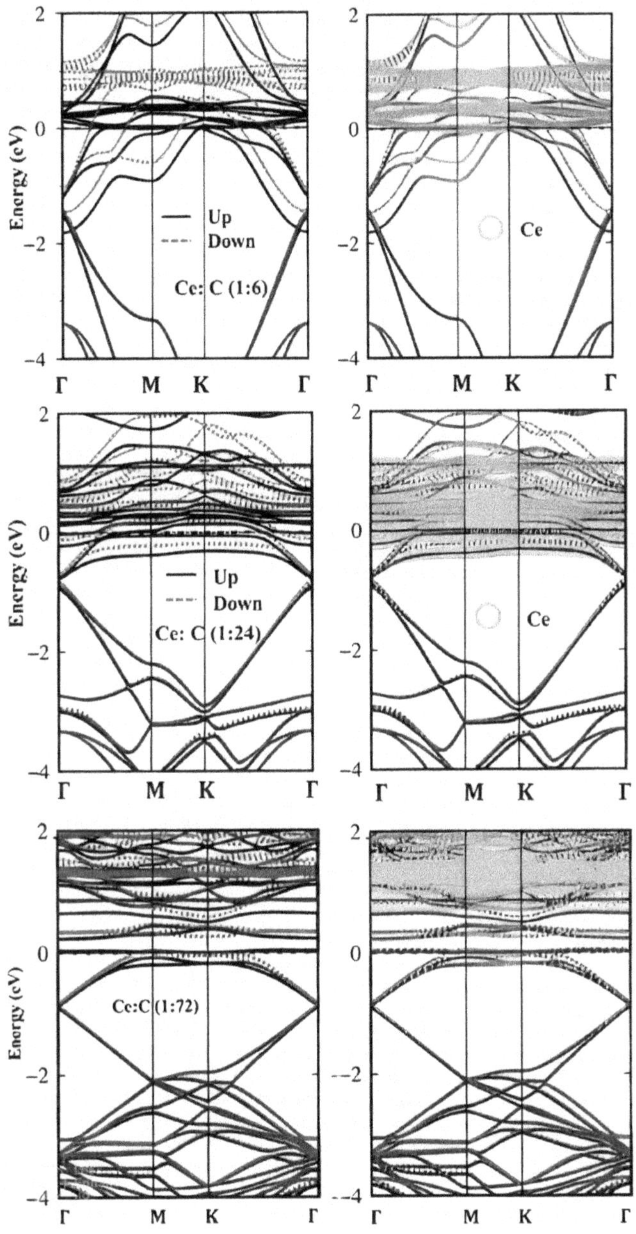

Fig. 4.4. Quasi-band structure of Ce-adsorbed graphene, corresponding to stable configurations: (1:6, 1:24, and 1:72). Project-quasi band structure of Ce atom-adsorbed graphene (green circles).

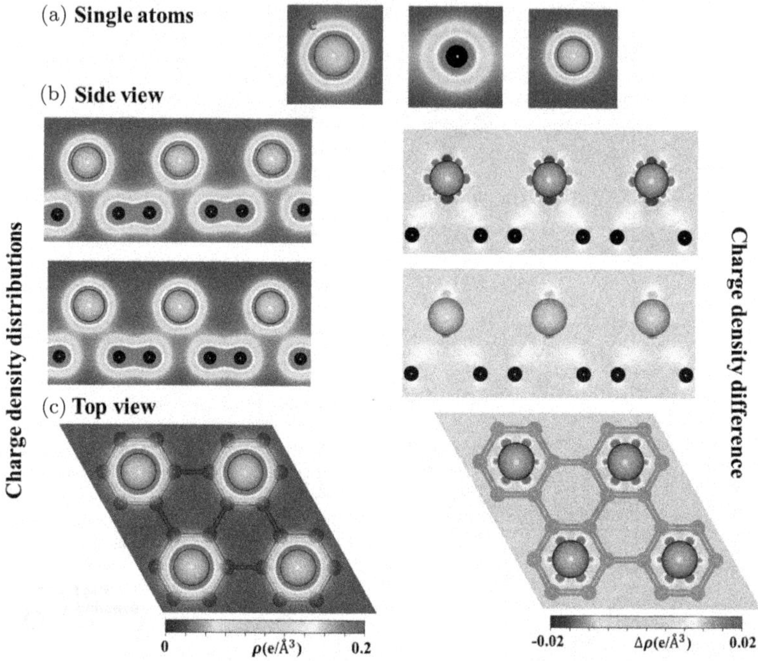

Fig. 4.5. (a) La and Ce single atoms and the (b) side and (c) top views of charge density distributions and charge density difference at the 16.67% highest concentration in La-/Ce-adsorbed graphene.

but rather important (5d_{yz}, 5d_{xy}, 5d_{z^2}, 5d_{xz}, 5$d_{x^2-y^2}$ – (5d_{yz}, 5d_{xy}, 5d_{z^2}, 5d_{xz}, 5$d_{x^2-y^2}$ and (4f_{xyz}, 4f_{yz^2}, 4f_{z^3}, 4f_{xz^2}, 4$f_{z(x^2-3y^2)}$, 4$f_{x(x^2-3y^2)}$, 4$f_{y(y^2-3x^2)}$ – 4f_{xyz}, 4f_{yz^2}, 4f_{z^3}, 4f_{xz^2}, 4$f_{z(x^2-3y^2)}$, 4$f_{x(x^2-3y^2)}$, 4$f_{y(y^2-3x^2)}$) orbitals in the La–La and Ce–Ce bonds, respectively, in sufficiently high concentrations.

Interestingly, spin distributions are indicated in three types of Ce-adsorbed graphene. The configurations for the spin-density distributions could provide additional information regarding their magnetic properties. All the Ce-doped graphene structures exhibit ferromagnetic behavior, as is clearly illustrated in Fig. 4.6. The spin-up magnetic moments (ratio of 1:6) almost dominate the spin-density arrangement, but the opposite is true for a ratio of 1:72, which is closely related to the adatom-dominated spin-split energy bands near the Fermi level in Fig. 4.4.

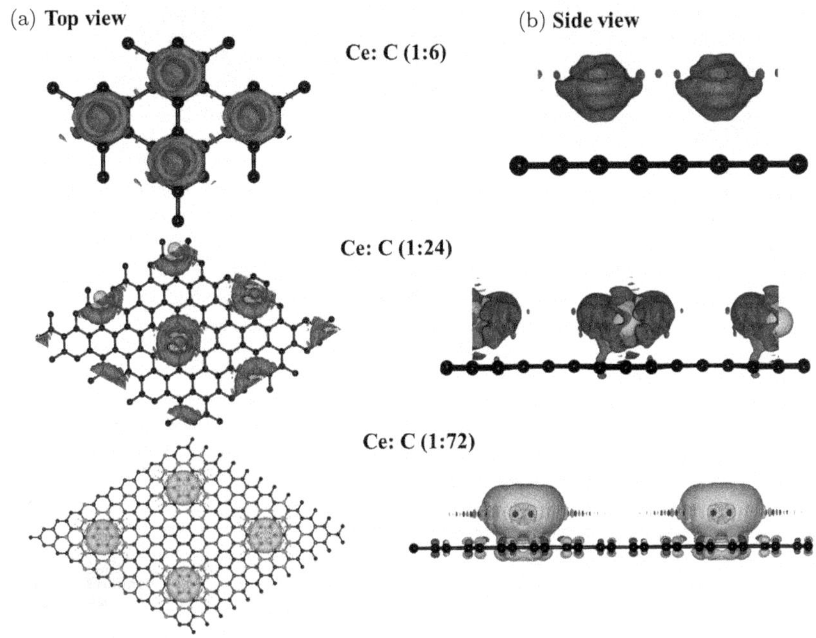

Fig. 4.6. Spin distributions (top and side views) of Ce-adsorbed graphene in different cases: 1:6, 1:24, and 1:72.

The DOS is determined from the energy dispersions of band-edge states, corresponding to the band structures; furthermore, the orbital-projected DOS directly reflects the features of the chemical bonds. The La-/Ce-adsorbed graphene, for the example case of 1:6, exhibits many 2D van Hove singularities, as seen in Fig. 4.7. The strong symmetric peaks and the shoulder structures, respectively, come from the linear bands, the saddle point, and the local extreme points. The special structures in $2p_z$-related DOS are dramatically changed by the orbital hybridization of 5d, 4f, and $2p_z$. The DOS due to the conduction band π^*-states is finite at the Fermi level, which is a result of the shift in the π band. The center of the linear E dependence, which is slightly dependent on the adatom concentration, is shifted downward. The good fit of the DOS of La and C are evident in the significant orbital hybridizations of C-$2p_z$ and La-5d/Ce-4f.

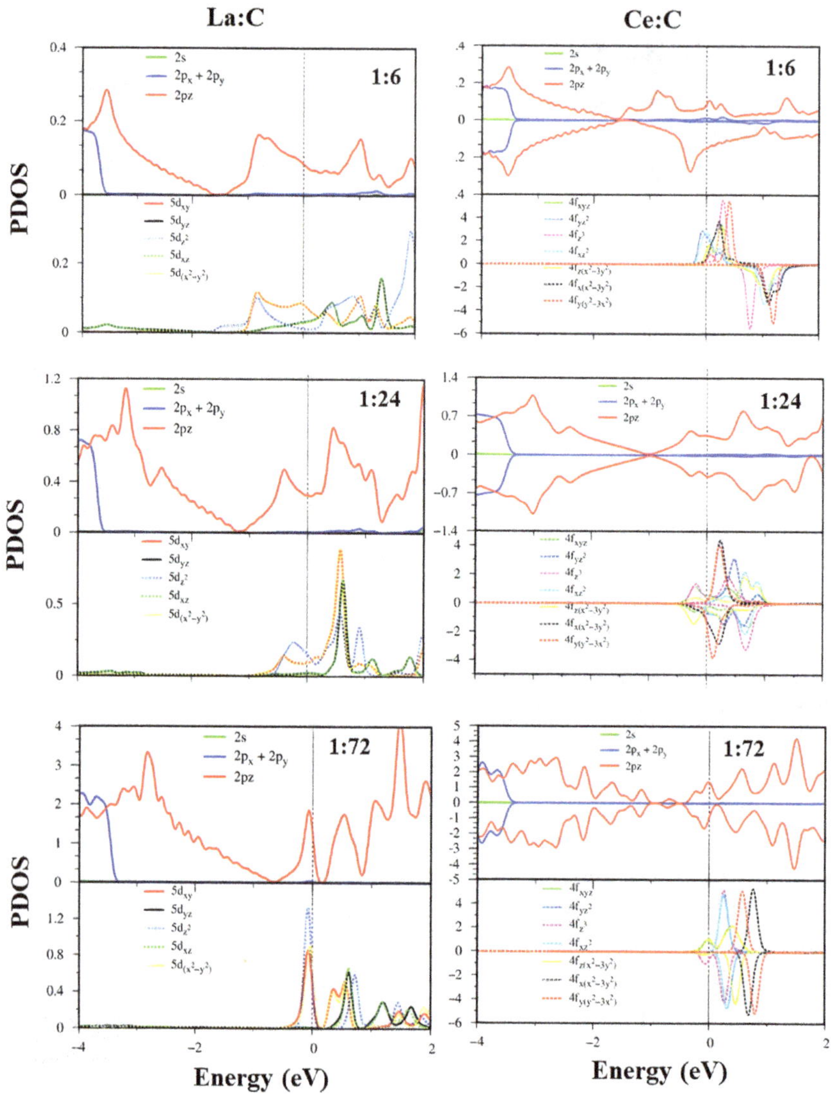

Fig. 4.7. Van hove singularities in the projected density of states of La, Ce, and C orbitals at different concentrations.

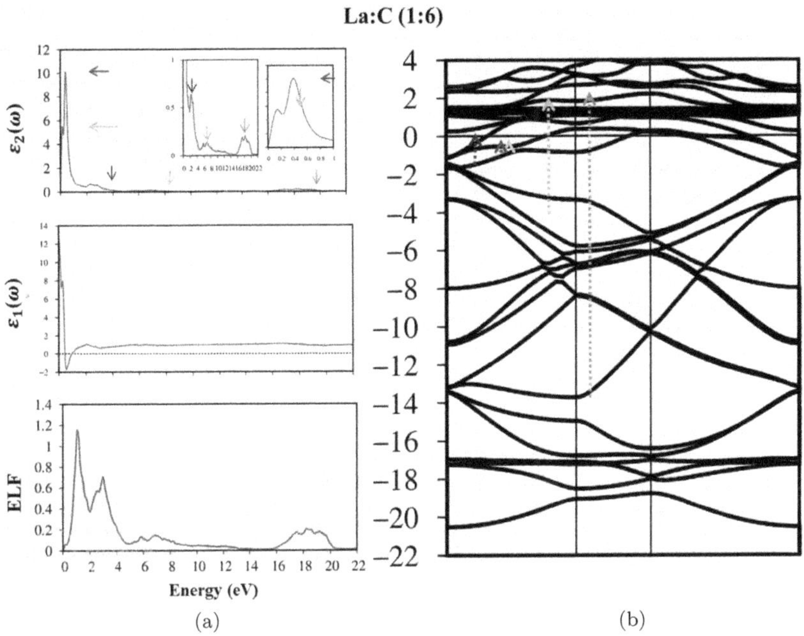

Fig. 4.8. (a) Imaginary and real parts of the dielectric functions, as well as the energy loss function (ELF). (b) Vertical excitation from occupied states to unoccupied ones denoted by arrows.

In addition to electronic properties, we also consider the optical excitations of La-/Ce-adsorbed graphene at the 16.67% highest concentrations, as indicated in Figs. 4.8 and 4.9. When the 2D pristine/adsorbed graphene is perturbed by an electromagnetic (EM) wave, the occupied electronic states and the unoccupied ones are vertical excitations each other. All the charge carriers are capable of dynamically screening this external field, thus generating an induced current density. Within the linear response, the screening ability of charges is well characterized by the transverse dielectric function at the long-wavelength limit [$\epsilon(q \rightarrow 0, \omega)$]. According to basic electrodynamics, this frequency-dependent function is very useful in understanding the main features of the energy loss function. The relevant physical quantities are consistent with one another when using the same theoretical framework. Furthermore, the single-particle electronic excitations (electron–hole pairs) and collective

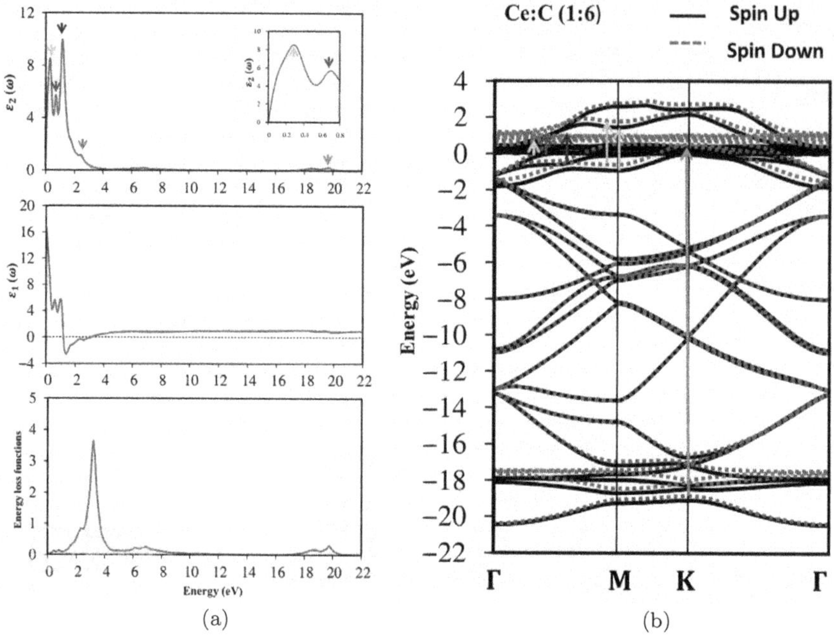

Fig. 4.9. (a) Imaginary and real parts of the dielectric functions, as well as the energy loss function (ELF). (b) Vertical excitation from occupied states to unoccupied ones denoted by arrows.

excitations (plasmon modes) are expected to dominate the couplings of charges and EM waves. Delicate analyses are done via direct linking among the orbital-dominated band-edge states (band structures), the joint Hove singularities (the large DOS), and the specific absorption frequency (optical spectra). Each prominent absorption structure is identified from the corresponding orbital hybridization.

The optical properties of La-/Ce-adsorbed graphene are exhibited over a wide energy range, up to 22 eV. Its frequency-dependent bare response function, which is due to the vertical interband electron–hole excitations, shows five prominent peaks at 0.10 (0.30), 0.30 (0.7), 2.20 (1.2), 7.0 (2.40), and 18 (19.5) eV with 16.67% concentration of La-(Ce-)adsorbed graphene (Tables 4.2 and 4.3). The featured results only reflect the different orbital hybridizations of 2D graphene honeycomb crystal arising from, respectively, the inter-π-band transitions inside a Dirac cone structure, the higher-energy

Table 4.2. Optical excitations related to orbital hybridizations without excitonic effect at the 16.67% highest concentration in La-adsorbed graphene.

Without excitonic effect (eV)	Arrow color	Excited orbitals
0.1	Green	($2p_z$) – C ($2p_z$)
0.3	Blue	C ($2p_z$) – C ($2p_z$)
2.2	Purple	C ($2p_z$) – C ($2p_z$)
		La ($5d_{xy}$, $5d_{xy}$, $5d_{yz}$, $5d_{z^2}$, $5d_{x^2-y^2}$) – La ($5d_{xy}$, $5d_{xy}$, $5d_{yz}$, $5d_{z^2}$, $5d_{x^2-y^2}$)
7.0	Yellow	La ($5d_{xy}$, $5d_{xy}$, $5d_{yz}$, $5d_{z^2}$, $5d_{x^2-y^2}$) – La ($5d_{xy}$, $5d_{xy}$, $5d_{yz}$, $5d_{z^2}$, $5d_{x^2-y^2}$)
18	Cyan	C (2s, $2p_x$, $2p_y$, $2p_z$) – La ($5d_{xy}$, $5d_{xy}$, $5d_{yz}$, $5d_{z^2}$, $5d_{x^2-y^2}$)

Table 4.3. Optical excitations related to orbital hybridizations without excitonic effect at the highest concentration of 16.67% in Ce-adsorbed graphene.

Without excitonic effect (eV)	Arrow color	Excited orbitals	Spin
0.3	Green	$4f_{yz^2}$–$4f_{z(x^2-3y^2)}$, $4f_{y(y^2-3x^2)}$	Spin up – spin up
0.7	Blue	$2p_z$–$4f_{z^3}$	Spin up – spin up
1.2	Purple	2s, $2p_x$, $2p_y$, $2p_z$–$4f_{z^3}$	Spin down – spin down
2.4	Yellow	2s, $2p_x$, $2p_y$, $2p_z$-2s, $2p_x$, $2p_y$, $2p_z$	Spin up – spin up
		2s, $2p_x$, $2p_y$, $2p_z$–$4f_{y(y^2-3x^2)}$, $4f_{x(x^2-3y^2)}$	Spin up – spin up
19.5	Cyan	2s, $2p_x$, $2p_y$–$4f_{yz^2}$, $4f_{xz^2}$, $4f_{x(x^2-3y^2)}$	Spin up – spin up

ones associated with the saddle M-point, and the composite excitations of the σ and π bands from the K and M points. To compare the imaginary parts of the dielectric functions, the real parts of the dielectric function are also indicated. Apparently, the real part $\varepsilon_1(\omega)$[59] of the dielectric function is related to the imaginary part $\varepsilon_2(\omega)$ of the dielectric function through the Kramers–Kronig relationship.[60] The real part of the dielectric functions is weakly dependent on the energy in the inactive region, e.g., the dielectric constant at zero point $\varepsilon_1(0)$ is approximately 12 (16) eV for La-(Ce-)doped graphene.

Following the charge screenings of all π and σ valence electrons, the energy loss functions (Figs. 4.8 and 4.9) clearly reveal the featured

Table 4.4. Collective excitations in energy loss functions at the 16.67% highest concentration in La-/Ce-adsorbed graphene.

Without excitonic effect	Orbitals
1 and 3 eV	Free carrier plasmon mode (La-5d)
7 eV	π-C(2p$_z$) + La (5d)
18.2 eV	$\pi + \sigma - C(2s, 2p_x, 2p_y, 2p_z) + La\,(5d)$
3 eV	Free carrier plasmon mode (Ce-4f)
7 eV	π-C(2p$_z$) + Ce (4f)
19.5 eV	$\pi + \sigma - C(2s, 2p_x, 2p_y, 2p_z) + Ce\,(4f)$

plasmon modes corresponding to the prominent peaks, the similar ones, and the drastic plasmon edges. From the calculated results of the former two, the π plasmon, the second π-electronic collective excitations, respectively, as evident in Table 4.4. Apparently, such energy-dependent plasmons belong to the quantization modes of the coherent charge oscillations under the long-wavelength limit, corresponding to the different orbital-dominated valence electrons.

4.4. Concluding Remarks

In this chapter, the geometric, electronic, and optical properties of La-/Ce-adsorbed graphene were determined using DFT based on VASP. As opposed to the zero-gap semiconductor behavior of monolayer graphene, all of the rare-earth-adsorbed cases belong to the negative-type doping, thus implying enhanced electronic conductivity behavior.

Besides, the La-/Ce-adsorbed graphene are different from each other in their fundamental properties. The main differences are the optimal lattice constants, the La–C and Ce–C chemical bond lengths, the modified Dirac cone structures, the spin-degenerate or spin-split energy spectra around the Fermi level, the adatom-enriched charge density distributions, the ferromagnetic or non-magnetic nature under different concentrations, the special van Hove singularities, and the single and collective optical excitations. Most importantly, the significant hybridizations of ($5d_{yz}$, $5d_{xy}$, $5d_{z^2}$, $5d_{xz}$, $5d_{z^2-y^2}$)/[$4f_{xyz}$, $4f_{yz^2}$, $4f_{z^3}$, $4f_{xz^2}$, $4f_{z(x^2-3y^2)}$, $4f_{x(x^2-3y^2)}$, $4f_{y(y^2-3x^2)}$] & $2p_z$ orbitals and ($5d_{yz}$, $5d_{xy}$, $5d_{z^2}$, $5d_{xz}$, $5d_{z^2-y^2}$)–($5d_{yz}$, $5d_{xy}$, $5d_{z^2}$, $5d_{xz}$, $5d_{z^2-y^2}$)/[$4f_{xyz}$, $4f_{yz^2}$, $4f_{z^3}$, $4f_{xz^2}$, $4f_{z(x^2-3y^2)}$, $4f_{x(x^2-3y^2)}$, $4f_{y(y^2-3x^2)}$]–[$4f_{xyz}$, $4f_{yz^2}$, $4f_{z^3}$, $4f_{xz^2}$, $4f_{z(x^2-3y^2)}$, $4f_{x(x^2-3y^2)}$, $4f_{y(y^2-3x^2)}$] have been identified in the La–C/Ce–C and La–La/Ce–Ce bonds.

On the theoretical side, van Hove singularities and optical properties could be examined through high-resolution measurements. For example, the angle-resolved photoemission spectroscopy (ARPECT)[61] and scanning tunneling spectroscopy (STS)[62] methods can be used to test the occupied states in quasi-band structures and van Hove singularities in the DOS, respectively.

References

[1] Levi, P.; Rosenthal, R. *The Periodic Table.* New York: Schocken Books, 1984.

[2] Scerri, E. R.; Worrall, J. Prediction and the periodic table. In *Selected Papers on the Periodic Table by Eric Scerri*, World Scientific, 2009, pp 45–90.

[3] Ambacher, O. Growth and applications of group III-nitrides. *Journal of physics D: Applied Physics* **1998**, *31* (20), 2653.

[4] Morelli, D.; Heremans, J.; Slack, G. Estimation of the isotope effect on the lattice thermal conductivity of group IV and group III-V semiconductors. *Physical Review B* **2002**, *66* (19), 195304.

[5] Brennan, J.; Siegrist, T.; Carroll, P.; Stuczynski, S.; Reynders, P.; Brus, L.; Steigerwald, M. Bulk and nanostructure group II-VI compounds from molecular organometallic precursors. *Chemistry of Materials* **1990**, *2* (4), 403–409.

[6] Mujica, A.; Rubio, A.; Munoz, A.; Needs, R. High-pressure phases of group-IV, III–V, and II–VI compounds. *Reviews of modern physics* **2003**, *75* (3), 863.

[7] Lin, S.-Y.; Tran, N. T. T.; Lin, M.-F. Diversified phenomena in metal- and transition-metal-adsorbed graphene nanoribbons. *Nanomaterials* **2021**, *11* (3), 630.

[8] Tran, N. T. T.; Nguyen, D. K.; Lin, S. Y.; Gumbs, G.; Lin, M. F. Fundamental properties of transition-metals-adsorbed graphene. *ChemPhysChem* **2019**, *20* (19), 2473–2481.

[9] Caragiu, M.; Finberg, S. Alkali metal adsorption on graphite: A review. *Journal of Physics: Condensed Matter* **2005**, *17* (35), R995.
[10] Shung, K. W.-K. Lifetime effects in low-stage intercalated graphite systems. *Physical Review B* **1986**, *34* (2), 1264.
[11] Li, N.; Wang, Z.; Zhao, K.; Shi, Z.; Gu, Z.; Xu, S. Large scale synthesis of N-doped multi-layered graphene sheets by simple arc-discharge method. *Carbon* **2010**, *48* (1), 255–259.
[12] Son, Y.-W.; Cohen, M. L.; Louie, S. G. Energy gaps in graphene nanoribbons. *Physical Review Letters* **2006**, *97* (21), 216803.
[13] Yang, X.; Dou, X.; Rouhanipour, A.; Zhi, L.; Räder, H. J.; Müllen, K. Two-dimensional graphene nanoribbons. *Journal of the American Chemical Society* **2008**, *130* (13), 4216–4217.
[14] Dutta, S.; Pati, S. K. Novel properties of graphene nanoribbons: A review. *Journal of Materials Chemistry* **2010**, *20* (38), 8207–8223.
[15] Dresselhaus, M. S.; Dresselhaus, G.; Eklund, P.; Rao, A. Carbon nanotubes. In *The Physics of Fullerene-based and Fullerene-related Materials*, Springer, 2000, pp. 331–379.
[16] Haddon, R. C. *Carbon Nanotubes*, Vol. 35, ACS Publications, 2002, pp. 997–997.
[17] Dresselhaus, M.; Dresselhaus, G.; Saito, R. Physics of carbon nanotubes. *Carbon* **1995**, *33* (7), 883–891.
[18] Curl, R. F.; Smalley, R. E. Fullerenes. *Scientific American* **1991**, *265* (4), 54–63.
[19] Taylor, R.; Walton, D. R. The chemistry of fullerenes. *Nature* **1993**, *363* (6431), 685–693.
[20] Khuong Dien, V.; Thi Han, N.; Nguyen, T. D. H.; Huynh, T. M. D.; Pham, H. D.; Lin, M.-F. Geometric and electronic properties of Li 2 GeO 3. *Frontiers in Materials* **2020**, *7*, 288.
[21] Han, N. T.; Dien, V. K.; Tran, N. T. T.; Nguyen, D. K.; Su, W.-P.; Lin, M.-F. First-principles studies of electronic properties in lithium metasilicate (Li 2 SiO 3). *RSC Advances* **2020**, *10* (41), 24721–24729.
[22] Thi, H. N.; Tran, N. T. T.; Dien, V. K.; Nguyen, D. K.; Lin, M.-F. Diversified Properties in 3D ternary oxide compound: Li2SiO3. In *Lithium-Ion Batteries and Solar Cells*, CRC Press, 2021, pp 79–101.
[23] Lonkar, S. P.; Deshmukh, Y. S.; Abdala, A. A. Recent advances in chemical modifications of graphene. *Nano Research* **2015**, *8* (4), 1039–1074.
[24] Zheng, H.; Duley, W. First-principles study of edge chemical modifications in graphene nanodots. *Physical Review B* **2008**, *78* (4), 045421.
[25] Kumar, R.; Tanwar, M. Effect of some physical perturbations and their interplay on Raman spectral line shapes in silicon: A brief review. *Journal of Raman Spectroscopy* **2021**, *52* (12), 2100–2118.

[26] Ding, Y.; Wang, Y. Structural, electronic, and magnetic properties of adatom adsorptions on black and blue phosphorene: A first-principles study. *The Journal of Physical Chemistry C* **2015**, *119* (19), 10610–10622.

[27] Tran, N. T. T.; Gumbs, G.; Nguyen, D. K.; Lin, M.-F. Fundamental properties of metal-adsorbed silicene: A DFT study. *ACS omega* **2020**, *5* (23), 13760–13769.

[28] Lin, S.-Y.; Lin, Y.-T.; Tran, N. T. T.; Su, W.-P.; Lin, M.-F. Feature-rich electronic properties of aluminum-adsorbed graphenes. *Carbon* **2017**, *120*, 209–218.

[29] Eder, E. Properties of substitutions and unifications. *Journal of Symbolic Computation* **1985**, *1* (1), 31–46.

[30] Abdelkader, A. M.; Kinloch, I. A.; Dryfe, R. A. Continuous electrochemical exfoliation of micrometer-sized graphene using synergistic ion intercalations and organic solvents. *ACS applied Materials & Interfaces* **2014**, *6* (3), 1632–1639.

[31] Chang, S.-L.; Lin, S.-Y.; Lin, S.-K.; Lee, C.-H.; Lin, M.-F. Geometric and electronic properties of edge-decorated graphene nanoribbons. *Scientific Reports* **2014**, *4* (1), 1–8.

[32] Yamazoe, N.; Tamaki, J.; Miura, N. Role of hetero-junctions in oxide semiconductor gas sensors. *Materials Science and Engineering: B* **1996**, *41* (1), 178–181.

[33] Kimura, W. D. What are electromagnetic waves. *Electromagnetic Waves and Lasers*, San Rafael, CA, USA: Morgan & Claypool Publishers, 2017.

[34] Alhaidari, A. Solution of the Dirac equation with position-dependent mass in the Coulomb field. *Physics Letters A* **2004**, *322* (1–2), 72–77.

[35] De, S.; Rosen, J.; Dagan, A.; Hannaford, B.; Swanson, P.; Sinanan, M. Assessment of tissue damage due to mechanical stresses. *The International Journal of Robotics Research* **2007**, *26* (11–12), 1159–1171.

[36] Shiu, Y.-T.; Weiss, J. A.; Hoying, J. B.; Iwamoto, M. N.; Joung, I. S.; Quam, C. T. The role of mechanical stresses in angiogenesis. *Critical Reviews in Biomedical Engineering* **2005**, *33* (5), 431–510.

[37] Huang, J.; Webb, W. Viscous damping of thermal excitations on the interface of critical fluid mixtures. *Physical Review Letters* **1969**, *23* (4), 160.

[38] Forslund, A.; Ruban, A. Ab initio surface free energies of tungsten with full account of thermal excitations. *Physical Review B* **2022**, *105* (4), 045403.

[39] Han, N. T.; Dien, V. K.; Lin, M.-F. Excitonic effects in the optical spectra of Li2SiO3 compound. *Scientific reports* **2021**, *11* (1), 1–10.

[40] Han, N. T.; Dien, V. K.; Lin, S.-Y.; Chung, H.-C.; Li, W.-B.; Tran, N. T. T.; Liu, H.-Y.; Pham, H. D.; Lin, M.-F. Comprehensive understanding of electronic and optical properties of Li2SiO3 compound.

[41] Dien, V. K.; Pham, H. D.; Tran, N. T. T.; Han, N. T.; Huynh, T. M. D.; Nguyen, T. D. H.; Fa-Lin, M. Orbital-hybridization-created optical excitations in Li2GeO3. *Scientific Reports* **2021**, *11* (1), 1–10.

[42] Han, N. T.; Tran, N. T. T.; Dien, V. K.; Nguyen, D. K.; Lin, M.-F. Diversified properties in 3D ternary oxide compound. In *Lithium-Ion Batteries and Solar Cells: Physical, Chemical, and Materials Properties*, 2021.

[43] Chung, N. T. T. T.; Han, N. T.; Liu, H.-Y.; Pham, H. D.; Lin, M.-F. Experimental measurements.

[44] Hanke, W.; Sham, L. Many-particle effects in the optical excitations of a semiconductor. *Physical Review Letters* **1979**, *43* (5), 387.

[45] Dodd, R.; Edwards, M.; Clark, C. W.; Burnett, K., Collective excitations of Bose-Einstein-condensed gases at finite temperatures. *Physical Review A* **1998**, *57* (1), R32.

[46] Parr, R. G. Density functional theory. *Annual Review of Physical Chemistry* **1983**, *34* (1), 631–656.

[47] Khuong Dien, V.; Thi Han, N.; Nguyen, T. D. H.; Huynh, T. M. D.; Pham, H. D.; Lin, M.-F. Geometric and Electronic Properties of Li2GeO3. *Frontiers in Materials* **2020**, *7*, 288.

[48] Sun, G.; Kürti, J.; Rajczy, P.; Kertesz, M.; Hafner, J.; Kresse, G., Performance of the Vienna ab initio simulation package (VASP) in chemical applications. *Journal of Molecular Structure: THEOCHEM* **2003**, *624* (1–3), 37–45.

[49] Perdew, J.; Burke, K.; Ernzerhof, M., Perdew, burke, and ernzerhof reply. *Physical Review Letters* **1998**, *80* (4), 891.

[50] Blöchl, P. E., Projector augmented-wave method. *Physical Review B* **1994**, *50* (24), 17953.

[51] Politzer, P.; Murray, J. S. The Hellmann-Feynman theorem: A perspective. *Journal of Molecular Modeling* **2018**, *24* (9), 1–7.

[52] Meservey, R.; Tedrow, P. Spin-polarized electron tunneling. *Physics Reports* **1994**, *238* (4), 173–243.

[53] Crépieux, A.; Bruno, P. Theory of the anomalous Hall effect from the Kubo formula and the Dirac equation. *Physical Review B* **2001**, *64* (1), 014416.

[54] Nagatomi, T.; Shimizu, R.; Ritchie, R. Energy loss functions for electron energy loss spectroscopy. *Surface Science* **1999**, *419* (2–3), 158–173.

[55] Uchida, H.; Huang, D.; Grey, F.; Aono, M., Site-specific measurement of adatom binding energy differences by atom extraction with the STM. *Physical Review Letters* **1993**, *70* (13), 2040.

[56] Kano, S.; Tada, T.; Majima, Y. Nanoparticle characterization based on STM and STS. *Chemical Society Reviews* **2015**, *44* (4), 970–987.

[57] Wong, S. L.; Huang, H.; Chen, W.; Wee, A. T. STM studies of epitaxial graphene. *MRS Bulletin* **2012**, *37* (12), 1195–1202.

[58] Novoselov, K. S.; Geim, A. K.; Morozov, S. V.; Jiang, D.; Katsnelson, M. I.; Grigorieva, I.; Dubonos, S.; Firsov, A. Two-dimensional gas of massless Dirac fermions in graphene. *Nature* **2005**, *438* (7065), 197–200.

[59] Bertsch, G. F.; Iwata, J.-I.; Rubio, A.; Yabana, K. Real-space, real-time method for the dielectric function. *Physical Review B* **2000**, *62* (12), 7998.

[60] Hutchings, D. C.; Sheik-ahae, M.; Hagan, D. J.; Van Stryland, E. W. Kramers-Krönig relations in nonlinear optics. *Optical and Quantum Electronics* **1992**, *24* (1), 1–30.

[61] Lv, B.; Qian, T.; Ding, H. Angle-resolved photoemission spectroscopy and its application to topological materials. *Nature Reviews Physics* **2019**, *1* (10), 609–626.

[62] Binnig, G.; Rohrer, H. Scanning tunneling microscopy. *Surface science* **1983**, *126* (1–3), 236–244.

CHAPTER 5

INTERCALATION OF 4D TRANSITION METALS INTO GRAPHITE

Thi My Duyen Huynh[*] and Ming-Fa Lin[*,†]

[*] *Department of Physics, College of Sciences, National Cheng Kung University, Taiwan*
[†] *Hierarchical Green Energy Materials, Hi-Research Center, National Cheng Kung University, Taiwan*

Abstract

Through the quasi-particle framework implemented using VASP calculations, the structural, electronic, and optical properties of 4d transition-metal intercalated graphite were investigated and discussed.

Keywords: Graphene, 4d transition metals, intercalation, electronic and optical properties.

5.1. Graphite: Structure, Properties, and Applications

Owing to their numerous applications in electronic devices,[1–3] energy storage,[4–6] and other fields,[7–11] carbon materials play a significant role in material development. Carbon remarkably presents the extraordinary ability to combine with itself and other chemical elements in various ways,[12–24] resulting in the carbon family[11,15] which covers a wide range of geometries, from three-dimensional (3D) to zero-dimensional (0D) ones. Abundantly available structures of carbon could be naturally found, such as in coal, diamond, and graphite[25,26] Besides, some other forms could be created from carbon, for example, synthetic diamond and graphite, cokes, carbon black, graphitic fibers, and diamond-like carbon. Among them, diamond and graphite are two common forms of carbon in nature that are related to sp^3 and sp^2 hybridizations[4,27,28] Graphite is a fundamental and crucial member of the carbon family, which is considered a raw material that could be easily expanded to obtain an expandable graphite system for more applications.

5.1.1. *Structural characteristics*

In the bulk phase, graphite has a layered structure in which elemental carbon has the lowest energy state at ambient temperature and pressure[9,29–31] Graphite consists of a stack of parallel carbon layers, called graphene sheets, with carbon tightly bonded in hexagonal rings. The crystal structure of pure hexagonal graphite is described in Fig. 5.1, corresponding to the hexagonal lattice with D^4_{6h} of the $P6_3/mmc$ space group. Due to its stacked structure, graphite forms distinct types of stackings, with AA (Fig. 5.1(a)) and AB (Fig. 5.1(b)) being the popular ones for construction and investigation. Apart from these two common stackings, different structural symmetries could be formed depending on the c-axis stacking[32,33] of graphene sheets, for example ABC, AAA, and ABAB, leading

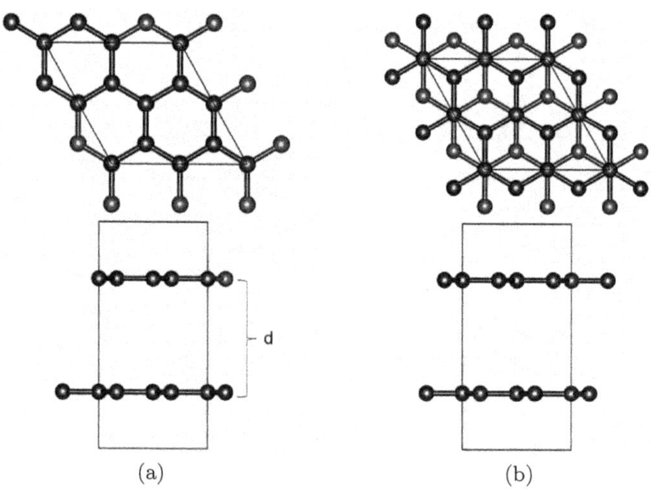

Fig. 5.1. The top and side views of (a) AA- and (b) AB-stacked graphite.

to numerous models, such as hexagonal, Bernal, and rhombohedral. Accordingly, the AA structure is formed by directly stacking two single sheets, one over the other. However, in the AB form, the second layer is experimentally found[34,35] to be a translated C–C bond length along the zigzag direction. This form presents the lowest ground-state energy, so it is the most favorable phase. Thus, AB graphite is the focus of this work aimed toward examining changes in its properties after chemical modifications. As shown in Fig. 5.1(b), the sp^2 bonding gives rise to planar structures of two graphene sheets with an interlayer distance d of 3.463 Å. This structure forms two different surfaces, corresponding to the flake-like morphology of natural graphite.[36,37] Layers in graphite are linked by weak van der Waals (vdW) forces, making graphite a slippery material that easily contributes to intercalation. They are stacked together by π–π interaction of the electronic network.[6] A uniform environment is found in this system due to the identical C–C bonds with a bond length of 1.42 Å. The unit cell is indicated by a black line, which is shown to contain eight carbon atoms per single layer when viewed from the top. This structural feature is fundamental to creating and naming a modified system.

5.1.2. *Electronic properties*

The electronic properties of graphite are determined using their band energy structures and densities of states (DOSs), which are dependent on their structural configurations formed through chemical bonds, interlayer distance, and orbital hybridization. The band energy spectra and orbital-projected DOSs of AB graphite are shown in Fig. 5.2, highlighting a close relationship between these features. In Fig. 5.2(a), the high-symmetric points are defined along Γ, M, K, A, L, and H in the first Brillouin zone (BZ) of the hexagonal lattice. A small overlap between the conduction and valence bands is found, which demonstrates the semi-metallic nature of AB graphite, corresponding to low-density free carriers. The occupied valence and unoccupied conduction bands are symmetric about the Fermi level. A Dirac-like linear dispersion is exhibited

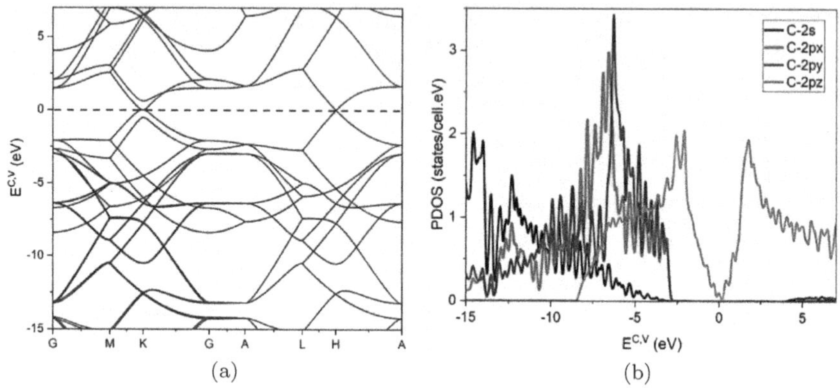

Fig. 5.2. (a) The band energy spectra and (b) orbital-projected DOS of AB-stacked graphite.

along the k_z direction, in which the Dirac cone locates at the K and H points. The superposition of the linear and parabolic dispersions is revealed in this band energy, corresponding to the growth of low-lying energy. Furthermore, saddle points appear along the M $\rightarrow$ L transformation due to van Hove singularities (vHs) rising from the band-edge state. Besides, the orbital-projected DOSs in Fig. 5.2(b) show the corresponding band edge, saddle points, and band energy dispersions, plotted in black, red, blue, and green. It should be noted that the two peaks around the Fermi level are distributed symmetrically between the occupied valence and unoccupied conduction bands. Hence, the V-shaped feature could be realized in this region, illustrating the quasi-localized state. Moreover, the semi-metallic feature is confirmed by a finite DOS at $E = 0$. Especially, the π bond based on the p_z orbital near the Fermi level arises from the strong interaction of C–C bonds, causing the thermal and electrical conduction properties. The sp^2 hybridized at low-lying energies due to the coexistence of s, p_x, and p_y orbitals that induce a strong σ bond, resulting in orbital hybridization that depends on the intralayer and interlayer interactions. Thus, graphite is anisotropic because of the difference between in-plane and out-of-plane C–C bonds.

In addition, the charge density further demonstrates that the bondings in graphite depend on the distribution of orbitals during

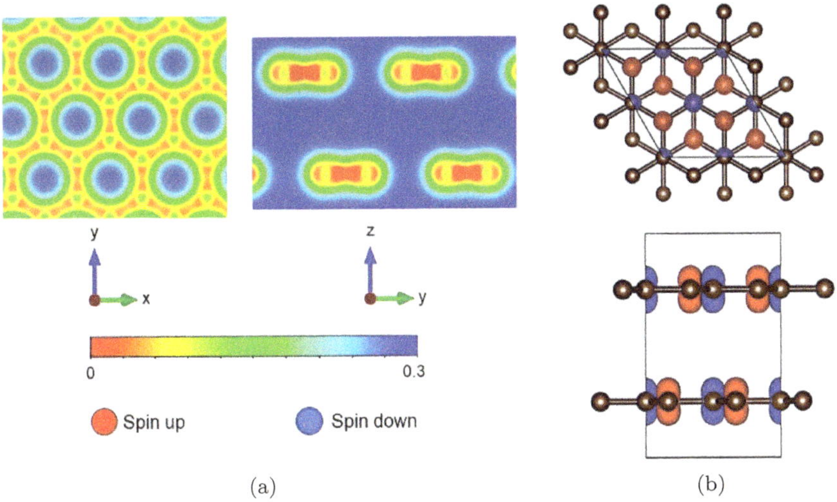

Fig. 5.3. (a) The charge density in (x, y) and (y, z) planes, and (b) spin distribution in the top and side views of AB-stacked graphite.

interactions, as shown in Fig. 5.3(a). Each layer indicates a stable plane due to the strong σ bond corresponding to the (2s, $2p_x$, $2p_y$) orbitals, as indicated by the red and yellow regions, while the strong π bond comes from the $2p_z$ orbital, which exhibits strong dominance in charge density, as indicated by the green regions. A weak but significant interaction between the two layers is found, which is dependent on the vdW forces between two p_z orbital parallel layers. On the other hand, the stacking effect induces the magnetic configuration in graphite, as shown in Fig. 5.3(b), in which the red and blue circles represent spin up and spin down, respectively. According to the spin distribution shown in this figure, AB graphite belongs to the antiferromagnetic–antiferromagnetic (AFM–AFM) configuration, which exhibits spin splitting (see the side view in Fig. 5.3(b)) and zero magnetic moment. The magnetic feature could be affected by modifying structural properties, as discussed later.

5.1.3. *Emergent and potential applications*

As mentioned above, a single layer in graphite is known as graphene, and graphite is the most common source of graphene. Based on graphite, a variety of 2D and 1D materials can be formed, such as nanotubes, fullerenes, and graphene nanoribbons. Graphite has been reported to show high electrical conductivity, low light absorbance, and high chemical resistance,[13] which makes it a unique and significant material for widely different fields. Furthermore, the preparation of graphene from graphite involves the utilization of more than 99.9% of its fixed carbon content[38] Hence, graphene with much enhanced properties and wider applications could be created. On the other hand, graphite can reportedly function as an anode material for lithium-ion batteries (LIBs)[30,39,40] and sodium-ion batteries (SIBs),[41,42] in which a lithium or sodium ion is intercalated into graphite. Additionally, untreated natural graphite is also used as an active electrode material in LIBs.[30,39,40] Apart from these two, both natural and synthetic graphites are interesting and promising materials for cathode plates. Therefore, graphite and graphite-based materials remarkably contribute to the development of batteries.[6,13,18,30,39,40,43,44] In addition, expanded graphite (EG) has shown promise for symmetrical supercapacitors based on its working mechanism as an electrode material.[27] It could improve the electrochemical performance of hybrid materials due to its interfacial interaction with a conductive polymer, which can accelerate the transmission of electrolyte ions and electrons during charging or discharging. The fabrication of graphite and EG batteries is expected to reach high capacities and long-term cycling stability. Thus, the modification of graphite has attracted interest from researchers for enhancing diverse high-energy stationary energy storage and adding to the huge commercial success of devices.

5.2. Modifications of Graphite

With sp^2-bonded carbon in a hexagonal or honeycomb lattice, graphite and graphite-based materials present high surface areas, high thermal conductivities, large charge carrier mobilities, and strong mechanical strengths. The development of graphite compounds has dramatically expanded, leading to the formation of graphite intercalation compounds (GICs). Furthermore, the weak vdW interaction between the layers in graphite creates a potential environment for intercalating ions or molecules, which exhibit excellent physical and chemical properties compared to pristine graphite.

5.2.1. *Graphite intercalation compounds*

GICs are formed by the insertion of atomic or molecular layers of different chemical elements between layers of graphite. GICs can be characterized by the stage, charge transfer, and number of species within the host structures. Among them, stage n is simply defined as the number of graphite layers between intercalate layers, which decreases with an increase in the intercalated concentration. GICs containing one substance are called binary compounds, while ternary compounds are GICs with two atomic or molecular species. Importantly, the interplanar spacing between graphene layers in graphite is expanded during intercalation without disrupting the carbon layers. Generally, GICs could be divided into two classes according to the character of the bonding, namely covalent GICs, such as graphite oxide and graphite fluoride, and ionic GICs, such as ion-intercalated graphite.[4] However, ionic GICs have received more attention than covalent ones because the π bonds in graphite can accept or donate electrons from or to the intercalation. Alkali materials are known as donor-type because of their electron-donating ability, whereas halogens, halide ions, and acids act as electron acceptors that intercalate with graphite to form acceptor-type materials.[25] For the preparation of GICs, several strategies have been applied, which form intercalated systems, labeled CX_m, where C is the intercalation host (graphite), X is the guest inserted into graphite, and m indicates electrons transferred from or to the carbon network. Accordingly, the reaction can be expressed as follows[13] based on the type of intercalation:

$$\mathrm{C} + m\mathrm{X} \rightarrow \mathrm{CX}_m \tag{5.1}$$

$$\mathrm{C}_x + \mathrm{X} \leftrightarrow \mathrm{C}_x^{m+} \cdot \mathrm{X}^{m-} \tag{5.2}$$

Up to now, a wide range of experimental techniques have been reported for synthesizing graphite through the intercalate process. The features of an intercalated system could be obtained from X-ray diffraction (XRD), transmission electron microscopy (TEM), nuclear magnetic resonance (NMR), and neutron diffraction.[12] Among them, XRD is good for analyzing the structural changes during ionic or solvent intercalation.[30,45,46] TEM can be used to observe the dynamic evolution of the interface, the electrochemical deposition process, and the internal structure, which are useful for processing in batteries.[47] While NMR could provide detailed information about the local features and chemical bonds of crystal structures, which are suitable for studying the chemical states in GICs.[48] Besides, spectroscopic techniques can directly probe the density of occupied or unoccupied states in GICs, e.g., electron energy loss spectroscopy (EELS), X-ray absorption spectroscopy (XAS), X-ray emission spectroscopy (XES),[4] and angle-resolved X-ray emission spectroscopy (ARXES). On the other hand, X-ray Raman scattering (XRS) could characterize the electronic properties[49] of the intercalated systems, especially lithium-intercalated graphite (LIG) in LIBs.

As one of the most promising and potential energy storage systems, LIBs have been synthesized and applied for smart devices due to their high energy density, long cycle life, low cost, and friendly environment.[13] Furthermore, GICs have been used for the growth of SIBs with similar chemical/physical properties but at lower costs.[13] As mentioned earlier, graphite has been successfully applied as an anode material for LIBs by intercalating lithium into graphite. Each lithium occupies the center of a hexagonal carbon along the c-axis, forming a stable intercalation.[50] Despite a great number of carbonaceous materials used in the energy storage field, LIG still remains to be the predominantly used anode material in LIBs. Apart from being applied as anode materials, GICs have other uses. For example, aluminum-intercalated graphite functions as an anion intercalation cathode in dual-ion batteries.[6] Additionally, sodium-intercalated graphite (NaIG) is used as an anode material with high energy for SIBs,[51,52] demonstrating the stable structure and high electrical conductivity of GICs. Along with the development of LIBs and SIBs, potassium-intercalated batteries (PIBs) have become more attractive because of their abundant natural resources and similar characteristics as LIBs and SIBs.[53] This battery system performs a successful extraction of KC_8 during the charge process. Hence, GICs have been realized as emergent systems in batteries and energy storage systems. Apart from these fields, GICs with metal atoms reveal potential superconductivity at transition temperatures.[14,54]

5.2.2. *Transition metal-intercalated graphite*

The electrons accepted or donated by the guest-atom intercalation can modify the electronic properties of graphite. Along with alkalis and aluminum, transition metals offer potential atomic intercalation into graphite, which is predicted to exhibit enriched and essential properties, including magnetic configurations, superconductivity, and high mechanical and thermal properties.[55] In fact, carbon compounds with magnetic 3d metals show high performance in spintronics, as indicated by the induced magnetic moment of Mn-/Fe-/Co-intercalated graphite.[56] This feature results in a range of peculiar spatial spin-density distributions that are significant for spintronic applications. The magnetic behavior could be governed by the p–d hybridization and variations in exchange interactions,[57] corresponding to the bonding of transition metals and carbon. In the transition metals' structural features, the d orbitals play an essential role in the compounds. The dominating p orbitals of carbon are strongly hybridized with d states in most MC_6 compounds, as observed in V- and Cr- intercalated graphite.[57] Thus, the d electron interacts with the bonding electrons, which reveals the correlating interaction between these two electrons. In a transition metal-intercalated graphite, the C atoms in graphene sheets have sp^2-hybridized bonds, and each C atom provides one π bond which is normal to the graphene sheet. The above-mentioned magnetic configurations might depend on the hybridization of the s and d orbitals of transition metals and the s and p orbitals of carbon in the interacted system. In the structural intercalation, each transition-metal atom shares six electrons of the carbon atoms in the neighboring graphene sheets.

On the other hand, the ideal graphene sheet is non-magnetic, while the presence of modified systems could induce magnetism.

Thus, GICs are essential from a two-dimensional magnetic system viewpoint. The transition-metal atoms in GICs could cause stable and tunable magnetism, leading to various technical effects. Even though there have been some reports of transition metal-intercalated graphite exhibiting magnetic behavior, detailed information about magnetic configurations still needs to be clarified for more applications. In particular, many transition metals, such as 4d and 5d configurations, have been absent the in intercalation of graphite. Inspired by GICs' applications and the potential features of transition metals in GICs, 4d transition metals, including zirconium (Zr) and niobium (Nb), are considered for intercalated graphite systems. In this work, the fundamental properties of Zr- and Nb-intercalated graphite, including geometric, electronic, and optical features, will be discussed considering various concentrations. Accordingly, the magnetism is also examined following these different structural systems to verify the induced magnetic configuration in the interaction between the d orbitals of 4d transition metals and the s and p orbitals of carbon. Furthermore, the electronic characteristics are expected to be affected and adjusted based on the intercalants (atom intercalates into the system), resulting in band-energy structures and DOSs. Simultaneously, the orbital hybridization of these orbitals through interaction between graphene sheets and intercalants could be proved via charge density and orbital-projected DOSs.

5.3. Zr-intercalated Graphite

As mentioned above, AB graphite is chosen for intercalation because of its theoretically stable configuration and probability of induced magnetism. Hexagonal graphite has a layered structure based on the large difference between in-plane and out-of-plane interactions. The vdW force is induced in graphite between the neighboring graphene sheets. A unit cell of eight carbon atoms in each layer in hexagonal form is considered, namely the C_8 graphite structure. The enlarged unit cell is also examined with a 2×2 supercell of C_8, labeled C_{16}, for varying concentrations. The concentration of intercalation is defined based on the number of intercalants and the number of carbon atoms in a single graphene sheet. Thus, the intercalated system of Zr into graphite is described in Fig. 5.4, with concentrations of (1:8), (2:8), and (1:16), corresponding to one, two, and one Zr atom intercalating into graphite C_8 and C_{16}, respectively. As shown in this figure, the intercalant locates at the center of the hexagonal ring of a honeycomb with the most stable intercalated position. This forms a binary Zr–GIC compound in hexagonal $P6_3/mmc$ crystalline structure, which could be called ZrC_8 and ZrC_{16} due to the unit cell. Especially, the (1:16) case induces the slight fluctuations in the planar graphene sheet, which might arise from the large unit cell and the strong interaction between the intercalant and the graphene sheet compared to the weak vdW on the other side without the intercalant.

Moreover, the lattice parameters indicate the modification of Zr-intercalated graphite through a change in the lattice constant, chemical bonds, and interlayer distance, as listed in Table 5.1. In the case of the C_8 structure, Zr intercalation increases the lattice constant with an increase in the concentration of Zr atoms. Besides, the C_{16} system increases the lattice constant because of the 2×2 supercell. All systems exhibit the changed bond length of the C–C

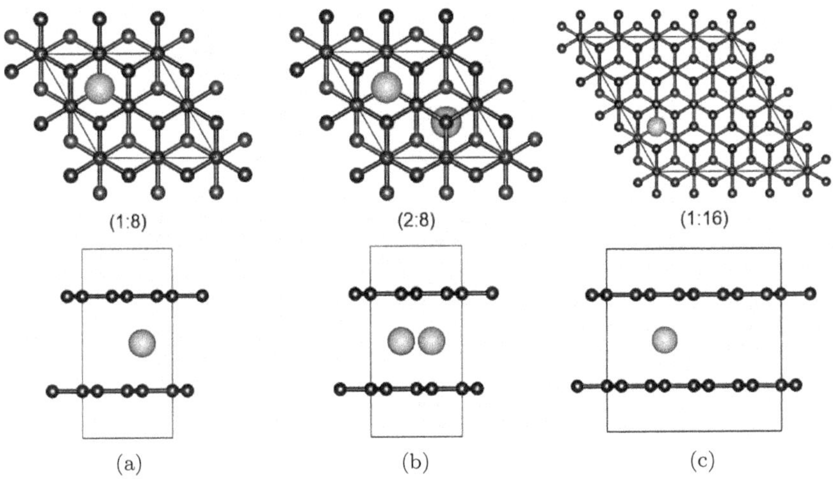

Fig. 5.4. Structural optimization as seen in the top and side views of Zr-intercalated graphite with varying concentrations of (a) (1:8), (b) (2:8), and (c) (1:16).

Table 5.1. Structural parameters of Zr-intercalated graphite.

System	Lattice constant (Å)	C–C bond length (Å) (1)	(2)	Interlayer distance (Å)	Magnetic moment
(1:8)	4.975	1.447	1.426	4.06	0
(2:8)	5.012	1.453	1.436	4.28	0
(1:16)	9.89	1.445	1.428	3.68	2.1

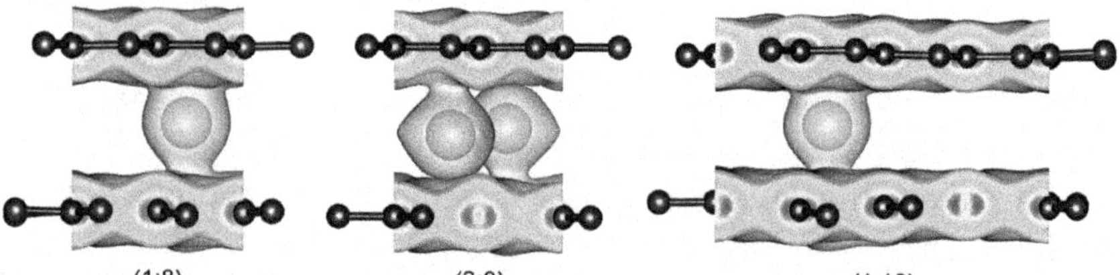

Fig. 5.5. The charge density of Zr-intercalated graphite.

bonds after intercalation, in which the bond length around the intercalant is larger than that of the faraway C–C bond, labeled (1) and (2) as shown in the table. For example, C–C (1) and C–C (2) are respectively 1.447 and 1.426 Åin the (1:8) case. The chemical bond length in the intercalated region is adjusted, while on the other side, the C-C bond length remains unchanged in the graphene sheet, indicating the effect of intercalated atoms on chemical bonding. Importantly, the interlayer distance increases after intercalating in all cases, in which C_8 systems reveal a dramatic change compared to pristine graphite. As a result, the interlayer distance in the (1:8) and (2:8) systems are 4.06 and 4.28 Å, respectively. In contrast, the interlayer distance in the (1:16) case is 3.68 Å, which is slightly larger than that of graphite. Hence, the concentration of intercalated systems could modify the structural features of graphite, probably leading to governing the electronic and optical properties. The same effect could be found in alkali-intercalated graphite[28] and 3d transition metal-intercalated GICs.[55,57] These bondings could be further verified through charge density, as shown in Fig. 5.5. In this figure, the strong interaction between Zr and the carbon graphene sheet is indicated by the yellow regions in the spacing between the graphene sheets and the green regions around the C–C bonds. The C–C bonds with sp^2 and π bonds due to the s, p orbitals of carbon are also affected by the d orbitals of Zr, as indicated by the red and green regions, compared to the pristine case (see Fig. 5.3). The hybridized bonds are discussed in detail using orbital-projected DOSs later to give more information.

In regard to magnetism, the spin distribution and magnetic moment are considered, as illustrated in Fig. 5.6 and Table 5.1, respectively. There is clear polarization in the spin up and spin down of hexagonal ZrC_8 and ZrC_{16}, which are identical. According to Table 5.1, the magnetic moments of both (1:8) and (2:8) systems are zero, corresponding to non-magnetic or antiferromagnetic behavior. Based on the spin distribution shown in Figs. 5.6(a) and 5.6(b), spin splitting of spin up and spin down is found in graphene sheets, corresponding to red and blue circles, which belongs to antiferromagnetic behavior. In contrast, the (1:16) system exhibits a ferromagnetic configuration based on the strong contribution of spin up compared to spin down, plotted in red and blue circles, respectively. The magnetic moment is 2.1 (see Table 5.1), which is

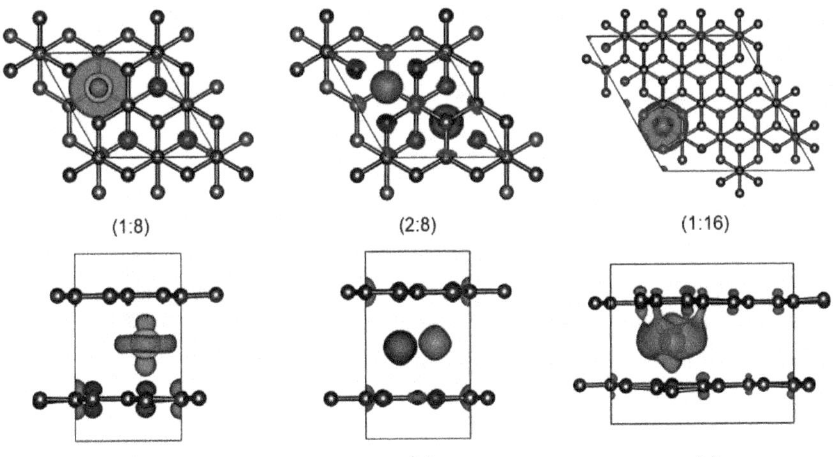

Fig. 5.6. Spin distribution of (a) (1:8), (b) (2:8), and (c) (1:16) Zr-intercalated graphite. Red and blue circles represent spin-up and spin-down distributions.

consistent with the ferromagnetic feature. The difference between C_8 and C_{16} structures is caused by the concentration of Zr atoms, demonstrating the influence of the intercalant in adjusting the magnetism of graphite. This finding elucidates the magnetic configurations in GICs, which are in good agreement with 3d transition metal-intercalated graphite.[55,57]

The electronic properties of Zr-intercalated graphite are further presented through band-energy structures and orbital-projected DOSs, as shown in Figs. 5.7 and 5.8. In Fig. 5.7, the band energy spectra of ZrC_8 and ZrC_{16} with the concentrations of (1:8), (2:8), and (1:16) are shown, considering the dominance of Zr atoms in the whole band. The high-symmetry K point along G–M–K–G–A–L–H–A of the first BZ of the hexagonal lattice is implemented, which is the same as with pristine graphite, as discussed above. All cases exhibit the overlap between conduction and valence bands around the Fermi level, corresponding to metallic behavior. The dispersion in the low-lying valence band is almost identical to that of the band structure of AB graphite (see Fig. 5.2). The Dirac cone at the K and H points in graphite is transformed into having a low energy in the valence bands of around $-2\,\mathrm{eV}$ due to the strong effect of the Zr atoms near the Fermi level. The partial flat bands are found in the low-lying energy, ranging from -7 to -8 and from -12 to $-13\,\mathrm{eV}$, along the G–A points of the (1:8) and (2:8) systems (see Fig. 5.7(a) and 5.7(b)), corresponding to the symmetric structure of graphite. A greater degeneracy of the flat bands revealed in the (1:16) case might come from the larger unit cell of the high-symmetric structure. As evident from the Zr atom dominance, indicated with green circles, the intercalant exhibits a major and strong effect on conduction bands and the energy band near the Fermi level, leading to an adjustment of the band features in these regions compared to graphite. In terms of orbital distribution, the orbital-projected DOS plotted in Fig. 5.8 provides detailed information about the orbital dominance and bonding in Zr-intercalated graphite. The orbital distribution of carbon and Zr in the compound is divided between the left- and right-hand sites with three corresponding systems. The strong contribution of C-p orbitals is indicated below the Fermi level,

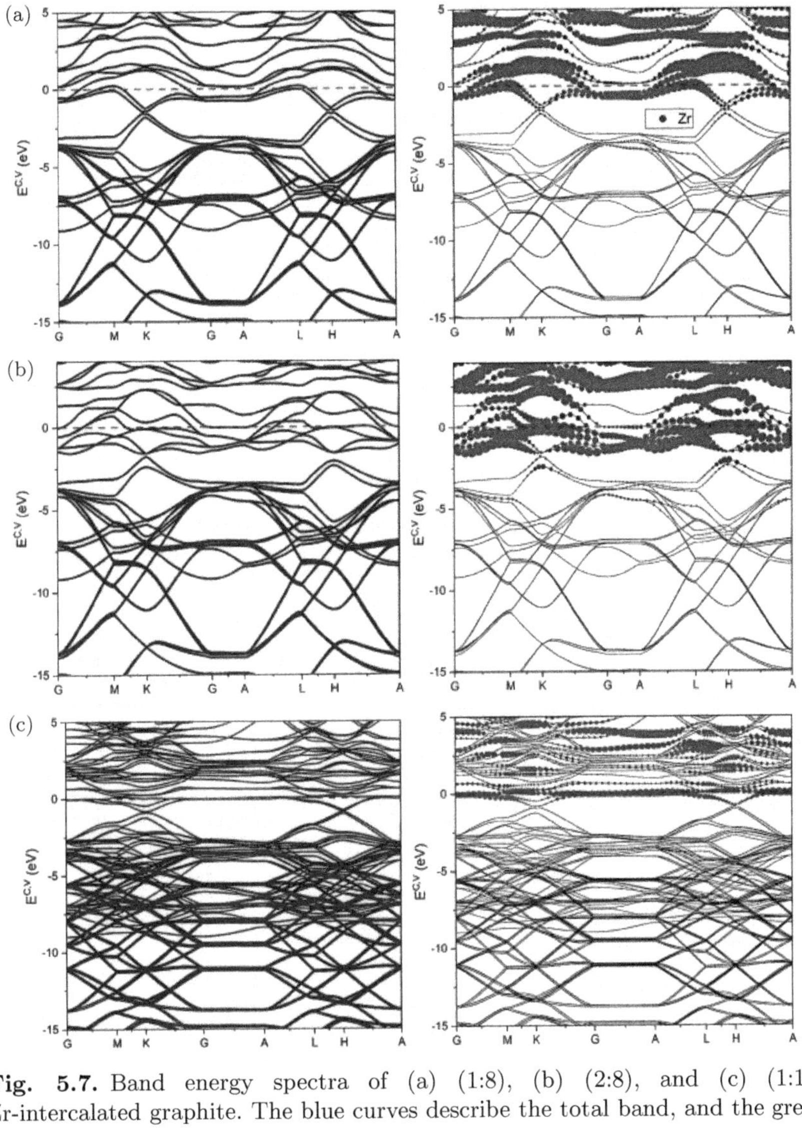

Fig. 5.7. Band energy spectra of (a) (1:8), (b) (2:8), and (c) (1:16) Zr-intercalated graphite. The blue curves describe the total band, and the green circles represent Zr atom dominance.

while the states just above the Fermi level are dominated by the Zr d-orbitals. The C atoms in graphite sheets have sp^2 hybridization, and each C has one π bond that is normal to the graphite sheet. These bonds result in the strong dominance of $2p_z$ and 2s, $2p_x$, $2p_y$

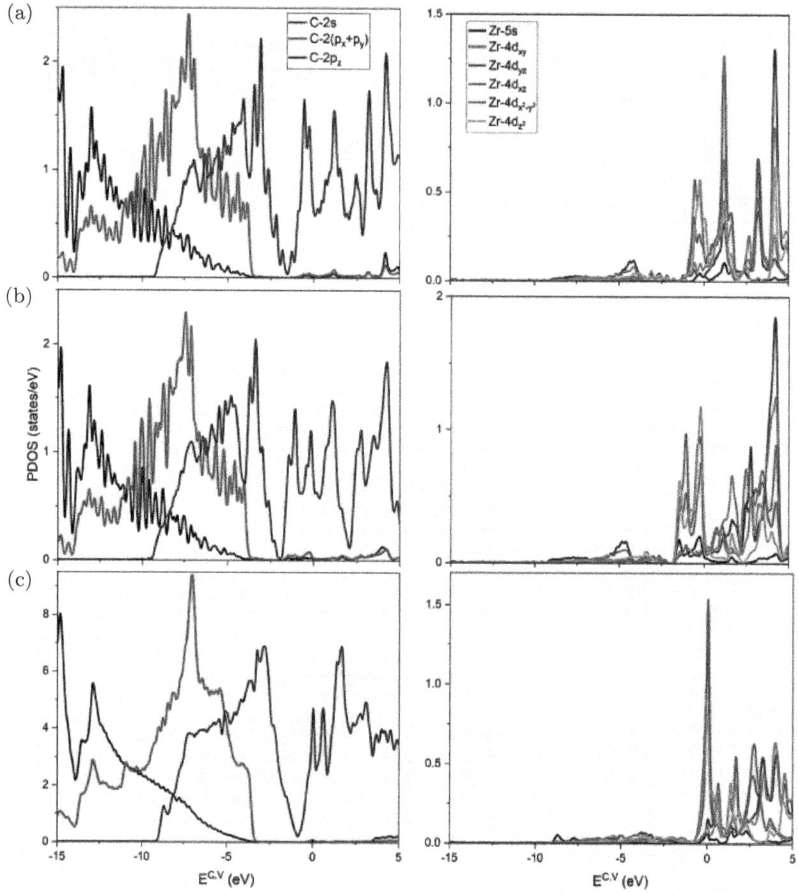

Fig. 5.8. Orbital-projected DOSs of (a) (1:8), (b) (2:8), and (c) (1:16) Zr-intercalated graphite.

in the conduction band, around the Fermi level, and the valence bands, respectively. Even though the C-s and p orbitals reveal similar features of contribution in the DOSs, the peaks and dispersions are different due to the effect of the intercalant. Along with the p_z orbital of carbon, the 4d orbitals of Zr make a significant contribution to the conduction band and the energy band around the Fermi level, in which $4d_{x^2-y^2}$ and $4d_{z^2}$ have strong dominance near the Fermi level, while $4d_{xy}$, $4d_{yz}$, $4d_{xz}$ strongly contribute to the conduction band. Hence, the dominant p orbitals of carbon are strongly hybridized with d states in all the compounds. Moreover, the s and d orbitals of Zr atoms in the intercalate sheet and the s and p orbitals of C in the graphite sheets form the hybridization, which is responsible for the induced magnetic configuration of this GIC. Similar characteristics had been reported for 3d transition metals (e.g., V, Cr, Mn, Fe, Co, and Ni)[57] intercalating into the C_6 graphite system with six carbon atoms per unit cell from the top side.

In addition, the optical properties, including their related factors of dielectric function, energy loss spectrum, absorption, refractivity, and reflectance, are shown in Fig. 5.9 to provide detailed information about the system. In Fig. 5.9(a), the two parts of the dielectric function are described with two remarkably strong peaks in the energy range of 0–1.5 eV and some weak peaks in both the real (ε_1) and imaginary (ε_2) parts, colored in green and wine-red curves, respectively. The corresponding relation between these two parts is illustrated via the Kramers–Kronig relationships,[58] which are explained through the following related formula:

$$n(\omega) - 1 = \frac{2}{\pi} P \int_0^{\infty} \frac{\omega' \kappa(\omega')}{\omega'^2 - \omega^2} d\omega' \tag{5.3}$$

The peaks of the energy loss function, as shown in Fig. 5.9(b), are consequently found in the vanished region of ε_1 in the xx direction. The loss function corresponds to broadened regions at approximately 7.5–13 and 15–18 eV in the xx and zz directions, respectively, relating to plasmons and forms of collective excitation. The vanishing real part of the dielectric function and a minimum of the imaginary part can also determine this plasmon. Thus, the plasmon modes are associated with the imaginary part that could propagate along the surface. Besides, small vibrations are observed with increasing energy in the range of 3–8 eV, indicated by the weak peaks of the two parts. It should be noted that the region above 10 eV cannot be interpreted in terms of classical oscillators. Simultaneously, the ultraviolet absorption of this system is caused by these transitions related to its band structure, indicating that the system starts to absorb photons in the ultraviolet regions at 1.5 eV, as shown in Fig. 5.9(c). The zz direction exhibits stronger

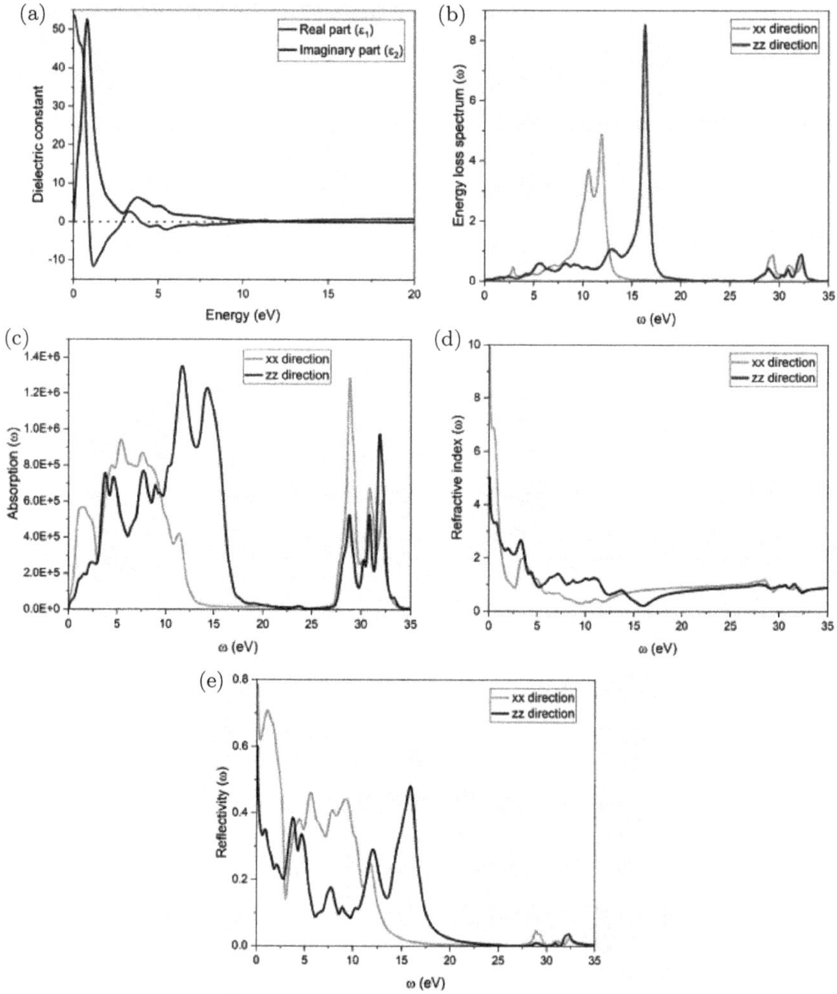

Fig. 5.9. Optical properties of (1:8) Zr-intercalated graphite: (a) the dielectric function with real and imaginary parts, (b) energy loss function, (c) absorption spectra, (d) refractive index, and (e) reflectivity in the xx and zz directions.

adsorption in the larger energy range of 10–17.5 eV compared to the xx/yy direction. The lowest limit of the conduction band and the upper valence band is defined by the valence excitation region, which is extended up to 17.5 eV. The real part behaves mainly as a classical oscillator, in which it vanishes in the energy range of

10–17.5 eV, corresponding to the maxima of absorption at these frequencies. Furthermore, the reflectance and refractive properties are calculated based on the dielectric function shown in Fig. 5.9(d) and 5.9(e). In Fig. 5.9(d), the maximum refractive index is found in the low-lying energies of approximately 0 and 2.5 eV in the two directions. This index accordingly decreases with increasing energy, as the wavelength increases. Especially, the change in the refractive index is produced from a change in absorption through the Kramers–Kronig relationships. These two factors are consistent with the characteristics of the dielectric function, indicating a close relation between dielectric function and linear optical properties. In addition, the minimum of strong reflectivity is determined by the imaginary part of the dielectric constant in the energy ranges of 2.5–3 and 7–10 eV in the zz and xx/yy directions, respectively, indicating a collective plasma resonance. Simultaneously, the broad maxima at 0 and 2 eV in the xx and zz directions, respectively, relate to the strong peaks in dielectric constant.

5.4. Nb-intercalated Graphite

Similarly, the optimized structures of Nb-intercalated graphite are similar to those of Zr cases, in which the Nb atom is intercalated at the center of a honeycomb ring of the graphite sheets, as shown in Fig. 5.10. The lattice constant, chemical bonds, and interlayer distance belonging to the structural parameters of Nb-intercalated graphite are mentioned in Table 5.2 for (1:8), (2:8), and (1:16) concentrations. The lattice constant increases with an increase in the concentration of the Nb atoms and supercell graphite, as in the case of (1:16). All systems exhibit a dramatic change in the bond length of the C–C bonds after intercalation, in which the bond length around the intercalant is larger than that of the faraway C–C bond, which is similar to the Zr cases. Accordingly, C–C (1) and C–C (2) are respectively 1.448 and 1.426 Åin the (1:8) case. Nb atoms cause the chemical bond length in the intercalated region to adjust, while on the other side, the C–C bonds of the graphene sheet remain unchanged. Importantly, the interlayer distance increases

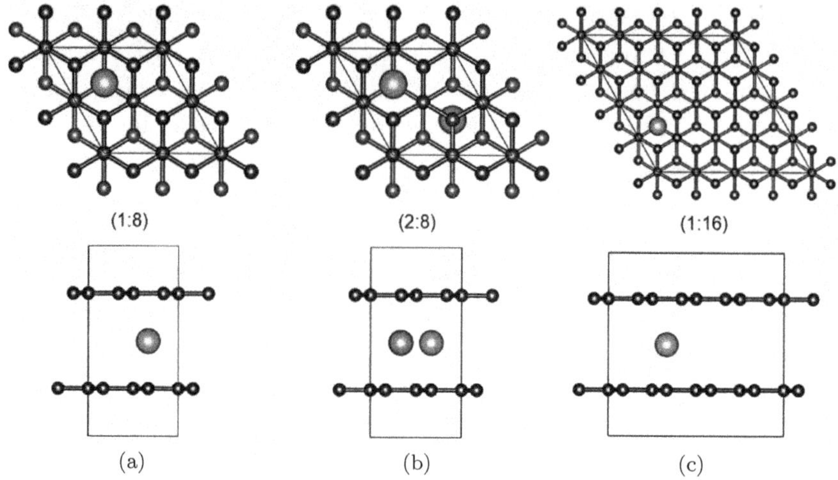

Fig. 5.10. Structural optimization as seen in the top and side views of Nb-intercalated graphite with varying concentrations of (a) (1:8), (b) (2:8), and (c) (1:16).

Table 5.2. Structural parameters of Nb-intercalated graphite.

System	Lattice constant (Å)	C–C bond length (Å) (1)	C–C bond length (Å) (2)	Interlayer distance (Å)	Magnetic moment
(1:8)	4.97	1.448	1.426	3.86	0
(2:8)	4.99	1.445	1.429	4.28	0
(1:16)	9.89	1.445	1.437	3.48	1.04

after intercalating in all cases because of the intercalant. As a result, the interlayer distances of the (1:8) and (2:8) systems are 3.86 and 4.28 Å, respectively. It should be noted that the Zr atom induces a larger interlayer distance in the case of (1:8) compared to the Nb case. While the interlayer distance of the (1:16) case is 3.48 Å, which is almost the same as that of graphite. Hence, the concentration of intercalated systems could modify the structural features of graphite, probably governing the electronic and optical properties. Nb-intercalated graphite not only reveals features similar to the Zr cases but also its distinct characteristics. To further clarify the bonding features, the charge density shown in Fig. 5.11 could

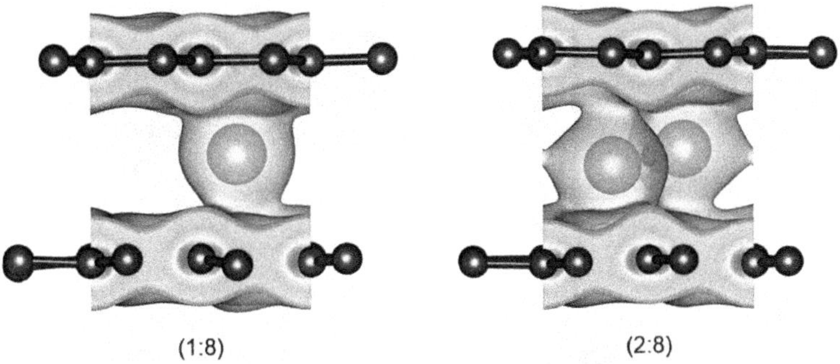

Fig. 5.11. The charge density of Nb-intercalated graphite.

provide detailed information about these bonds. In this figure, the strong interaction between Nb and the carbon graphene sheet is indicated by the yellow regions in the spacing between the graphene sheets and the green regions around the C–C bonds that are stronger than the interaction between Zr and the graphite sheets. The C–C bonds with sp^2 and π bonds due to the s and p orbitals of carbon are also affected by the d orbitals of Nb, as indicated by the red and green regions. Because of the similarities in their outermost 4d orbital structures, both Zr and Nb exhibit similar properties of bondings after intercalation.

The spin distribution and magnetic moment are considered for illustrating the induced magnetism in Nb-intercalated graphite, as shown in Fig, 5.12 and Table 5.2, respectively. There is clear polarization in the spin up and spin down of hexagonal ZrC_8 and ZrC_{16}, which are identical. According to Table 5.2, the magnetic moments of both (1:8) and (2:8) systems are zero, corresponding to non-magnetic or antiferromagnetic behavior, which is the same as in the Zr cases. Based on the spin distribution shown in Figs. 5.12(a) and 5.12(b), the spin splitting of spin up and spin down is found in the graphene sheets and two Nb atoms in the (2:8) case, as indicated by red and blue circles, which represents antiferromagnetic behavior. The magnetic moment of (1:16) is 1.04 (see Table 5.2), which is consistent with the ferromagnetic feature. Even though the induced magnetism is similar, the Nb cases demonstrate a difference from the Zr cases due to the spin distributions of spin up and spin down at

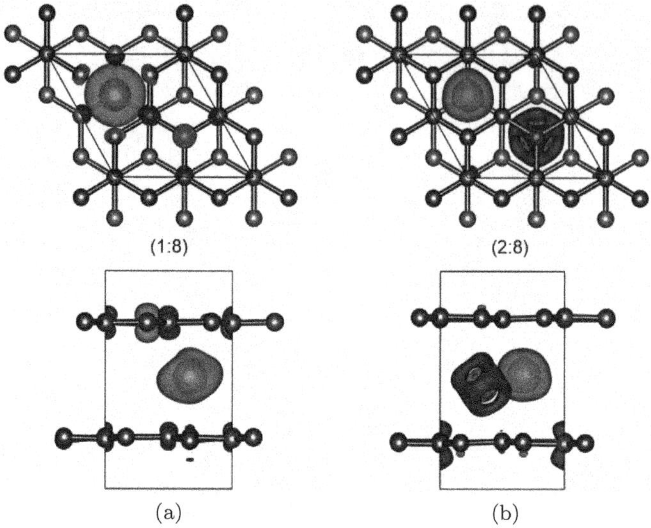

Fig. 5.12. Spin distribution of Nb-intercalated graphite.

the same concentration. The electronic properties of Nb-intercalated graphite are further presented through band-energy structures, as shown in Fig. 5.13. In this figure, the band-energy spectra of NbC_8 with the concentrations of (1:8) and (2:8) are presented, considering the atom dominance of Nb atoms in the whole band along G–M–K–G–A–L–H–A of the first BZ of the hexagonal. These bands reveal features similar to those of the Zr-intercalated graphite, where the overlap between the conduction and valence bands around the Fermi level is exhibited, corresponding to metallic behavior. The Dirac cone at the K and H points in graphite is transformed into low energy in the valence bands at around -2 eV due to the strong effect of Nb atoms near the Fermi level. The partial flat bands are found in low-lying energy, ranging from -7 to -8 and from -12 to -13 eV along the G–A points of the (1:8) and (2:8) systems (see Figs. 5.13(a) and 5.13(b)), corresponding to the symmetric structure of graphite. The Nb atom dominance, indicated by cyan circles, shows a larger energy range dominance (from -5 to 5 eV) compared to the Zr cases. The intercalant also exhibits a major and strong effect in the conduction bands and the energy band near the Fermi

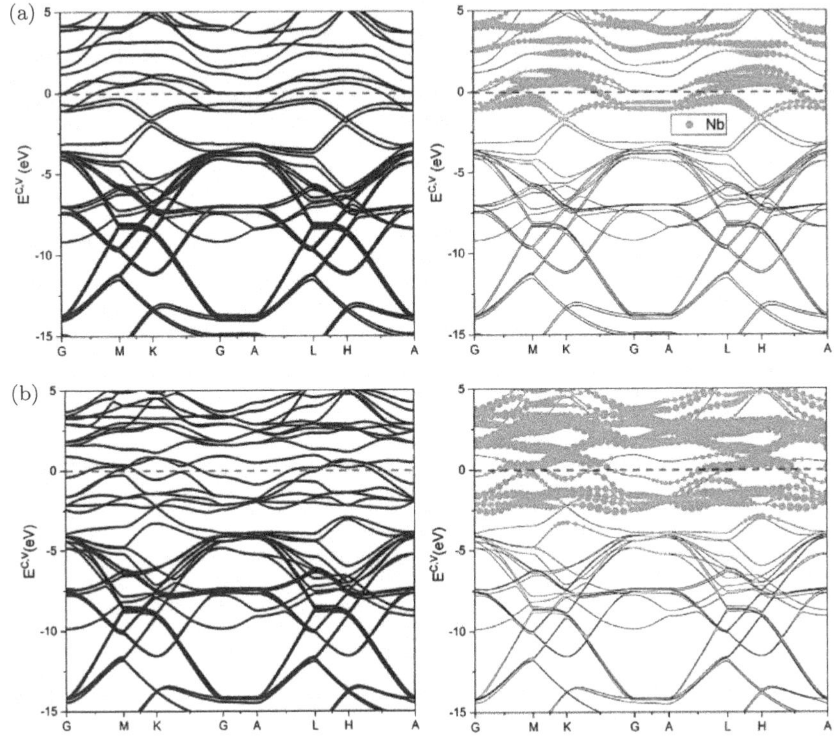

Fig. 5.13. Band energy spectra of (a) (1:8) and (b) (2:8) Nb-intercalated graphite. The blue curves describe the total band, and the light-cyan circles represent Nb atom dominance.

level, leading to adjusted band features in these regions compared to graphite.

In regard to orbital distribution, the orbital-projected DOS plotted in Fig. 5.14 provides detailed information about the orbital dominance and bonding in Nb-intercalated graphite. The orbital distribution of carbon and Nb in the compounds of the (1:8) and (2:8) systems is displayed, which results in the strong hybridization of these orbitals. The C atoms in graphite sheets have sp^2 hybridization, and each C has one π bond that is normal to the graphite sheet, corresponding to the strong contributions of C-p and Nb-d orbitals. These bonds result in the strong dominance of $2p_z$ and 2s, $2p_x$, $2p_y$ in the conduction band, around the Fermi level, and valence

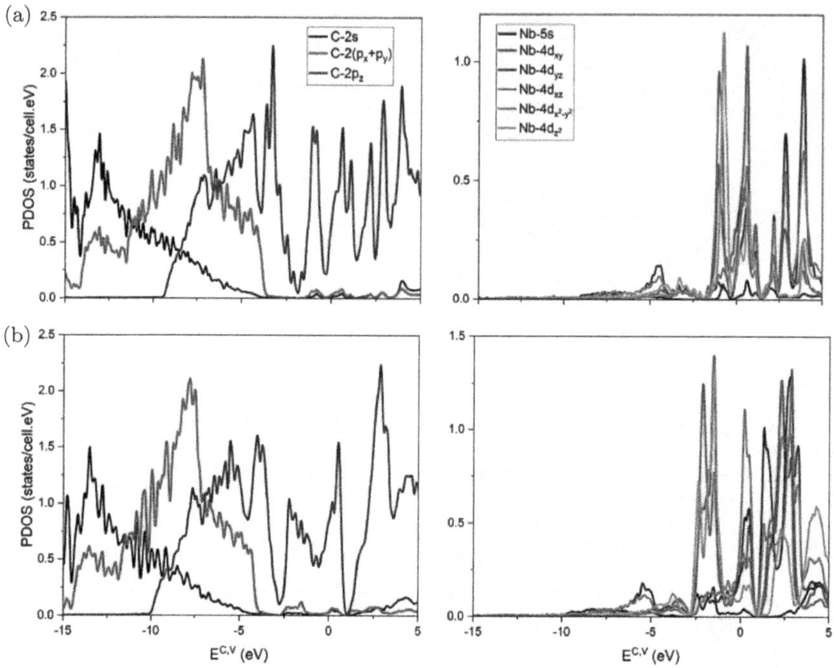

Fig. 5.14. Orbital-projected DOSs of (a) (1:8) and (b) (2:8) Nb-intercalated graphite.

bands, respectively. The effect of the intercalant is also indicated in the differences between the (1:8) and (2:8) cases in the peaks of DOSs. The 4d orbitals of Nb make a significant contribution to the conduction band and the energy band around the Fermi level, which is consistent with the band-energy spectra, in which $4d_{x^2-y^2}$ and $4d_{z^2}$ show strong dominance near the Fermi level, while $4d_{xy}$, $4d_{yz}$, and $4d_{xz}$ strongly contribute to the conduction band. Hence, the hybridized bondings between C-s,p and Nb-s,d are demonstrated through this contribution. Furthermore, the Nb cases exhibit higher DOSs compared to the Zr systems, even though they have the same outermost electron configuration of the 4d orbitals. In short, Nb- and Zr-intercalated graphites show similarities and distinctions in their electronic properties, as indicated by their band energy spectra and orbital-projected DOSs.

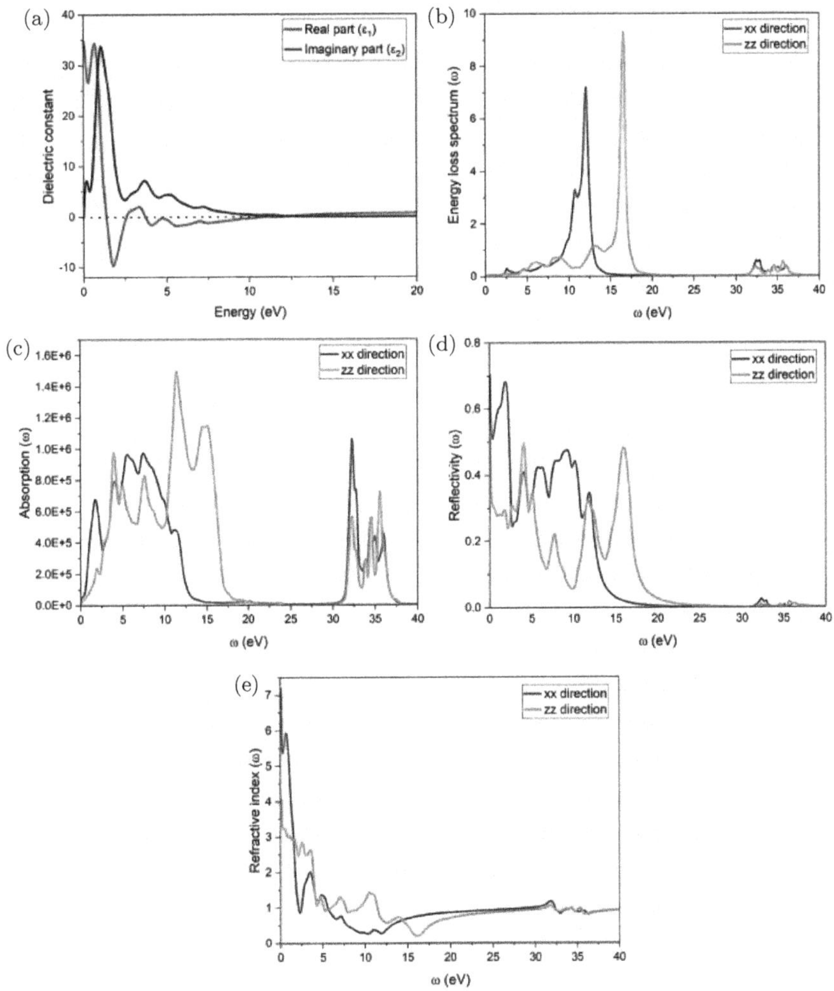

Fig. 5.15. Optical properties of Nb-intercalated graphite in the (1:8) system: (a) the dielectric function with real and imaginary parts, (b) energy loss function, (c) absorption spectra, (d) reflectivity, and (e) refractive index in the xx and zz directions.

The optical properties, dielectric function, energy loss spectrum, absorption, refractivity, and reflectance are shown in Fig. 5.15 to provide detailed information about the system. Most features resemble those of the Zr-intercalated graphite system. In Fig. 5.15(a), the two parts of the dielectric function are described with two remarkably strong peaks in the energy range of 0–1.5 eV and some weak peaks in both the real (ε_1) and imaginary (ε_2) parts, colored in red and blue curves, respectively. The peaks of the energy loss function, as shown in Fig. 5.15(b), are consequently found in the vanished region of ε_1 in the xx direction. The loss function relating to plasmons and forms of collective excitation corresponds to regions broadened at approximately 9–13 and 15–18 eV in the xx and zz directions, respectively. Besides, small vibration is observed with an increase in energy in the range of 3–8 eV, as indicated by the weak peaks of the two parts, which are similar to the weak peaks of the dielectric function in the Zr case. The region above 10 eV (see Fig. 5.15(a)) cannot be interpreted in terms of classical oscillators. Simultaneously, the ultraviolet absorption of this system is caused by these transitions related to its band structure, indicating that the system starts to absorb photons in the ultraviolet regions at 1.5 eV, as shown in Fig. 5.15(c). The zz direction also exhibits stronger adsorption in the larger energy range of 10–17.5 eV compared to the xx/yy direction. The real part behaves mainly as a classical oscillator in which it vanishes in an energy range of 10–17.5 eV, corresponding to the maxima of absorption at these frequencies. Furthermore, the reflectance and refractive properties are based on the dielectric function shown in Figs. 5.15(d) and 5.15(e), with properties similar to those in the Zr case. In Fig. 5.15(d), the minimum of strong reflectivity is determined by the imaginary part of the dielectric constant in the energy range of 0–5 eV for both the zz and xx/yy directions, respectively, indicating a collective plasma resonance. Simultaneously, the broad maxima at 0–2 eV in the xx direction relate to the strong peak in the dielectric constant, while the maximum refractive index is found in the low-lying energies of approximately 0 and 2.5 eV in the xx direction, as shown in Fig. 5.15(e). This index accordingly decreases with an increase in energy, as the wavelength increases, which is caused by the change in absorption. These two factors are consistent with the characteristics of the dielectric function and the results of the Zr-intercalated graphite, indicating a close relation between the dielectric function and the linear optical properties of two 4d transition metals.

In general, the electronic structures of GICs with Zr- and Nb-intercalated graphites are modified after intercalation due to hybridization of the carbon-p orbital and the d orbitals of a transition metal. As a result, all intercalated systems reveal a metallic configuration based on the 4d transition metal (Zr, Nb). The strong interaction between these orbitals of C and Zr or Nb was indicated by the charge density and orbital-projected DOSs. Furthermore, the induced magnetism in hexagonal configurations is demonstrated, in which the C_8 systems show antiferromagnetic behavior and the C_{16} cases exhibit ferromagnetic one. It indicates that the concentration of the intercalated atoms could define the fundamental properties of GICs. The hybridization and spin distribution based on the intercalated magnetic 4d transition-metal atoms appear to be favorable for the realization of magnetism in the system. Besides, the type of intercalant atom also affects the electronic and optical properties of GICs, resulting in the differences between transition-metal and alkali intercalations.

References

[1] Geng, Z. S.; Hahnlein, B.; Granzner, R.; Auge, M.; Lebedev, A. A.; Davydov, V. Y.; Kittler, M.; Pezoldt, J.; Schwierz, F. Graphene nanoribbons for electronic devices. *Annalen Der Physik* **2017**, *529* (11).

[2] Celis, A.; Nair, M. N.; Taleb-Ibrahimi, A.; Conrad, E. H.; Berger, C.; de Heer, W. A.; Tejeda, A. Graphene nanoribbons: Fabrication, properties and devices. *Journal of Physics D-Applied Physics* **2016**, *49* (14).

[3] Nishad, A. K.; Sharma, R. Lithium-intercalated graphene interconnects: Prospects for on-chip applications. *IEEE Journal of the Electron Devices Society* **2016**, *4* (6), 485–489.

[4] Zhang, H.; Yang, Y.; Ren, D. S.; Wang, L.; He, X. M. Graphite as anode materials: Fundamental mechanism, recent progress and advances. *Energy Storage Materials* **2021**, *36*, 147–170.

[5] Shuvo, M. A. I.; Khan, M. A. R.; Karim, H.; Morton, P.; Wilson, T.; Lin, Y. R. Investigation of modified graphene for energy storage

applications. *ACS Applied Materials & Interfaces* **2013,** *5* (16), 7881–7885.

[6] Zhang, M.; Song, X. H.; Ou, X. W.; Tang, Y. B. Rechargeable batteries based on anion intercalation graphite cathodes. *Energy Storage Materials* **2019**, *16*, 65–84.

[7] Luo, X. C.; Chugh, R.; Biller, B. C.; Hoi, Y. M.; Chung, D. D. L. Electronic applications of flexible graphite. *Journal of Electronic Materials* **2002**, *31* (5), 535–544.

[8] Zhang, T. T.; Wu, S.; Yang, R.; Zhang, G. Y. Graphene: Nanostructure engineering and applications. *Frontiers of Physics* **2017**, *12* (1).

[9] Wissler, M. Graphite and carbon powders for electrochemical applications. *Journal of Power Sources* **2006**, *156* (2), 142–150.

[10] Bonal, J. P.; Gosmain, L. Graphites for nuclear applications. *Actualite Chimique* **2006**, 23–27.

[11] Tong, X.; Wei, Q. L.; Zhan, X. X.; Zhang, G. X.; Sun, S. H. The new graphene family materials: Synthesis and applications in oxygen reduction reaction. *Catalysts* **2017**, *7* (1).

[12] Li, N.; Su, D. In-situ structural characterizations of electrochemical intercalation of graphite compounds. *Carbon Energy* **2019**, *1* (2), 200–218.

[13] Xu, J. T.; Dou, Y. H.; Wei, Z. X.; Ma, J. M.; Deng, Y. H.; Li, Y. T.; Liu, H. K.; Dou, S. X. Recent progress in graphite intercalation compounds for rechargeable metal (Li, Na, K, Al)-ion batteries. *Advanced Science* **2017**, *4* (10).

[14] Weller, T. E.; Ellerby, M.; Saxena, S. S.; Smith, R. P.; Skipper, N. T. Superconductivity in the intercalated graphite compounds C(6)Yb and C(6)Ca. *Nature Physics* **2005**, *1* (1), 39–41.

[15] Kawai, S.; Saito, S.; Osumi, S.; Yamaguchi, S.; Foster, A. S.; Spijker, P.; Meyer, E. Atomically controlled substitutional boron-doping of graphene nanoribbons. *Nature Communications* **2015**, *6*.

[16] Jiang, J. K.; Kang, J. H.; Cao, W.; Xie, X. J.; Zhang, H. J.; Chu, J. H.; Liu, W.; Banerjee, K. Intercalation doped multilayer-graphene-nanoribbons for next-generation interconnects. *Nano Letters* **2017**, *17* (3), 1482–1488.

[17] Terrones, H.; Lv, R. T.; Terrones, M.; Dresselhaus, M. S. The role of defects and doping in 2D graphene sheets and 1D nanoribbons. *Reports on Progress in Physics* **2012**, *75* (6).

[18] Zhang, H.; Guo, G. L.; Adenusi, H.; Qin, B. S.; Li, H. H.; Passerini, S.; Huang, W. Advances and issues in developing intercalation graphite cathodes for aqueous batteries. *Materials Today* **2022**, *53*, 162–172.

[19] Zhao, X. X.; Song, P.; Wang, C. C.; Riis-Jensen, A. C.; Fu, W.; Deng, Y.; Wan, D. Y.; Kang, L. X.; Ning, S. C.; Dan, J. D.;

Venkatesan, T.; Liu, Z.; Zhou, W.; Thygesen, K. S.; Luo, X.; Pennycook, S. J.; Loh, K. P. Engineering covalently bonded 2D layered materials by self-intercalation. *Nature* **2020**, *581* (7807), 171-+.
[20] Daukiya, L.; Nair, M. N.; Cranney, M.; Vonau, F.; Hajjar-Garreau, S.; Aubel, D.; Simon, L. Functionalization of 2D materials by intercalation. *Progress in Surface Science* **2019**, *94* (1), 1–20.
[21] Yamada, Y.; Usui, K.; Chiang, C. H.; Kikuchi, K.; Furukawa, K.; Yamada, A. General observation of lithium intercalation into graphite in ethylene-carbonate-free superconcentrated electrolytes. *ACS Applied Materials & Interfaces* **2014**, *6* (14), 10892–10899.
[22] Silva, C. C.; Cai, J. Q.; Jolie, W.; Dombrowski, D.; Zum Hagen, F. H. F.; Martinez-Galera, A. J.; Schlueter, C.; Lee, T. L.; Busse, C. Lifting epitaxial graphene by intercalation of alkali metals. *Journal of Physical Chemistry C* **2019**, *123* (22), 13712–13719.
[23] Algdal, J.; Balasubramanian, T.; Breitholtz, M.; Kihlgren, T.; Wallden, L. Thin graphite overlayers: Graphene and alkali metal intercalation. *Surface Science* **2007**, *601* (4), 1167–1175.
[24] Ahmad, S.; Miro, P.; Audiffred, M.; Heine, T. Tuning the electronic structure of graphene through alkali metal and halogen atom intercalation. *Solid State Communications* **2018**, *272*, 22–27.
[25] Sengupta, R.; Bhattacharya, M.; Bandyopadhyay, S.; Bhowmick, A. K. A review on the mechanical and electrical properties of graphite and modified graphite reinforced polymer composites. *Progress in Polymer Science* **2011**, *36* (5), 638–670.
[26] Jara, A. D.; Betemariam, A.; Woldetinsae, G.; Kim, J. Y. Purification, application and current market trend of natural graphite: A review. *International Journal of Mining Science and Technology* **2019**, *29* (5), 671–689.
[27] Zhang, D.; Tan, C.; Zhang, W. Z.; Pan, W. J.; Wang, Q.; Li, L. Expanded graphite-based materials for supercapacitors: A review. *Molecules* **2022**, *27* (3).
[28] Takahara, I.; Mizoguchi, T. First principles study on formation and migration energies of sodium and lithium in graphite. *Physical Review Materials* **2021**, *5* (8).
[29] Wallace, P. R. The band theory of graphite. *Physical Review* **1947**, *71* (9), 622–634.
[30] Zhao, H. P.; Ren, J. G.; He, X. M.; Li, J. J.; Jiang, C. Y.; Wan, C. R. Modification of natural graphite for lithium ion batteries. *Solid State Sciences* **2008**, *10* (5), 612–617.
[31] Wilhelm, H. A.; L'Heureux, J. Natural and synthetic graphite powders: Production and main industrial uses. *Actualite Chimique* **2006**, 19–22.

[32] Okada, S.; Kobayashi, T. Electronic properties of graphite with rotational stacking arrangement. *Japanese Journal of Applied Physics* **2009**, *48* (5).

[33] Yang, Y. P.; Zou, Y. C.; Woods, C. R.; Shi, Y. M.; Yin, J.; Xu, S. G.; Ozdemir, S.; Taniguchi, T.; Watanabe, K.; Geim, A. K.; Novoselov, K. S.; Haigh, S. J.; Mishchenko, A. Stacking order in graphite films controlled by van der Waals technology. *Nano Letters* **2019**, *19* (12), 8526–8532.

[34] Ho, J. H.; Lu, C. L.; Hwang, C. C.; Chang, C. P.; Lin, M. F. Coulomb excitations in AA- and AB-stacked bilayer graphites. *Physical Review B* **2006**, *74* (8).

[35] Lu, C. L.; Chang, C. P.; Huang, Y. C.; Ho, J. H.; Hwang, C. C.; Lin, M. F. Electronic properties of AA- and ABC-stacked few-layer graphites. *Journal of the Physical Society of Japan* **2007**, *76* (2).

[36] Wohner, N.; Lam, P.; Sattler, K. Energetic stability of graphene nanoflakes and nanocones. *Carbon* **2014**, *67*, 721–735.

[37] Barnard, A. S.; Snook, I. K. Modelling the role of size, edge structure and terminations on the electronic properties of graphene nano-flakes. *Modelling and Simulation in Materials Science and Engineering* **2011**, *19* (5).

[38] Wu, Z. S.; Ren, W. C.; Gao, L. B.; Liu, B. L.; Jiang, C. B.; Cheng, H. M. Synthesis of high-quality graphene with a predetermined number of layers. *Carbon* **2009**, *47* (2), 493–499.

[39] Chae, S.; Choi, S. H.; Kim, N.; Sung, J.; Cho, J. Integration of graphite and silicon anodes for the commercialization of high-energy lithium-ion batteries. *Angewandte Chemie-International Edition* **2020**, *59* (1), 110–135.

[40] Hu, D. Z.; Chen, L.; Tian, J.; Su, Y. F.; Li, N.; Chen, G.; Hu, Y. L.; Dou, Y. S.; Chen, S.; Wu, F. Research progress of lithium plating on graphite anode in lithium-ion batteries. *Chinese Journal of Chemistry* **2021**, *39* (1), 165–173.

[41] Liu, Q.; Xu, R. G.; Mu, D. B.; Tan, G. Q.; Gao, H. C.; Li, N.; Chen, R. J.; Wu, F. Progress in electrolyte and interface of hard carbon and graphite anode for sodium-ion battery. *Carbon Energy* **2022**, *4* (3), 458–479.

[42] Park, J.; Xu, Z. L.; Kang, K. Solvated ion intercalation in graphite: Sodium and beyond. *Frontiers in Chemistry* **2020**, *8*.

[43] Verma, P.; Sasaki, T.; Novak, P. Chemical surface treatments for decreasing irreversible charge loss and preventing exfoliation of graphite in Li-ion batteries. *Electrochimica Acta* **2012**, *82*, 233–242.

[44] Luo, P.; Zheng, C.; He, J. W.; Tu, X.; Sun, W. P.; Pan, H. G.; Zhou, Y. P.; Rui, X. H.; Zhang, B.; Huang, K. Structural engineering in graphite-based metal-ion batteries. *Advanced Functional Materials* **2022**, *32* (9).

[45] Aleksensky, A. E.; Baidakova, M. V.; Yagovkina, M. A.; Siklitsky, V. I.; Vul, A. Y.; Naramoto, H.; Lavrentiev, V. I. Nanodiamonds intercalated with metals: Structure and diamond-graphite phase transitions. *Diamond and Related Materials* **2004**, *13* (11–12), 2076–2080.

[46] Onuma, H.; Kubota, K.; Muratsubaki, S.; Ota, W.; Shishkin, M.; Sato, H.; Yamashita, K.; Yasuno, S.; Komaba, S. Phase evolution of electrochemically potassium intercalated graphite. *Journal of Materials Chemistry A* **2021**, *9* (18), 11187–11200.

[47] Wu, X. Y.; Li, S. M.; Yang, B.; Wang, C. M. In situ transmission electron microscopy studies of electrochemical reaction mechanisms in rechargeable batteries. *Electrochemical Energy Reviews* **2019**, *2* (3), 467–491.

[48] Letellier, M.; Chevallier, F.; Beguin, F. In situ Li-7 NMR during lithium electrochemical insertion into graphite and a carbon/carbon composite. *Journal of Physics and Chemistry of Solids* **2006**, *67* (5–6), 1228–1232.

[49] Nonaka, T.; Kawaura, H.; Makimura, Y.; Nishimura, Y. F.; Dohmae, K. In situ X-ray Raman scattering spectroscopy of a graphite electrode for lithium-ion batteries. *Journal of Power Sources* **2019**, *419*, 203–207.

[50] Dahn, J. R. Phase-diagram of LIXC6. *Physical Review B* **1991**, *44* (17), 9170–9177.

[51] David, L.; Singh, G. Reduced graphene oxide paper electrode: Opposing effect of thermal annealing on Li and Na cyclability. *Journal of Physical Chemistry C* **2014**, *118* (49), 28401–28408.

[52] Guan, Z. R. X.; Shen, X.; Yu, R. C.; Wang, Z. X.; Chen, L. Q. Chemical intercalation of solvated sodium ions in graphite. *Electrochimica Acta* **2016**, *222*, 1365–1370.

[53] Tossici, R.; Berrettoni, M.; Rosolen, M.; Marassi, R.; Scrosati, B. Electrochemistry of KC8 in lithium-containing electrolytes and its use in lithium-ion cells. *Journal of the Electrochemical Society* **1997**, *144* (1), 186–192.

[54] Flores-Livas, J. A.; Sanna, A. Superconductivity in intercalated group-IV honeycomb structures. *Physical Review B* **2015**, *91* (5).

[55] Singh, R.; Prakash, S. First-principles investigation into structural and magnetic properties of binary graphite 3d-transition metal

intercalated compounds (XC6; X = Cr, Mn, Fe). *Carbon* **2010**, *48* (5), 1341–1344.

[56] Kaloni, T. P.; Kahaly, M. U.; Schwingenschlogl, U. Induced magnetism in transition metal intercalated graphitic systems. *Journal of Materials Chemistry* **2011**, *21* (46), 18681–18685.

[57] Grechnev, G. E.; Lyogenkaya, A. A.; Kolesnichenko, Y. A.; Prylutskyy, Y. I.; Hayn, R. Electronic structure and magnetic properties of graphite intercalated with 3d-metals. *Low Temperature Physics* **2014**, *40* (5), 450–453.

[58] Peiponen, K. E.; Vartiainen, E. M. Kramers-Kronig relations in optical-data inversion. *Physical Review B* **1991**, *44* (15), 8301–8303.

CHAPTER 6

INTERCALATION OF 5D RARE-EARTH ELEMENTS INTO GRAPHITE

Wang Yu-Ming and Ming-Fa Lin

Department of Physics, National Cheng Kung University, Tainan, Taiwan

Abstract

The rare-earth metal atoms of Pd and Pu with 5f orbitals are outstanding candidates for fully exploring the significant decoration effects owing mainly to their seven types of probability functions. The highly complicated orbital hybridizations and spin configurations closely related to the C–Pd/C–Pu, C–C, and Pd–Pd/Pu–Pu bonds are fully examined under a unified framework.

Keywords: Graphite, intercalation, 5d elements.

A carbon atom has an atomic number of 6 in the period table and presents four active orbitals of (2s, $2p_x$, $2p_y$, $2p_z$) for any condensed-matter system. To greatly enhance the chemical bondings, two fully occupied 2s orbitals in each isolated atom can transform into a half-occupied configuration of (2s, $2p_z$). In general, there exist sp^3, sp^2, and sp bondings, depending on the neighboring-atom interactions. The 3D diamond[1] and layered graphite,[2] respectively, reveal the first and second kinds, with four and three nearest neighbors. The former belongs to the largest-gap insulator. The latter possesses a hexagonal honeycomb crystal, in which the well-characterized stacking configurations are AB,[3] ABC,[4] and AA.[5] The interlayer atomic interactions are purely due to the $2p_z$–$2p_z$ orbital hybridizations, and they are perpendicular to the σ bonding of (2s, $2p_x$, $2p_y$) – (2s, $2p_x$, $2p_y$) under a planar honeycomb crystal. Such a layered structure clearly exhibits a dimensionality crossover behavior. The low-dimensional systems are very close to one another in terms of geometric and electron properties, e.g., few-layer graphenes with different stackings,[6] 1D carbon nanotubes,[7] graphene nanoribbons,[8] and carbon tori.[9] A single-walled carbon nanotube and a finite-width graphene nanoribbon could be regarded, respectively, as a rolled-up graphitic sheet in the cylindrical form and a straightly cut graphene along the longitudinal directions. The π-electronic states of the former are sampled from those of the latter under a boundary condition. Three types of geometry-dominated carbon nanotubes are identified from the curvature effects. The strong relations between radius, chirality, and bandgap are further verified from the combinational measurements of STM and STS.[10,11] Some unusual results are also obtained for graphene nanoribbons with open edges and finite-width quantum confinements.[12] The low-lying valence and conduction states are mainly determined by the π-electronic bondings. The current research clearly shows that the position-dependent atomic interactions are well established with empirical formulas. These analytic equations are reliable and convenient for the tight-binding model. The various Hamiltonians of graphene-related materials are successful for a wider development of a grand quasi-particle framework (details available in previously published books).[13] Their successes represent similar possibilities for other atom-based materials and compounds.

The chemical modifications, which include intercalations/deintercalations,[14] adsorptions,[15] substitutions,[16] decorations,[17] and heterojunctions,[18] are very powerful in generating a number of mainstream compounds. Graphite possesses a uniform interlayer spacing of $\sim$3.35 Å, where the interlayer van der Waals interactions are weak but very significant for the creation of the popular AB-stacking Bernal.[19] The planar honeycomb clearly indicates the perpendicular π and σ bondings, leading to well-defined electronic states in the absence of anticrossing valence subbands.[20] This system frequently serves as the anode/cathode material for the positive-/negative-ion-based batteries, owing to the available graphitic spacings for adatom intercalations/deintercalations during the charging and discharging processes. For example, graphite lithium compounds, with periodic graphitic sheets and well-behaved intercalant layers,[21] exhibit metallic behavior under a blueshift of the Fermi level.[22] On the other hand, a pristine system behaves as a semi-metal due to the weak overlaps of the valence and conduction energy subbands.[23] The n- or p-type dopings are predicted to arise from the interlayer carbon–intercalant chemical bondings; furthermore, they are responsible for the unusual quasi-particle behaviors. The free conduction electrons or valence holes are capable of creating the greatly enhanced quasi-particle phenomena, e.g., the intercalant-induced energy subbands,[24] the density-dependent optical gaps,[25] the Coulomb-field excitations in the presence of coexistent and electron–hole pairs and plasmon modes,[26] the featured Landau subbands with non-crossing/crossing/anticrossing behaviors, the rich magneto-optical spectra (the magneto-selection rules),[27] the outstanding electrical conductivities,[28] and superconductivity temperature.[29] Their concise pictures of physical, chemical, and material sciences strongly rely on the phenomenological models. The previous theoretical predictions only use the rigid-band model (the Dirac cone structure of a monolayer graphene).[30] This is not enough to characterize the electronic properties of graphite intercalation compounds since the carbon-intercalant chemical bondings play a critical role in the above-mentioned quasi-particle properties. The establishment of reliable Hamiltonians has become a focus of study in recent investigations. This strategy is very helpful in developing a grander quasi-particle framework.

To greatly comprehend the intercalation effects and develop theoretical models, the rare-earth metal atoms are considered outstanding candidates for intercalant guest ones.[31] Only a few of them are predicted to survive in the earth, while they play critical roles in common industrial products and the mineral economy. How to efficiently utilize their characteristics has become a significant current issue. The featured quasi-particle phenomena are fully dominated by seven types of 4f or 5f orbitals: (1) f_{x^3}, (2) f_{xz^2}, (3) f_{yz^2}, (4) f_{xyz}, (5) $f_{z(x^2-y^2)}$, (6) $f_{x(x^2-y^2)}$ and (7) $f_{x(x^2-3y^2)}$, with the spin-up and spin-down configurations.[32] More interestingly, these orbital-based crystals frequently exhibit rich and unique quasi-particle phenomena, such as high-temperature superconductors (e.g., YbCuO),[33] heavy Fermion compounds,[34] and intermetallic compounds, and chemical adsorptions on graphene,[35] silicene,[36] germanene,[37] tinene,[38] and plumbene.[39] Graphite rare-earth metal compounds are chosen for a model study. Such materials are expected to possess rather complicated quasi-particles because more multi-orbital hybridizations and rich ferromagnetic spin modulations simultaneously come to exist in carbon–intercalant and intercalant–intercalant bonds. The unusual intercalation effects could be very sensitive to the various intercalants (the occupied configurations of distinct rare-earth metal atoms)[31] relative concentrations,[40] and symmetric distributions.[41] The critical mechanisms can be explained only through delicate VASP simulations and analyses (detailed in Sections 6.1–6.5,)[42] but not using the tight-binding model.[43]

The VASP simulations are very suitable for studying the quantum quasi-particles of graphite Pa/U intercalation compounds[42] to achieve concise pictures of multi-/single-orbital hybridizations and spin configurations. The results calculated agree with one another under a unified framework, as clearly illustrated in previously published books. The bonding-enriched crystal symmetries,[44] the net magnetic momentum of atom- and orbital-decomposed contributions,[45] the featured band structures and wave functions with atom and spin dominances,[46] the spatial charge / spin-density distributions, the atom- orbital- and spin-projected van Hove singularities,[47] and the prominent absorption peaks and optical gaps in the presence of excitonic effects.[48] The highly anisotropic environment of graphitic sheets can drive the chemical interactions/ de-intercalations very efficiently during the charging/discharging processes in lithium-ion-based batteries. The semi-metallic Bernal graphite, due to the interlayer van der Waals interactions,[49] becomes a metallic compound. The blueshift of the Fermi level can be estimated from the low-energy van Hove singularity and an optical gap. All essential properties are likely to be very sensitive to the intercalant concentrations and arrangements. The stage-n layered structure is available for characterizing a close geometric relation between the host and intercalant atoms.[50] The intralayer σ bonding of C–C bonds, the very complicated 5f–5f orbital hybridizations of Pa–Pa/U–U bonds, the interlayer $2p_z$–$2p_z$ hybridizations in neighboring graphenes, and the interlayer $3p_z$–5f multi-orbital mixings need to be thoroughly examined using the delicate data. There exist certain important differences among graphite Pa/U/Li/Na/K/Rb/Cs/$AlCl_4$ intercalation compounds. The various experimental examinations are discussed in detail. How to build a tight-binding model from these critical mechanisms is a very interesting research question. This is very helpful in achieving an expanded grand quasi-particle framework.

6.1. The Optimal Crystals of Graphite and Graphite Pa/U Intercalation Compounds

The unusual features of nitrogen substitutions in silicene are investigated in detail by using the VASP software.[51,52] The exchange and correlation energies due to many-particle electron–electron interactions are calculated using the Perdew–Burke–Ernzerhof (PBE) functional under the generalized gradient approximation.[53] Furthermore, the electron–ion interactions can be characterized by the projector augmented wave (PAW) pseudopotentials.[54] A space of 15 Å width is inserted between the silicene planes to avoid their interaction. In general, a plane-wave basis set with a maximum kinetic energy cutoff of 600 eV is chosen to expand the wave function. The first Brillouin zone is sampled with $5 \times 5 \times 5$ and $40 \times 40 \times 5$ k-point meshes within the Monkhorst–Pack scheme for further calculations of the electronic and magnetic properties. The energy convergence is set to be 10^{-8} eV between two consecutive steps, and the maximum Hellmann–Feynman force acting on each atom is less than 0.01 eV Å^{-1} during the ionic relaxations.

The delicate VASP simulations can present the optimal crystal structures of the Bernal graphite (Figs. 6.1(a) and 6.1(b))[19] and the stage-1, stage-2, stage-3, and stage-4 graphite Pa/U intercalation compounds (Figs. 6.1(c) and 6.1(d), Figs. 6.1(e) and 6.1(f), Figs. 6.1(g) and 6.1(h), and Figs. 6.1(i) and 6.1(j), respectively).[50] Carbon host atoms and rare-earth metal guest ones form the planar honeycomb lattices and the intercalant layers, respectively. This obviously indicates that the very strong σ bondings of C-(2s, $2p_x$, $2p_y$)–(2s, $2p_x$, $2p_y$) hybridizations almost remain the same after the metal-type interactions.[55] The stage-n compounds have n graphitic sheets between periodic Pa/U intercalant layers. There exist the intralayer C–C and Pa–Pa/U–U bonds and the interlayer C–C and C–Pa/C–U ones, in which the van der Walls interactions disappear under the $n - 1$ case. The significant geometric parameters and the interlayer distances between the periodic intercalant layers, two neighboring graphenes, or Pa–C or U–C layers are clearly presented in Table 6.1. Their features are closely related to the interlayer bonding phenomena and thus determine the band property near the Fermi level (the metallic and semi-metallic behaviors[56]) and optical transitions (Figs. 6.6(a)–6.6(x)).[57]

The graphite Pa and U intercalation compounds can exhibit stable stage-n crystal structures. Similar geometry properties are clearly revealed in the other 5f[58]/4f[59] rare-earth metal atoms and even in the 3d/4d/5d/6d transition-metal ones.[60] Moreover, the analytic formulas, which characterize the concentration relationships between the host and guest atoms, are identical for $C_{8n}R$, $C_{8n}T$, and $C_{8n}Na$[61]/C_{8n}[62] K/$C_{8n}Rb$[63]/$C_{8n}Cs$[64] but comparable for $C_{6n}Li$.[65] Graphite lithium intercalation compounds possess the smallest interlayer distances, the highest adatom concentrations, and the strongest chemical bondings (Table 6.1). These indicate the prominent interlayer multi-orbital hybridizations of metallic carbon–intercalant bonds. The donor-type dopings of conduction electrons are expected to dominate the low-energy quasi-particle properties. How to evaluate the blueshifts of the Fermi level will be discussed in Sections 6.4 and 6.5. In addition, there also exist acceptor-type graphite intercalation compounds,[66] which lead to

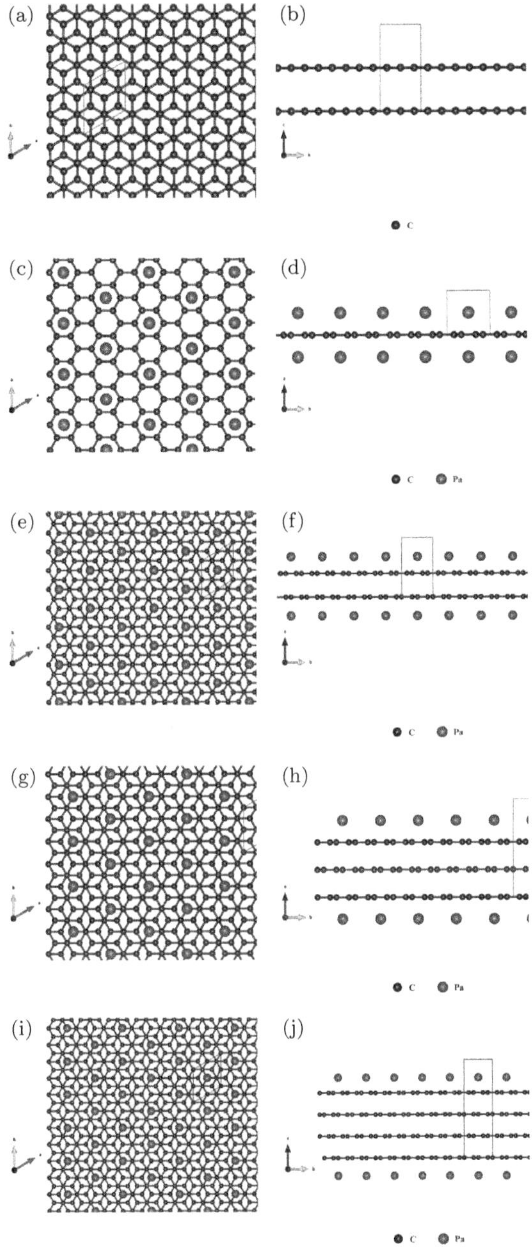

Fig. 6.1. The top/side views of the geometric structures for (a)/(b) Bernal graphite, and the (c)/(d) stage-1, (e)/(f) stage-2, (g)/(h) stage-3, and (i)/(j) stage-4 graphite Pa intercalation compounds.

Table 6.1. The geometric parameters, namely the interlayer distances of graphene–graphene, intercalant–intercalant, and graphene–intercalant for stage-n graphite Pa/U interaction compounds.

	Graphene–graphene (Å)	**Intercalant–intercalant (Å)**	**Graphene–intercalant (Å)**
PaC8	4.36	5.02	2.62
PaC16	3.11	4.97	2.63
PaC24	4.75	4.95	2.77
PaC32	4.97	4.95	2.72
UC8	4.43	5.04	2.65
UC16	4.73	4.96	2.67
UC24	4.83	4.95	2.81
UC32	5.05	4.94	2.76

the creation of free valence holes. Such systems, which have been successfully synthesized in experimental laboratories,[67] consist of large intercalants with negative charges, e.g., $AlCl_4$ and Al_2Cl_7 in ion-based batteries.[68] Both donor- and acceptor-type systems present the rich and unique quasi-particle metals in terms of crystal symmetries, blue-/redshifts of E_F, 3D optical plasmons coexistence with electron–hole excitations,[69] and finite optical gaps (which are discussed latter).[70]

The theoretical predictions on the position-dependent chemical bonds or crystal symmetries could be verified by the prominent patterns of X-ray scattering[71] and reflection low-energy electron diffraction (RLEED).[72] The diffraction peaks are purely due to elastic scatterings between photons and electrons, in which a periodic charge density distribution is responsible for the diffraction rule: The wave-vector change of the incident particles equals a specific reciprocal lattice vector. According to the theoretical calculations, the scattering amplitude is very sensitive to the lattice points and the charge density distributions around them. As a result, high-precision X-ray measurements of diffraction peaks are available for estimating the active chemical bondings within a unit cell, e.g., the ionic bondings of 3D sodium chloride.[73] The graphite Pa/U intercalation compounds simultaneously have covalent, metallic, and van der Waals bondings.[49] Their superposition leads to highly non-uniform and anisotropic carrier density distributions, as clearly shown in Figs. 6.3(a)–6.3(f). The strong cooperation between the experimental examinations and the VASP simulations might be able to solve this open issue. In addition, the high-resolution observations of scanning tunneling microscopy (STM)[10] is only suitable for 2D few-layer materials, but not for bulk graphite intercalation compounds because of the very weak quantum currents.[74]

6.2. Unusual Band Structures with Atom and Spin Dominances

Monolayer graphene,[6] graphite,[2] and graphite Pa/U intercalation compounds[42] present diverse electronic structures, in which they, respectively, exhibit the zero-gap,[75] semi-metallic,[76] and metallic behaviors near the Fermi level. A graphitic sheet possesses a planar honeycomb crystal and is able to create a well-defined Dirac cone because of the uniform strength of three nearest-neighboring $2p_z$-orbital hopping integrals. The linear and isotropic valence and conduction energy subbands just intersect at the K/K′ valleys (the corners of the hexagonal first Brillouin zone).[77] The vanishing density of states at E_F shows the unique behavior of semiconducting quasi-particles.[78] The π- and σ-electronic states, which are well characterized along the KMΓ and ΓMK paths, respectively, with band widths of 7.20 and 7.05 eV. Their subband anti-crossings are absent in the wave-vector dispersion because of the non-deformed crystal. The perpendicular feature remains the same after the interlayer van der Walls interactions and intercalant effects, mainly owing to the very stable honeycomb crystals or the strong π bondings.[79] Through the weak but significant interlayer $2p_z$–$2p_z$ orbital hybridizations, the AB-/ABC-/AA-stacked graphite systems display somewhat overlapping valence and conduction subbands along the KH paths (Fig. 6.2(a)). The simultaneous existences of valence holes and conductions electrons can greatly diversify the various quasi-particle properties, e.g., the stacking-, temperature-,[80] polarization-direction-,[81] and magnetic-field-dependent electron–hole pair excitations,[82] plasmon modes[83]/magneto-plasmons,[84] Landau subands,[85] optical

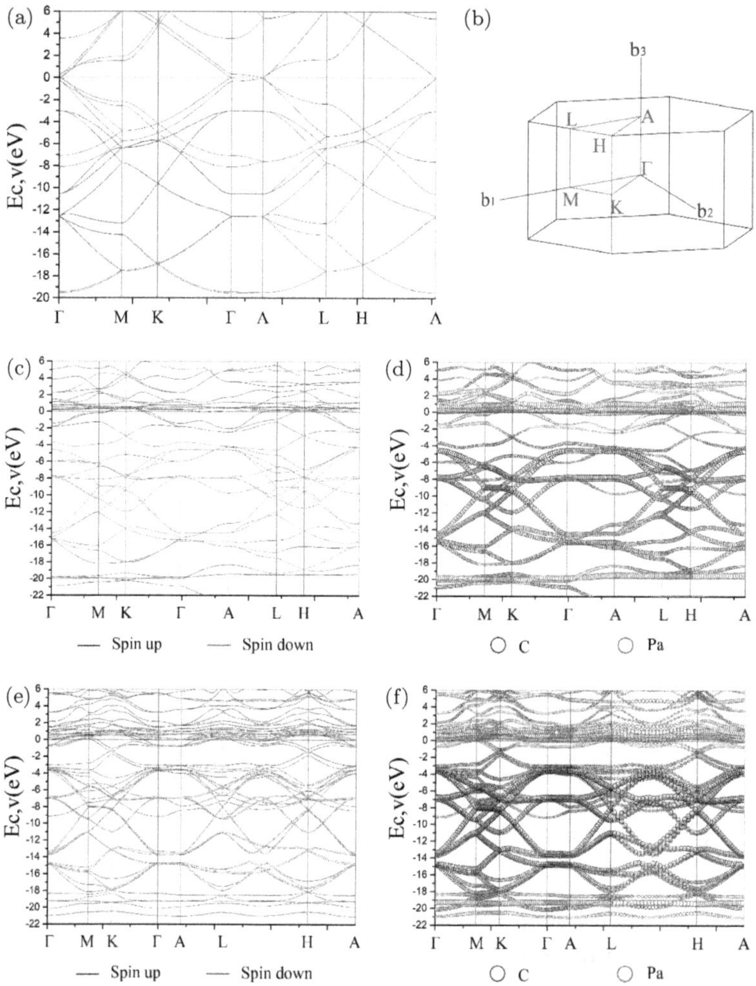

Fig. 6.2. The band structures with carbon/intercalant dominances for (a) pristine graphite (b) within the first Brillouin zone, the (c)/(d) stage-1, (e)/(f) stage-2, (g)/(h) stage-3, and (i)/(j) stage-4 graphite Pa compounds, and the (k)/(l) stage-1, (m)/(n) stage-2, (o)/(p) stage-3, and (q)/(r) stage-4 graphite U compounds.

absorption spectra,[86] and magneto-selection rules. Of course, the dimensionality crossover phenomenon between monolayer graphene and graphite should be a very interesting research topic for the modulation strategy.

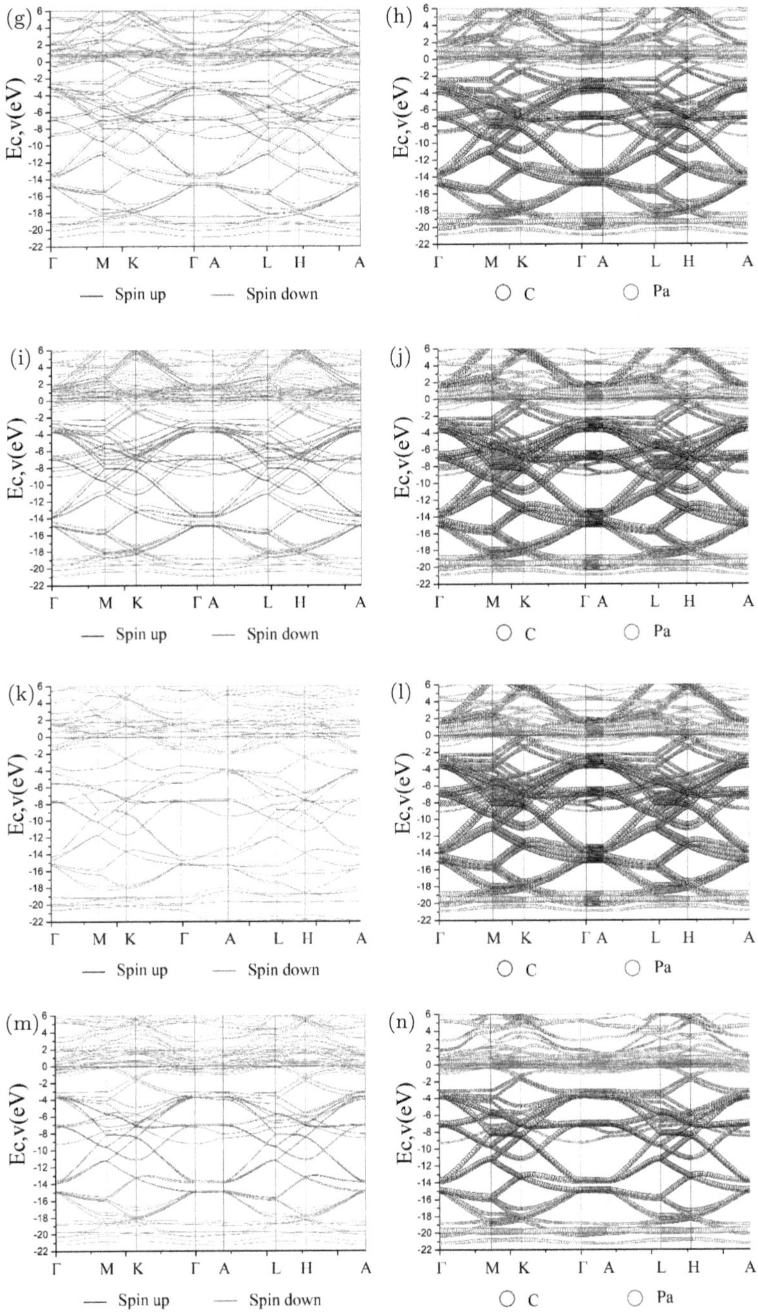

Fig. 6.2. (*Continued*)

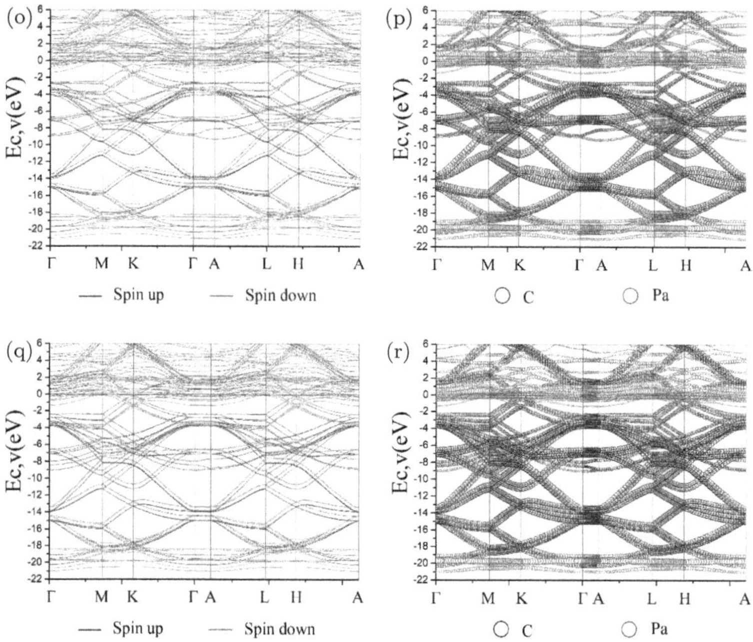

Fig. 6.2. (*Continued*)

The rare-earth metal intercalations[58] totally change the electronic energy spectra and wave functions, as clearly shown in Fig. 6.2: band structures with carbon/intercalant dominances for the (c)/(d) stage-1, (e)/(f) stage-2, (g)/(h) stage-3, and (i)/(j) stage-4 graphite Pa compounds and the (k)/(l) stage-1, (m)/(n) stage-2, (o)/(p) stage-3, and (q)/(r) stage-4 graphite U compounds. The diversified quasi-particle behaviors cover: (1) more energy subbands with an increase in n, mainly owing to an enlarged unit cell (the zone-folding effects),[87] (2) the greatly enhanced asymmetry of valence holes and conduction electrons about the Fermi level $E_F = 0$, (3) the blueshift of E_F due to the semimetal–metal transition,[88] (4) the reduced free carrier density with a decrease in intercalant concentrations (larger values of n),[89] (5) the modified Dirac cone structure at the K valley of $E^v =\sim -3.10$ eV (see also the density of states in Fig. 6.5),[90] (6) the π-band width along KMΓ and their partners due to the zone-folding effects,[87] (7) the well-characterized σ-electronic valence states initiated from the Γ point at $E^v = -4.10$ eV, (8) the spin split of certain subbands across E_F (the multi-spin-related Fermi surfaces),[91] (9) the Pa/U dominances within a specific energy range of $-3.0 < E^{c,v} < 6.0$ eV above the modified Dirac cone structure and $-22.0 < E^{c,v} < -19.0$ eV (the red balls),[30] (10) the carbon dominances in the π- and σ-electronic valence states over $-19.0 < E^{c,v} < -3.0$ eV (the blue balls),[92] and (11) a plenty of critical points in the energy–wave-vector space with different curvatures (various energy dispersions and van Hove singularities, as shown in Fig. 6.5).[47]

The featured quasi-particle properties directly reflect the total chemical bondings, as follows: (I) the intralayer σ and π bondings in a planar carbon-honeycomb crystal,[93] (II) the interlayer $2p_z$–$2p_z$ hybridizations (the semi-metallic van der Waals interactions),[49] (III) the interlayer metallic C–Pa/C–U bondings, and (IV) the intralayer metallic Pa–Pa/U–U ones. Such multi-orbital hybridizations are responsible for the energy-dependent atom dominances in the above-mentioned (9) and (10). The rare-earth metal intercalants can drive the ferromagnetic spin configurations and thus create ferromagnetic metals. Most of the spin-split free carriers are associated with the intercalation effects. It is well known that the Fermi surfaces lead to certain unusual quasi-particle phenomena, such as the long-range Friedel oscillations due to the Coulomb-field charged impurity (with a period of π/k_F[94]), the Fermi-momentum-induced electron–hole pair excitation boundaries,[82] the Fermi-energy-dominated plasmon modes,[26] the optical threshold frequencies,[95] and the Kohn anomaly of phonon spectra related to free carrier screening.[96] More interestingly, the multi-Fermi surfaces in coaxial carbon nanotubes in few-layer graphene systems have been predicted to exhibit the beating behavior with an unusual superposition of certain prominent Friedel oscillations.[97] As for the blueshift of the Fermi level, its value can be estimated from the modified Dirac points,[30] and the optical gaps.[25]

The high-resolution measurements of angle-resolved photoemission spectroscopy (ARPES)[98] are very powerful for observing the occupied quasi-particle energy spectra and lifetimes.[99] Such examinations have been successfully done on the 3D Bernal graphite,[100] few-layer graphene systems with different stacking configurations and layer numbers,[101] and 1D graphene nanoribbons,[102] but not on carbon nanotubes because of the non-straight deformations.[103] A graphite sample has prominent surface quantum confinement during many inelastic scatterings, leading to the specific non-conservation of the perpendicular momentum transfer. This issue could be solved by using the featured energy dispersions along the k_z direction. Similar behaviors arise in graphite intercalation compounds. How to overcome them depends on the strong cooperation between experimental[67] and theoretical researches.[104] The blueshift of E_{F} and the redshifts of the initial π and σ valence subbands become smaller with a decrease in the rare-earth metal atom concentration. These could be examined through the ARPES observations,[98] while the opposite is true for the complicated spin-split Fermi surfaces, according to the VASP simulations (Fig. 6.2).

6.3. Non-uniform Charge- and Spin-density Distributions

Both graphite and graphite rare-earth metal intercalation compounds display the featured charge density distributions. Interestingly, figs. 6.3(a)–6.3(f) clearly illustrate the coexistence of four kinds of chemical bondings. Each isolated carbon atom exhibits four active orbitals of 2s and (2p$_x$, 2p$_y$, 2p$_z$), in which the former and the latter, respectively, correspond to the inner red and the outer yellow/green regions. Similarly, an independent Pa/U atom reveals the spherical probability distributions of the two-6s and seven-5f orbitals. Their dramatic transformations occur after the formation of 3D layered graphites[105] and graphite Pa/U intercalation compounds. The 2D carbon-honeycomb crystals, which survive under a planar structure, should possess perpendicular σ and π bondings through the (2s, 2p$_x$, 2p$_y$) – (2s, 2p$_x$, 2p$_y$) and 2p$_z$–2p$_z$ orbital hybridizations (Fig. 6.3(a)) [Refs]. Any two neighboring carbon atoms have quite strong interactions, leading to the dog-bond-like red regions, as seen from the top view, and the most stable crystal in the world. Along the z direction, the weak but significant

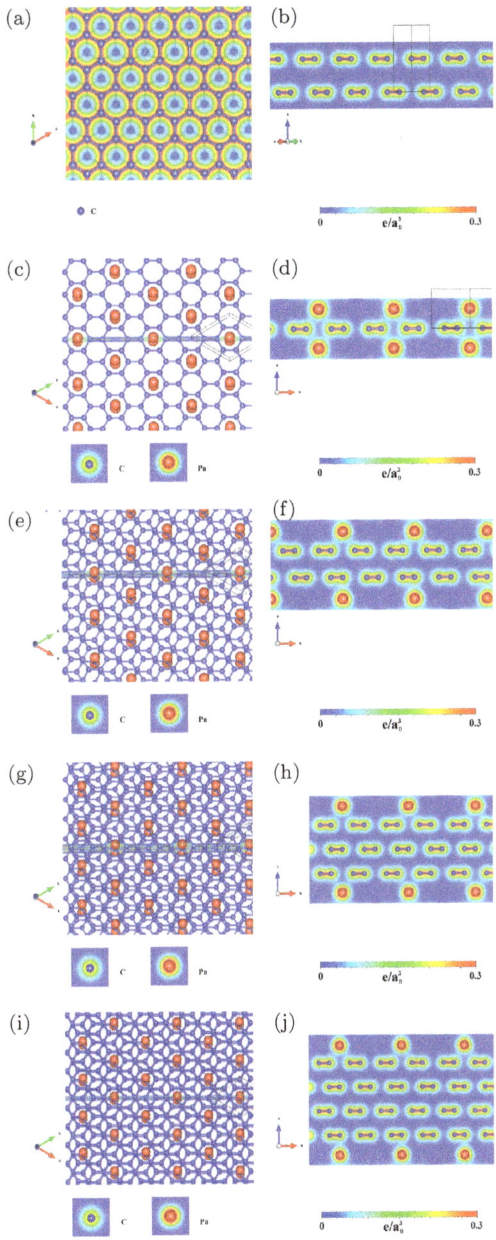

Fig. 6.3. The top/side views of the spatial charge density distributions for (a)/(b) Bernal graphite and the (c)/(d) stage-1 and (e)/(f) stage-2, (g)/(h) stage-3, and (i)/(j) stage-4 graphite Pa compounds. Also shown for comparison are those of isolated carbon and Pa atoms.

interlayer van der Waals interactions of $2p_z$ orbitals are responsible for the attractions of neighboring graphitic sheets and the semi-metallic behavior. Of course, the position-dependent carbon atom interactions are very sensitive to the stacking configurations (AA,[5] AB,[3] or ABC[4] stackings). Such featured bondings also appear in the $n \geq 2$ intercalation compounds (Figs. 6.3(e) and 6.3(f)). Most importantly, the rare-earth metal intercalations can present four kinds of chemical bondings on distinct planes simultaneously: (I) the intralayer σ and π orbital hybridizations on honeycomb lattices,[93] (II) the intralayer intercalant–intercalant metallic bondings of ($5f_{x^3}$, $5f_{xz^2}$, $5f_{yz^2}$, $5f_{xyz}$, $5f_{z(x^2-y^2)}$, $5f_{x(x^2-y^2)}$, $5f_{x(x^2-3y^2)}$) – ($5f_{x^3}$, $5f_{xz^2}$, $5f_{yz^2}$, $5f_{xyz}$, $5f_{z(x^2-y^2)}$, $5f_{x(x^2-y^2)}$, $5f_{x(x^2-3y^2)}$),[106] (III) the interlayer $2p_z$–$2p_z$ interactions between two neighboring graphene atoms, and (IV) the interlayer host–guest metallic orbital mixings about $2p_z$– ($5f_{x^3}$, $5f_{xz^2}$, $5f_{yz^2}$, $5f_{xyz}$, $5f_{z(x^2-y^2)}$, $5f_{x(x^2-y^2)}$, $5f_{x(x^2-3y^2)}$).[107] These concise pictures are consistent with a number of merged van Hove singularities (as discussed in Figs. 6.5(a)–6.5(p)).[47] The rich phenomena of chemical bondings are very helpful in establishing the tight-binding model[43] or the generalized tight-binding model[108] with suitable hopping integrals and on-site ionization energies. Whether the predicted charge density distributions could be examined through the measured diffraction peaks of elastic scatterings,[109] as done for ionic crystals, which require strong cooperation among the VASP simulations,[42] phenomenological models,[110] and experimental observations.[67]

The spatial spin-density arrangements along the top/side views are clearly illustrated in Fig. 6.4 under four intercalation cases: the (a)/(b) and (c)/(d) stage-1 and the (e)/(f) and (g)/(h) stage-2 graphite rare-earth metal Pa and U compounds. The intercalant Pa and U layers can generate rather strong ferromagnetic configurations through the giant red volumes, while the opposite is true for the neighboring carbon atoms through the greatly reduced blue ones. The prominent ferromagnetic spin configurations and negligible antiferromagnetic ones are fully consistent with the atom- and orbital-projected net magnetic moments (as detailed in Table 6.2). Also, note that Pa and U can even induce magnetic moments of

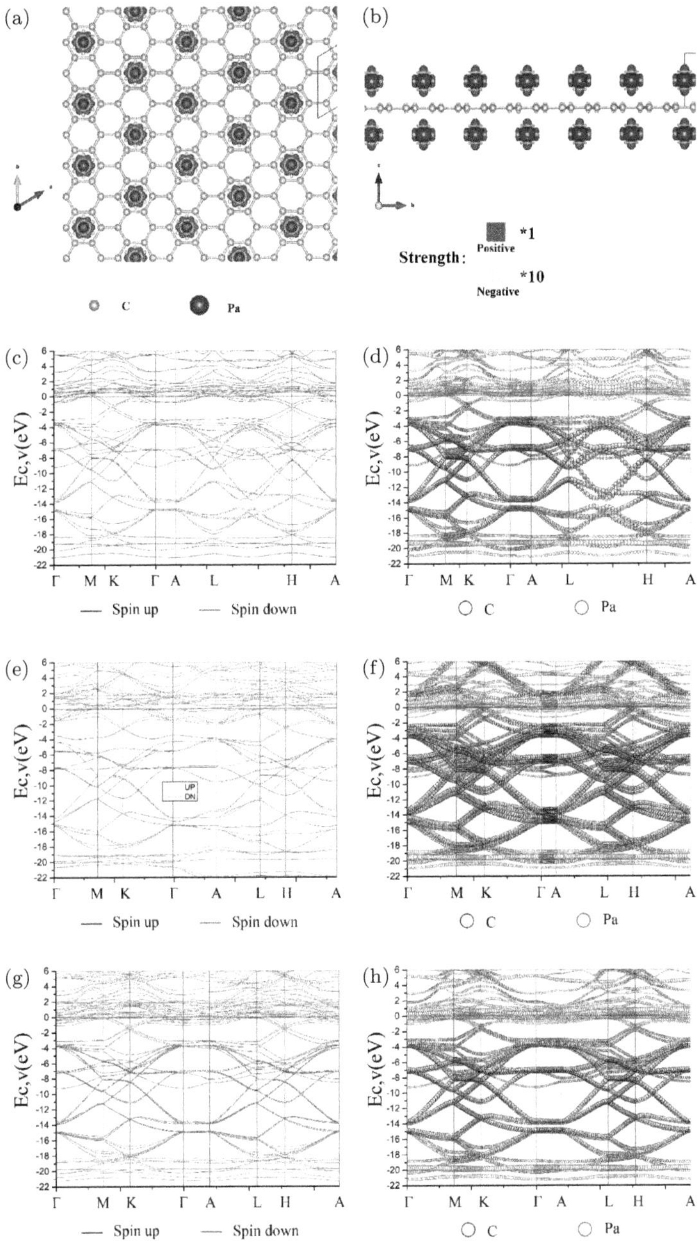

Fig. 6.4. The spin-charge distributions along the (a) top and (b) side views of stage-2 graphite rare-earth-metal compounds. The band energy spectra and the atom-dominated band structure for (c)/(d) State-I, (e)/(f) state-II, and (g)/(h) state-III Pa intercalate graphite.

Table 6.2. The atom- and orbital-decomposed magnetic moments for stage-1, stage-2, stage-3, and stage-4 graphite Pa and U compounds.

	Magnetic moment (μ_B)
PaC8	0.10
PaC16	0.20
PaC24	0.74
PaC32	1.71
UC8	2.58
UC16	2.98
UC24	3.28
UC32	3.46

0.064 and 2.345 μ_B per unit cell, respectively, regardless of C–Pa/C–U bond lengths (2.62/2,65 Å). The magnetic strength ratio between the host carbon atoms and the guest rare-earth metal ones is roughly smaller than 2%/0.6% under the Pa/U intercalation cases (the enhanced blue balls). Most of the carbon atoms display very weak spin-down distributions, except for the first and fourth ones. Under the spin-dependent phenomenological model,[110] only the ferromagnetic configurations of Pa/U intercalant atoms, but not the host carbon atoms, need to be taken into account through the Heisenberg-like Hamiltonian. Such interactions are quite different from the Hubbard-like on-site electron–electron Coulomb ones.[94] Obviously, the extremely magnetic configurations frequently appear after the rare-earth metal intercalations.[58] The metallic Pa/U–C bonds can greatly promote the carrier mobility,[111] while they cannot effectively mediate the ferromagnetic spin distributions. The above-mentioned information is very helpful in establishing the spin-based ferromagnetic Hamiltonians with reliable parameters and thus expanding the quasi-particle framework.[13] These unusual magnetic behaviors could be verified from the SQUID[112] or inelastic neutron scatterings[113] and the spin-polarized STM of the spin-related surface morphologies.[10] Both charge- and spin-dependent interactions are combined together to establish the reliable parameters of the tight-binding model, clearly indicating an expanded quasi-particle framework.

6.4. Atom-, Orbital-, and Spin-decomposed Densities of States

After the delicate calculations, the energy-dependent 3D/3D/1D density of states is defined by the constant-energy surface/loop/point integration of the inverse magnitude of group velocity, except for the 0D quantum dots[114] that exhibit the delta-function-like symmetric peaks with a broadening factor.[115] In general, the critical points, which occur in the energy–wave-vector space,[116] display abnormal group velocities,[117] present the van Hove singularities,[47] and account for the special structures under physical perturbations, e.g., the strong electron–hole excitations[69] and plasmon modes[83] through the external time-dependent Coulomb fields[118] and the prominent absorption peaks due to the joint van Hove singularities under electromagnetic-wave perturbations.[119] In general, it is relatively easy to obtain the analytic forms of the 2D/1D/0D singular density of states (e.g., the details of 1D van Hove singularities given in Chapter 10),[47] while the opposite is true for the 3D ones. For example, the 3D parabolic and partially flat energy dispersions, with isotropic features near the band-edge states, could exhibit the square-root shoulder structures[120] and the delta-function-like symmetric peaks,[115] respectively. The 2D, 1D, and 0D VASP predictions on van Hove singularities are directly examined through high-resolution STS measurements.[11] However, the quantum is negligible for 3D samples with enough widths. Their van Hove singularities only reflect the available channels during the external field perturbations (as detailed in Section 6.5]. As for 3D pristine graphites[121] and graphite rare-earth metal intercalation compounds,[58] the orbital- and atom-/orbital-/spin-decomposed density of states are available for thoroughly exploring the critical multi-orbital hybridizations[122] and the featured Fermi surfaces, respectively, through the merged van Hove singularities and the minimum value below $E_F = 0$ (blueshift).[47] Both of them are very sensitive to the stage-n crystal structures.[123]

The atom-, orbital-, and spin-decomposed densities of states can reveal the multi-orbital hybridizations of the four types of chemical

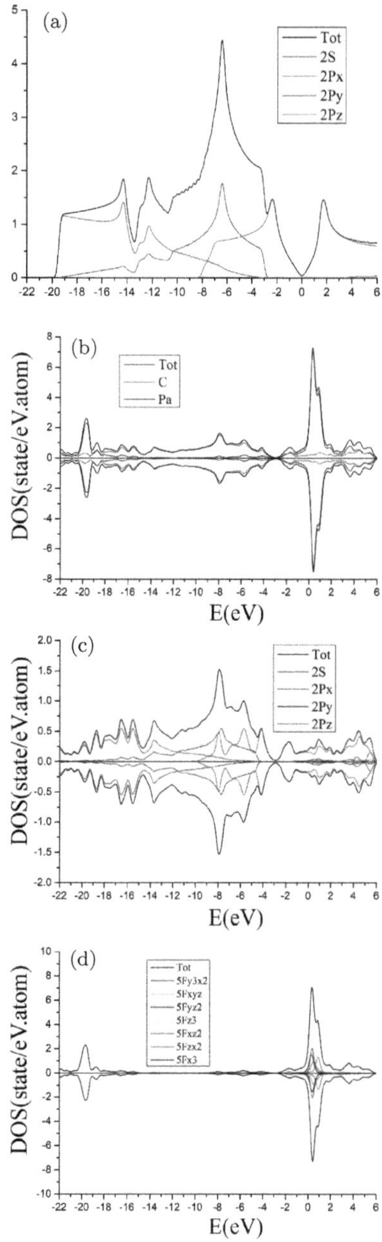

Fig. 6.5. The atom-, orbital-, and spin-decomposed density of states for (a) pristine graphite, the (b)/(c)/(d) stage-1, (e)/(f)/(g) stage-2, (h)/(i)/(j) stage-3, and (k)/(l)/(m) stage-4 graphite Pa compounds, and the (n)/(o)/(p) stage-1 graphite U compound.

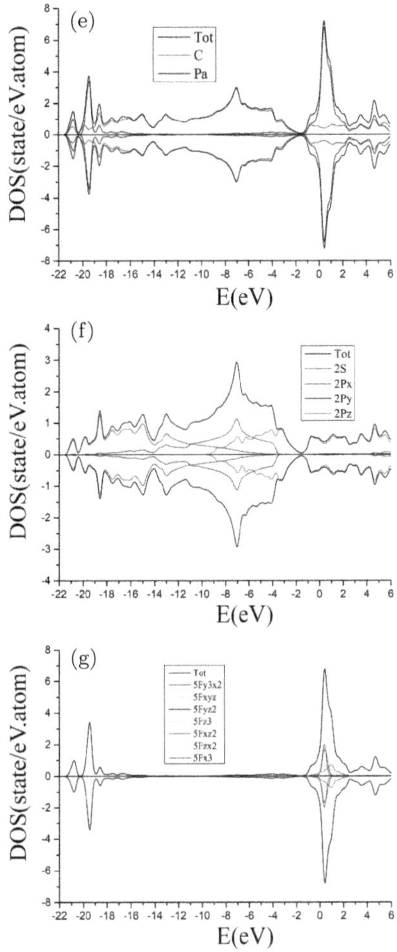

Fig. 6.5. (*Continued*)

bondings in graphite rare-earth metal intercalation compounds, as shown in Fig. 6.5 under (a) a pristine Bernal graphite, (b)/(c)/(d) stage-1, (e)/(f)/(g) stage-2, (h)/(i)/(j) stage-3, and (k)/(l)/(m) stage-4 Pa-intercalant systems, and (n)/(o)/(p) stage-1 U-guest-atom ones. The van Hove singularities in 3D graphite only behave as broadening peak structures. The π- and σ-electronic states, which, respectively, come from the intralayer $2p_z$–$2p_z$ (the red curve)[124] and ($2s$, $2p_x$, $2p_y$) – ($2s$, $2p_x$, $2p_y$) orbital hybridizations

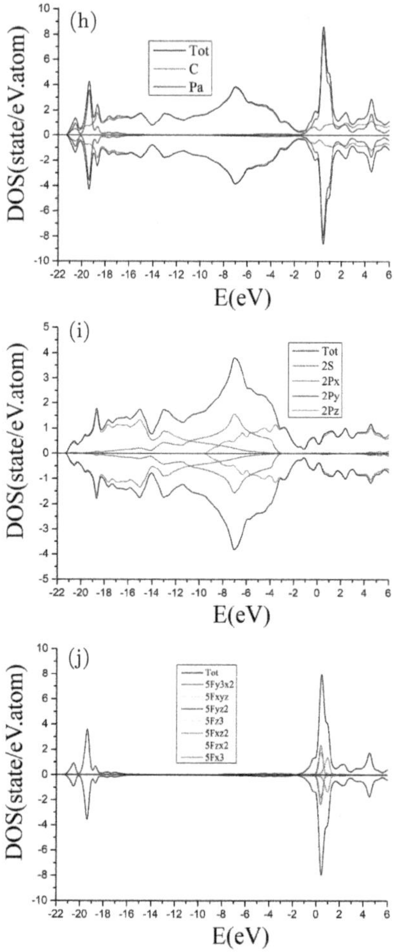

Fig. 6.5. (*Continued*)

(the red, green, and blue curves).[125] These two kinds of peaks cannot merge together, which clearly illustrate the perpendicular chemical bondings within a planar carbon-honeycomb crystal. At the Fermi level of $E_F = 0$, the density of states is finite but small under a weak band overlap of valence and conduction subbands (Fig. 6.2(a)). That is, the semi-metallic behavior is purely due to the interlayer van der Waals interactions (interlayer $2p_z$–$2p_z$ mixings).[124] The former and the latter are well characterized within

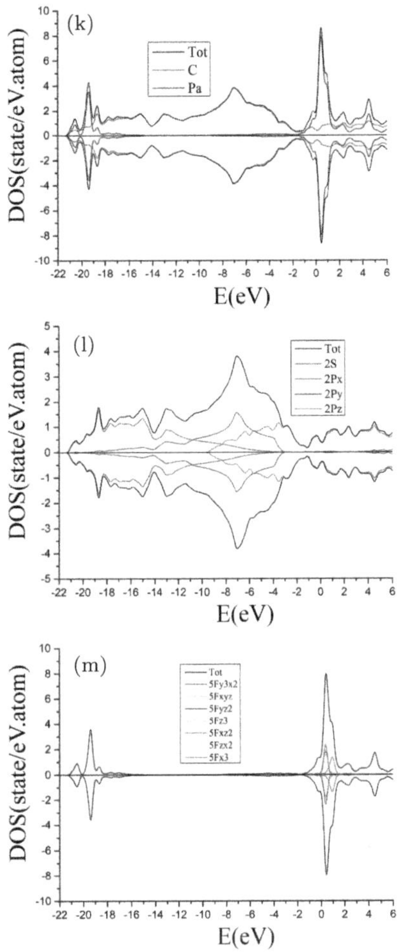

Fig. 6.5. (*Continued*)

$-3.0 < E^{c,v} < 3.0$ eV and deeper energies, respectively. In the rare-earth metal interactions, the minimum density of state due to the π valence subbands presents a redshift of $\sim$$-3.1$ eV. The strong C–Pt and Pt–Pt bonds dominate within $-3.1 < E^{c,v} < 6.0$ eV, covering part of the carbon-2p$_z$ contributions through the metallic and semi-metallic bonds, The seven-5f contributions of (f_{x^3}, f_{xz^2}, f_{yz^2}, f_{xyz}, $f_{z(x^2-y^2)}$, $f_{x(x^2-y^2)}$ and $f_{x(x^2-3y^2)}$) are comparable through the spin-split metallic ferromagnetism, in which their magnitudes

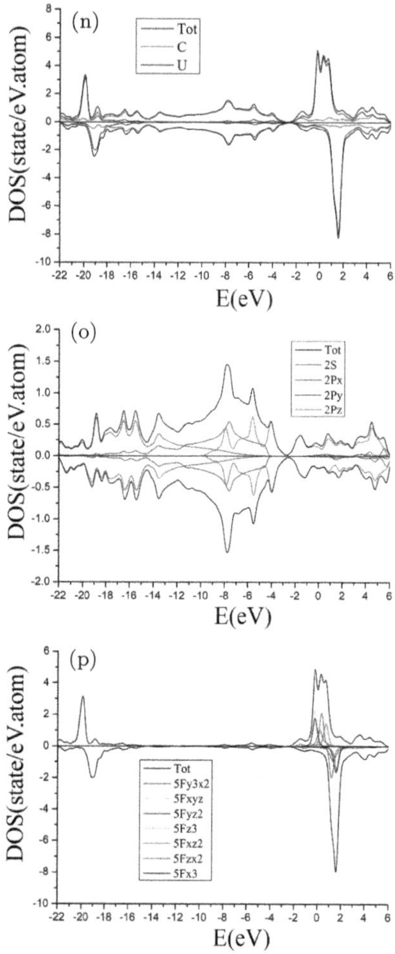

Fig. 6.5. (*Continued*)

cannot be well distinguished from the high-resolution STS/SP-STS [Refs]. The Pt-atom dominances also survive within $-22.0 < E^{c,v} < -19.0$ eV. On the other hand, the carbon dominances of the π- and σ-electronic valence states appear within $-22.0 < E^{c,v} < -3.10$ eV in the absence of spin splitting.[126] The above-mentioned van Hove singularities clearly illustrate four kinds of critical chemical bondings in graphite Pa/U intercalation compounds, which are consistent with those examined from the band structures (Fig. 6.2).[127] In addition, similar results are revealed in few-layer graphene intercalation compounds.[6]

The high-precision STS measurements are available for examining the strong relationships between the featured van Hove singularities and the multi-orbital hybridizations or ferromagnetic spin configurations. Such examinations have been very successful for the geometry-enriched π-electronic properties, covering few-layer graphene systems with different stacking configurations and layer numbers, 1D single-walled/coaxial carbon nanotubes with various radii and chiralities, and graphene nanoribbons in the presence of quantum-confinement effects[128] and open-edge structures.[129] The carbon honeycomb-related systems exhibit the metallic, semi-metallic, and finite-gap behaviors, which could be directly observed from a finite or zero density of state at $E_F = 0$. The π- and σ-electronic features could be observed within $-3.0 < E < 3.0$ eV and deeper energies. After the rare-earth-metal Pa/U intercalations,[58] this specific range undergoes a large redshift (larger than 3.0 eV) with enhanced metallic behavior of C–Pa/C–U and Pa–Pa/U–U bonds. The spin-polarized STS is urgently required for the spin-split ferromagnetic metals. Also, note that the minimum density of states, corresponding to a modified Dirac cone structure, arises at high valence energies with an increase in n (a decrease in intercalant concentration)[30] Interestingly, the concise pictures of chemical, physical, and material science could be achieved from the experimental and theoretical studies.[104] However, achieving simultaneous progress among the VASP simulations,[42] experimental observations,[67] and phenomenological models,[110] has become a tough challenge since it is almost impossible for the proper establishment of the 14-5f orbital- and spin-dependent Hamiltonians.[130]

6.5. Unusual Dielectric Functions, Energy Loss Spectra, Reflectances, and Absorption Coefficients

The vertical optical transitions, which come from the physical perturbations of electromagnetic waves,[131] can exhibit unusual quasi-particle semi-metallic phenomena in graphite materials and graphite Pa/U compounds, mainly owing to the semi-metallic[56] and ferromagnetic-metallic[132] behaviors, respectively. The incident photons carry energies but are almost without momenta, so that the initial and final states possess the same wave vectors. Their joint density of states and electric dipole matrix elements, respectively, determine the available excitation channels and the scattering absorption probabilities. When the former is associated with band-edge states, the prominent absorption structures come to exist under a finitely induced dipole moment. The semi-metallic systems and ferromagnetic metals are able to attain the orbital- and atom-/orbital-spin-projected electronic states, which are useful for developing composite quasi-particles (a grander framework).[13] Each absorption structure would be attributed to the atom, orbital, and spin dominances of the initial and final states under delicate analyses. Furthermore, it might be associated with the significant superposition of different chemical bondings under the same spin configurations. These behaviors have never been reported in previously published works. Whether the optical threshold frequencies of stage-n donor-type metals could be estimated using the value of the Fermi-level blueshift is worthy of detailed discussion. The single-particle electron–hole excitations and collective excitation modes,[69] which, respectively, are revealed in the dielectric functions[133] and energy loss spectra, display the diversified quasi-particles. There exist certain important differences between the π- and σ-electronic plasmons,[26] mainly owing to their perpendicular features. Their close relations with reflectance spectra and absorption coefficients could be verified from the high-precision optical measurements.[134] In addition, the many-body excitons cannot survive because the screening effects are too strong to present the attractive Coulomb potentials between the excited valence holes and conduction electrons.

Figure 6.6 clearly shows the VASP simulations of the imaginary- and real-part dielectric functions (the solid and dashed curves) / energy loss spectra / reflectances / absorption coefficients for (a)/(b)/(c)/(d) pristine graphite, the (e)/(f)/(g)/(h) stage-1, (i)/(j)/(k)/(l) stage-2, (m)/(n)/(o)/p) stage-3, (q)/(r)/(s)/(t), and stage-4 graphite Pa compounds, and the (u)/(v)/(w)/(x) stage-1 graphite U

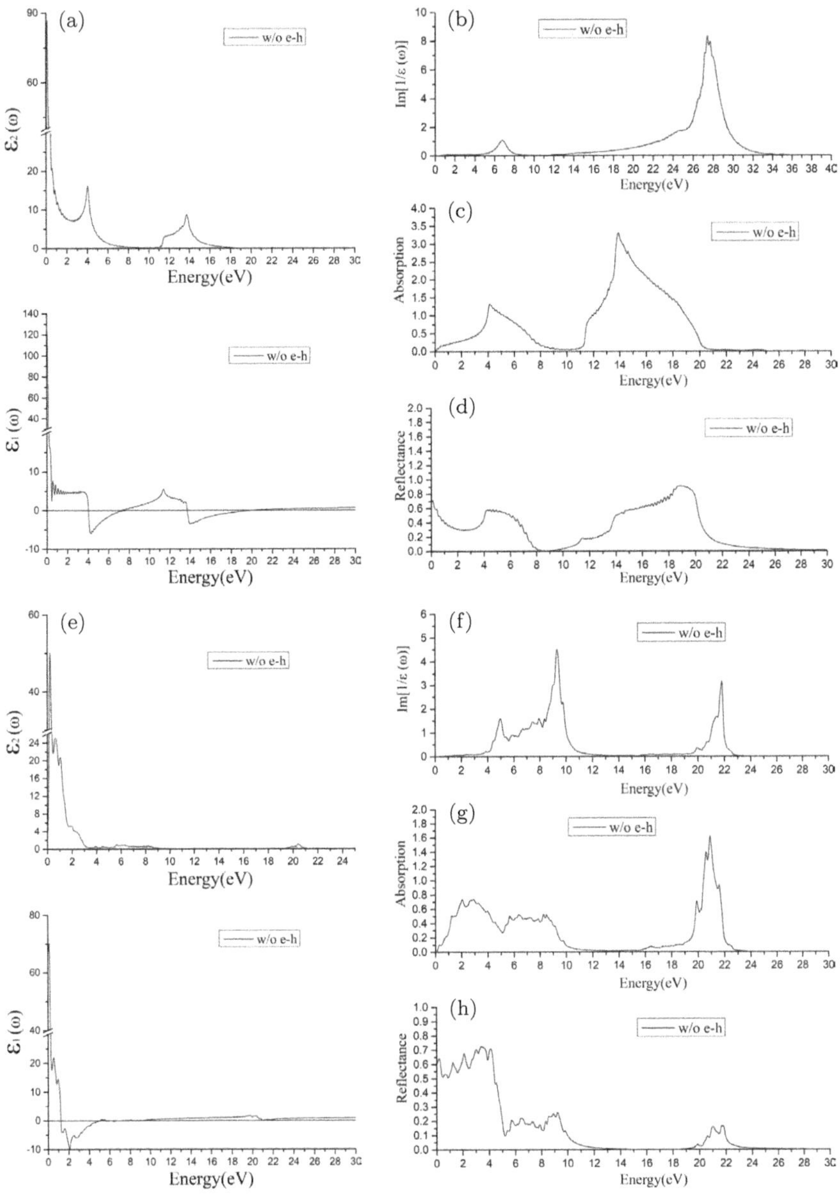

Fig. 6.6. The dielectric functions/energy loss spectra/ absorption / reflectances coefficients for (a)/(b)/(c)/(d) a pristine graphite, the (e)/(f)/(g)/(h) stage-1, (i)/(j)/(k)/(l) stage-2, (m)/(n)/(o)/p) stage-3, (q)/(r)/(s)/(t) stage-4 graphite Pa compounds, and the (u)/(v)/(w)/(x) stage-1 graphite U compound.

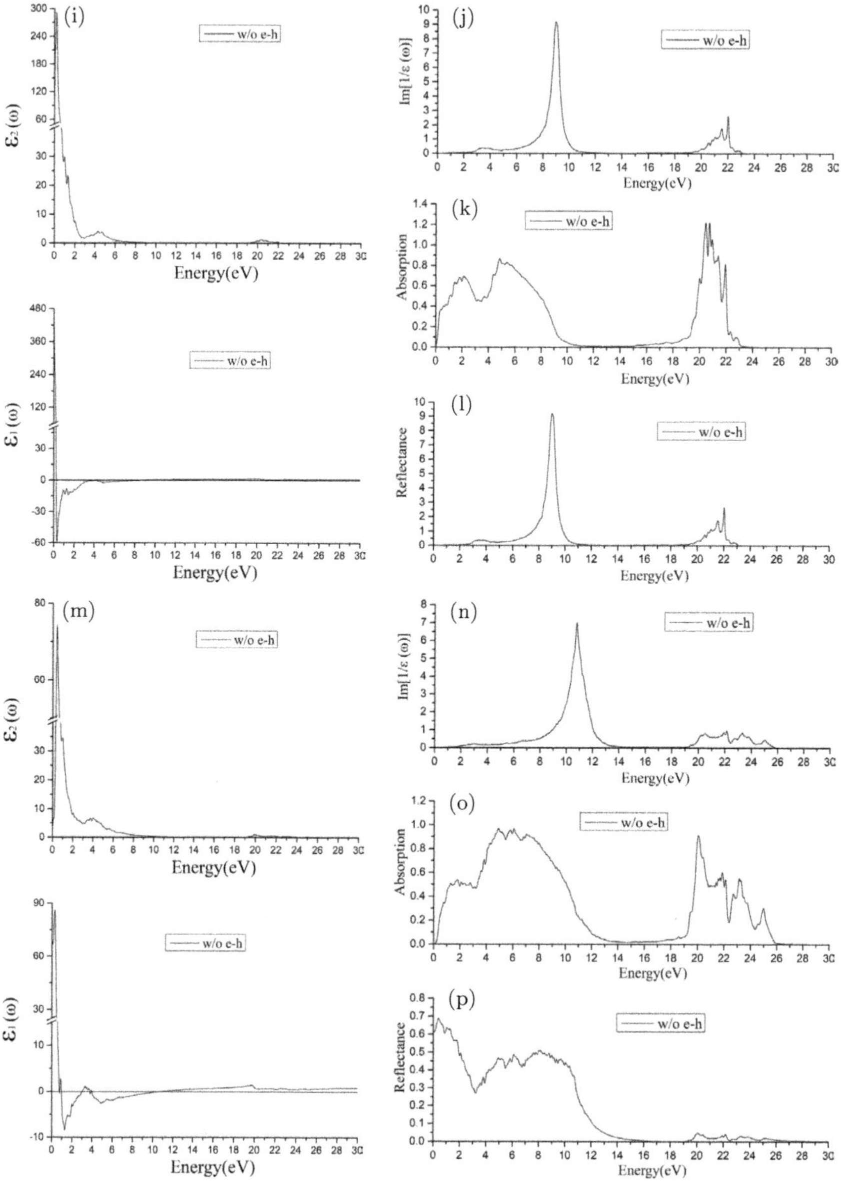

Fig. 6.6. (*Continued*)

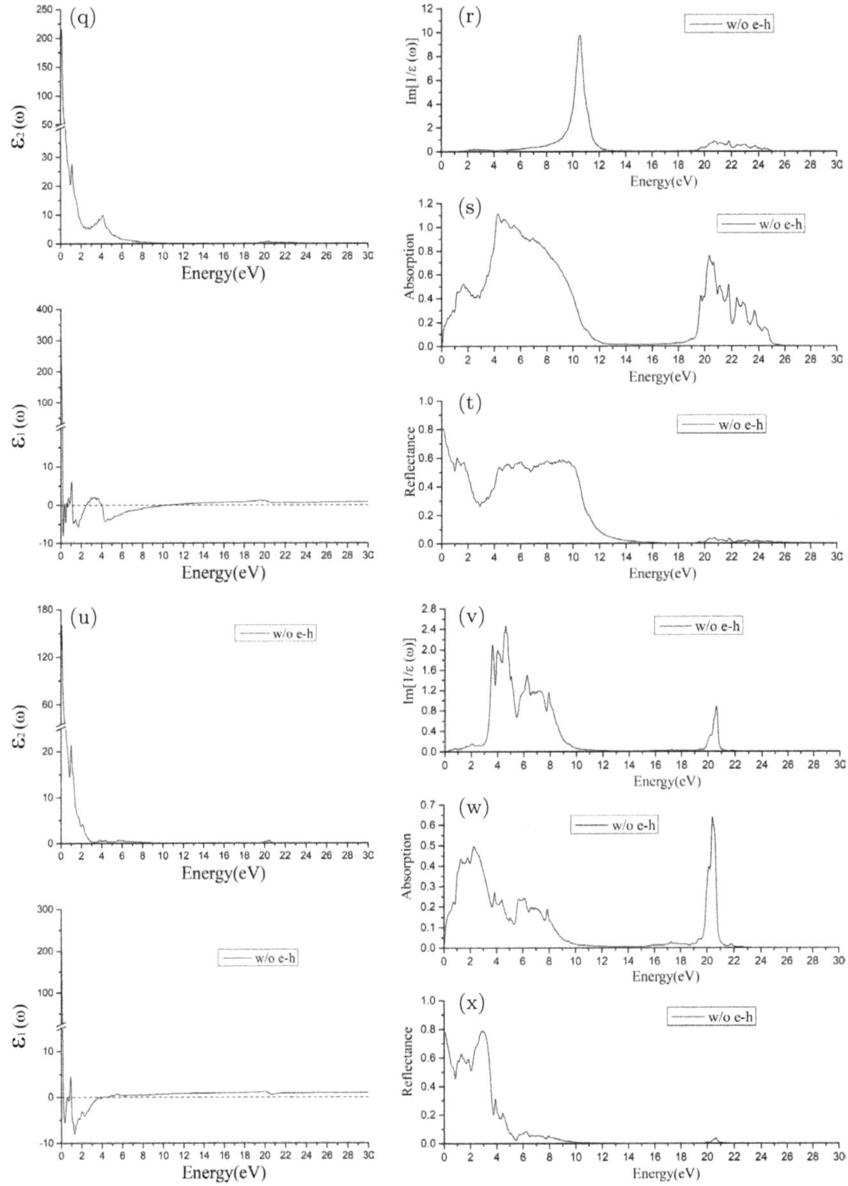

Fig. 6.6. (*Continued*)

one. Bernal graphite presents very low densities of free valence holes and conduction electrons through the semi-metallic van der Waals interactions (as shown earlier in Fig. 6.2(a)).[135] The band-overlap-created carriers can create the polarization- and temperature-dependent optical excitations. This work is focused on the planar electric fields at zero temperature because of the strong anisotropy and the vanishing of the complicated electron–phonon scatterings. The low-frequency excitation peak appears at ~0.01 eV, as indicated by the solid curve in Fig. 6.6(a). Most of the π-electronic states, which are dominated by intralayer $2p_z$–$2p_z$ bondings, are associated with the saddle point (the M point). These very prominent van Hove singularities are able to induce very strong charge screenings in the bare (Fig. 6.6(a)) and screened response functions (Fig. 6.6(b)). The π-electronic single-particle and collective excitations, respectively, occur at ~5.00 and 7.00 eV. These two kinds of electronic excitations in any condensed-matter system will accompany each other in their related phase diagrams.[136] As for $\sigma + \pi$ valence electrons, which possess higher carrier density and deeper energy, they contribute significantly to electron–hole pair excitations and plasmon modes appearing at 14.0 eV and 24.0 eV, respectively. In short, a 3D semi-metallic graphite clearly illustrates the low-frequency π and $\sigma + \pi$ plasmons at distinct frequency ranges.

There are a number of high-resolution reflectance measurements made for graphite[137] and graphite intercalation compounds.[66] The frequency-dependent bare[138] and screened[139] response functions can be achieved with the full support of the Kramers–Kronig relations.[140] Both Taft and Philipp have successfully obtained the frequency-dependent reflectance spectra with a range of up to 26 eV in 1965, covering the low-, middle-, and high-frequency charge screenings of carbon-(2s, $2p_x$ $2p_y$, $2p_z$) orbitals. Through delicate analyses, the estimated values of dominating absorption structures in the imaginary-part dielectric function and energy loss function, corresponding to free conduction electrons and valence holes,[141] π valence electrons,[142] and $\pi + \sigma$ ones[143] are ~0.02, ~5.00, and ~15 eV and ~0.44, ~7.00, and ~25 eV, respectively. [PRB 16. 2896(1977)] The diversified electronic excitations are consistent with those obtained in the current work (Figs. 6.6(a)–6.6(d)). Their main features also directly reflect the Coulomb-field perturbations[144] through the momentum- and frequency-dependent excitation phase diagrams. More interestingly, the free-carrier Coulomb excitations are very sensitive to the direction and magnitude of momentum transfer, as clearly verified through high-precision measurements of reflection electron energy loss spectroscopy (REELS).[145] As to donor- or acceptor-type graphite intercalation compounds, their optical reflectance measurements are frequently assisted by the rigid-band approximations.[146] This estimates that the red- or blueshift of the Fermi level is half of the threshold absorption frequency. However, the graphite rare-earth metal compounds present a modified Dirac cone structure at ~ -3.10 eV (Figs. 6.2(c) and 6.2(d)), so that such a viewpoint becomes meaningless. Their optical properties are strongly affected by the intralayer C–C covalent bondings, the interlayer semi-metallic van der Waals interactions, and the metallic C–Pa/C–U and Pa–Pa/U–U ones. The experimental examinations shown in Figs.6(e)–6(x), especially for the low-frequency composite quasi-particles, are very important in developing a wider framework.

6.6. Summary

The delicate VASP simulations and analyses are successful in thoroughly exploring the chemical intercalation effects.[147] The rare-earth metal graphite intercalation compounds have clearly illustrated the diverse quasi-particle phenomena, which can be utilized to realize a grand framework. A 2D monolayer graphene and a 3D layered graphite, respectively, exhibit the zero-gap semiconducting[148] and semi-metallic behaviors,[56] directly reflecting the hexagonal honeycomb symmetry[79] and the weak but important van der Waals interactions.[49] The latter are very sensitive to AB, ABC, and AA stacking configurations.[3–5] The σ and π bondings of (2s, $2p_x$, $2p_y$) – (2s, $2p_x$, $2p_y$) and $2p_z$–$2p_z$ are perpendicular to each other on a planar crystal. This property almost remains unchanged even in the presence of intercalant intercalations. Some of the van der Waals interactions are replaced by the interlayer host–guest and intralayer intercalant–intercalant metallic bondings for the $n \geq 2$ modifications, except for the total absence in the $n = 1$ case. Five kinds of chemical bondings will coexist in honeycomb lattices, the intercalant ones, and the interlayer spacings close to the host and guest atoms. The layered crystals of graphite/graphite Pa/U intercalation compounds are responsible for all charge- and spin-enriched quasi-particles. The concise pictures of chemistry and physics are achieved from the featured VASP results, including the layered crystal symmetries before and after the Pa/U intercalations,[149] the ferromagnetic metals with more energy subbands and band-edge states (the superposition of zone-folding and intercalation effects),[87] the atom- and orbital-dominated magnetic moments, the Pa-/U-initiated spin arrangement modulations,[150] the atom-, orbital- and spin-projected van Hove singularities, and the spin-related absorption peaks and optical gaps in the absence of excitonic effects.

More interestingly, the rich excitation quasi-particles are well characterized by the complicated superposition of atom-, orbital-, and spin-dependent initial and final subband states, that is, such redefined electron–photon couplings represent an enlarged quasi-particle framework (as similarly concluded in Chapter 10). The first-principles simulations are available for deriving the reliable parameters of the tight-binding model, e.g., the complicated orbital-dependent hopping integrals from the metallic bonds of C–Pa/C–U of $2p_z$– ($5f_{x^3}$, $5f_{xz^2}$, $5f_{yz^2}$, $5f_{xyz}$, $5f_{z(x^2-y^2)}$, $5f_{x(x^2-y^2)}$ and $5f_{x(x^2-3y^2)}$) and the metallic Pa–Pa/U–U bonds of ($5f_{x^3}$, $5f_{xz^2}$, $5f_{yz^2}$, $5f_{xyz}$, $5f_{z(x^2-y^2)}$, $5f_{x(x^2-y^2)}$, $5f_{x(x^2-3y^2)}$) – ($5f_{x^3}$, $5f_{xz^2}$, $5f_{yz^2}$, $5f_{xyz}$, $5f_{z(x^2-y^2)}$, $5f_{x(x^2-y^2)}$, $5f_{x(x^2-3y^2)}$).

The first-principles predictions on the rare-earth metal intercalation effects could be verified using high-precision experimental observations. The X-ray diffractions[151] and RLEED[72] are powerful in examining the planar stage-n crystal structures, in which the carbon honeycomb lattices and the intercalant layers present the periodic distances. The measured results, which are fully supported by the VASP simulations, might be useful in estimating the spatial charge density distributions,[152] as done for ionic crystals. Both STM and STS become meaningless in determining the geometry-enriched band properties across the Fermi level since the quantum tunneling currents are almost negligible for sufficiently thick 3D samples. The spin-polarized ARPES is able to observe the spin-split occupied energy subbands below the Fermi level[22] and the quasi-particle lifetimes[153] only through very strong cooperation with the numerical simulations, such as more energy dispersions and critical points due to the zone-folding and intercalation effects.[87] The spin-up and spin-down Fermi surfaces are well separated in an observable window, clearly indicating their cooperation or competition partnerships. For example, this will lead to the beating behaviors of the long-range Friedel oscillations in the presence of static Coulomb impurity screenings[154] or very complicated Coulomb decay rates.[155] The optical reflectance,[156] absorption,[157] and transmission[158] spectroscopies provide sufficient information about the frequency-dependent dielectric functions, screened response functions, and reflectance spectra/absorption coefficients in the absence of excitonic effects, respectively.[48] The blueshift of the Fermi level can be estimated from the energy spacing between the negative energy of the minimum density of states and $E_F = 0$. In short, the simultaneous progress among the first-principles simulations,[159] the phenomenological models,[110] and the experimental observations is clearly illustrated by the rare-earth metal chemical intercalations.[58] This strategy is worthwhile for developing the binary/ternary layered semiconductor compounds, e.g., GaS,[160] GaSe,[161] and GaTe[162] and their In-related counterparts, and their electromagnetic-wave[119] and Coulomb-field[144] composite quasi-particles.[13]

References

[1] Bachmair, F.; *et al.* A 3D diamond detector for particle tracking. *Nuclear Instruments and Methods in Physics Research Section A: Accelerators, Spectrometers, Detectors and Associated Equipment* **2015**, *786*, 97–104.

[2] Kelly, B. T. Physics of graphite. 1981.

[3] Jeong, H.-K.; *et al.* Evidence of graphitic AB stacking order of graphite oxides. *Journal of the American Chemical Society* **2008**, *130* (4), 1362–1366.

[4] Charlier, J.-C., Gonze, X.; Michenaud, J.-P. First-principles study of the stacking effect on the electronic properties of graphite (s). *Carbon* **1994**, *32* (2), 289–299.
[5] Chiu, C. W.; *et al.* Absorption spectra of AA-stacked graphite. *New Journal of Physics* **2010**, *12* (8), 083060.
[6] Gunes, F.; *et al.* Layer-by-layer doping of few-layer graphene film. *ACS Nano* **2010**, *4* (8), 4595–4600.
[7] Xu, Q.; *et al.* Function-driven engineering of 1D carbon nanotubes and 0D carbon dots: Mechanism, properties and applications. *Nanoscale* **2019**, *11* (4), 1475–1504.
[8] Cai, J.; *et al.* Graphene nanoribbon heterojunctions. *Nature Nanotechnology* **2014**, *9* (11), 896–900.
[9] Meunier, V.; Lambin, P.; Lucas, A. A. Atomic and electronic structures of large and small carbon tori. *Physical Review B* **1998**, *57* (23), 14886.
[10] Tersoff, J.; Hamann, D. R. Theory of the scanning tunneling microscope. *Physical Review B* **1985**, *31* (2), 805.
[11] Feenstra, R. M. Scanning tunneling spectroscopy. *Surface Science* **1994**, *299*, 965–979.
[12] Joly, V. L. J.; *et al.* Observation of magnetic edge state in graphene nanoribbons. *Physical Review B* **2010**, *81* (24), 245428.
[13] Jorio, A.; *et al. Raman Spectroscopy in Graphene Related Systems.* John Wiley & Sons, 2011.
[14] Sethuraman, V. A.; *et al.* Surface structural disordering in graphite upon lithium intercalation/deintercalation. *Journal of Power Sources* **2010**, *195* (11), 3655–3660.
[15] Giannozzi, P.; Car, R.; Scoles, G. Oxygen adsorption on graphite and nanotubes. *The Journal of Chemical Physics* **2003**, *118* (3), 1003–1006.
[16] Dos Santos, M. C.; Alvarez, F. Nitrogen substitution of carbon in graphite: Structure evolution toward molecular forms. *Physical Review B* **1998**, *58* (20), 13918.
[17] Kakaei, K. One-pot electrochemical synthesis of graphene by the exfoliation of graphite powder in sodium dodecyl sulfate and its decoration with platinum nanoparticles for methanol oxidation. *Carbon* **2013**, *51*, 195–201.
[18] Zhang, W.; *et al.* Supramolecular linear heterojunction composed of graphite-like semiconducting nanotubular segments. *Science* **2011**, *334* (6054), 340–343.
[19] Yin, L.-J.; *et al.* Landau quantization in graphene monolayer, Bernal bilayer, and Bernal trilayer on graphite surface. *Physical Review B* **2015**, *91* (11), 115405.

[20] Lin, C.-Y.; *et al. Electronic and Optical Properties of Graphite-related Systems*. CRC Press, 2017.

[21] Yazami, R.; Touzain, P. A reversible graphite-lithium negative electrode for electrochemical generators. *Journal of Power Sources* **1983**, *9* (3), 365–371.

[22] Gerischer, H. An interpretation of the double layer capacity of graphite electrodes in relation to the density of states at the Fermi level. *The Journal of Physical Chemistry* **1985**, *89* (20), 4249–4251.

[23] Shyu, F.-L.; Lin, M.-F. Plasmons and optical properties of semimetal graphite. *Journal of the Physical Society of Japan* **2000**, *69* (12), 3781–3784.

[24] Tang, X.; *et al.* Robust above-room-temperature ferromagnetism in few-layer antimonene triggered by nonmagnetic adatoms. *Advanced Functional Materials* **2019**, *29* (15), 1808746.

[25] Huang, H.; *et al.* Oxygen density dependent band gap of reduced graphene oxide. *Journal of Applied Physics* **2012**, *111* (5), 054317.

[26] Smolyaninov, I. I.; *et al.* Single-photon tunneling via localized surface plasmons. *Physical Review Letters* **2002**, *88* (18), 187402.

[27] Ho, Y.-H.; *et al.* Magneto-optical selection rules in bilayer Bernal graphene. *ACS Nano* **2010**, *4* (3), 1465–1472.

[28] Aoki, H.; Ando, T. Effect of localization on the Hall conductivity in the two-dimensional system in strong magnetic fields. *Solid State Communications* **1981**, *38* (11), 1079–1082.

[29] Mineev, V. P.; Samokhin, K.; Landau, L. D. *Introduction to Unconventional Superconductivity*. CRC Press, 1999.

[30] Ochiai, T.; Onoda, M. Photonic analog of graphene model and its extension: Dirac cone, symmetry, and edge states. *Physical Review B* **2009**, *80* (15), 155103.

[31] Perri, J. A., Binder, I.; Post, B. Rare Earth "Disilicides". *The Journal of Physical Chemistry* **1959**, *63* (4), 616–619.

[32] Brooks, M. S. S.; Kelly, P. J. Large orbital-moment contribution to 5 f band magnetism. *Physical Review Letters* **1983**, *51* (18), 1708.

[33] Rupich, M. W.; *et al.* Metalorganic deposition of YBCO films for second-generation high-temperature superconductor wires. *MRS Bulletin* **2004**, *29* (8), 572–578.

[34] Honda, F.; *et al.* Novel electronic states of heavy fermion compound YbCo2Zn20. *Journal of the Physical Society of Japan* **2014**, *83* (4), 044703.

[35] Li, F.; Su, Y.; Zhao, J. Shuttle inhibition by chemical adsorption of lithium polysulfides in B and N co-doped graphene for Li–S batteries. *Physical Chemistry Chemical Physics* **2016**, *18* (36), 25241–25248.

[36] Feng, J.-W.; *et al.* Gas adsorption on silicene: A theoretical study. *Computational Materials Science* **2014**, *87*, 218–226.
[37] Xia, W.; *et al.* A first-principles study of gas adsorption on germanene. *Physical Chemistry Chemical Physics* **2014**, *16* (41), 22495–22498.
[38] Nguyen, D. K.; *et al.* Rich essential properties of Si-doped graphene. In *Green Energy Materials Handbook*. CRC Press, 2019, pp. 83–107.
[39] Sharma, M.; Sharma, R. Plumbene: A next generation hydrogen storage medium. *International Journal of Hydrogen Energy* **2021**, *46* (66), 33197–33205.
[40] Datsyuk, V.; *et al.* Chemical oxidation of multiwalled carbon nanotubes. *Carbon* **2008**, *46* (6), 833–840.
[41] Heydarpour, Y.; Aghdam, M. M.; Malekzadeh, P. Free vibration analysis of rotating functionally graded carbon nanotube-reinforced composite truncated conical shells. *Composite Structures* **2014**, *117*, 187–200.
[42] Hafner, J. Ab-initio simulations of materials using VASP: Density-functional theory and beyond. *Journal of Computational Chemistry* **2008**, *29* (13), 2044–2078.
[43] Sutton, A. P.; *et al.* The tight-binding bond model. *Journal of Physics C: Solid State Physics* **1988**, *21* (1), 35.
[44] Touhara, H. K. Y. N.; *et al.* On the structure of graphite fluoride. *Zeitschrift für anorganische und allgemeine Chemie* **1987**, *544* (1), 7–20.
[45] Liu, W.; *et al.* High precision measurements of the ground state hyperfine structure interval of muonium and of the muon magnetic moment. *Physical Review Letters* **1999**, *82* (4), 711.
[46] Karpan, V. M.; *et al.* Graphite and graphene as perfect spin filters. *Physical Review Letters* **2007**, *99* (17), 176602.
[47] Li, G.; *et al.* Observation of Van Hove singularities in twisted graphene layers. *Nature Physics* **2010**, *6* (2), 109–113.
[48] Mele, E. J..; Ritsko, J. J. Fermi-level lowering and the core exciton spectrum of intercalated graphite. *Physical Review Letters* **1979**, *43* (1), 68.
[49] Zaremba, E.; Kohn, W. Van der Waals interaction between an atom and a solid surface. *Physical Review B* **1976**, *13* (6), 2270.
[50] Dresselhaus, M. S.; Dresselhaus, G. Light scattering in graphite intercalation compounds. In *Light Scattering in Solids III*. 1982, pp. 3–57.
[51] Kresse, G.; Furthmuller, J. *Physical Review B – Condensed Matter and Materials Physics*, **1996**, *54*, 11169.

[52] Kresse, G.; Joubert, D. *Physical Review B – Condensed Matter and Materials Physics*, **1999**, *59*, 1758.
[53] Perdew, J. P.; Burke, K.; Ernzerhof, M. *Physical Review Letters*, **1996**, *77*, 3865.
[54] Blöchl, P. E. *Physical Review B – Condensed Matter and Materials Physics* **1994**, *50*, 17953.
[55] Yürüm, Y.; Taralp, A.; Nejat Veziroglu, T. Storage of hydrogen in nanostructured carbon materials. *International Journal of Hydrogen Energy* **2009**, *34* (9), 3784–3798.
[56] Bubnova, O.; *et al.* Semi-metallic polymers. *Nature Materials* **2014**, *13* (2), 190–194.
[57] Taft, E. A.; Philipp, H. R. Optical properties of graphite. *Physical Review* **1965**, *138* (1A), A197.
[58] Petiau, J.; *et al.* Delocalized versus localized unoccupied 5f states and the uranium site structure in uranium oxides and glasses probed by x-ray-absorption near-edge structure. *Physical Review B* **1986**, *34* (10), 7350.
[59] El Hachimi, A. G.; *et al.* First-principles prediction of the magnetism of 4f rare-earth-metal-doped wurtzite zinc oxide. *Journal of Rare Earths* **2014**, *32* (8), 715–721.
[60] Khomskii, D. *Transition Metal Compounds.* Cambridge University Press, 2014.
[61] Nijamudheen, A.; *et al.* Impact of surface modification on the lithium, sodium, and potassium intercalation efficiency and capacity of few-layer graphene electrodes. *ACS Applied Materials & Interfaces* **2020**, *12* (17), 19393–19401.
[62] Weitz, I. S.; Rabinovitz, M. The application of C_8K for organic synthesis: reduction of substituted naphthalenes. *Journal of the Chemical Society, Perkin Transactions* **1993**, *1* (1), 117–120.
[63] Kobayashi, M.; *et al.* Superconductivity in the first stage rubidium graphite intercalation compound C_8Rb. *Synthetic Metals* **1985**, *12* (1–2), 341–346.
[64] Gunasekara, N.; *et al.* Electronic band structure of C_8Cs studied by highly-angle-resolved ultraviolet photoelectron spectroscopy. *Journal of the Physical Society of Japan* **1987**, *56* (7), 2581–2589.
[65] Umegaki, I.; *et al.* Li-ion diffusion in Li intercalated graphite C_6Li and $C_{12}Li$ probed by μ+ SR. *Physical Chemistry Chemical Physics* **2017**, *19* (29), 19058–19066.
[66] Sorokina, N. E.; *et al.* Acceptor-type graphite intercalation compounds and new carbon materials based on them. *Russian Chemical Bulletin* **2005**, *54* (8), 1749–1767.

[67] Savoskin, M. V.; *et al.* Carbon nanoscrolls produced from acceptor-type graphite intercalation compounds. *Carbon* **2007**, *45* (14), 2797–2800.

[68] Bhauriyal, P.; Mahata, A.; Pathak, B. The staging mechanism of $AlCl_4$ intercalation in a graphite electrode for an aluminium-ion battery. *Physical Chemistry Chemical Physics* **2017**, *19* (11), 7980–7989.

[69] Huber, R.; *et al.* How many-particle interactions develop after ultrafast excitation of an electron–hole plasma. *Nature* **2001**, *414* (6861), 286–289.

[70] Stauber, T.; San-Jose, P.; Brey, L. Optical conductivity, Drude weight and plasmons in twisted graphene bilayers. *New Journal of Physics* **2013**, *15* (11), 113050.

[71] Mansur, H. S.; Oréfice, R. L.; Mansur, A. A. P. Characterization of poly (vinyl alcohol)/poly (ethylene glycol) hydrogels and PVA-derived hybrids by small-angle X-ray scattering and FTIR spectroscopy. *Polymer* **2004**, *45* (21), 7193–7202.

[72] Pendry, J. B. The application of pseudopotentials to low-energy electron diffraction II: Calculation of the reflected intensities. *Journal of Physics C: Solid State Physics* **1969**, *2* (12), 2273.

[73] Zheng, L.; *et al.* Effect of sodium chloride solution on quality of 3D-printed samples molded using wheat starch gel. *Food Hydrocolloids* **2022**, *123*, 107197.

[74] Dávila, M. E.; Le Lay, G. Few layer epitaxial germanene: a novel two-dimensional Dirac material. *Scientific Reports* **2016**, *6* (1), 1–9.

[75] Ho, J.-H.; *et al.* Electron decay rates in a zero-gap graphite layer. *Physics Letters A* **2006**, *357* (4–5), 401–406.

[76] Liu, J.; *et al.* A new 3D Dirac nodal-line semi-metallic graphene monolith for lithium ion battery anode materials. *Journal of Materials Chemistry A* **2018**, *6* (28), 13816–13824.

[77] Gonella, S.; Ruzzene, M. Analysis of in-plane wave propagation in hexagonal and re-entrant lattices. *Journal of Sound and Vibration* **2008**, *312* (1–2), 125–139.

[78] Hybertsen, M. S.; Louie, S. G. First-principles theory of quasiparticles: Calculation of band gaps in semiconductors and insulators. *Physical Review Letters* **1985**, *55* (13), 1418.

[79] Allen, M. J.; Tung, V. C.; Kaner, R. B. Honeycomb carbon: A review of graphene. *Chemical Reviews* **2010**, *110* (1), 132–145.

[80] Bowman, J. C.; Krumhansl, J. A. The low-temperature specific heat of graphite. *Journal of Physics and Chemistry of Solids* **1958**, *6* (4), 367–379.

[81] Lin, C.-Y.; *et al. Electronic and Optical Properties of Graphite-related Systems.* CRC Press, 2017.

[82] Yan, J.; *et al.* Observation of magnetophonon resonance of Dirac fermions in graphite. *Physical Review Letters* **2010**, *105* (22), 227401.

[83] Shung, K. W.-K. Dielectric function and plasmon structure of stage-1 intercalated graphite. *Physical Review B* **1986**, *34* (2) 979.

[84] Berman, O. L.; Gumbs, G.; Lozovik, Y. E. Magnetoplasmons in layered graphene structures. *Physical Review B* **2008**, *78* (8), 085401.

[85] Ho, C.-H.; Chang, C.-P.; Lin, M.-F. Landau subband wave functions and chirality manifestation in rhombohedral graphite. *Solid State Communications* **2014**, *197*, 11–15.

[86] Marinopoulos, A. G.; *et al.* Ab initio study of the optical absorption and wave-vector-dependent dielectric response of graphite. *Physical Review B* **2004**, *69* (24), 245419.

[87] Dresselhaus, M. S.; *et al.* Lattice vibrations in graphite and intercalation compounds of graphite. *Materials Science and Engineering* **1977**, *31*, 141–152.

[88] Panich, A. M.; Nakajima, T. Physical properties and CF bonding in fluorine-graphite intercalation compounds as seen by NMR. *Molecular Crystals and Liquid Crystals Science and Technology. Section A. Molecular Crystals and Liquid Crystals* **2000**, *340* (1), 77–82.

[89] Pfluger, P.; *et al.* Electronic properties of graphite intercalation compounds. *Physica B+C* **1980**, *99* (1–4), 395–400.

[90] Chun-Lei, W.; *et al.* High density gas state at water/graphite interface studied by molecular dynamics simulation. *Chinese Physics B* **2008**, *17* (7), 2646.

[91] Anderson, J. R.; *et al.* Effect of pressure on the Fermi surface of graphite. *Physical Review* 164.3 (1967): 1038.

[92] Charlier, J.-C.; Michenaud, J.-P.; Gonze, X. First-principles study of the electronic properties of simple hexagonal graphite. *Physical Review B* **1992**, *46* (8) 4531.

[93] Moosa, A.; Abed M. Graphene preparation and graphite exfoliation. *Turkish Journal of Chemistry* **2021**, *45* (3), 493–519.

[94] Taen, T.; Uchida, K.; Osada, T. Thickness-dependent phase transition in graphite under high magnetic field. *Physical Review B* **2018**, *97* (11), 115122.

[95] Chen, R.-B.; Chiu, Y.-H.; Lin, M.-F. A theoretical evaluation of the magneto-optical properties of AA-stacked graphite. *Carbon* **2013**, *54*, 268–276.

[96] Piscanec, S.; *et al.* Kohn anomalies and electron-phonon interactions in graphite. *Physical Review Letters* **2004**, *93* (18), 185503.

[97] Lin, M. F.; Shung, K. W.-K. Screening of charged impurities in graphite intercalation compounds. *Physical Review B* **1992**, *46* (19) 12656.
[98] Hermanson, J. Final-state symmetry and polarization effects in angle-resolved photoemission spectroscopy. *Solid State Communications* **1977**, *22* (1), 9–11.
[99] Fujimori, S.-I.; *et al.* Direct observation of a quasiparticle band in Ce Ir In 5: An angle-resolved photoemission spectroscopy study. *Physical Review B* **2006**, *73* (22), 224517.
[100] Schneider, J. M.; *et al.* Consistent interpretation of the low-temperature magnetotransport in graphite using the Slonczewski-Weiss-McClure 3D band-structure calculations. *Physical Review Letters* **2009**, *102* (16), 166403.
[101] Graf, D.; *et al.* Spatially resolved Raman spectroscopy of single-and few-layer graphene. *Nano Letters* **2007**, *7* (2), 238–242.
[102] Lin, Y.-M.; *et al.* Electrical observation of subband formation in graphene nanoribbons. *Physical Review B* **2008**, *78* (16), 161409.
[103] Bian, J.; Nicola, L. On the lubrication of rough copper surfaces with graphene. *Tribology International* **2021**, *156*, 106837.
[104] Tatar, R. C.; Rabii, S. Electronic properties of graphite: A unified theoretical study. *Physical Review B* **1982**, *25* (6), 4126.
[105] Li, J.; Gao, T.; Luo, J. Superlubricity of graphite induced by multiple transferred graphene nanoflakes. *Advanced Science* **2018**, *5* (3), 1700616.
[106] DiVincenzo, D. P.; Mele, E. J. Self-consistent effective-mass theory for intralayer screening in graphite intercalation compounds. *Physical Review B* **1984**, *29* (4), 1685.
[107] Inoue, M.; Hughes, H. P.; Yoffe, A. D. The electronic and magnetic properties of the 3d transition metal intercalates of TiS2. *Advances in Physics* **1989**, *38* (5), 565–604.
[108] Hancock, Y.; *et al.* Generalized tight-binding transport model for graphene nanoribbon-based systems. *Physical Review B* **2010**, *81* (24), 245402.
[109] Tamtögl, A.; *et al.* Graphene on Ni (111): Electronic corrugation and dynamics from helium atom scattering. *The Journal of Physical Chemistry C* **2015**, *119* (46), 25983–25990.
[110] Dodaro, J. F.; *et al.* Phases of a phenomenological model of twisted bilayer graphene. *Physical Review B* **2018**, *98* (7), 075154.
[111] Mnatsakanov, T. T.; *et al.* Carrier mobility model for GaN. *Solid-State Electronics* **2003**, *47* (1), 111–115.

[112] Clarke, J.; Braginski, A. I. *The SQUID Handbook*, Vol. 1. Weinheim: Wiley-Vch, 2004.
[113] Doster, W.; Cusack, S.; Petry, W. Dynamical transition of myoglobin revealed by inelastic neutron scattering. *Nature* **1989**, *337* (6209), 754–756.
[114] Angizi, S.; *et al.* A comprehensive review on planar boron nitride nanomaterials: From 2D nanosheets towards 0D quantum dots. *Progress in Materials Science* **2022**, *124*, 100884.
[115] Ho, J. H.; *et al.* Magnetoelectronic properties of a single-layer graphite. *Journal of the Physical Society of Japan* **2006**, *75* (11), 114703.
[116] Lin, M. F.; Huang, C. S.; Chuu, D. S. Plasmons in graphite and stage-1 graphite intercalation compounds. *Physical Review B* **1997**, *55* (20), 13961.
[117] Kriz, R. D.; Ledbetter, H. M. Elastic representation surfaces of unidirectional graphite/epoxy composites. In *Recent Advances in Composites in the United States and Japan*. ASTM International, 1985.
[118] Marinopoulos, A. G.; *et al.* Anisotropy and interplane interactions in the dielectric response of graphite. *Physical Review Letters* **2002**, *89* (7), 076402.
[119] Li, S.; *et al.* Bifunctional two-dimensional nanocomposite for electromagnetic wave absorption and comprehensive anti-corrosion. *Carbon* **2022**, *186*, 520–529.
[120] Lu, C.-L.; *et al.* Absorption spectra of trilayer rhombohedral graphite. *Applied Physics Letters* **2006**, *89* (22), 221910.
[121] Yin, H.; *et al.* Self-assembly of 3D-graphite block infiltrated phase change materials with increased thermal conductivity. *Journal of Cleaner Production* **2019**, *235*, 359–368.
[122] Wei-Bang, L.; *et al.* Geometric and electronic properties on stage-1 FeCl3-graphite intercalation compounds. arXiv preprint arXiv:2012.14648, **2020**.
[123] Fischer, J. E.; Te, T. Graphite intercalation compounds. 1978.
[124] Li, W.-Ba.; *et al.* Diverse fundamental properties in stage-n graphite alkali-intercalation compounds: Anode materials of Li+-based batteries. arXiv preprint arXiv:2001.02042, **2019**.
[125] Nagashima, A.; *et al.* Electronic states of monolayer graphite formed on TiC (111) surface. *Surface Science* **1993**, *291* (1–2), 93–98.
[126] Chico, L.; López-Sancho, M. P.; Munoz, M. C. Spin splitting induced by spin-orbit interaction in chiral nanotubes. *Physical Review Letters* **2004**, *93* (17), 176402.

[127] Slonczewski, J. C.; Weiss, P. R. Band structure of graphite. *Physical Review* **1958**, *109* (2), 272.
[128] Ghatak, K. P.; Biswas S. N. Influence of quantum confinement on the heat capacity of graphite. *Fizika A* **1994**, *3* (1), 7–24.
[129] Saad, M.; *et al.* Mesoscopic scale rearrangements of graphite nanoflake open edges under mild annealing treatments. *Vacuum* **2022**, *199*, 110977.
[130] Dyall, K. G. An exact separation of the spin-free and spin-dependent terms of the Dirac–Coulomb–Breit Hamiltonian. *The Journal of Chemical Physics* **1994**, *100* (3), 2118–2127.
[131] Yeh, C. Perturbation approach to the diffraction of electromagnetic waves by arbitrarily shaped dielectric obstacles. *Physical Review* **1964**, *135* (5A), A1193.
[132] Li, G.; *et al.* Competition between ferromagnetic metallic and paramagnetic insulating phases in manganites. *Journal of Applied Physics* **2002**, *92* (3), 1406–1410.
[133] Penn, D. R. Wave-number-dependent dielectric function of semiconductors. *Physical Review* **1962**, *128* (5), 2093.
[134] Evans, M.; *et al.* Thermo-optic noise in coated mirrors for high-precision optical measurements. *Physical Review D* **2008**, *78* (10), 102003.
[135] Chen, J.; *et al.* III–VI van der Waals heterostructures for sustainable energy related applications. *Nanoscale* **2019**, *11* (13), 6431–6444.
[136] Connolly, J. A. D.; Petrini, K. An automated strategy for calculation of phase diagram sections and retrieval of rock properties as a function of physical conditions. *Journal of Metamorphic Geology* **2002**, *20* (7), 697–708.
[137] Rantitsch, G. *et al.* On the discrimination of semi-graphite and graphite by Raman spectroscopy. *International Journal of Coal Geology* **2016**, *159*, 48–56.
[138] Wehling, T. O., *et al.* Strength of effective coulomb interactions in graphene and graphite. *Physical review letters* 106.23 (2011): 236805.
[139] Mao, C.; *et al.* Selecting the best graphite for long-life, high-energy Li-ion batteries. *Journal of the Electrochemical Society* **2018**, *165* (9), A1837.
[140] Lucarini, V.; *et al.* *Kramers-Kronig Relations in Optical Materials Research*, Vol. 110. Springer Science & Business Media, 2005.
[141] Bardyszewski, W. The dielectric function of holes in semiconductors of zinc-blende structure. *Solid State Communications* **1986**, *57* (11), 873–876.

[142] Pichler, T.; *et al.* The dielectric function of dimerised C59N. *Synthetic Metals* **1997**, *86* (1–3), 2313–2314.
[143] Mohan, B.; Kumar, A.; Ahluwalia, P. K. A first principle calculation of electronic and dielectric properties of electrically gated low-buckled mono and bilayer silicene. *Physica E: Low-dimensional Systems and Nanostructures* **2013**, *53*, 233–239.
[144] Gorshkov, V. G. On relativistic perturbation theory for a Coulomb field. *Soviet Physics — JETP* **1961**, *13*, 1037.
[145] Wang, Z. L.; Cowley, J. M. Reflection electron energy loss spectroscopy (reels): A technique for the study of surfaces. *Surface Science* **1988**, *193* (3), 501–512.
[146] Lee, M.-S.; Mahanti, S. D. Validity of the rigid band approximation in the study of the thermopower of narrow band gap semiconductors. *Physical Review B* **2012**, *85* (16), 165149.
[147] Hao, Y.; *et al.* Pillaring effect of K ion anchoring for stable V2O5-based zinc-ion battery cathodes. *ChemNanoMat* **2020**, *6* (5), 797–805.
[148] Baskey, M.; Saha, S. K. A graphite-like zero gap semiconductor with an interlayer separation of 2.8 Å. *Advanced Materials* **2012**, *24* (12), 1589–1593.
[149] Du, L.; *et al.* Engineering symmetry breaking in 2D layered materials. *Nature Reviews Physics* **2021**, *3* (3), 193–206.
[150] Zheng, H.; *et al.* Simultaneous modulation of magnetic and dielectric transition via spin-crossover-tuned spin arrangement and charge distribution. *Angewandte Chemie International Edition* **2018**, *57* (28), 8468–8472.
[151] Warren, B. E. *X-ray Diffraction.* Courier Corporation, 1990.
[152] Klein, B. J.; Pack, G. R. Calculations of the spatial distribution of charge density in the environment of DNA. *Biopolymers: Original Research on Biomolecules* **1983**, *22* (11), 2331–2352.
[153] Abrahams, E.; *et al.* Quasiparticle lifetime in disordered two-dimensional metals. *Physical Review B* **1981**, *24* (12), 6783.
[154] Pereira, V. M.; Kotov, V. N.; Castro Neto, A. H. Supercritical Coulomb impurities in gapped graphene. *Physical Review B* **2008**, *78* (8), 085101.
[155] Chiu, C.-W.; *et al.* Coulomb decay rates in monolayer doped graphene. *RSC Advances* **2020**, *10* (4), 2337–2346.
[156] Forker, R.; Gruenewald, M.; Fritz, T. Optical differential reflectance spectroscopy on thin molecular films. *Annual Reports Section C (Physical Chemistry)* **2012**, *108*, 34–68.

[157] Jeong, S. H.; *et al.* Optical absorption spectroscopy for determining carbon nanotube concentration in solution. *Synthetic Metals* **2007**, *157* (13–15) 570–574.

[158] Kremers, M.; *et al.* Optical transmission spectroscopy of switchable yttrium hydride films. *Physical Review B* **1998**, *57* (8), 4943.

[159] Segall, M. D.; *et al.* First-principles simulation: Ideas, illustrations and the CASTEP code. *Journal of Physics: Condensed Matter* **2002**, *14* (11), 2717.

[160] Shen, G.; *et al.* Vapor– solid growth of one-dimensional layer-structured gallium sulfide nanostructures. *ACS Nano* **2009**, *3* (5), 1115–1120.

[161] Fernelius, N. C. Properties of gallium selenide single crystal. *Progress in Crystal Growth and Characterization of Materials* **1994**, *28* (4), 275–353.

[162] Shenoy, U. S.; *et al.* Electronic structure and properties of layered gallium telluride. *Chemical Physics Letters* **2016**, *651*, 148–154.

CHAPTER 7

FEATURED PROPERTIES OF 5D TRANSITION METAL SUBSTITUTIONS INTO GRAPHENE

Thi Dieu Hien Nguyen[*], Wen-Dung Hsu[*] and Ming-Fa Lin[†]

[*] *Department of Materials Science and Engineering, National Cheng Kung University, Tainan, Taiwan*
[†] *Department of Physics, National Cheng Kung University, Tainan, Taiwan*

Abstract

The 5d transition metal guest-atom substituted graphene systems extensively change the featured properties of condensed-matter systems. Significant chemical bonds of C–C and C–M (where M is W, Os, Ir, or Pt) are formed by single and multi-hybridizations. W-, Os-, and Pt-substituted graphenes form narrow-bandgap semiconductors, with a range of 0.21–0.4 eV, while Ir substitution results in a semi-metal with a bandgap of 0.04 eV. Moreover, the magnetic properties are exposed to W- and Ir-substituted graphenes. The geometric, electronic, magnetic, and optical characteristics are essentially discussed in this book chapter.

Keywords: Density functional theory, orbital hybridizations, transition metal substitutions, electronic properties, optical properties, magnetic configurations.

7.1. Introduction

Carbon possesses four active orbital configurations, namely 2s, $2p_x$, $2p_y$, and $2p_z$, which determine all the essential properties of related compounds. Diverse structures of C-related materials, covering zero-dimensional (0D) carbon/graphene quantum dots,[1–3]

C_{60} fullerenes,[4,5] one-dimensional (1D) carbon nanotubes,[6–12] 1D graphene nanoribbons,[13–16] two-dimensional (2D) layered graphene,[17] and three-dimensional (3D) graphite systems, show chiefly a significant contribution of sp^2 hybridization. Among them, 2D graphene shows promise as a potential candidate for further research in the condensed-matter field, which has been discussed in both experimental[18] and theoretical analyses.[19,20] Remarkably, from an experimental perspective, since the first successful isolation of monolayer graphene by Novoselov in 2004,[18] interest in graphene has been particularly prominent due to many exceptional essential properties that are characteristic of this material. Interestingly, the Nobel Prize in Physics in 2010 was awarded for outstanding groundbreaking research on graphene, which might have presented both benefits and challenges for scientists when working on graphene.

Chemical modifications include adsorptions, substitutions, intercalations, decorations, and heterojunctions, which are efficient approaches to extensively enhancing the various rich behaviors of materials and their applications. Guest-atom-substituted graphene has been applied to a wide range of elements on the periodic table, e.g., alkalies, halogens, transition metals, and rare-earth elements. Among them, transition metals in diverse groups, such as 3d, 4d, 5d, and 6d, might make great contributions to basic and applied science. Previous studies have reported theoretical[21–23] and experimental measurements of substituted systems, with some of them showing rich magnetic moments, which result in the splitting of spin up and spin down over an energy range. For example, manganese (Mn) with the $3d^5$ orbital configuration possesses a massive magnetic moment of $3\mu_B$ at diluted guest-atom concentrations.[23] This might be useful for comprehensive investigations of the innovative behavior of graphene through chemical modification methods. In this chapter, we focus on 5d transition metals as guest atoms for the substitution process in graphene.

The substitution systems of tungsten (W), osmium (Os), iridium (Ir), and platinum (Pt) are investigated under the theory of hybridization in the chemical bonds, which is seldom discussed in the literature. 5d transition metal-substituted graphene compounds are systematically examined using first-principles analyses in combination with spin interaction and the crucial chemical relations of π, σ, sp^2, and sp^3 for planar and buckling structures. The current work shows highly non-uniform structures under the diverse C–C, W–C, Os–C, Ir–C, and Pt–C chemical bonds. Furthermore, several rich characteristics come into existence, such as the strongly ruined or modified Dirac cone on the band energy; the semimetal, semiconducting, or metallic behaviors; magnetic configurations; and optical spectra. Theoretical calculations are significantly tested by experimental investigations, such as scanning tunneling microscopy/transmission electron microscopy (STM/TEM)[24] for geometric structures, angle-resolved photoemission spectroscopy (ARPES)[24–26] for wave-vector-dependent energy bands, scanning tunneling spectroscopy (STS) for van Hove singularities, the superconductor quantum interference device (SQUID) for magnetic behaviors, and optical spectroscopies for optical properties.

7.2. Computational Techniques

The investigations for the featured optimal structures, electronic characteristics, magnetic configurations, and optical properties of 5d transition metal-substituted enriched graphene structures are based on the density functional theory (DFT)[27] using the Vienna Ab-initio Simulation Package (VASP). The important exchange-correlation energy is calculated using the generalized gradient approximation[28] and the Perdew–Burke–Ernzernhof functional,[29] while the ion–electron interaction is calculated by the projected augmented wave (PAW) approach.[30] In this model, we also add 15Å in the z-direction to neglect the interaction in this direction. The k-point meshes are applied on a wave-vector basis using the Monkhorst–Pack scheme as $7 \times 7 \times 1$ for optimal structures and $15 \times 15 \times 1$ for other physical and chemical behaviors. The energy of 10^{-5}eV is utilized for two consecutive steps to check the convergence. Moreover, the limit for the Hellmann–Feynman force of each atom is likely less than 0.01 eV·Å^{-1} in the relaxational process.

7.3. Optimal Stability

Owing to their high atomic numbers, the 5d transition group elements show complicated electron configurations, as depicted in Fig. 7.1. Therefore, after substituting W, Os, Ir, or Pt into graphene, these structures come into existence with unique characteristics (Fig. 7.2(a) and 7.2(b) and Table 7.1). The modulation of chemical bonds creates highly non-uniform structures which are totally different from those of pristine graphene. Two mainstream chemical interactions cover C–C and C–M, where M denotes the guest-atom substitution, including W, Os, Ir, and Pt. It is noted that W and Ir represent magnetic properties, so the spin-dependent many and single particles need to be taken into account in the numerical calculations. The geometric structures are considered in a 1:31 concentration ratio for the four guest-atom-substituted graphenes, while the 1:49 case is considered for W and Os. The higher concentrations, for example, 1:18 and 1:8, might be difficult to form in this structure due to the heavy atomic number of 5d transition metals. Pristine graphene presents a well-known C–C chemical length of 1.42 Å, while the substituted modification displays a slight modulation of 1.40–1.43 for the specific W, Os, Ir, and Pt atoms, as shown in Table 7.1. Furthermore, the stable structures present buckling, which mainly appears at the substituting position of the guest atoms. The wide difference spread ranging from 1.60 (Ir) to 1.75 (Os). These unusual features might play a substantial role in fundamentally determining the essential properties of modulated systems, which might refer to the contribution of sp^3 hybridization instead of the well-defined identification of σ and π bonds in graphene. The observation of C–C and C–M fluctuation and buckling behavior might predict a rich and complicated charge density, which leads to an interesting understanding of other electronic and optical properties.

7.4. Wave Vector-independent Band Characteristics

Rich band structures for 2D systems reveal rich and unique electronic features. In our model analyses, we use the supercell atoms of 4×4 (31 C atoms and 1 impure atom) and 5×5 (49 C atoms

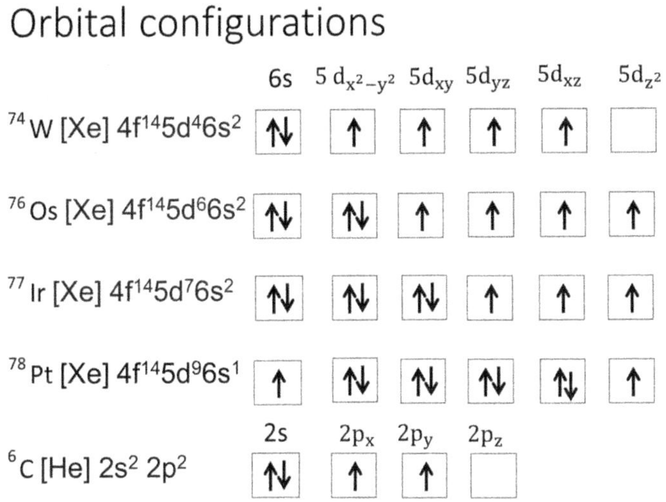

Fig. 7.1. The ground-state orbital configurations of W, Os, Ir, Pt, and C.

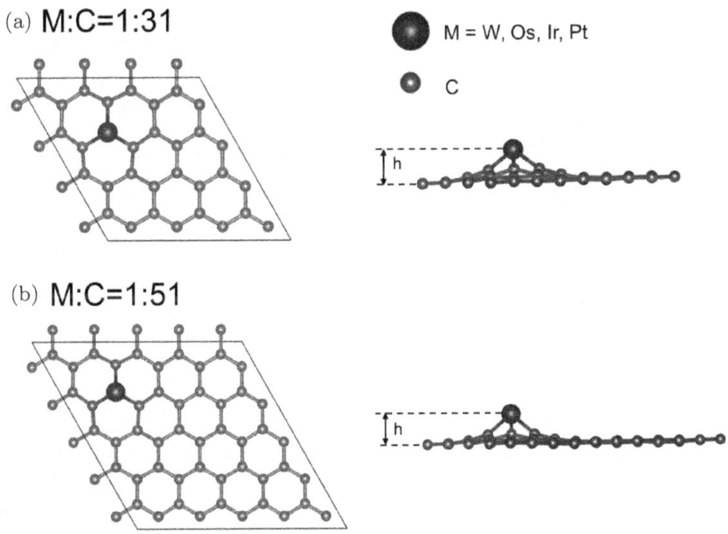

Fig. 7.2. Geometric structures of 5d transition metal-substituted graphene: (a) 1:31 and (b) 1:49.

Table 7.1. Optimal structures with diverse features: chemical bonds, height, total energy, magnetic moment, and band gap.

	Concentration	C–M	C–C	Height (h)	E_0	Magnetic moment	E_g
C	1:31 (4 × 4)	—	1.42	—	−298.02	—	0
Pristine	1:49 (5 × 5)	—	1.42	—	−465.66	—	0
W	1:31 (4 × 4)	1.94	1.42–1.43	1.73	−294.28	1.51	0.30
	1:49 (5 × 5)	1.95	1.42–1.43	1.72	−462.02	−1.51	0.21
Os	1:31 (4 × 4)	1.89	1.42–1.44	1.65	−293.52	0	0.40
	1:49 (5 × 5)	1.89	1.42–1.43	1.75	−461.23	0	0.34
Ir	1:32 (4 × 4)	1.90	1.41–1.44	1.60	−291.91	0.56	0.04
Pt	1:32 (4 × 4)	1.93	1.40–1.43	1.64	−288.97	0	0.22

and 1 impure atom) so that the zone folding effect can be applied with the K-path of Γ-M-K-Γ (Fig. 7.3(a)). The eventual Dirac cone at the K point (Fig. 7.3(b)) demonstrates the significant role of $2p_z$ hybridization (π bands) for pristine graphene. Many parabolic dispersions come into existence frequently, and the specific saddle point appears at M points. Additionally, the crossing/non-crossing features result from the orthogonal of the $\sigma(sp^2–sp^2)$ and π bands. The substitution configurations produce plenty of subbands because of the large atomic numbers in the supercell and the diverse single and multi-hybridizations, as shown in Figs. 7.3(c)–7.3(h). In most available cases with low concentrations, the transition metal-dominated strong free electrons or holes are near the Fermi level, leading to the distortion of Dirac cones in substitution systems. A typical feature of four atoms in 5d transition is that there exists strongly asymmetric behavior between the occupied hole bands and unoccupied electron bands. This phenomenon illustrates complicated hopping integrals and energy in phenomenological models, which present strong interaction of atom-dependent orbitals. The dramatic change in band structures is clearly visible in the W, Os, Ir, and Pt atoms. Remarkably, the spin-split band characteristics with spin up and spin down, are shown by the black and dashed-red lines for W and Ir in Figs. 7.3(c) and 7.3(g), respectively. It is noted that the magnetic moment with a value of 1.51 creates strong splitting near

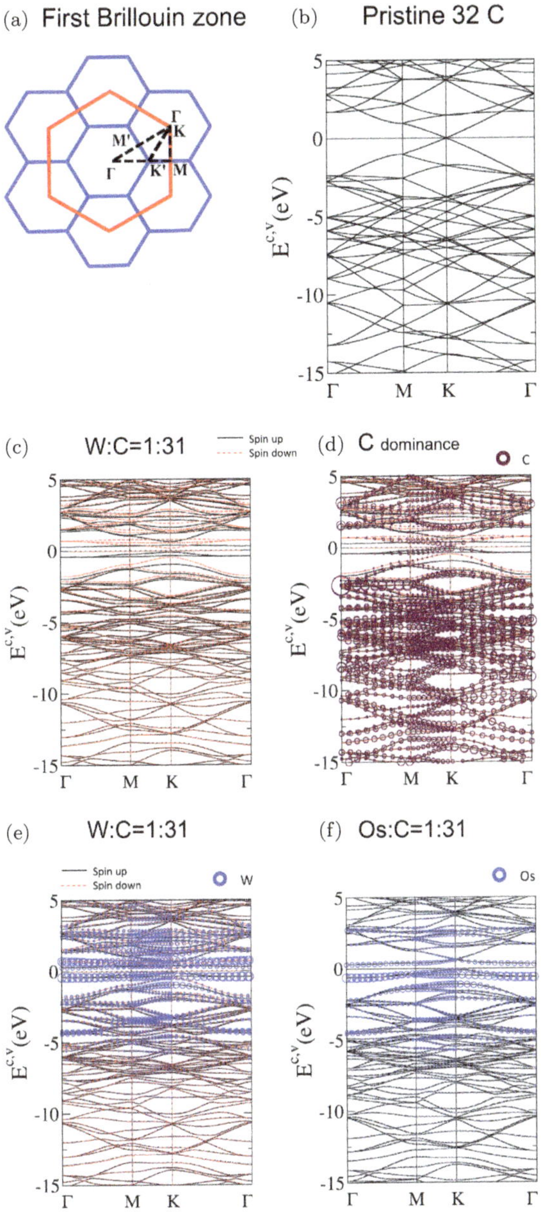

Fig. 7.3. (a) The first Brillouin zone under the zone folding effect, and the band structures along with the high symmetric point for 1:31 concentration of (b) pristine graphene, (c) W with spin up and spin down, (d) C dominance in W-substituted graphene, (e) W dominance in W-substituted graphene, (f) Os and its dominance, (g) Ir, and (h) Pt.

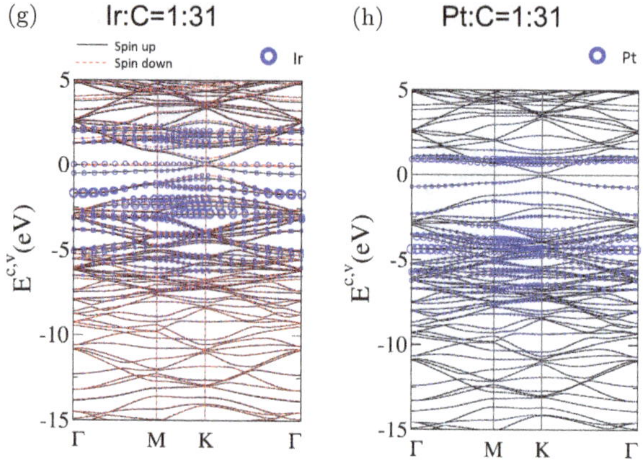

Fig. 7.3. (*Continued*)

the Fermi level (E_F= 0), while in the deeper energy range, undistinguished spin separations are observed. The band structures are sensitive to the kind of guest atom and the M (W, Os, Ir, or Pt) concentration. The case of 1:31 for W, Os, and Pt presents a narrow bandgap semiconductor of 0.3, 0.4, and 0.22 eV, respectively, while Ir demonstrates a 0.04 eV semi-metal gap. These gaps are measured from the highest occupied states to the lowest unoccupied states, for example, W:C (3.23%) possesses a direct bandgap at the K point with unique ferromagnetic spin configurations (Fig. 7.3(c)). A similar phenomenon can be observed with 1:49 of W- and Os-substituted graphene, with an increasing number of subbands based on the Moiré superlattice unit cell. Transition metals dominate in the range of −6.0 to 2.0 eV, as indicated by the blue circles, and C atoms have higher arrangements in the whole energy band of −15.0 to 5.0 eV, as indicated in deep-purple color, but not near the Fermi level (Fig. 7.3(d)). Apparently, the complicated orbital hybridizations of transition metal–C essentially determine the chemical interactions between them. The co-dominance of (W, Os, Ir, or Pt) and C is located at $-6.0 \leq E^{v,c} \leq 2.0$ eV, revealing the interaction of W–C, Os–C, Ir–C, and Pt–C. Remarkably, there exist interesting flat bands near the Fermi level at −0.4 and 0.3 eV, which might be related to the C 2p$_z$ orbitals, rather than C 2p$_x$ and 2p$_y$ (detailed in the DOS sections). Moreover, the energy bandwidths are quite large up to 2.0 eV, which will strongly modify the diverse energy range, along with the high-symmetry points Γ-M-K- Γ. The non-monotonous wave-vector-dependent behaviors appear between two high-symmetric points, with the band-edge states frequently created in the energy bands.

For experimental measurements, ARPES is a useful technique to measure the wave-vector dependence of the valence bands close to the Fermi level for non-magnetic structures, while spin-polarized ARPES is used for examining the magnetic moment bands of nanomaterials.

7.5. Rich Atom- and Orbital-decomposed van Hove Singularities

The significant densities of states (DOSs) play a substantial role in the comprehensive understanding of the electronic properties of substituted systems, as shown in Figs. 7.4(a)– 7.4(f). The chemical bonds of W–C, Os–C, Ir–C, Pt–C, and C–C can be demonstrated by the atom, orbital, and spin DOSs with critical orbital hybridizations. In general, DOSs might come into existence in diverse forms, such as asymmetric V-shapes, specific delta-function-like distinctive peaks, and logarithmic structures, which are associated with the band structures. In comparison with the pristine case, the Dirac cone in the band structure will be slightly modified, leading to a change in the V-shape in the case of substitutions of 5d transition metals at low concentrations of 3.23% and 1.96%. The various band-edge states result in diverse van Hove singularities for all available configurations. In the energy range of −1.1 to 1.1 eV, the C-2p_z (light-green line) and M-(6s, 5$d_{x^2-y^2}$, 5d_{xy}, 5d_{yz}, 5d_{xz}, 5d_{z^2}) (solid yellow, green, red, dashed pink, purple, blue, respectively) show strong contributions. The common 5d orbitals appear in the DOSs of impure atoms at identical positions with diverse intensities, as shown in Figs. 7.5(a)–7.5(d) and Figs. 7.6(a) and 7.6(b). The spin-decomposed DOSs of W in Figs. 7.5(a) and 7.6(a) depict the different peak positions in spin up and spin down, which exhibit the strong split-spin configurations of ferromagnetic structures. In these cases, the van Hove singularities

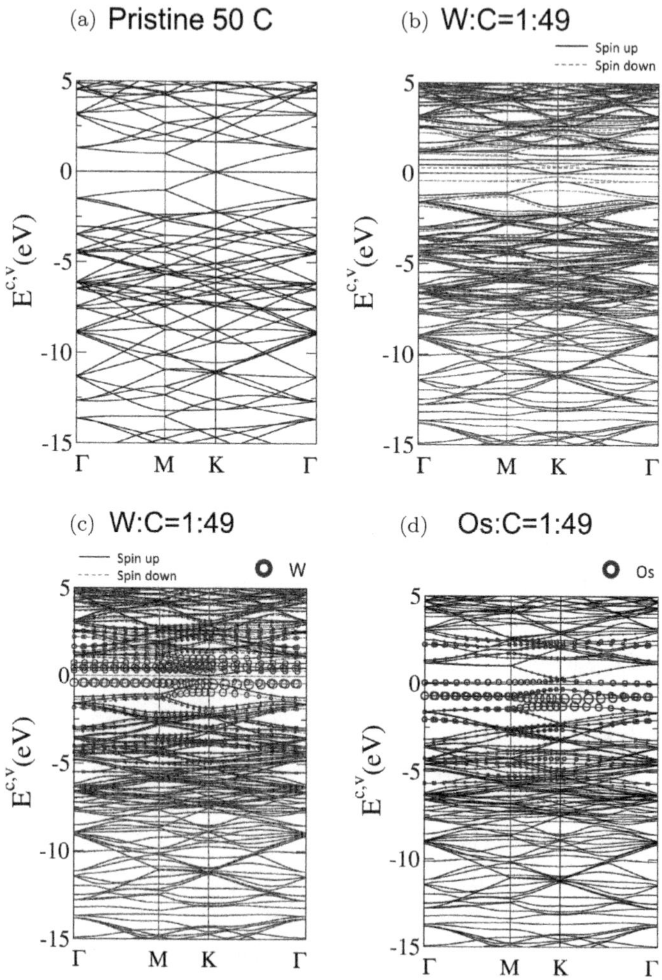

Fig. 7.4. Wave vector-independent band structures along with the high-symmetric point for 1:49 concentration of (a) pristine graphene, (b) W with spin up and spin down, (c) W dominance in W-substituted graphene, and (d) Os dominance in Os-substituted graphene.

are extended over the entire energy spectrum, but they seem to be difficult to classify as pure π or σ bonds (sp^2–sp^2). As a result, the C–C bonds are caused by the mixing of (sp^2–sp^2) and (sp^3–sp^3)of 2s, $2p_x$, $2p_y$, and $2p_z$. The essential chemical bonds of C–W are originated from (2s, $2p_x$, $2p_y$, $2p_z$) – (6s, $5d_{x^2-y^2}$, $5d_{xy}$, $5d_{yz}$,

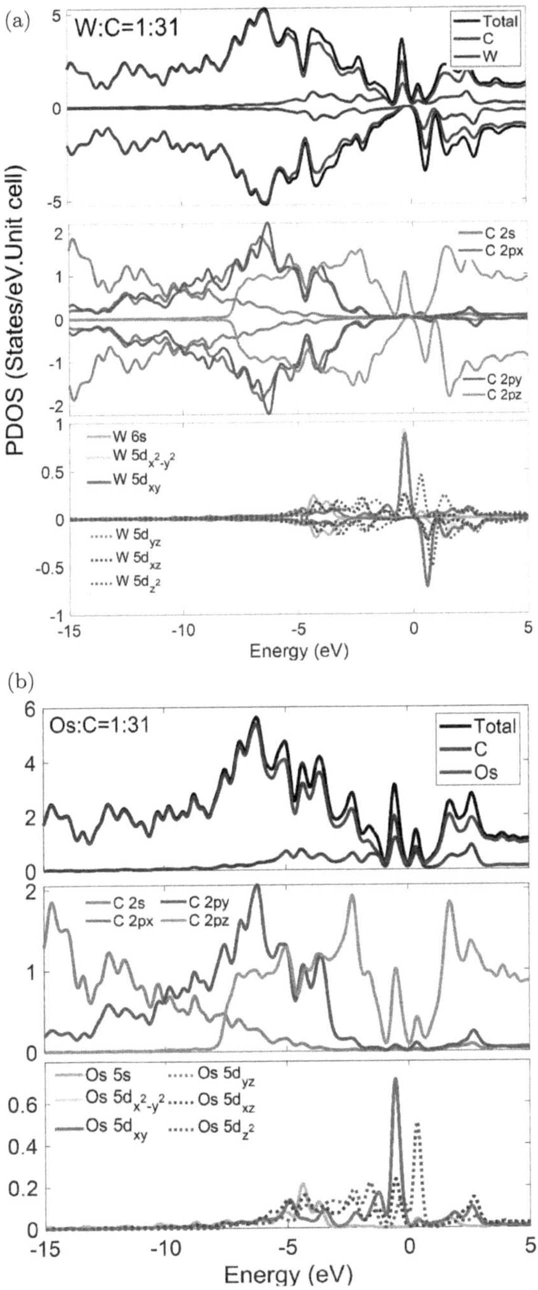

Fig. 7.5. The atom and orbital densities of states for M-substituted systems with 1:31 concentration for (a) W, (b) Os, (c) Ir, and (d) Pt.

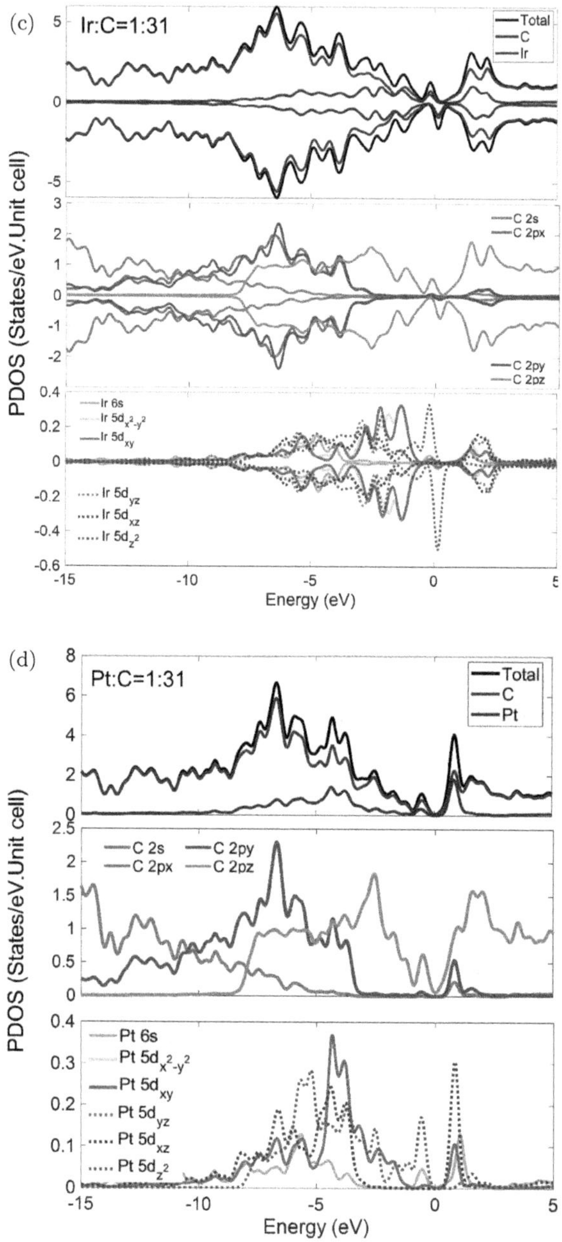

Fig. 7.5. (*Continued*)

(a)

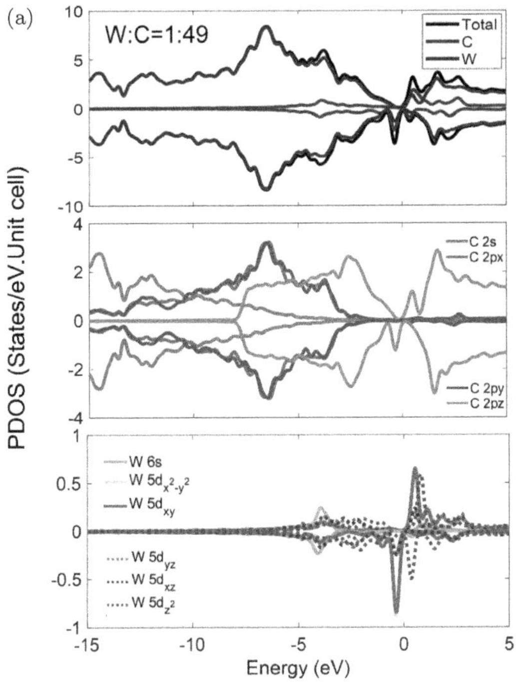

(b)

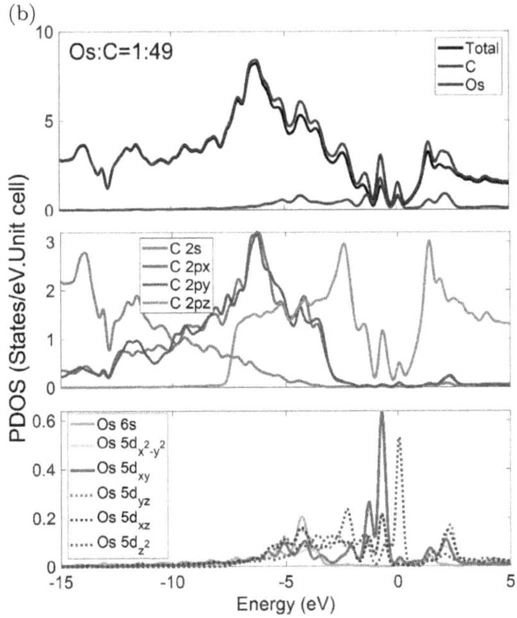

Fig. 7.6. The atom and orbital densities of states for M-substituted systems with 1:49 concentration for (a) W and (b) Os.

$5d_{xz}$, $5d_{z^2}$). It should be pointed out that the mixing of six-orbital hybridizations of 5d transition metals and four-orbital hybridizations of carbon might appear at identical overlapping positions, leading to the creation of physical and chemical interactions. Regarding the spin-split energy, the ferromagnetic properties are mainly featured by the energy bands and the van Hove singularities about the Fermi level. The DOSs of Ir also possess magnetic properties, but $5d_{z^2}$ contributes higher intensities in comparison to W–guest atoms with $5d_{x^2-y^2}$ and $5d_{xy}$. For Os and Pt, the spin-up and spin-down states might simultaneously exist, which might destroy the magnetic properties in these materials. On the other hand, the mixing of six orbitals (6s, $5d_{x^2-y^2}$, $5d_{xy}$,$5d_{yz}$, $5d_{xz}$, $5d_{z^2}$) extends to the occupied states as deep as -10 eV, while Os shows less participation in the lower energy range.

From an experimental perspective, van Hove singularities are popularly identified using STS or SP-STS. The principal methods for these techniques involve measuring the differential (dI/dV), which is proportional to the DOS. To date, a few successful experiments have been conducted for identifying the van Hove singularities of graphene nanoribbons, few-layer graphene systems (monolayer graphene, AA-, AB-, and ABC-stacked graphites, and 2D trilayer graphene systems in AAA, ABA, ABC, and AAB stackings). Additionally, high-resolution spin-polarized spectroscopies are available to determine the spin-split magnetic properties and the DOS structures near the Fermi level.

7.6. Spatial Charge Densities and Spin Configurations

W-, Os-, Ir-, and Pt-substituted graphenes exhibit specific charge distribution properties. It is noted that the isolated carbon atoms reveal the contributions of four activated orbitals, including (2s, $2p_x$, $2p_y$, and $2p_z$) by the spherical-like shape. In Fig. 7.7(a), the appearance of the red core of 2s in red color might partially take part in the chemical interaction of C–C or C–M atoms (where M is W, Os, Ir, or Pt), while the yellow-green region is classified

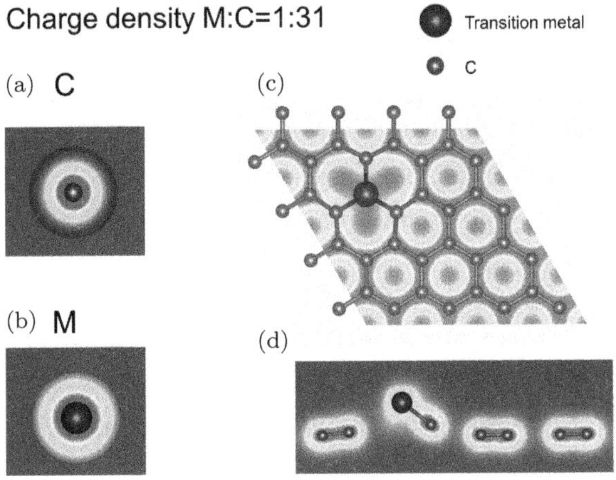

Fig. 7.7. Charge density for the M-substituted carbon system with 1:31 concentration: (a) isolated M, (b) C, (c) top view, and (d) side view.

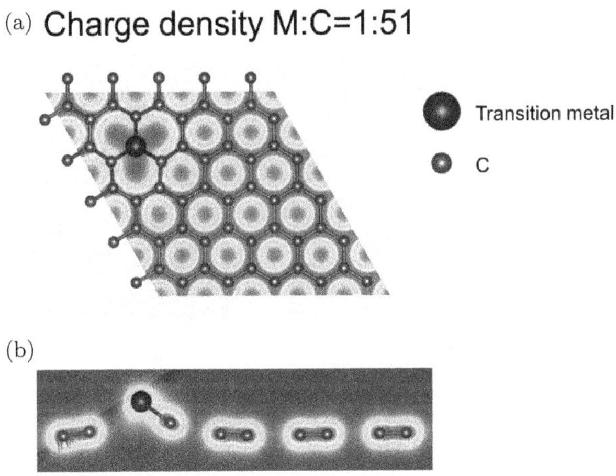

Fig. 7.8. Charge density for M-substituted carbon system with 1:49 concentration: (a) top view and (b) side view.

as ($2p_x$, $2p_y$, $2p_z$) orbitals. Figure 7.7(b) illustrates the isolated transition metal atom, comprising the W, Os, Ir, or Pt guest atoms. The red core is specific to the 6s orbital, while the yellow-green region describes the contribution of M-($5d_{x^2-y^2}$, $5d_{xy}$, $5d_{yz}$, $5d_{xz}$, $5d_{z^2}$). Figures 7.7(c) and 7.7(d) and Figs. 7.8(a) and 7.8(b) show the charge densities of the substitution systems, which belong to the 1:31 and 1:49 concentration cases, respectively. The chemical bondings of C–C and C–M determine the shape change of the charge density based on the contribution of diverse hybridization, resulting from C (2s, 2p$_x$, 2p$_y$, 2p$_z$) – C (2s, 2p$_x$, 2p$_y$, 2p$_z$) / C (2p$_z$) – C (2p$_z$) / C(2s, 2p$_x$, 2p$_y$) – C (2s, 2p$_x$, 2p$_y$), and C (2s, 2p$_x$, 2p$_y$, 2p$_z$) – M (5d$_{x^2-y^2}$, 5d$_{xy}$, 5d$_{yz}$, 5d$_{xz}$, 5d$_{z^2}$).

The ferromagnetic properties take place in the W and Ir, which are revealed in the spin-density distribution shown in Figs.7.9(a)–7.9(c) and Figs. 7.10(a)–7.10(c). In general, there exist two kinds of spin configurations, comprising the Hubbard and Heisenberg models. It is interesting that the W and Ir spin moments belong to the latter, which presents the contributions of both spin up and spin down. Figures 7.9(a)–7.9(c) and Figs. 7.10(a)–7.10(c) show the spin configurations for two concentrations of 1:31 and 1:49,

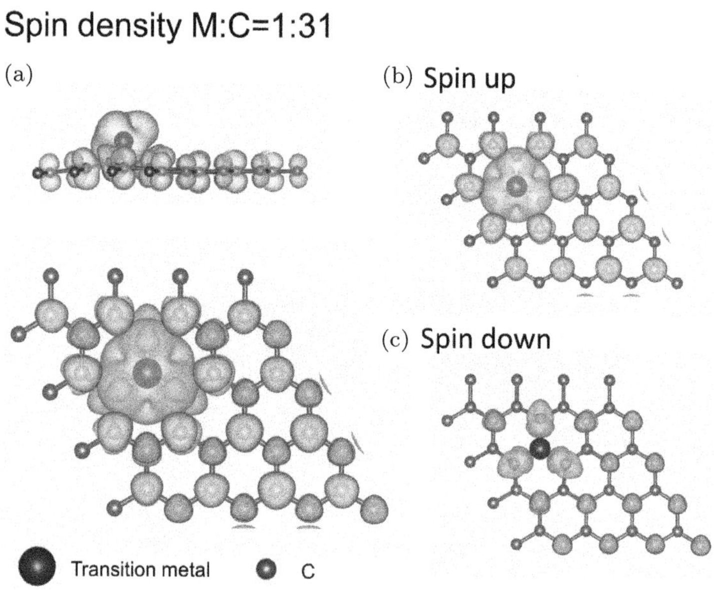

Fig. 7.9. Spin configuration for 1:31 M-substituted graphene: (a) whole system, (b) spin up, and (c) spin down.

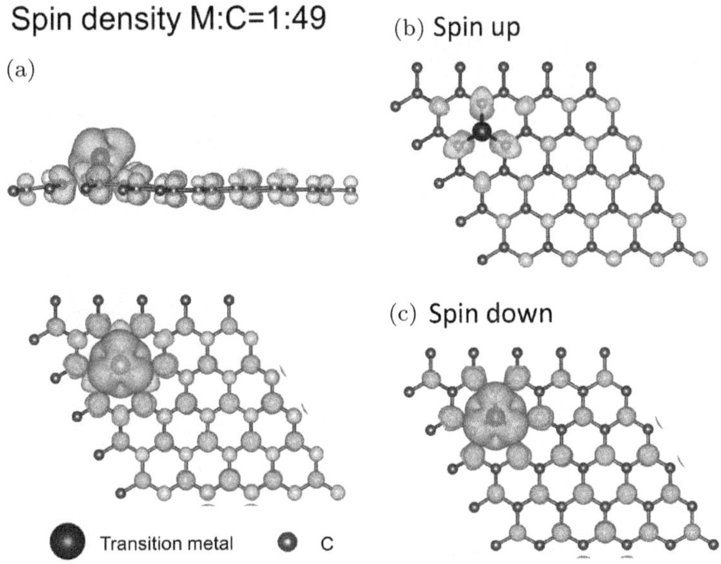

Fig. 7.10. Spin configuration for 1:49 M-substituted graphene: (a) whole system, (b) spin up, and (c) spin down.

indicated by yellow and cyan, respectively, for spin-up and spin-down configurations. The impure guest atoms (the blue atom) show the spin-up (1:31 case) or spin-down (1:49) density in the positive half-plane of $z > 0$. The spin distributions present highly asymmetric characteristics, which play a substantial role in determining the net magnetic moments of the ferromagnetic systems. The spin-split magnetic moments near the Fermi level in the valence band might be due to compatible spin-up contributions of $2p_z$ and $(5d_{x^2-y^2}, 5d_{xy}, 5d_{yz}, 5d_{xz}, 5d_{z^2})$. Remarkably, the absolute values of spin up and spin down are listed in Tables 7.2, 7.3, and 7.4 based on the distribution of the magnitude of the spin moment in each case. The dominance of spin up is more likely stronger in the 1:31 case than in the 1:49 one for W. The total magnetic moments are 1:52 (1:31) and -1.51 (1:49) for W and 0.56 for Ir, which are helpful for crucial magnetic applications and engineering integrations.

Table 7.2. List of magnetizations of 1:31 W.

# of ion	**of**	**Ion**	s	p
1	0.002	0.028	−0.001	0.029
2	0.006	0.054	−0.001	0.059
3	0.006	0.057	−0.001	0.061
4	0.002	0.029	−0.001	0.03
5	0.002	0.039	−0.001	0.041
6	0.006	0.054	−0.002	0.058
7	0.006	0.057	−0.001	0.061
8	0.002	0.044	−0.001	0.045
9	0.002	0.045	−0.001	0.046
10	0.006	0.046	−0.001	0.05
11	0.006	0.046	−0.001	0.05
12	0.002	0.024	−0.001	0.025
13	0.002	0.04	−0.001	0.042
14	0.002	0.029	−0.001	0.03
15	0.002	0.024	−0.001	0.024
16	−0.001	−0.021	0.002	−0.021
17	−0.002	−0.025	0.002	−0.024
18	−0.007	−0.054	0.008	−0.053
19	−0.002	−0.023	0.002	−0.023
20	−0.002	−0.019	0.002	−0.018
21	−0.002	−0.019	0.002	−0.018
22	−0.007	−0.057	0.008	−0.057
23	−0.007	−0.057	0.008	−0.056
24	−0.002	−0.018	0.002	−0.018
25	−0.002	−0.02	0.002	−0.019
26	−0.002	−0.019	0.002	−0.018
27	−0.002	−0.025	0.002	−0.026
28	−0.001	−0.017	0.001	−0.017
29	−0.002	−0.018	0.002	−0.017
30	−0.002	−0.018	0.002	−0.018
31	−0.001	−0.018	0.001	−0.018
32	0.027	−0.004	1.267	1.29
Total	0.03	0.18	1.3	1.52

Table 7.3. List of magnetizations of 1:49 W.

#	of	Ion	s	p
1	−0.001	−0.015	0.001	−0.015
2	−0.001	−0.018	0.001	−0.019
3	−0.005	−0.043	0.001	−0.047
4	−0.005	−0.039	0.001	−0.043
5	−0.001	−0.015	0.001	−0.016
6	−0.001	−0.017	0.001	−0.017
7	−0.001	−0.018	0.001	−0.019
8	−0.005	−0.049	0.001	−0.053
9	−0.006	−0.043	0.001	−0.047
10	−0.002	−0.032	0.001	−0.033
11	−0.001	−0.026	0.001	−0.027
12	−0.002	−0.035	0.001	−0.036
13	−0.005	−0.055	0.001	−0.059
14	−0.005	−0.053	0.001	−0.057
15	−0.001	−0.024	0.001	−0.024
16	−0.001	−0.028	0.001	−0.028
17	−0.001	−0.026	0.001	−0.027
18	−0.001	−0.023	0.001	−0.023
19	−0.001	−0.022	0.001	−0.023
20	−0.001	−0.017	0.001	−0.018
21	−0.001	−0.02	0.001	−0.02
22	−0.002	−0.023	0.001	−0.024
23	−0.001	−0.017	0.001	−0.017
24	−0.001	−0.017	0.001	−0.017
25	0.001	0.013	−0.001	0.013
26	0.001	0.012	−0.001	0.012
27	0.002	0.022	−0.002	0.022
28	0.007	0.056	−0.008	0.055
29	0.002	0.023	−0.001	0.023
30	0.001	0.014	−0.001	0.014
31	0.001	0.012	−0.001	0.012
32	0.001	0.014	−0.002	0.013
33	0.007	0.052	−0.008	0.051
34	0.007	0.053	−0.008	0.053
35	0.001	0.015	−0.002	0.014
36	0.001	0.013	−0.001	0.013
37	0.001	0.013	−0.001	0.013
38	0.001	0.014	−0.002	0.013
39	0.002	0.02	−0.002	0.02

(Continued)

Table 7.3. (*Continued*)

#	of	Ion	s	p
40	0.001	0.012	−0.001	0.012
41	0.001	0.012	−0.001	0.011
42	0.001	0.014	−0.001	0.014
43	0.001	0.012	−0.001	0.012
44	0.001	0.011	−0.001	0.011
45	0.001	0.013	−0.001	0.013
46	0.001	0.013	−0.001	0.014
47	0.001	0.015	−0.001	0.014
48	0.001	0.014	−0.001	0.014
49	0.001	0.013	−0.001	0.014
50	−0.026	0.004	−1.253	−1.275
Total	−0.03	−0.20	−1.28	−1.51

Table 7.4. List of magnetizations of 1:31 Ir.

#	of	Ion	s	p
1	−0.001	−0.012	0.002	−0.012
2	−0.001	−0.014	0.002	−0.014
3	−0.001	−0.014	0.002	−0.014
4	−0.001	−0.013	0.002	−0.012
5	−0.001	−0.01	0.001	−0.009
6	−0.001	−0.014	0.002	−0.013
7	−0.001	−0.014	0.002	−0.013
8	−0.001	−0.01	0.001	−0.01
9	−0.001	−0.01	0.001	−0.01
10	−0.001	−0.013	0.002	−0.013
11	−0.001	−0.014	0.002	−0.013
12	−0.001	−0.01	0.001	−0.01
13	−0.001	−0.01	0.001	−0.01
14	−0.001	−0.009	0.001	−0.009
15	−0.001	−0.012	0.002	−0.012
16	0.002	0.044	−0.001	0.045
17	0.003	0.038	−0.001	0.04
18	0.003	0.04	−0.001	0.042
19	0.003	0.03	−0.001	0.032
20	0.001	0.017	0	0.018
21	0.001	0.015	0	0.016
22	0.003	0.058	−0.001	0.06
23	0.003	0.053	−0.001	0.055
24	0.001	0.011	−0.001	0.011

(*Continued*)

Table 7.4. (*Continued*)

#	of	Ion	s	p
25	0.001	0.02	−0.001	0.021
26	0.001	0.013	0	0.013
27	0.002	0.045	−0.001	0.047
28	0.002	0.019	−0.001	0.02
29	0.001	0.011	−0.001	0.012
30	0.001	0.014	0	0.015
31	0.002	0.028	−0.001	0.029
32	0.019	0.04	0.198	0.257
Total	0.03	0.32	0.21	0.56

7.7. Optical Properties

The optical properties of condensed-matter systems can be fully examined through dielectric functions, absorption, energy loss function (ELF), and reflectivity index. Among them, the dielectric functions are specific to the screening abilities, which are mainly caused by the incident electromagnetic waves. In this case, the energy and kinetic energy need to be conserved, while dielectric functions also need to satisfy the Pauli exclusion principle, the square root of dipole moments, and joint DOS. In general, the imaginary dielectric function can be identified by the formula

$$[\mathrm{Im}(\epsilon(\omega))] = \frac{4\pi e^2}{m^2\omega^2}\int_{BZ}^{\square} \frac{2}{(2\pi)^3}|\langle i|\hat{e}\cdot\vec{p}|f\rangle|^2 f(E^C) \times(1-f(E^C))\delta(E_f-E_i-\omega)d^3k.$$

It is noted that the complex dielectric functions include imaginary and real parts in Figs. 7.11(a) and 7.11(b) and have strong combinations using the K–K relation. This is calculated by the integration in the range of $-\infty$ to ∞ with the principal value $\varepsilon_1(\omega) = 1 + \frac{2}{\pi}\mathrm{P}\int_0^{\infty} \square\frac{\omega'\varepsilon_2(\omega)}{\omega'^2-\omega^2}d\omega$. In this system, xx and yy overlap with each other (shown in red color), and the zz direction is shown in blue. Nine prominent peaks come into existence in the imaginary part, which are denoted by P1, P2, P3, P4, P5, P6, P7, P8, and P9. The transitions are marked at the band structures in Fig.7.11(c), which is based on the flat band or band-edge state. In this case, P1 might be created due to the inner transitions of the valence or

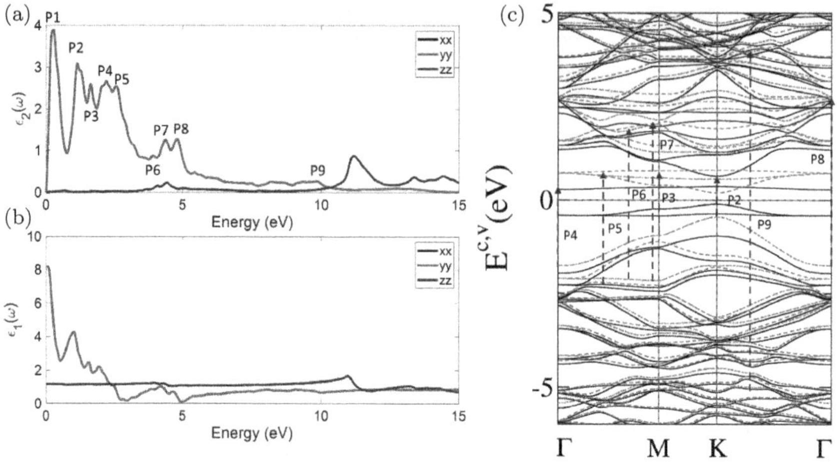

Fig. 7.11. Complex dielectric function: (a) Imaginary part $\epsilon_2(\omega)$ and (b) real part $\epsilon_1(\omega)$.

Table 7.5. Physical peaks along with band transitions, orbitals, and spins.

Peak numbers	Energy (eV)	Band transitions	Orbitals and spins
P1	0.252	—	—
P2	1.134	K	W (↑)-($5d_{x^2-y^2}$, $5d_{xy}$, $5d_{yz}$, $5d_{xz}$) – W (↓) ($5d_{x^2-y^2}$, $5d_{xy}$, $5d_{yz}$, $5d_{xz}$)
P3	1.617	M	W (↑)-($5d_{x^2-y^2}$, $5d_{xy}$, $5d_{yz}$, $5d_{xz}$) – W (↓) ($5d_{x^2-y^2}$, $5d_{xy}$, $5d_{yz}$, $5d_{xz}$)
P4	2.185	Γ	W (↑)-($5d_{x^2-y^2}$, $5d_{xy}$, $5d_{yz}$, $5d_{xz}$) – W (↑) ($5d_{x^2-y^2}$, $5d_{xy}$, $5d_{yz}$, $5d_{xz}$)
P5	2.563	Γ–M	W (↑)-($5d_{xz}$, $5d_{Z^2}$) – W (↑) (6s)
P6	3.907	Γ–M	W (↓)-($5d_{Z^2}$) – W (↑) ($5d_{x^2-y^2}$, $5d_{xy}$)
P7	4.369	Γ–M	W (↑)-($5d_{Z^2}$) – W (↓) ($5d_{x^2-y^2}$, $5d_{xy}$, $5d_{yz}$, $5d_{xz}$)
P8	4.789	Γ	W (↑) ($5d_{x^2-y^2}$, $5d_{xy}$, $5d_{yz}$, $5d_{xz}$) – W (↑) ($5d_{x^2-y^2}$, $5d_{xy}$, $5d_{yz}$, $5d_{xz}$)
P9	9.83	K – Γ	C (↑/↓) (2s, $2p_x$, $2p_y$, $2p_z$) – C (↓) (2s, $2p_x$, $2p_y$, $2p_z$)

conduction bands. Table 7.5 lists the specific peaks associated with their band positions and joint van Hove singularities. For instance, at the energy of 1.134 eV, the transition occurs at the K point and is classified as W (↑)-($5d_{x^2-y^2}$, $5d_{xy}$, $5d_{yz}$, $5d_{xz}$) – W (↓)-($5d_{x^2-y^2}$,

$5d_{xy}$, $5d_{yz}$, $5d_{xz}$), where W (↑) and W (↓) illustrate spin up and spin down, respectively. Similar analysis can be done for other peaks that appear in the imaginary part.

Other optical characteristics provide a full understanding of the substituted system. The ELF is known as the Coulomb excitation under the collective pictures, as shown in Fig.7.12(a). This physical quantity results in the plasmon modes in the diverse energy range. For the xx case, the highest peak is observed at 5.21 eV, which is known as the plasmon mode. In general, the ELF can be directly calculated by ELF $= \frac{\epsilon_2(\omega)}{\epsilon_1^2(\omega)+\epsilon_2^2(\omega)}$, which is mainly specific to the screening response abilities. Other collective modes come

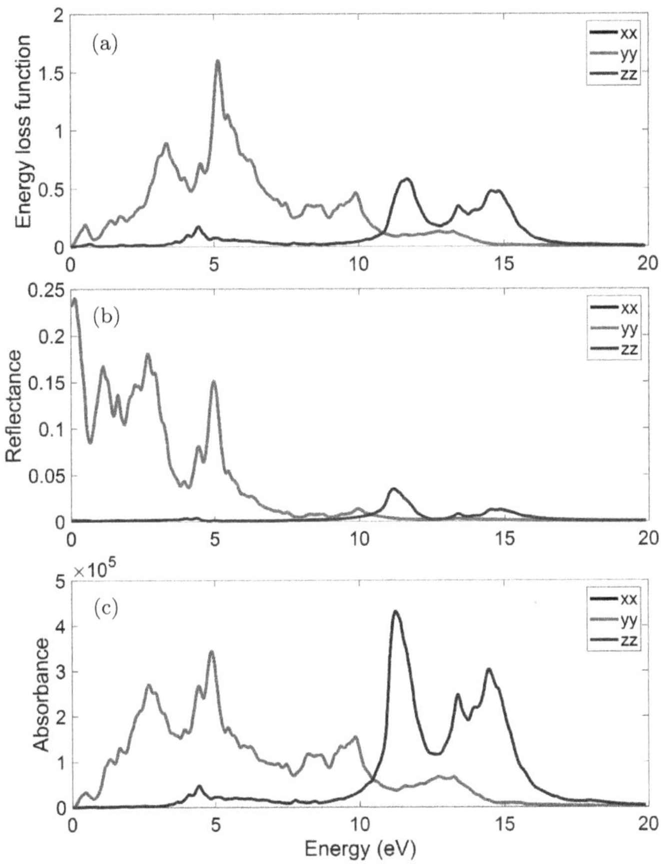

Fig. 7.12. Optical properties: (a) energy loss function, (b) reflectance, and (c) absorbance.

into existence due to the diverse valence and conduction subbands. The reflectance and absorbance are presented in Figs. 7.12(b) and 7.12(c). At low frequencies, this reflectance has a larger value than at higher frequencies, which might be due to the declination of an electromagnetic wave in the forbidden frequency range. Both reflectance and absorption are inversely proportional to the decay rate, thus leading to the appearance of the diverse peak. The reflectivity coefficient occurs with a value smaller than 1 because it is calculated using the ratio of the incident light flux and the outgoing component. These optical properties can be thoroughly tested through high-resolution spectroscopies.

7.8. Concluding Remarks and Future Perspectives

The extensive optimal stabilities and the electronic, magnetic, and optical properties of 5d transition metal-substituted graphene have been comprehensively investigated using first-principles simulations. The crucial chemical bonds of C–C, C–M (with M denoting W, Os, Ir, or Pt) are mainly determined by (2s, $2p_x$, $2p_y$, $2p_z$)-(2s, $2p_x$, $2p_y$, $2p_z$) and (2s, $2p_x$, $2p_y$, $2p_z$)-(6s, $5d_{x^2-y^2}$, $5d_{xy}$, $5d_{yz}$, $5d_{xz}$, $5d_{z^2}$). It is important to know that when we substitute the 5d guest atoms into the host atoms, a strong buckling is created due to hybridized characteristics. The spin configurations dominate in W- and Ir-based chemical modification systems, which belong to the Heisenberg model. The magnetic moments of W-(1:31), W-(1:49), and Ir-(1:31) are 1.52, −1.51, and 0.56, respectively. We also investigated the optical properties of W, including complex dielectric functions, absorption spectra, reflectance, and energy loss functions. The significant combination of the prominent peaks of the imaginary part, the band transitions, the joint DOSs, and spin configurations is clearly identified in this book chapter. We can use this strategy for chemical modifications of other group elements in the periodic table. Furthermore, many low-dimensional condensed-matter systems, especially the critical determination of prominent peaks in the optical spectra, can be systematically investigated.

References

[1] Nielsen, E.; Rahman, R.; Muller, R. P. A many-electron tight binding method for the analysis of quantum dot systems. *Journal of Applied Physics* **2012**, *112*, 114304.

[2] Li, L.; Zhu, X. Enhanced photocatalytic hydrogen evolution of carbon quantum dot modified 1D protonated nanorods of graphitic carbon nitride. *ACS Applied Nano Materials* **2018**, *1*, 5337–5344.

[3] Grushevskaya, H.; Krylov, G.; Kruchinin, S.; Vlahovic, B.; Bellucci, S. Electronic properties and quasi-zero-energy states of graphene quantum dots. *Physical Review B* **2021**, *103*, 235102.

[4] Zhang, X.; Huang, Y.; Wang, Y.; Ma, Y.; Liu, Z.; Chen, Y. Synthesis and characterization of a graphene–C60 hybrid material. *Carbon* **2009**, *47*, 334–337.

[5] Sanchís, J.; Milačič, R.; Zuliani, T.; Vidmar, J.; Abad, E.; Farré, M.; Barceló, D. Occurrence of C60 and related fullerenes in the Sava River under different hydrologic conditions. *Science of the Total Environment* **2018**, *643*, 1108–1116.

[6] He, X.; Léonard, F.; Kono, J. Uncooled carbon nanotube photodetectors. *Advanced Optical Materials* **2015**, *3*, 989–1011.

[7] Reznik, D.; Olk, C.; Neumann, D.; Copley, J. X-ray powder diffraction from carbon nanotubes and nanoparticles. *Physical Rreview B* **1995**, *52*, 116.

[8] Komatsu, N.; Shimawaki, T.; Aonuma, S.; Kimura, T. Ultrasonic isolation of toroidal aggregates of single-walled carbon nanotubes. *Carbon* **2006**, *44*, 2091–2092.

[9] Booshehri, L. G. Ultrafast and magneto-optical spectroscopy of excitons and phonons in carbon nanotubes, 2010.

[10] Liu, L.; Zhao, J. Toroidal and coiled carbon nanotubes. *Syntheses and Applications of Carbon Nanotubes and Their Composites* **2013**, 257–282.

[11] Chiu, C.; Chiu, Y.; Shyu, F.; Chang, C.; Chuu, D.; Lin, M.-F. Temperature-dependent carrier dynamics in metallic carbon nanotubes. *Physics Letter A* **2005**, *346*, 347–354.

[12] Dillon, A. C.; Jones, K.; Bekkedahl, T.; Kiang, C.; Bethune, D.; Heben, M. Storage of hydrogen in single-walled carbon nanotubes. *Nature* **1997**, *386*, 377–379.

[13] Kimouche, A.; Ervasti, M.; Drost, R.; Halonen, S.; Harju, A.; Joensuu, P. M.; Sainio, J.; Liljeroth, P. Ultra-narrow metallic armchair graphene nanoribbons. *Nature Communications* **2015**, *6*, 1–6.

[14] Ribeiro Jr, L. A.; da Cunha, W. F.; Fonseca, A. L. D. A.; e Silva, G. M.; Stafström, S. Transport of polarons in graphene nanoribbons. *The Journal of Physical Chemistry Letters* **2015**, *6*, 510–514.

[15] Lin, S. Y.; Tran, N. T. T.; Chang, S. L.; Su, W. P.; Lin, M. F. *Structure- and Adatom-Enriched Essential Properties of Graphene Nanoribbons.* CRC Press, 2018.
[16] Chung, H.; Huang, Y.; Lee, M.; Chang, C.; Lin, M. Quasi-landau levels in bilayer zigzag graphene nanoribbons. *Physica E: Low-dimensional Systems and Nanostructures* **2010**, *42*, 711–714.
[17] Tran, N. T. T.; Lin, S.-Y.; Lin, C.-Y.; Lin, M.-F. *Geometric and Electronic Properties of Graphene-related Systems: Chemical Bonding Schemes.* CRC Press, 2017.
[18] Geim, A. K.; Novoselov, K. S. *Nanoscience and Technology: A Collection of Reviews from Nature Journals.* World Scientific, 2010, pp. 11–19.
[19] Giesbers, A.; Ponomarenko, L.; Novoselov, K.; Geim, A.; Katsnelson, M.; Maan, J.; Zeitler, U. Gap opening in the zeroth Landau level of graphene. *Physical Review B* **2009**, *80*, 201403.
[20] Castro Neto, A. H.; Guinea, F.; Peres, N. M. R.; Novoselov, K. S.; Geim, A. K. The electronic properties of graphene *RvMP* **2009**, *81*, 109–162.
[21] Krasheninnikov, A.; Lehtinen, P.; Foster, A. S.; Pyykkö, P.; Nieminen, R. M. Embedding transition-metal atoms in graphene: Structure, bonding, and magnetism. *Physical Review Letters* **2009**, *102*, 126807.
[22] Santos, E. J.; Ayuela, A.; Sánchez-Portal, D. First-principles study of substitutional metal impurities in graphene: Structural, electronic and magnetic properties. *New Journal of Physics* **2010**, *12*, 053012.
[23] Alonso-Lanza, T.; Ayuela, A.; Aguilera-Granja, F. Substitutional 4d and 5d impurities in graphene *Physical Chemistry Chemical Physics* **2016**, *18*, 21913–21920.
[24] Bussolotti, F.; Chi, D.; Goh, K. J..; Huang, Y. L.; Wee, A. T. *2D Semiconductor Materials and Devices.* Elsevier, 2020, pp. 199–220.
[25] Damascelli, A. Probing the electronic structure of complex systems by ARPES. *Physica Scripta* **2004**, *2004*, 61.
[26] Hwang, E.; Sarma, S. D. Quasiparticle spectral function in doped graphene: Electron-electron interaction effects in ARPES. *Physical Review B* **2008**, *77*, 081412.
[27] Grimme, S.; Antony, J.; Ehrlich, S.; Krieg, H. A consistent and accurate ab initio parametrization of density functional dispersion correction (DFT-D) for the 94 elements H-Pu. *The Journal of Chemical Physics* **2010**, *132*, 154104.
[28] Perdew, J. P.; Burke, K.; Ernzerhof, M. Generalized gradient approximation made simple. *Physical Review Letters* 1996, *77*, 3865.

[29] Ernzerhof, M.; Scuseria, G. E. Assessment of the Perdew–Burke–Ernzerhof exchange-correlation functional. *The Journal of Chemical Physics* **1999**, *110*, 5029–5036.

[30] Rangel, T.; Caliste, D.; Genovese, L.; Torrent, M. A wavelet-based projector augmented-wave (PAW) method: Reaching frozen-core all-electron precision with a systematic, adaptive and localized wavelet basis set. *Computer Physics Communications* **2016**, *208*, 1–8.

CHAPTER 8

SUBSTITUTIONS OF 4F RARE-EARTH ELEMENTS INTO GRAPHENE

Nguyen Thanh Tuan and Ming-Fa Lin

Department of Physics, National Cheng Kung University, Tainan, Taiwan

Abstract

This study investigates the electronic and magnetic properties of 4f-rare earth metal substitutions in graphene monolayers using first-principles calculations. The incorporation of rare earth metals into the graphene lattice modifies its electronic structure and introduces a significant bandgap, enabling potential semiconductor device applications. The presence of 4f electrons leads to localized magnetic moments, offering possibilities for spintronic applications. Hybridized states between the carbon atoms and 4f orbitals generate unique electronic and magnetic phenomena, including magnetic anisotropy and spin polarization. This research provides insights into the design of graphene-based materials with tailored functionalities, facilitating their implementation in future electronic and spintronic devices.

Keywords: Graphene monolayer, 4f rare-earth metals, substitutions.

Each element in the periodic table exhibits a unique atomic configuration. The properties of these elements enable them to play critical roles in the basic and applied sciences. The various chemical bindings found in these elements decide both the structures of well-defined and irregular crystals. According to current scientific research, the concise orbital hybridization mechanisms and the ferromagnetic, antiferromagnetic, or non-magnetic spin configurations are closely related to the following three types of active orbitals, which corresponds to atomic configurations: (1) (s, p_x, p_y, p_z) orbitals about Group-I–Group-VIII condensed-matter systems, (2) (d_{z^2}, $d_{x^2-y^2}$, d_{xy}, d_{yz}, d_{xz}) orbitals of transition-metal atoms (e.g., Sc, Ti, Fe, Co, Ni, Cu, Ag, and Au with 3d or 4d orbitals, and (3) (f_{x^3}, f_{xz^2}, f_{yz^2}, f_{xyz}, $f_{z(x^2-y^2)}$, $f_{x(x^2-3y^2)}$, $f_{y(3x^2-y^2)}$) orbitals of rare-earth metal atoms (e.g., La, Ce, Pr, Sm, Eu, Gd, Yb, and Lu with 4f orbitals). Generally, the rich single- or multi-orbital hybridizations, which create whole condensed-matter systems, mainly cover the pure and impure bindings for sp^3–sp^3, sp^3–d^5, sp^3–f^7, d^5–d^5, d^5–f^7, and f^7–f^7. Besides, the reduced orbital mixings could generate the diversified high-symmetry crystal, such as s–s (e.g., Group-I elements: alkali metals), s–sp^3 (Group I, alkali metal-based compounds, such as hydrogenated graphene,[1] silicene,[2] germanene,[3] tinene,[4] and plumbene[5]), p_z–p_z, and (s, p_x, p_y, p_z) – (s, p_x, p_y, p_z) (e.g., graphene-related systems); similarly, we have s–d^5, s–f^7, p–d^5, and p–f^7. In addition, the former might create prominent spin configurations of ferromagnetism or antiferromagnetism, where the d^5 and f^7 orbitals play a major role. Obviously, the strong relationships between charges/spins and composite quantum quasi-particles have been successfully showed in previously published books, and they will be greatly expanded in the current and near-future studies.

Specifically, the four active orbitals (2s, $2p_x$, $2p_y$, $2p_z$) present in carbon atoms are capable of forming a number of graphene-related 3D,[6] 2D,[7] or 1D[8] materials through sp^2-dominated chemical bindings, such as the 3D AA-,[9] AB-,[10] or ABC-[11] stacked bulk graphite, 2D layered graphene systems with AAA,[12] ABA,[13] ABC,[14] or AAB[15] stacking configurations, and 1D coaxial carbon nanotubes in the presence of rotation symmetries,[16] graphene nanoribbons under the finite-width quantum confinements and the chiral,[17] achiral open edges, carbon fullerenes with closed surfaces, carbon disks, and carbon chains,[18] rings,[18] or tori.[19] Most importantly, the low-energy quasi-particle behaviors have been identified as being driven by the carbon-$2p_z$ orbitals because of their almost perpendicularities with the deeper-energy σ-electronic states of (2s, $2p_x$, $2p_y$, $2p_z$). These features have simultaneously led to great progress through first-principles simulations, phenomenological models, and experimental observation, including the largely diversified phenomena pertaining to many physical properties, such as geometric, electronic, magnetic, Coulomb excitation and decay, optical, transport, mechanical, and thermal properties. Furthermore, highly non-uniform and anisotropic orbital-mixing environments are clearly exhibited by the above-mentioned materials; therefore, chemical modifications could be a very powerful technique for creating rich and unique composite quasi-particles. Besides, the following efficient modulations have been achieved through state-of-the-art experimental syntheses: (I) chemisorption (e.g., alkalizations,[20] hydrogenations,[21] oxidizations,[22] and halogenations[23]); (II) substitutions (SiC,[24] GeC,[25] and SnC[26]); (III) intercalations,[27] de-intercalations,[28] (donor- or acceptor-type graphite intercalation compounds), decorations (planar, curved, folded, or scrolled graphene nanoribbons[29]); (V) bilayer composites (C–BN/ –Si/ –Ge/ –Sn/ –Pb);[30] and (VI) unusual heterojunctions.[31] Importantly, the diversified chemical bindings, corresponding to the host–host, host–guest, and guest–guest bonds, are expected to exhibit a lot of unique quasi-particle behaviors through delicate examinations.

For the rare-earth elements, there are seven kinds of active f orbitals that can exhibit unique quasi-particle phenomena. Many well-known compounds are synthesized with these elements, including high-temperature superconductors, notably quaternary YbBaCuO compounds with a critical temperature of ~90 K.[32] Fermion compounds with massive low-temperature heat capacities are also a prominent example (the ratio of the effective and bare mass is ~1000 in Cu_6/ $CeAl_3$/ $CeCu_2Si_2$/ $YbAl_3$/ UBe_{13}/ UPt_3.[33–35] Besides that, there are intermetallic systems, high-performance materials for ion-based batteries (such as the quaternary $LiLaTiO_4$ electrolyte compound[36]). These above-mentioned mainstream materials have been successfully synthesized in experimental laboratories and are able to provide outstanding platforms for fully exploring unusual scientific phenomena, the most prominent inconsistency is the magnetic field configuration, so many critical devices are built up and down magnet/ super-high magnetic trains in transports and SQUID for measurements. These are produced by seven kinds of f^7multi-orbital hybridizations and their rich configurations, which lead to the above-mentioned quasi-particle behaviors. Interestingly, the composite quasi-particle behaviors are expected to emerge after chemical absorptions, decorations, heterojunctions, superpositions, or substitutions, even when these properties are combined with other elements. In a specific way, the rare-earth adatoms can exhibit effective adsorption,[37] absorption,[38] such as oxidization[39] and hydrogenations[40] on/in graphene systems; on the other hand, C–C, C–R, and R–R bonds (R: rare-earth elements) play an important role in explaining essential properties, such as crystal symmetry, the features of energy band structure, the merged van Hove singularities, the covalent charge density distribution, the net magnetic moment within a primitive unit cell, the unusual spin-up and/or spin-down arrangements, and optical transitions with excitonic effects. The VASP simulation package[41] is very useful for calculations and establishing the phenomenological models of material systems based on the density function theory,[42] specifically, the reliable parameters of the orbital and spin dependence in the Hubbard-like tight-binding model.[43]

La- and Gd-enriched quasi-particle behaviors are created through chemical absorptions, adsorptions, or substitutions on monolayer graphene, mainly owing to the very complicated chemical bonding, such as C–C, C–La/C–Gd, and La–La/Gd–Gd bonds. The first-principles simulations are available for thoroughly examining the concise orbital mixings and spin configurations. Besides, the calculated results are consistent with one another under a unified quasi-particle framework, as clearly illustrated in previously published books. They consist of the optimal absorbed structures with hollow, bridge, or top sides and C–C, C–La/C–Gd, and La–La/Gd–Gd bond lengths; the featured energy band structure and wave functions of, say, the semiconductor or metal type; the existence or disappearance of the well-defined π and σ bands; the modification, distortion, or absence of the Dirac cone structures; the host- and guest-atom dominances at specific energies; the spatial charge density distributions of the observable variations on the (x, z)/(y, z)/(x, y) planes through the host–guest bonds with C-2p_z and La-d^5 or C-2p_z and Gd-f^7; the atom-, orbital-, and spin-projected van Hove singularities with many merged structures from the significant orbital couplings;

the modulate spin configurations shown through ferromagnetic or antiferromagnetic arrangements on the guest and/or honeycomb lattices; and the prominent absorption spectra of the dielectric function, energy loss function, reflectance, and absorption coefficient, with close relationships between photons and the dominating orbitals of the initial and final states in the presence of excitonic effects.

8.1. La- and Gd-adatom Substitutions in Graphene Monolayer

A graphene monolayer has a hexagonal lattice structure. The size of the unit cell is 2.46 Å, and the bonding length of C–C in the lattice is 1.42 Å;[44] therefore, the graphene structure is a planar sheet with sp^2 hybridization of the carbon atoms (Figs. 8.1(a) and 8.1(b)). The calculations pertaining to the graphene structure and the La- and Gd-adatom substitutions in graphene sheets are based on the density functional theory (DFT) and performed through the VASP simulation package[45] via first principles.[46] For this chapter, the rare-earth atoms substituted in graphene are La and Gd, both of which exhibit strong magnetism, which is a characteristic feature of rear-earth metal elements. The substitutions are performed for two concentrations, 1:32 and 1:50, in the supercells of graphene. With the increase in radius, the structure of pristine graphene undergoes a significant change that is exhibited through a change in lattice parameters and bond lengths.

For the first case of La-atom substitution in a graphene sheet, the results are shown in Table 8.1 and Figs. 8.1(c)–8.1(f). The absorption of the large-size La atoms into the graphene structure at different concentrations leads to a change in the lattice constant of the hexagonal lattice and the C–C bond lengths. In the 1:32 concentration case, the lattice parameter is 10.24 Å, the bond length of La–C is 2.02 Å, and that of C–C is in the range of 1.37–1.48 Å, and there is no buckling in this case; therefore, the σ and π bondings are orthogonal. Besides, the sp^2 hybridization prevails due to the post-substitution graphene remaining in the planar form. Similarly, for the 1:50 concentration case, the substitutions in the graphene structure show differences in size compared to the graphene monolayer: the lattice

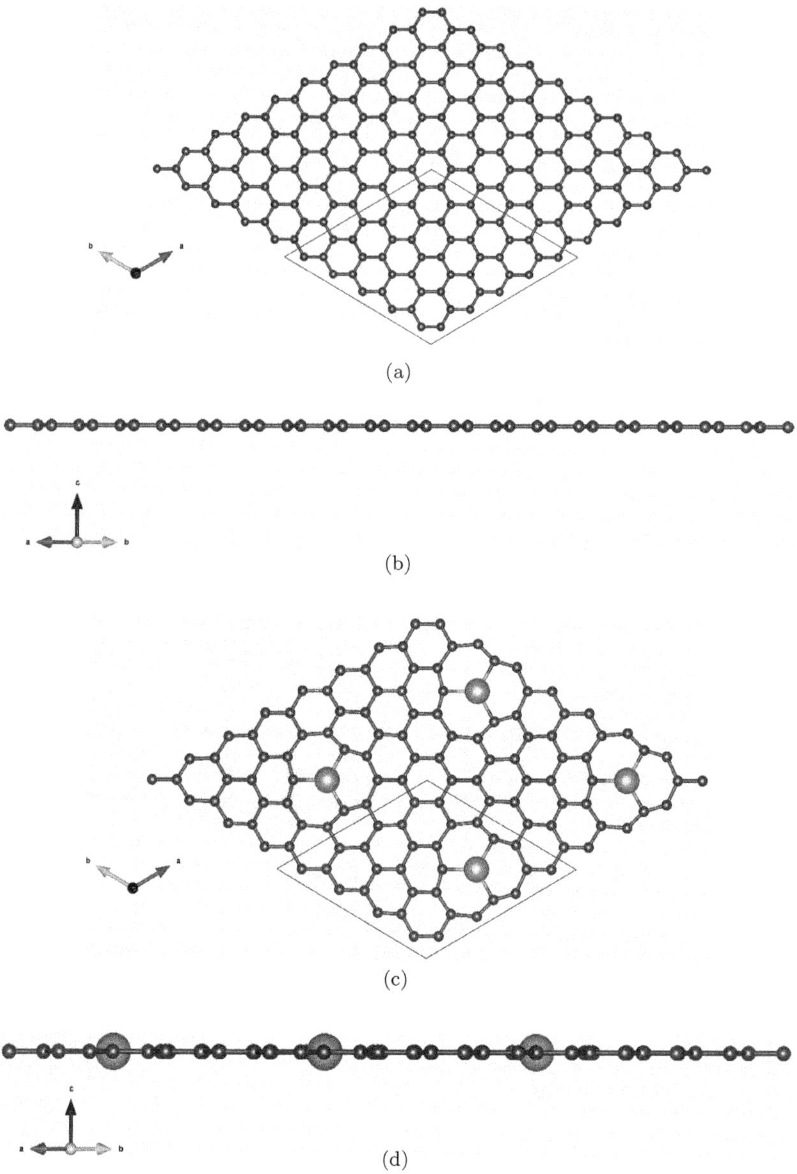

Fig. 8.1. (a) Top view and (b) side view of graphene monolayer. (c) Top view and (d) side view of La-atom-substituted graphene with La:C = 1:32. (e) Top view and (f) side view of La-atom-substituted graphene with La:C = 1:50. (g) Top view and (h) side view of Gd-atom-substituted graphene with Gd:C = 1:32. (i) Top view and (j) side view of Gd-atom-substituted graphene with Gd:C = 1:50.

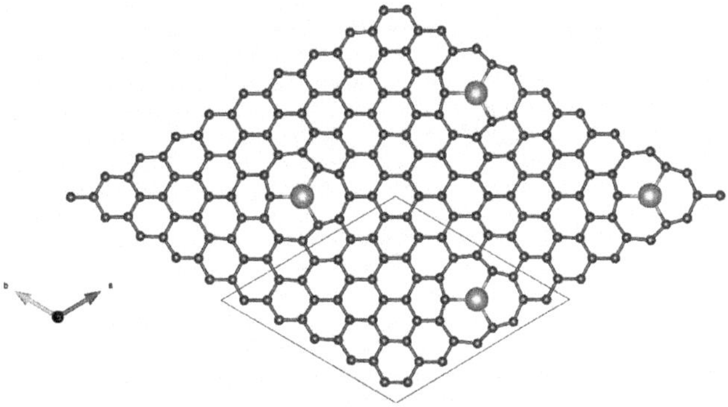

(e)

(f)

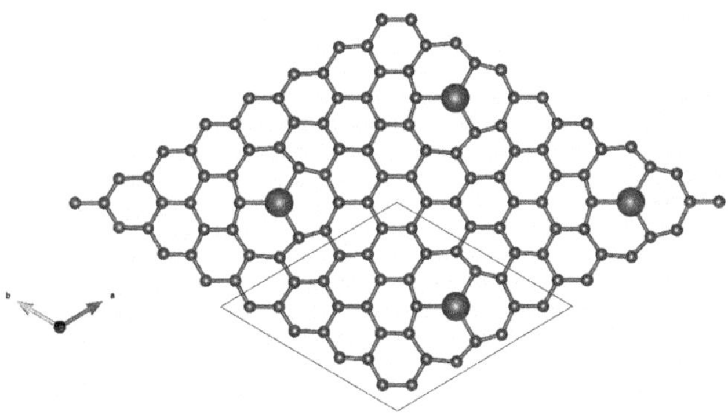

(g)

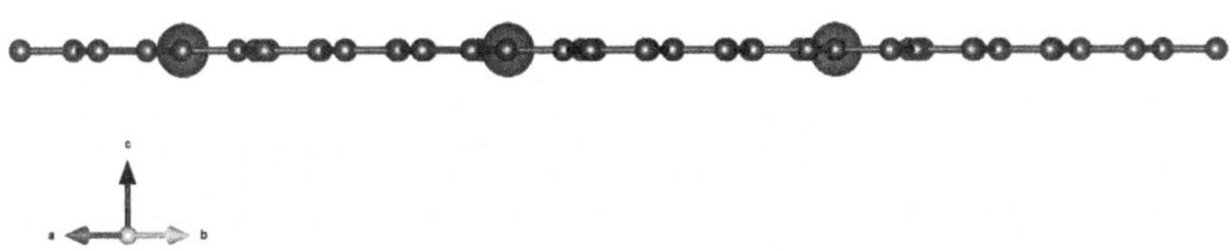

(h)

Fig. 8.1. (*Continued*)

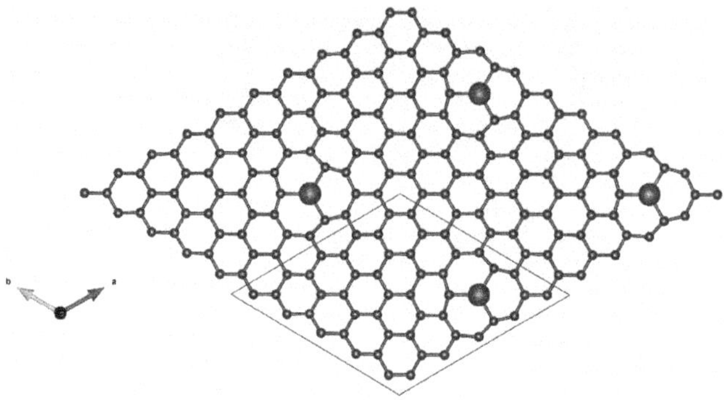

(i)

(j)

Fig. 8.1. (*Continued*)

Table 8.1. Geometric structure data after optimization of graphene and rare-earth atom substituted in graphene.

	Lattice parameters	Buckle	Bond lengths (C–C) (Å)	Bond lengths (R–C) (Å)
Graphene	2.46	0	1.42	
La–Graphene 1:32	10.24	0	1.37–1.48	2.02 Å
La–Graphene 1:50	12.63	0	1.37–1.46	2.01 Å
Gd–Graphene 1:32	10.22	0	1.38–1.48	1.98 Å
Gd–Graphene 1:50	12.58	0	1.37–1.46	1.91 Å

parameter increases to 12.63 Å, as with the 1:32 case; the structure of pristine graphene is also significantly changed, with the La–C bond length becoming 2.01 Åand the C–C bond length becoming 1.37–1.46 Å. In general, the carbon atoms near the La atoms tend to be pushed away due to the size of the rare-earth element atoms, which are much larger than that of the carbon atoms.

Thus, the pristine structure of graphene is strongly influenced by the La atoms.

For the case of Gd atoms, the hexagonal structure of the supercell is maintained for both concentrations, but the pristine structure of graphene is changed. After substitution, the planar sheet remains flat without the buckling effect. Therefore, the π- and σ-bonds are orthogonal; this structure is usually relatively sturdy for flat structures, such as in graphene. However, the σ-bonds generally vary in strength, corresponding to the bond lengths of C–C according to their positions in the lattice. The Gd-atom substitution in graphene corresponds to a 4×4 supercell without a buckled structure. The C–C bond lengths are in the range of 1.38–1.48 Å, and the Gd–C bond length is 1.98 Å. The nearest carbon atoms to the Gd atom are pushed away, resulting in the distribution of carbon atoms and a significant alteration of the length of the C–C bond. Besides, in the case of a 5×5 supercell of graphene substituted with Gd atoms, the C–C bond lengths are in the range of 1.37–1.46 Å, and the Gd–C bond length is 1.91 Å. Also, the weak and strong C–C bonds are different depending on their bond lengths due to the large size of rare-earth atoms, resulting in a significant change from pristine graphene (Figs. 8.1(g)–8.1(j)).

8.2. Featured Band Structures

Graphene has a planar structure with a hexagonal unit cell, leading to the reciprocal lattice also having a hexagonal structure; therefore, the band energy is plotted based on the high-symmetry points, Γ, M, and K.[47] Besides, the high-symmetry structure, the band diagram features a linear isotropic Dirac cone around the K point.[48] In particular, the conduction and valence bands of this material coincide with the Fermi level at the K point, which shows that the bandgap of a graphene monolayer is zero, thus causing it to behave as a zero-gap semiconductor. For the valence band, the π-band is constructed from the $2p_z$ orbital that builds the π-bonds in graphene. The σ-band is built from three σ-bands, namely 2s, $2p_x$, and $2p_y$. The π and σ bands exhibit crossing behavior because the graphene monolayer has a planar sheet structure. The group velocity is zero at the M point, which is a critical point for the band-edge state, due to the strong Van Hove singularity at this point (Fig. 8.2(a)).

After substituting the rare-earth La atom in the graphene monolayer, the hexagonal lattice structure remains the same, but the lattice parameter and bond length of C–C increase because the rare-earth atoms are bigger than the C atoms and the buckling of

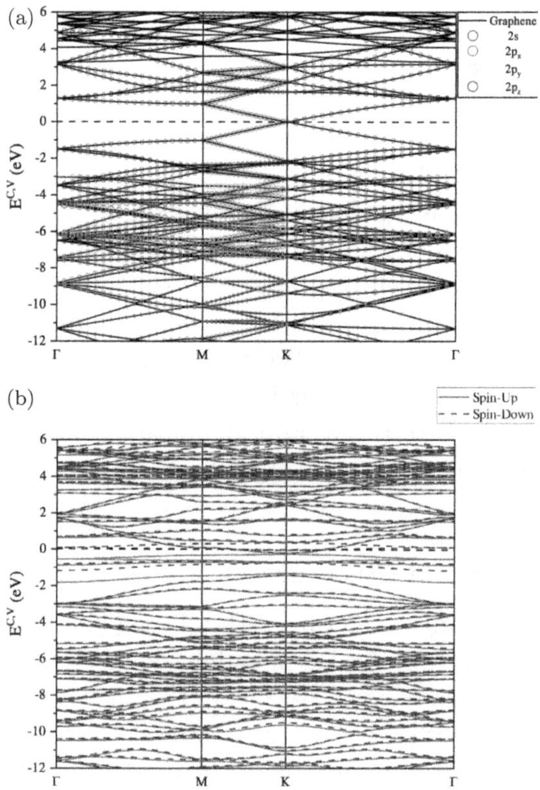

Fig. 8.2. Quasi-particle band structure of (a) graphene, (b) La atoms substituted in graphene – La:C = 1:32, (c) C-atom-dominated energy spectrum – La:C = 1:32, (d) La-atom-dominated energy spectrum – La:C = 1:32, (e) La atoms substituted in graphene – La:C = 1:50, (f) C-atom-dominated energy spectrum – La:C = 1:50, (g) La-atom-dominated energy spectrum – La:C = 1:50, (h) Gd-atoms substituted in graphene – Gd:C = 1:32, (i) C-atom-dominate energy spectrum – Gd:C = 1:32, (j) Gd-atom dominate energy spectrum – Gd:C = 1:32, (k) Gd-atom-substituted in graphene – Gd:C = 1:50, (l) C-atom-dominated energy spectrum – Gd:C = 1:50, and (m) Gd-atom-dominate energy spectrum – Gd:C = 1:50.

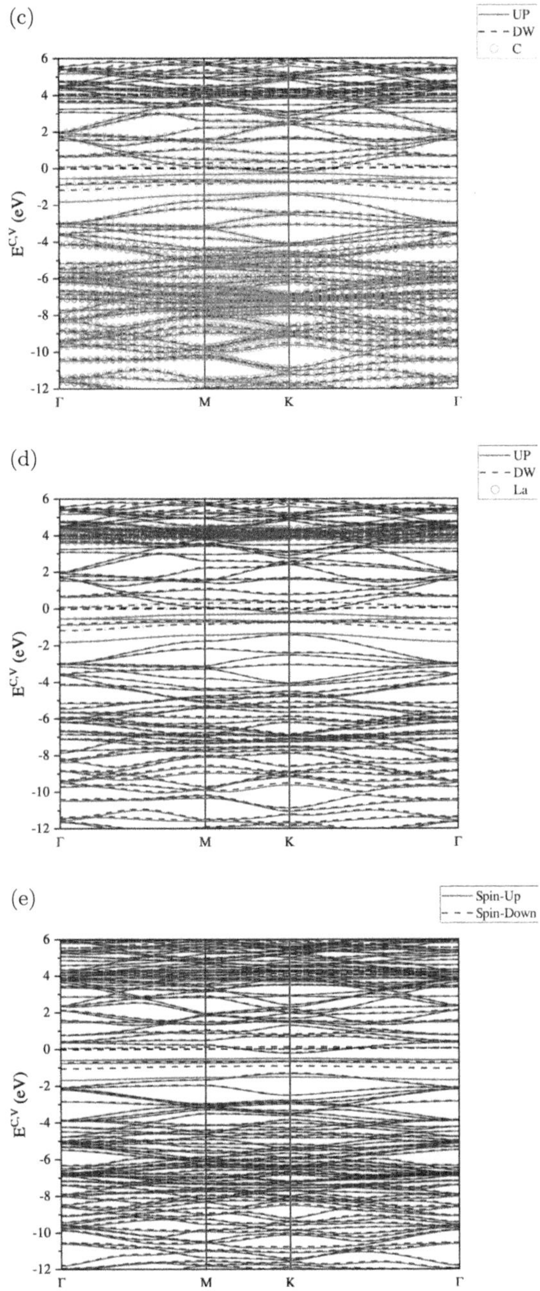

Fig. 8.2. (*Continued*)

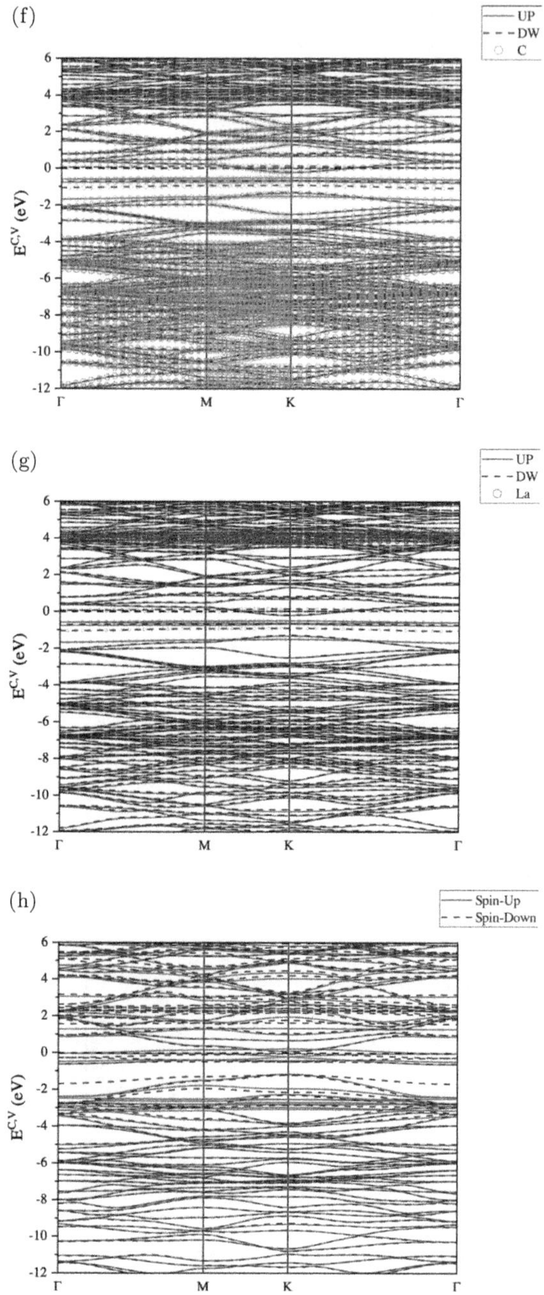

Fig. 8.2. (*Continued*)

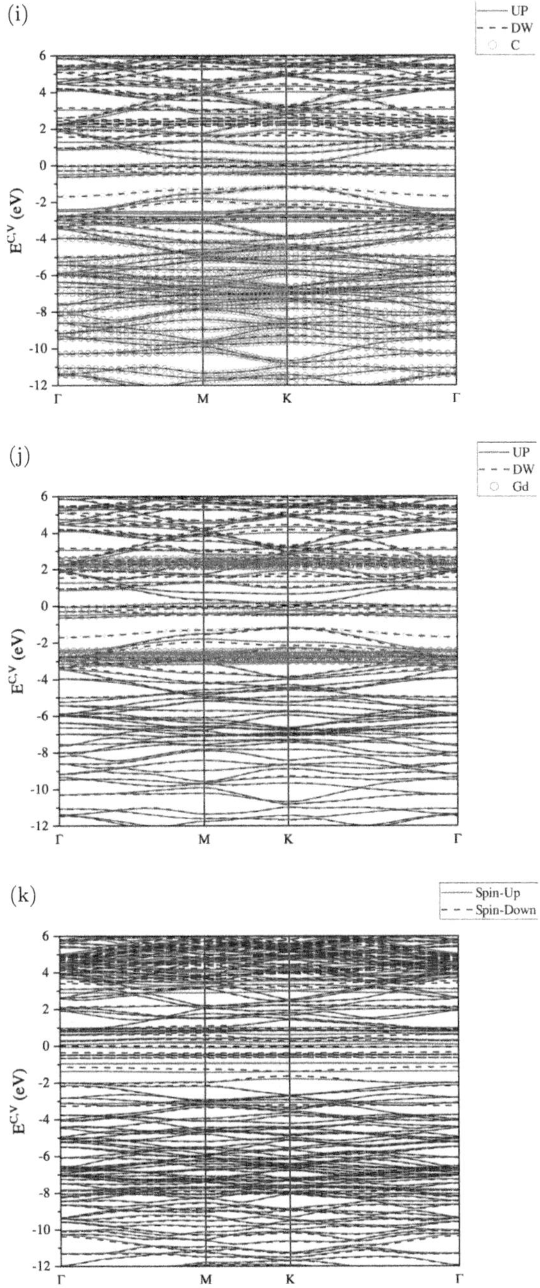

Fig. 8.2. (*Continued*)

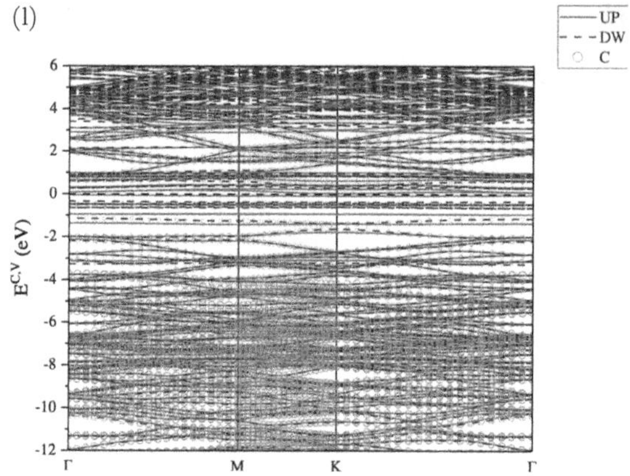

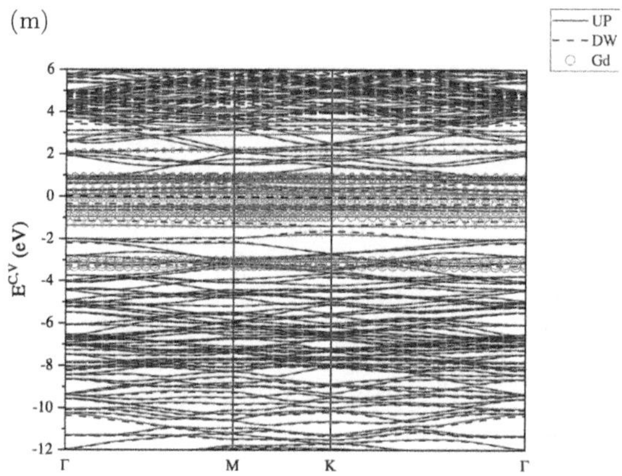

Fig. 8.2. (*Continued*)

the layer is zero; therefore, the σ- and π-binding are orthogonal, and they remain well defined and exhibit crossing behavior. However, the C–C bond length in different materials has different strengths and weaknesses corresponding to the bond length. After substituting the La atom, there appear many energy subbands over a large range of the band structure because of the contributions of electrons in the 6s and 5d orbitals of the La atom. The σ-bands undergo many modifications due to chemical modification. With the large diameter and the complexity of the distribution of electrons and orbitals, the La atoms are substituted in graphene, leading to the breaking of Dirac cones at the K points and causing the band structure of this material to be significantly changed compared to the energy band of pristine graphene. That means the graphene monolayer becomes metallic after substitution with La atoms. Besides, because La is a rare-earth element, magnetization is considered one of the outstanding properties of this element. This leads to the splitting of the bands, with the red color denoting spin up and the blue color spin down. Significant splitting around the Fermi level occurs due to the principal domination of the La atoms around this region and the lead zone. Additionally, for the 1:32 concentration case, the energy spectrum dominated by the quasi-particle band structure of the C atoms shows clearly the dominance of the 2s orbital over a wide range from −18 to −11 eV, the $2p_x$ and $2p_y$ orbitals from −12 to −4 eV, and the $2p_z$ orbital from −6 to 6 eV in the valence band. These orbitals support the building of the σ- and π-bonding. With La atoms, their quasi-particle band structure dominates the energy spectrum, with the 5d and 6s orbitals significantly dominating the distribution in the conduction band because La is a rare-earth metal that has a complex electronic configuration. For the 1:50 case, the energy spectrum dominated by the quasi-particle band structure of the C atoms clearly exhibits the dominance of the 2s orbital over a wide range from −18 to −11 eV , the $2p_x$ and $2p_y$ orbitals from −12 to −4 eV, and the $2p_z$ orbital from −6 to 6 eV in the valence band. These orbitals support the building of the σ- and π-bondings. Also, the quasi-particle band structure of the La atoms in the energy band of the 5d and 6s orbitals is significantly distributed in the conduction band in the range of 3–5 eV and around the Fermi level (Fig. 8.2(b)–8.2(g)).

For the Gd-atom substitution in a graphene monolayer, due to the contribution of orbitals ($5d_{z^2}$, $5d_{x^2-y^2}$, $5d_{xy}$, $5d_{yz}$, $5d_{xz}$) and ($4f_{x^3}$, $4f_{xz^2}$, $4f_{yz^2}$, $4f_{xyz}$, $4f_{z(x^2-y^2)}$, $4f_{x(x^2-3y^2)}$, $4f_{y(3x^2-y^2)}$) of the rare-earth element, many subbands are formed, which are clearly exhibited in the band structure. Also, the supercells with hexagonal lattice that are characteristic of graphene remain the same, although the Gd atom is substituted, so the band energy of Gd–graphene is still plotted along the high-symmetry points, Γ, M, and K in the first Brillouin zone, as in the graphene monolayer. Because Gd is a magnetic element, there are two colored lines in the band diagram, with the red color corresponding to the spin-up and the blue color to the spin down of the electrons in the atom. The metallic feature is clearly demonstrated through the breaking of the Dirac cone due to the large diameter and the major contribution of the orbitals in the 4f and 5d subshells of the Gd atom. There are three σ-bands constructed by the 2s, $2p_x$, and $2p_z$ orbitals of the C atom in the material, which form σ-bonding in the graphene sheet. Besides, the π-band belongs to the $2p_z$ orbital of the C atom, which builds the π-bonding. Because the planar structure of graphene is preserved, the π- and σ-bands exhibit crossing behavior. With the 1:32 concentration, in the C atom, the energy spectrum dominated by the quasi-particle band structure of carbon elements exhibits the contribution of the orbitals of the C atoms in the material: 2s in the range from -12 to -5 eV, $2p_x$ and $2p_y$ from -11 to 4 eV, $2p_z$ from -6 to 6 eV in the energy band. Moreover, the orbitals ($5d_{z^2}$, $5d_{x^2-y^2}$, $5d_{xy}$, $5d_{yz}$, $5d_{xz}$) and ($4f_{x^3}$, $4f_{xz^2}$, $4f_{yz^2}$, $4f_{xyz}$, $4f_{z(x^2-y^2)}$, $4f_{x(x^2-3y^2)}$, $4f_{y(3x^2-y^2)}$) of Gd also contribute significantly the around Fermi level in the range of -2 to 2 eV. Also, for the 1:50 concentration, the domination of the C atoms in the energy band also distribute in the depths of the energy range for orbital 2s, because which is deeper than the orbital $2p_x$, $2p_y$ and $2p_z$ outermost of atom. For the $2p_x$ and $2p_y$ orbitals, the domination is in the range of -12 to -4 eV, for the $2p_z$ orbital, it is the range of -8 to 6 eV. Besides, the dominance of Gd also contributes significantly around the Fermi level in the range of -4 to 2 eV (Fig. 8.2(h)–8.2(m)).

8.3. Charge and Spin-density Distributions

For the La-atom substituted graphene monolayer in the 1:32 concentration case, the multi-orbital hybridizations in the C–C bonds are directly revealed in the spatial charge density distributions. The π- and σ- bonds of the carbon atoms and La–C bonds are shown clearly in the top and side views. The top view reveals a significant hybridization of the 2s and ($2p_x$, $2p_y$) orbitals, which form sp^2 hybridization and construct the σ-bonds. The σ-bonds, denoted by red triangles, showed a weakening of the bond after being substituted, which was caused by the lengthening of the graphene sheet because La has a large atom. Besides, the overlap of the orbitals C-p_z and La-($5d_{z^2}$, $5d_{x^2-y^2}$, $5d_{xy}$, $5d_{yz}$, $5d_{xz}$) creates a very weak metallic C–La bond. For the same concentration, the charge density distribution of the Gd-atom-substituted graphene monolayer also clearly exhibits the same results. In general, the change in lattice parameters after rare-earth atoms are substituted into the pristine graphene sheet leads to strengthening or weakening depending on the change in the σ-bond. On the other hand, there is overlapping of the orbitals C-$2p_z$ and Gd-6s-($5d_{z^2}$, $5d_{x^2-y^2}$, $5d_{xy}$, $5d_{yz}$, $5d_{xz}$) – ($4f_{x^3}$, $4f_{xz^2}$, $4f_{yz^2}$, $4f_{xyz}$, $4f_{z(x^2-y^2)}$, $4f_{x(x^2-3y^2)}$, $4f_{y(3x^2-y^2)}$). This causes the metallic bonds to be very weak compared to the C–C bonds, making the starting material transform from a zero-gap semiconductor to metallic (Figs. 8.3(a)–8.3(f)).

For the concentration of 1:50, the substitution in graphene monolayer with the lanthanum element also exhibits similar results. The top and side views of charge density distributions clearly show the π- and σ-bonds of carbon, which characterize the monolayer graphene structure. The sp^2 hybridization is formed by the characteristic hybridization of the orbitals of the C atom, i.e., 2s with $2p_x$, $2p_y$ orbitals, which is presented in the top view of the charge distribution picture. Besides, the σ-bonds produced by sp^2 hybridization, denoted by the green area between the carbon and lanthanum atoms, are very weak compared to the σ-bonds of graphene monolayer due to the fact that lanthanum is a rare-earth element with a very large size. Also, the side view of the charge density picture also clearly shows significant weakening of the σ-bond as well as the formation of the carbon–lanthanum metallic bond made by the overlap of the orbitals C-$2p_z$ and La-6s-($5d_{z^2}$, $5d_{x^2-y^2}$, $5d_{xy}$, $5d_{yz}$, $5d_{xz}$). Also, in the Gd-atom-substituted graphene monolayer in the 1:50 concentration case, similar to the other concentration case, the σ-bonds of the C atoms in the graphene sheet change significantly after the Gd atom is substituted due to the low charge density between the C atoms. Besides, the lattice parameter of graphene after substitution with rare-earth atoms changes, leading to corresponding changes in the σ-bonds. Moreover, the Gd–C bonding is also constructed by the

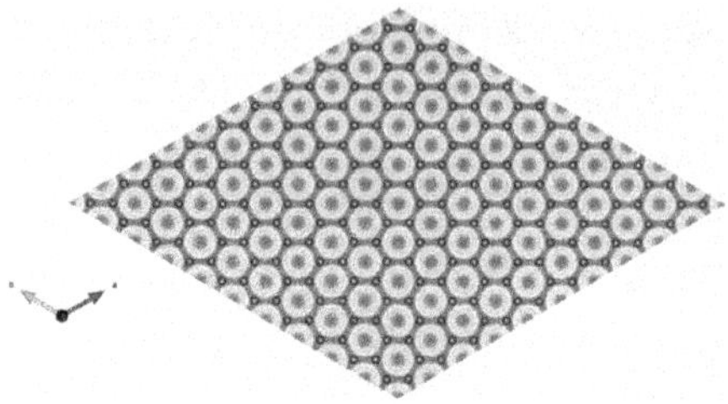

(a)

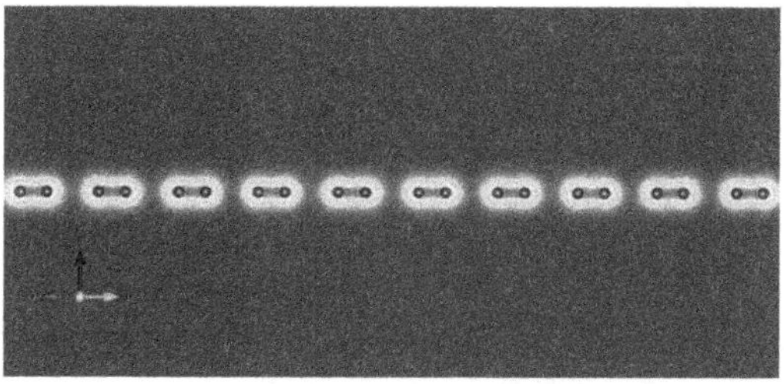

(b)

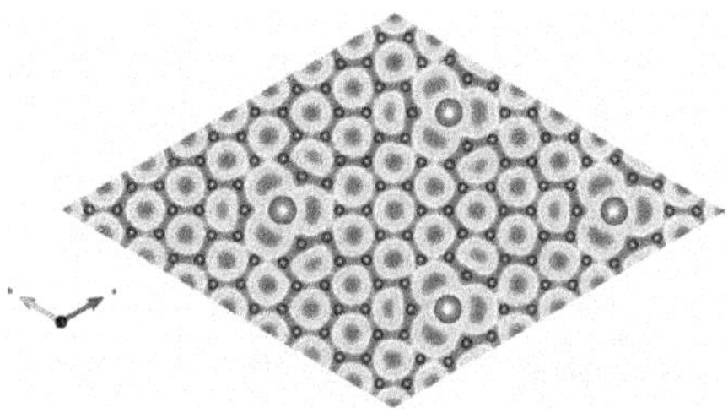

(c)

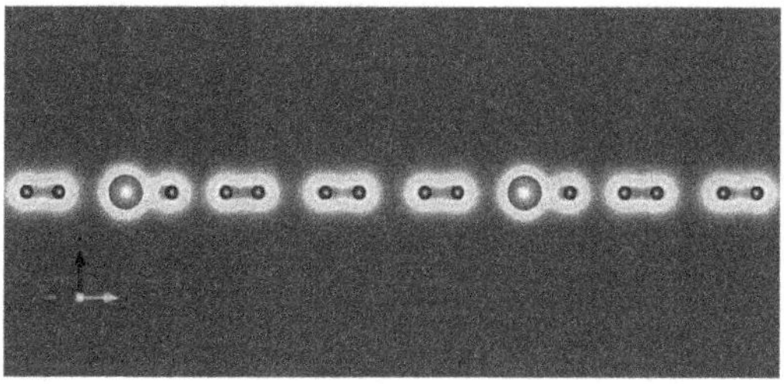

(d)

Fig. 8.3. Charge density of graphene (a) top view and (b) side view. Charge density of La atoms substituted in graphene single-side – La:C = 1:32: (c) top view and (d) side view. Charge density of La atoms substituted in graphene single-side: (e) top view and (f) side view. Charge density of Gd atoms substituted in graphene single-side – Gd:C = 1:32: (g) top view and (h) side view. Charge density of Gd atoms substituted in graphene single-side – Gd:C = 1:50: (i) top view and (j) side view.

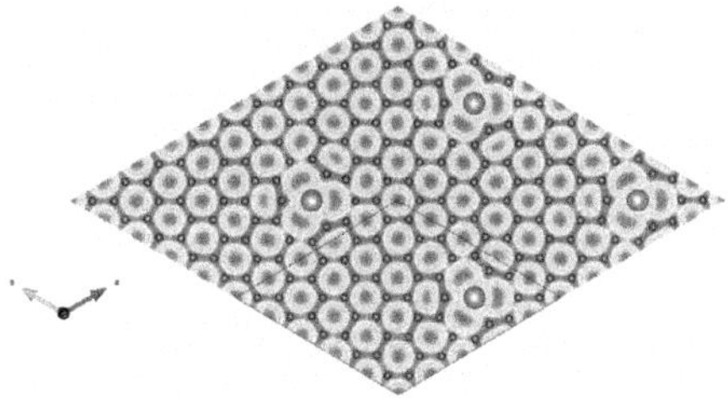

(e)

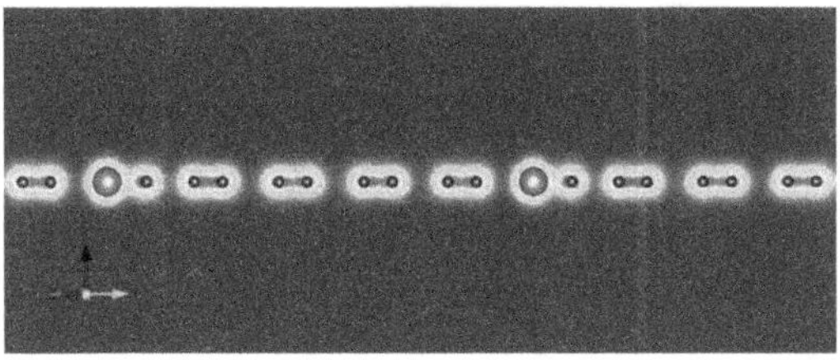

(f)

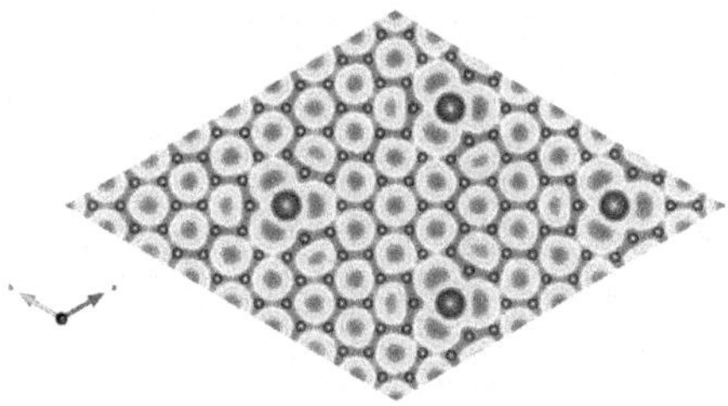

(g)

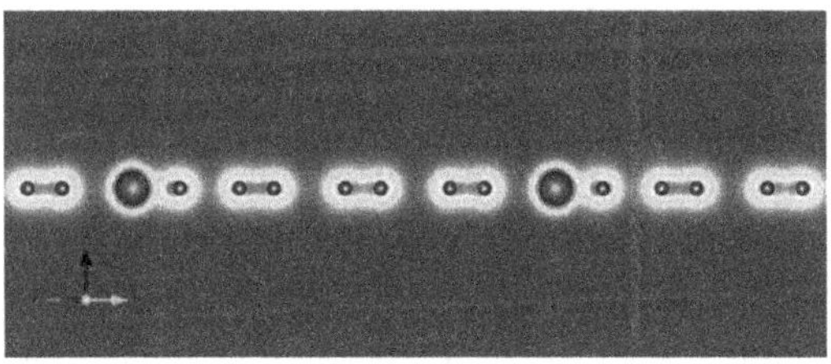

(h)

Fig. 8.3. (*Continued*)

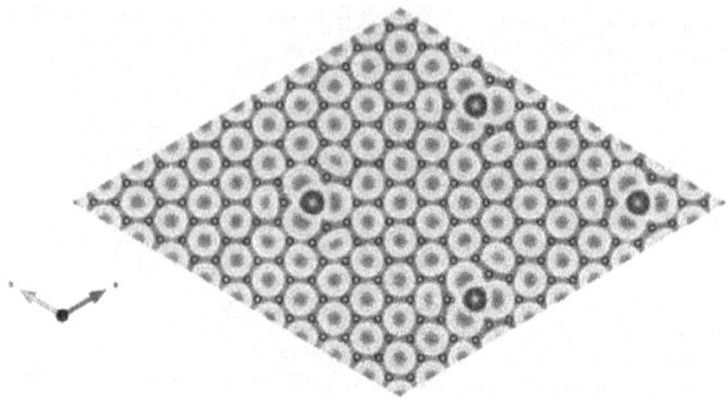

(i)

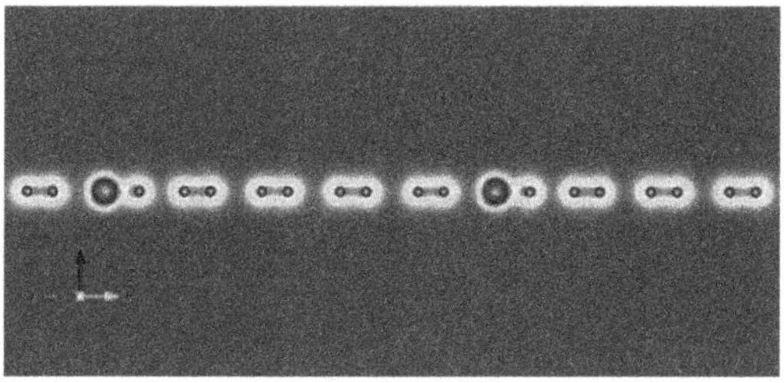

(j)

Fig. 8.3. (*Continued*)

overlap of the orbitals C-2p_z and Gd-6s-(5d_{z^2}, 5$d_{x^2-y^2}$, 5d_{xy}, 5d_{yz}, 5d_{xz}) –(4f_{x^3}, 4f_{xz^2}, 4f_{yz^2}, 4f_{xyz}, 4$f_{z(x^2-y^2)}$, 4$f_{x(x^2-3y^2)}$, 4$f_{y(3x^2-y^2)}$), with these interactions between the orbitals resulting in metallic Gd–C bonds (Figs. 8.3(g)–8.3(j)).

8.4. Van Hove Singularities

The energy-dependent density of states[49] is closely related to the state number between E and $E + dE$ for the closed wave-vector surfaces, specific even points for system material in 3D, 2D, or 1D, or constant-energy loops. The calculation formulas are mainly determined by the inverse of group velocities; therefore, the critical points in band structures clearly exhibit the well-known van Hove singularities because of the zero- or singular-group velocities. Generally, such points cover the maximum, minimum, or saddle points, such as the Dirac points of linearly particular valence and conduction bands. Their singular behaviors are very sensitive to changes in the band-edge states and dimensionalities. The current theoretical predictions of the analytic energy dependencies of van Hove singularities clearly present it to be $\sqrt{E}$,[50] constant plateaus,[51] delta-function-like structures,[52] $\ln|E|$,[53] V-valley,[54] or $\frac{1}{\sqrt{E}}$,[53] and delta functions correspond to the characteristic form as being 3D, 2D, or 1D parabolic,[55] linear,[56] constant-energy loop,[57] flat energy bands,[58] flat and 0D discrete states,[13] V-shaped,[59] partially flat energy dispersions,[60] or saddle points.[61] The low-dimensional van Hove singularities[62] could be directly verified using the high-precision STS measurements, except for those of 3D materials.[63] It is inevitable that all band-edge states have an important relationship with many of the essential properties.

The featured van Hove singularities of rare-earth atoms substituted in a graphene monolayer, as clearly illustrated in Figs. 8.4(a)–8.4(m), are available for determining the concise pictures of the orbital hybridizations of the C–C bonds. For the La atoms substituted in a graphene monolayer with a 1:32 concentration, the dominance of the orbitals of La and C atoms is clearly shown, with the orbitals of the C atom supporting in the valence band and the La atom supporting in the conduction band. Generally, the V-shape of the Dirac cone in this material is broken, which is different from that in the band structure of graphene. Besides, the density of states shows an asymmetry between spin up and spin down at a certain point. As in the graphene monolayer, the $2p_z$ orbital contributes mainly to the conduction band and partly to the valence band. Around the Fermi level, the $2p_x$ and $2p_y$ orbitals are not symmetric due to the influence of the La atoms that show strong magnetism. In addition, the support of the d orbitals of the La atoms is asymmetric around the zero point of energy. The dominance of the 6s orbital is very weak because this subshell lies deep in the atom. Generally, for the C atom, the projected density of state primarily ranges from a deep energy level to −4 eV for the orbital 2s, −12 to −4 eV for the orbitals $2p_x$ and $2p_y$, and −6 to 6 eV for the orbital $2p_z$. Besides, the orbitals ($5d_{z^2}$, $5d_{x^2-y^2}$, $5d_{xy}$, $5d_{yz}$, $5d_{xz}$) and 6s of the La atom dominate in the conduction band and partly in the valence band, which clearly present the metallic properties. Similarly, for the 1:50 concentration, the density of state asymmetry occurs around the Fermi level, and the Dirac cone is broken due to the influence of the La atom after substitution. Also, the support of the C atom in the valence band range is mainly from the 2s orbital, ranging from a deep energy level to −4, $2p_x$ and $2p_y$ from −12 to −4 eV, and −6 to 6 eV for $2p_z$. The domination of the La atom support is specific to the conduction band, which shows the properties of the metallic rare-earth elements.

For the van Hove singularities of Gd-atom substituted graphene monolayer with the 1:32 concentration, the density of states is significantly different from that of pristine graphene. The Dirac cone is broken, there is asymmetry around the Fermi level pool, and there is a significant contribution of the Ga atoms in the conduction band. The step function from parabolic bands because of the contribution of the orbitals from the carbon atoms. Besides, with the special

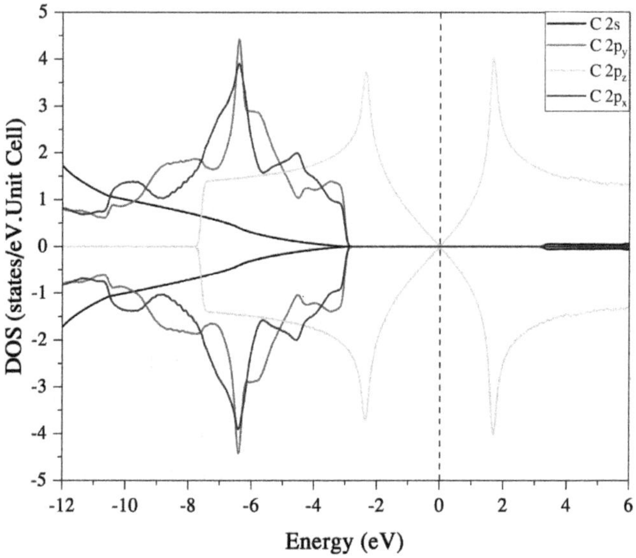

(a)

Fig. 8.4. Density of states of (a) graphene, (b) La atoms substituted in graphene – La:C = 1:32, (c) the C-(2s, $2p_x$, $2p_y$, $2p_z$) dominance – La:C = 1:32, (d) the La-6s-($5d_{z^2}$, $5d_{x^2-y^2}$, $5d_{xy}$, $5d_{yz}$, $5d_{xz}$) dominance – La:C = 1:32, (e) La atoms substituted in graphene – La:C = 1:50, (f) the C-(2s, $2p_x$, $2p_y$, $2p_z$) dominance – La:C = 1:50, (j) the La-6s-($5d_{z^2}$, $5d_{x^2-y^2}$, $5d_{xy}$, $5d_{yz}$, $5d_{xz}$) dominance – La:C = 1:50, (h) Gd atoms substituted in graphene – Gd:C = 1:32, (i) the C-(2s, $2p_x$, $2p_y$, $2p_z$) dominance – Gd:C = 1:50, (j) the Gd-6s-($5d_{z^2}$, $5d_{x^2-y^2}$, $5d_{xy}$, $5d_{yz}$, $5d_{xz}$)-($4f_{x^3}$, $4f_{xz^2}$, $4f_{yz^2}$, $4f_{xyz}$, $4f_{z(x^2-y^2)}$, $4f_{x(x^2-3y^2)}$, $4f_{y(3x^2-y^2)}$) dominance – Gd:C = 1:32, (k) Gd atoms substituted in graphene – Gd:C = 1:50, (l) the C-(2s, $2p_x$, $2p_y$, $2p_z$) dominances – Gd:C = 1:50, and (m) the Gd-6s-($5d_{z^2}$, $5d_{x^2-y^2}$, $5d_{xy}$, $5d_{yz}$, $5d_{xz}$)-($4f_{x^3}$, $4f_{xz^2}$, $4f_{yz^2}$, $4f_{xyz}$, $4f_{z(x^2-y^2)}$, $4f_{x(x^2-3y^2)}$, $4f_{y(3x^2-y^2)}$) dominance – Gd:C = 1:50.

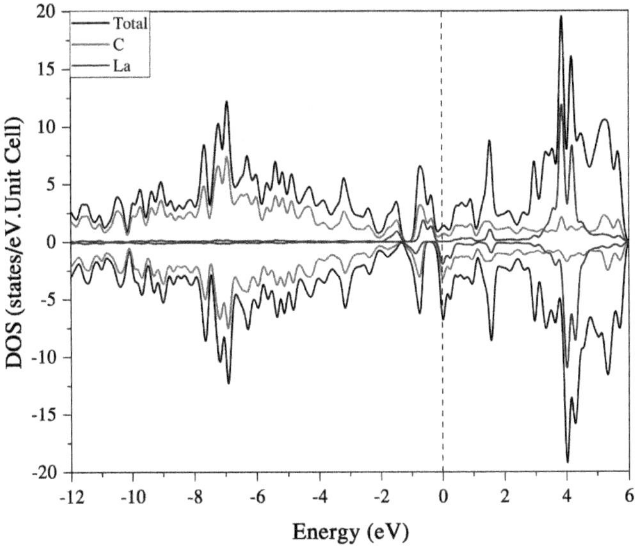

(b)

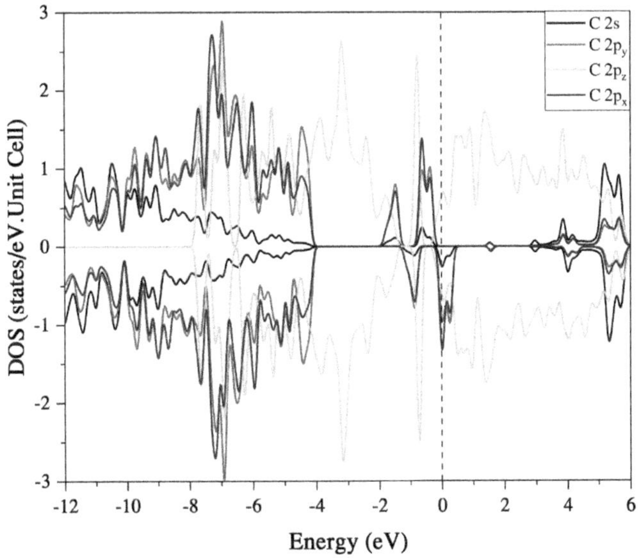

(c)

Fig. 8.4. (*Continued*)

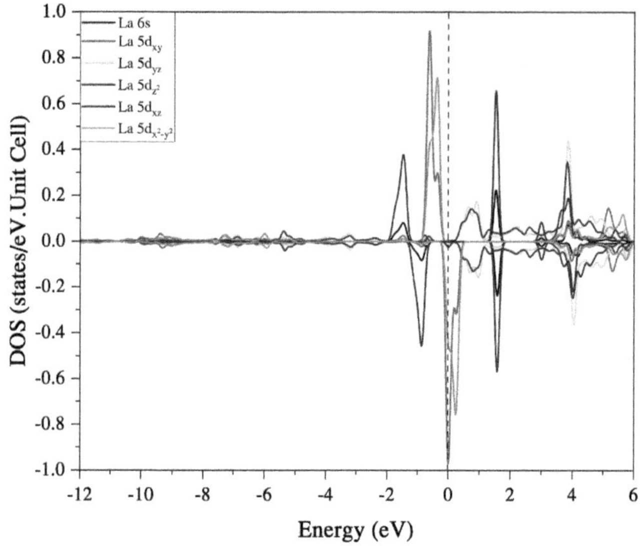

(d)

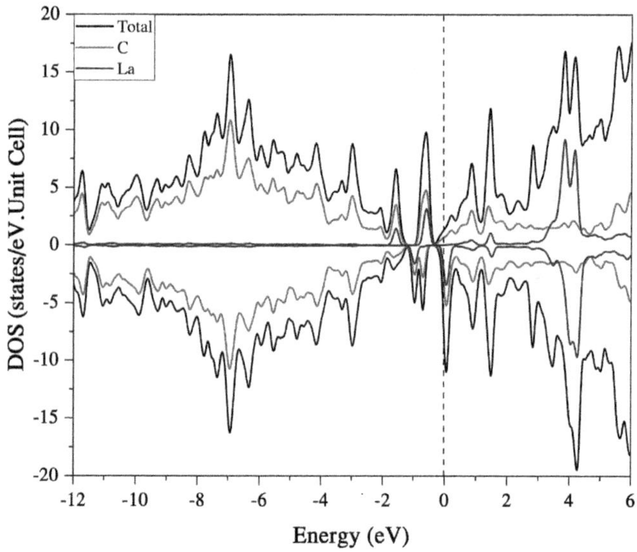

(e)

Fig. 8.4. (*Continued*)

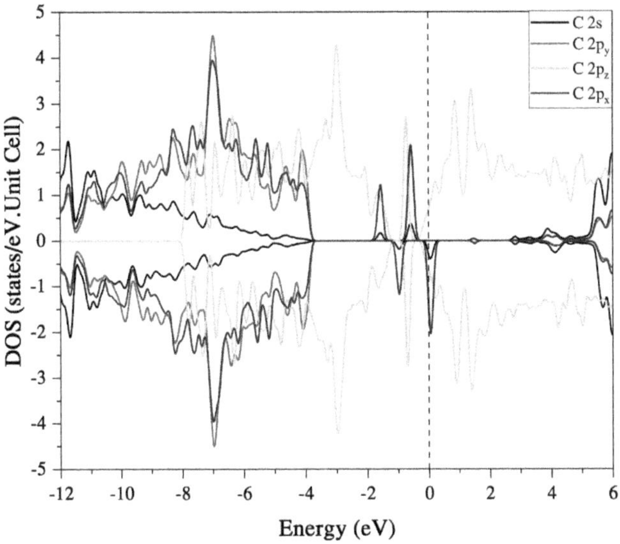

(f)

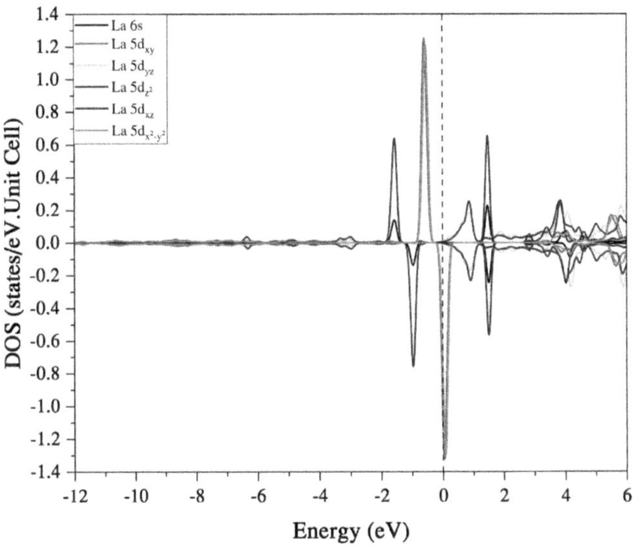

(g)

Fig. 8.4. (*Continued*)

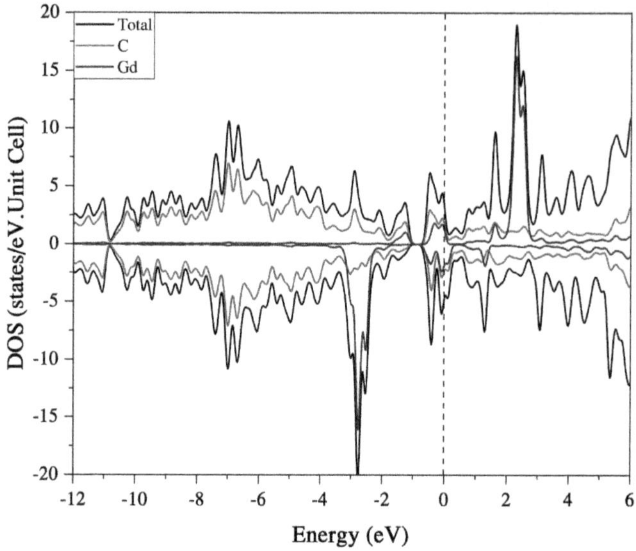

(h)

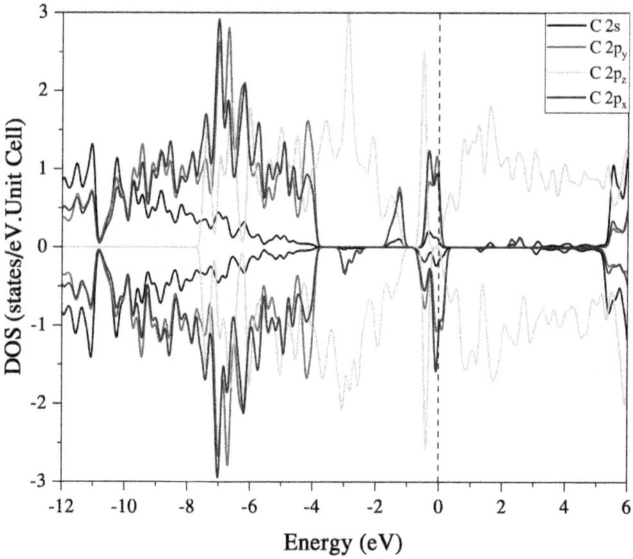

(i)

Fig. 8.4. (*Continued*)

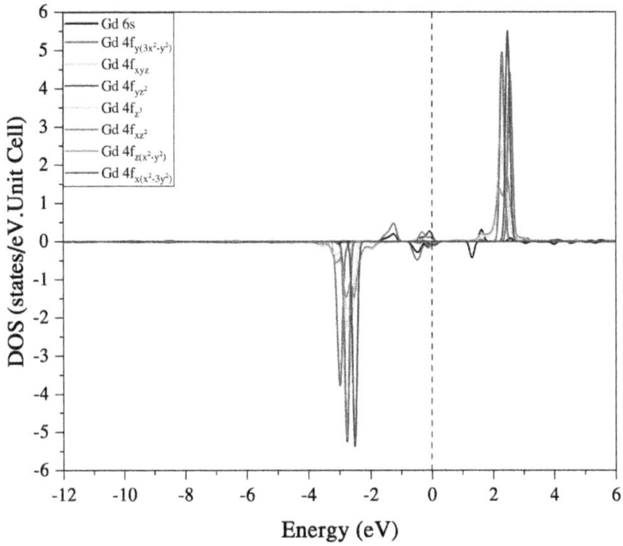

(j)

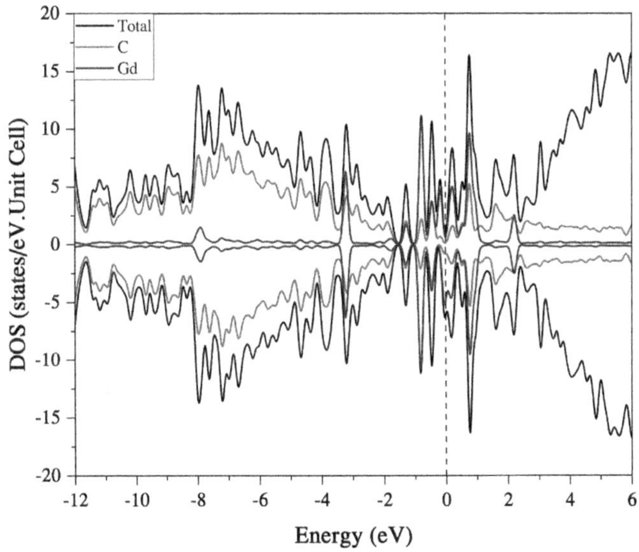

(k)

Fig. 8.4. (*Continued*)

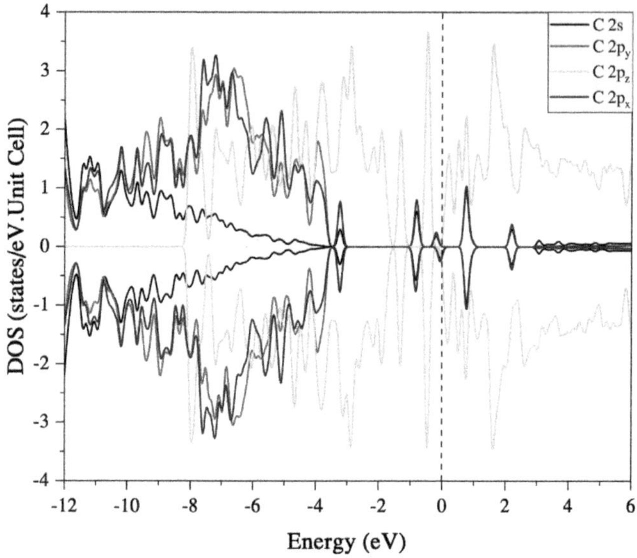

(l)

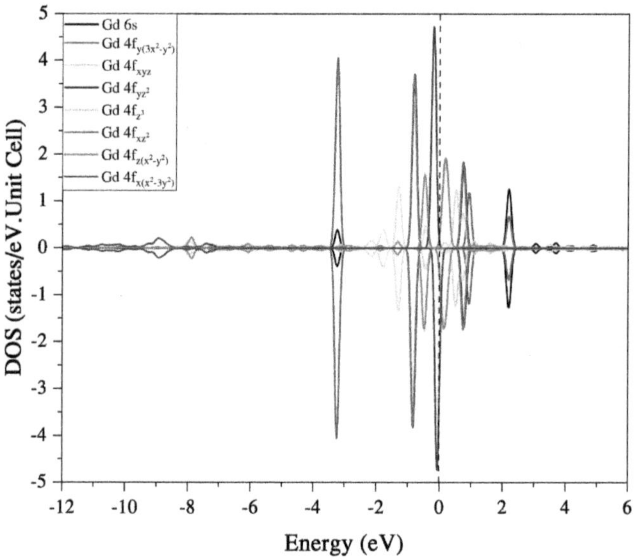

(m)

Fig. 8.4. (*Continued*)

properties of rare-earth elements, the density of states of the Gd atom splits into spin up and spin down in the diagram, which are asymmetric around the Fermi level. Additionally, Gd is a rare-earth metal with many orbitals in the atom: ($5d_{z^2}$, $5d_{x^2-y^2}$, $5d_{xy}$, $5d_{yz}$, $5d_{xz}$) and ($4f_{x^3}$, $4f_{xz^2}$, $4f_{yz^2}$, $4f_{xyz}$, $4f_{z(x^2-y^2)}$, $4f_{x(x^2-3y^2)}$, $4f_{y(3x^2-y^2)}$), so the main contribution of the Gd atom is in the conduction band for spin up and in the valence band for spin down. For the C atom, the projected density of states is in a deep energy range for the orbital 2s, from −12 to −4 eV for the orbitals $2p_x$ and $2p_y$, and from −6 to 6 eV for the distribution of the orbital $2p_z$ in the density of states diagram. Besides, the 1:50 concentration of the Gd atom in the graphene layer also exhibits the support of the orbitals of atoms in the material. Similarly, the support of the C atom is in a deep energy range for the orbital 2s, from −12 to −4 eV for the contributions of the orbitals $2p_x$ and $2p_y$, and a large area from the valence band to the conductor band for the orbital $2p_z$.

8.5. Spin-density Distributions

The carbon atoms in graphene have a non-magnetic characteristic. However, after substituting rare-earth elements, spin distribution occurs due to the influence of the rare-earth elements (Figs. 8.5(a)–8.5(d)). In the case of La atoms substituted in a graphene monolayer, the spin density is distributed mainly at the La element; besides, the carbon atoms around La also have a relatively small spin-density distribution. The higher the rare-earth concentration in graphene, the more distributed the spin density. Spin splitting is also clearly exhibited through the band structure and density of states. For the case of the gadolinium atoms, the spin-density distributions also present similar results for band structure and density of states, which show strong magnetism of this material through the splitting of spin up and spin down after substituting the rare-earth element. Moreover, spin-density distributions also appear around the C atoms in a graphene sheet because of the strong hybridization between the orbitals of Gadolinium atoms after substitution, although C is essentially a non-magnetic, non-metallic element. The magnetic moments of the orbitals are detailed in Tables 8.2(a)–8.2(d).

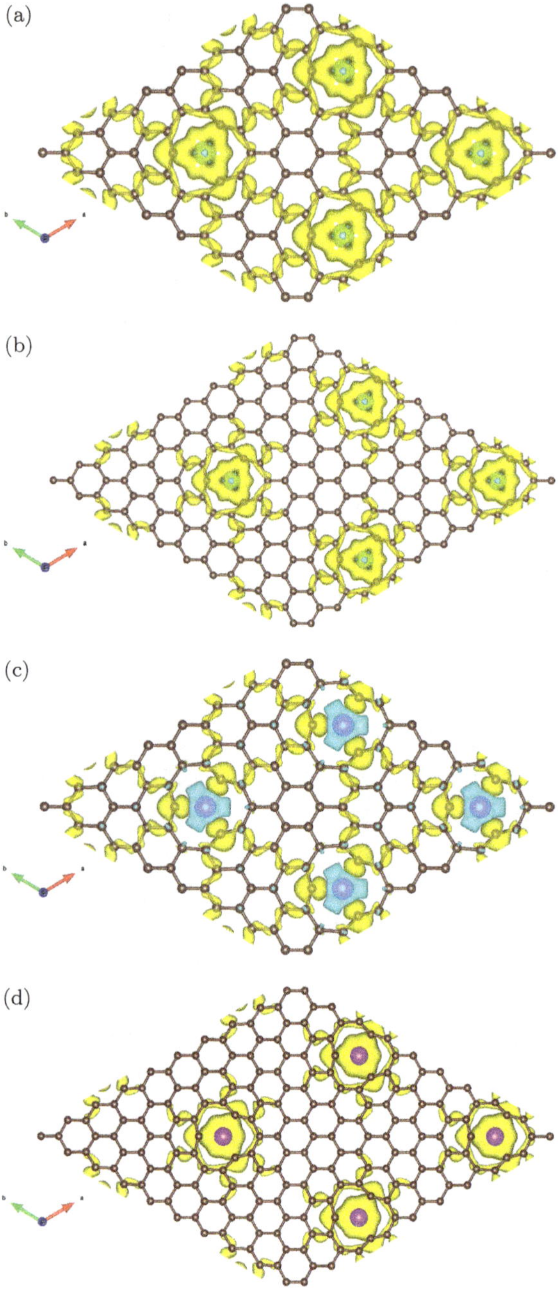

Fig. 8.5. Spin density distributions of La-atoms substituted in graphene: (a) La:C = 1:32, (b) La:C = 1:50, (c) Gd:C = 1:32, and Gd:C = 1:50.

Table 8.2. Magnetic moment (B) of La atom substituted in graphene with (a) La:C = 1:32 and (b) La:C = 1:50 and that of Gd atom substituted in graphene with (c) Gd:C = 1:32 and (d) Gd:C = 1:50.

(a)

# of ion	s	p	d	f	Total
1	0.000	−0.001	0.000	0.000	−0.001
2	0.001	0.008	0.000	0.000	0.009
3	0.000	−0.002	0.000	0.000	−0.002
4	0.002	0.008	0.000	0.000	0.009
5	0.002	0.009	0.000	0.000	0.011
6	0.002	0.007	0.000	0.000	0.009
7	0.000	−0.003	0.000	0.000	−0.003
8	0.000	−0.002	0.000	0.000	−0.002
9	0.002	0.003	0.000	0.000	0.005
10	0.002	0.004	0.000	0.000	0.005
11	0.001	0.007	0.000	0.000	0.008
12	0.000	0.000	0.000	0.000	0.000
13	0.002	0.005	0.000	0.000	0.008
14	0.002	0.007	0.000	0.000	0.010
15	0.000	0.000	0.000	0.000	−0.001
16	0.000	0.001	0.000	0.000	0.000
17	0.003	0.015	0.000	0.000	0.018
18	0.004	0.017	0.000	0.000	0.021
19	0.000	0.001	0.000	0.000	0.001
20	0.000	0.009	0.000	0.000	0.008
21	0.004	0.019	0.000	0.000	0.023
22	0.000	0.001	0.000	0.000	0.000
23	0.004	0.020	0.000	0.000	0.024
24	0.011	0.178	0.000	0.000	0.189
25	0.011	0.187	0.000	0.000	0.198
26	0.003	0.014	0.000	0.000	0.018
27	0.003	0.013	0.000	0.000	0.016
28	0.000	0.007	0.000	0.000	0.007
29	0.010	0.160	0.000	0.000	0.170
30	0.000	0.007	0.000	0.000	0.007
31	0.000	0.000	0.000	0.000	−0.001
La	0.003	0.060	0.380	0.062	0.505
Total	0.074	0.757	0.380	0.062	1.273

(Continued)

Table 8.2. (*Continued*)

(b)

# of ion	s	p	d	f	Total
1	0.000	0.000	0.000	0.000	0.000
2	0.001	0.003	0.000	0.000	0.004
3	0.000	−0.001	0.000	0.000	−0.001
4	0.001	0.004	0.000	0.000	0.004
5	0.000	0.000	0.000	0.000	0.000
6	0.000	−0.001	0.000	0.000	−0.001
7	0.001	0.004	0.000	0.000	0.005
8	0.000	−0.001	0.000	0.000	−0.001
9	0.000	−0.001	0.000	0.000	−0.001
10	0.001	0.004	0.000	0.000	0.005
11	0.002	0.006	0.000	0.000	0.009
12	0.002	0.006	0.000	0.000	0.008
13	0.001	−0.002	0.000	0.000	−0.001
14	0.000	−0.001	0.000	0.000	−0.001
15	0.000	−0.002	0.000	0.000	−0.001
16	0.002	0.003	0.000	0.000	0.005
17	0.002	0.004	0.000	0.000	0.005
18	0.001	0.004	0.000	0.000	0.005
19	0.001	0.004	0.000	0.000	0.004
20	0.000	0.000	0.000	0.000	0.000
21	0.002	0.004	0.000	0.000	0.007
22	0.002	0.005	0.000	0.000	0.008
23	0.000	0.000	0.000	0.000	0.000
24	0.000	0.000	0.000	0.000	0.000
25	0.000	0.000	0.000	0.000	0.000
26	0.003	0.013	0.000	0.000	0.017
27	0.004	0.014	0.000	0.000	0.018
28	0.000	0.000	0.000	0.000	0.000
29	0.000	−0.001	0.000	0.000	−0.001
30	0.000	0.001	0.000	0.000	0.001
31	0.001	0.003	0.000	0.000	0.003
32	0.000	0.000	0.000	0.000	0.000
33	0.001	0.003	0.000	0.000	0.004
34	0.000	0.000	0.000	0.000	0.000
35	0.000	0.010	0.000	0.000	0.010
36	0.004	0.015	0.000	0.000	0.019
37	0.000	0.001	0.000	0.000	0.001
38	0.000	0.001	0.000	0.000	0.001
39	0.004	0.015	0.000	0.000	0.020

(*Continued*)

Table 8.2. (*Continued*)

(b)

# of ion	s	p	d	f	Total
40	0.011	0.181	0.000	0.000	0.192
41	0.012	0.189	0.000	0.000	0.200
42	0.004	0.013	0.000	0.000	0.017
43	0.001	0.003	0.000	0.000	0.003
44	0.003	0.013	0.000	0.000	0.016
45	0.000	0.007	0.000	0.000	0.007
46	0.011	0.164	0.000	0.000	0.175
47	0.000	0.008	0.000	0.000	0.008
48	0.000	0.001	0.000	0.000	0.001
49	0.000	0.001	0.000	0.000	0.001
La	0.003	0.065	0.392	0.062	0.522
Total	0.080	0.760	0.392	0.062	1.294

(c)

# of ion	s	p	d	f	Total
1	0.000	0.002	0.000	0.000	0.002
2	0.000	−0.004	0.000	0.000	−0.004
3	0.000	0.004	0.000	0.000	0.004
4	−0.001	−0.004	0.000	0.000	−0.005
5	−0.001	−0.007	0.000	0.000	−0.008
6	0.000	−0.006	0.000	0.000	−0.007
7	0.000	0.003	0.000	0.000	0.003
8	0.000	0.003	0.000	0.000	0.003
9	0.001	0.001	0.000	0.000	0.001
10	0.001	−0.001	0.000	0.000	0.000
11	0.000	−0.004	0.000	0.000	−0.004
12	0.001	0.002	0.000	0.000	0.003
13	0.000	0.000	0.000	0.000	−0.001
14	−0.001	−0.003	0.000	0.000	−0.004
15	0.001	0.002	0.000	0.000	0.003
16	0.000	0.000	0.000	0.000	0.000
17	−0.001	−0.009	0.000	0.000	−0.010
18	−0.002	−0.011	0.000	0.000	−0.013
19	0.000	0.000	0.000	0.000	0.000
20	0.002	−0.003	0.000	0.000	−0.001

(*Continued*)

Table 8.2. (*Continued*)

(c)

# of ion	s	p	d	f	Total
21	−0.002	−0.014	0.000	0.000	−0.017
22	0.000	0.000	0.000	0.000	−0.001
23	−0.003	−0.015	0.000	0.000	−0.018
24	−0.009	−0.123	0.000	0.000	−0.132
25	−0.009	−0.138	0.000	0.000	−0.147
26	−0.001	−0.011	0.000	0.000	−0.011
27	0.000	−0.008	0.000	0.000	−0.009
28	0.002	−0.002	0.000	0.000	0.001
29	−0.008	−0.084	0.000	0.000	−0.092
30	0.002	−0.002	0.000	0.000	0.000
31	0.000	0.001	0.000	0.000	0.001
Gd	0.035	−0.040	−0.132	6.551	6.415
Total	0.006	−0.471	−0.132	6.551	5.955

(d)

# of ion	s	p	d	f	Total
1	0.000	−0.002	0.000	0.000	−0.002
2	0.000	−0.001	0.000	0.000	−0.001
3	0.000	−0.003	0.000	0.000	−0.003
4	0.000	−0.002	0.000	0.000	−0.001
5	0.000	−0.003	0.000	0.000	−0.003
6	0.000	−0.003	0.000	0.000	−0.003
7	0.000	−0.003	0.000	0.000	−0.003
8	0.000	−0.005	0.000	0.000	−0.005
9	0.000	−0.006	0.000	0.000	−0.006
10	0.000	−0.002	0.000	0.000	−0.002
11	0.000	−0.008	0.000	0.000	−0.008
12	0.000	−0.006	0.000	0.000	−0.007
13	0.000	−0.004	0.000	0.000	−0.004
14	0.000	−0.002	0.000	0.000	−0.002
15	0.000	−0.006	0.000	0.000	−0.006
16	−0.001	−0.002	0.000	0.000	−0.002
17	−0.001	−0.003	0.000	0.000	−0.003
18	0.000	−0.002	0.000	0.000	−0.002
19	0.000	−0.002	0.000	0.000	−0.002
20	0.000	−0.002	0.000	0.000	−0.003

(*Continued*)

Table 8.2. (*Continued*)

(d)

# of ion	s	p	d	f	Total
21	0.000	−0.002	0.000	0.000	−0.002
22	0.000	−0.004	0.000	0.000	−0.003
23	0.000	−0.003	0.000	0.000	−0.003
24	0.000	−0.001	0.000	0.000	−0.001
25	0.000	−0.001	0.000	0.000	−0.001
26	0.001	0.001	0.000	0.000	0.002
27	0.001	0.003	0.000	0.000	0.004
28	0.000	−0.001	0.000	0.000	−0.001
29	0.000	0.001	0.000	0.000	0.001
30	0.000	−0.001	0.000	0.000	−0.001
31	0.000	0.001	0.000	0.000	0.001
32	0.000	0.000	0.000	0.000	0.000
33	0.000	0.001	0.000	0.000	0.001
34	0.000	0.000	0.000	0.000	0.000
35	0.000	0.000	0.000	0.000	0.001
36	0.001	0.004	0.000	0.000	0.004
37	0.000	0.000	0.000	0.000	0.000
38	0.000	0.000	0.000	0.000	0.000
39	0.001	0.004	0.000	0.000	0.005
40	0.004	0.038	0.000	0.000	0.042
41	0.004	0.040	0.000	0.000	0.044
42	0.000	0.002	0.000	0.000	0.003
43	0.000	0.001	0.000	0.000	0.001
44	0.000	0.001	0.000	0.000	0.002
45	0.001	0.000	0.000	0.000	0.001
46	0.004	0.034	0.000	0.000	0.038
47	0.001	−0.001	0.000	0.000	0.000
48	0.000	0.000	0.000	0.000	0.000
49	0.000	−0.001	0.000	0.000	−0.001
Gd	0.007	0.008	0.045	−1.640	−1.579
Total	0.024	0.060	0.045	−1.640	−1.511

References

[1] Pumera, M.; Wong, C. H. A. Graphane and hydrogenated graphene. *Chemical Society Reviews* **2013**, *42* (14), 5987–5995.

[2] Osborn, T. H.; Farajian, A. A.; Pupysheva, O. V.; Aga, R. S.; Voon, L. L. Y. Ab initio simulations of silicene hydrogenation. *Chemical Physics Letters* **2011**, *511* (1–3), 101–105.

[3] Houssa, M.; Scalise, E.; Sankaran, K.; Pourtois, G.; Afanas' Ev, V.; Stesmans, A. Electronic properties of hydrogenated silicene and germanene. *Applied Physics Letters* **2011**, *98* (22), 223107.

[4] Matthes, L.; Pulci, O.; Bechstedt, F. Optical properties of two-dimensional honeycomb crystals graphene, silicene, germanene, and tinene from first principles. *New Journal of Physics* **2014**, *16* (10), 105007.

[5] Mahdavifar, S.; Tagani, M. B. Electronic and mechanical properties of Plumbene monolayer: A first-principle study. *Physica E: Low-dimensional Systems and Nanostructures* **2021**, 134, 114837.

[6] Jorio, A.; Dresselhaus, M. S.; Saito, R.; Dresselhaus, G. *Raman Spectroscopy in Graphene Related Systems.* John Wiley & Sons, 2011.

[7] Bonaccorso, F.; *et al.* Graphene, related two-dimensional crystals, and hybrid systems for energy conversion and storage. *Science* **2015**, *347* (6217), 1246501.

[8] Lin, C.-Y.; Wu, J.-Y.; Chiu, C.-W.; Lin, M.-F. *Coulomb Excitations and Decays in Graphene-Related Systems.* CRC Press, 2019.

[9] Chiu, C.-W.; Shyu, F.-L.; Lin, M.-F.; Gumbs, G.; Roslyak, O. Anisotropy of π-plasmon dispersion relation of AA-stacked graphite. *Journal of the Physical Society of Japan* **2012**, *81* (10), 104703.

[10] Ho, J.-H.; Lu, C.; Hwang, C.; Chang, C.; Lin, M.-F. Coulomb excitations in AA-and AB-stacked bilayer graphites. *Physical Review B* **2006**, *74* (8), 085406.

[11] Ho, C.-H.; Chang, C.-P.; Lin, M.-F. Evolution and dimensional crossover from the bulk subbands in ABC-stacked graphene to a three-dimensional Dirac cone structure in rhombohedral graphite. *Physical Review B* **2016**, *93* (7), 075437.

[12] Lin, C.-Y.; Yang, C.-H.; Chiu, C.-W.; Chung, H.-C.; Lin, S.-Y.; Lin, M.-F. Orbital hybridizations and Coulomb couplings of coaxial carbon nanotubes.

[13] Li, W.-B.; Tran, N. T. T.; Lin, S.-Y.; Lin, M.-F. Diverse fundamental properties in stage-n graphite alkali-intercalation compounds: Anode materials of Li+-based batteries. *arXiv preprint arXiv:2001.02042,* **2019**.

[14] Tran, N. T. T.; Lin, S.-Y.; Lin, C.-Y.; Lin, M.-F. *Geometric and Electronic Properties of Graphene-related Systems: Chemical Bonding Schemes.* CRC Press, 2017.

[15] Li, W.-B.; Tran, N. T. T.; Lin, S.-Y. Diverse phenomena in stage-n graphite alkali-intercalation compounds. In *Lithium-Ion Batteries and Solar Cells.* CRC Press, 2021, pp. 19–43.

[16] Lin, C.-Y.; Yang, C.-H.; Chiu, C.-W.; Chung, H.-C.; Lin, S.-Y.; Lin, M.-F. Theories for electronic excitations and de-excitations in 1D-3D carbon nanotube systems.

[17] Golor, M.; Lang, T. C.; Wessel, S. Quantum Monte Carlo studies of edge magnetism in chiral graphene nanoribbons. *Physical Review B* **2013**, *87* (15), 155441.

[18] Dresselhaus, M. S.; Dresselhaus, G.; Eklund, P. C. *Science of Fullerenes and Carbon Nanotubes: Their Properties and Applications.* Elsevier, 1996.

[19] Terrones, M. Carbon nanotubes: Synthesis and properties, electronic devices and other emerging applications. *International Materials Reviews* **2004**, *49* (6), 325–377.

[20] Li, W.-B.; Wang, Y.-M.; Chung, H.-C.; Lin, M.-F. 1D finite-width graphene nanoribbon systems: Alkalization and hydrogenation. *arXiv preprint arXiv:2206.11162,* **2022**.

[21] Ryu, S.; *et al.* Reversible basal plane hydrogenation of graphene. *Nano Letters* **2008**, *8* (12), 4597–4602.

[22] Yan, J.-A.; Xian, L.; Chou, M. Structural and electronic properties of oxidized graphene. *Physical Review Letters* **2009**, *103* (8), 086802.

[23] Tran, N. T. T.; Nguyen, D. K.; Glukhova, O. E.; Lin, M.-F. Coverage-dependent essential properties of halogenated graphene: A DFT study. *Scientific Reports* **2017**, *7* (1), 1–13.

[24] Mavrandonakis, A.; Froudakis, G. E.; Schnell, M.; Mühlhäuser, M. From pure carbon to silicon — carbon nanotubes: An ab-initio study. *Nano Letters* **2003**, *3* (11), 1481–1484.

[25] Sekkal W.; Zaoui, A. Predictive study of thermodynamic properties of GeC. *New Journal of Physics* **2002**, *4* (1), 9.

[26] Benzair, A.; Bouhafs, B.; Khelifa, B.; Mathieu, C.; Aourag, H. The ground state and the bonding properties of the hypothetical cubic zinc-blende-like GeC and SnC compounds. *Physics Letters A* **2001**, *282* (4–5), 299–308.

[27] Petrović, M.; *et al.* The mechanism of caesium intercalation of graphene. *Nature communications* **2013**, *4* (1), 1–8.

[28] Li, W.-B.; *et al.* $FeCl_3$ graphite intercalation compounds: Iron-ion-based battery cathodes. *First-Principles Calculations for Cathode* **2021**, 6–1.

[29] Chang, S.-L.; Lin, S.-Y.; Lin, S.-K.; Lee, C.-H.; Lin, M.-F. Geometric and electronic properties of edge-decorated graphene nanoribbons. *Scientific Reports* **2014**, *4* (1), 1–8.

[30] Kim, S. K.; Yoo, H. J. Formation of bilayer Ni–SiC composite coatings by electrodeposition. *Surface and Coatings Technology* **1998**, *108*, 564–569.

[31] Zhang, Y.-F.; *et al.* Sulfur-doped graphene nanoribbons with a sequence of distinct band gaps. *Nano Research* **2017**, 10 (10), 3377–3384.

[32] Soh, D.; Fan, Z. Study on YbBaCuO superconductor of Yb_2O_3–$BaCO_3$–CuO system. *Physica C: Superconductivity* **2000**, *337* (1–4), 292–296.

[33] Bauer, E.; Gratz, E.; Maikis, M.; Kirchmayr, H.; Roy, S.; Coles, B. Magnetoresistance of (CexGd1− x) Cu6 compounds. *Physica B: Condensed Matter* **1993**, *186*, 586–589.

[34] Berton, A.; Chaussy, J.; Cornut, B.; Lasjaunias, J.; Odin, J.; Peyrard, J. Specific heat of Ce3Al11 and CeAl3 compounds. *Journal of Magnetism and Magnetic Materials* **1980**, *15*, 379–380.

[35] Aliev, F.; Brandt, N.; Moshchalkov, V.; Chudinov, S. Superconductivity in $CeCu_2Si_2$. *Solid State Communications* **1983**, *45* (3), 215–218.

[36] Sebastian, L.; Gopalakrishnan, J. Lithium ion mobility in metal oxides: A materials chemistry perspective *Journal of Materials Chemistry* **2003**, *13* (3), 433–441.

[37] Liu, X.; *et al.* Metals on graphene: Correlation between adatom adsorption behavior and growth morphology. *Physical Chemistry Chemical Physics* **2012**, *14* (25), 9157–9166.

[38] Singha, A.; *et al.* 4f occupancy and magnetism of rare-earth atoms adsorbed on metal substrates. *Physical Review B* **2017**, *96* (22), 224418.

[39] Zhang, K.-C.; Li, Y.-F.; Liu, Y.; Zhu, Y.; Shi, L.-B. Giant magnetic anisotropy of rare-earth adatoms and dimers adsorbed by graphene oxide. *Physical Chemistry Chemical Physics* **2017**, *19* (20), 13245–13251.

[40] Álvarez-Rodríguez, J.; Rodríguez-Ramos, I.; Guerrero-Ruiz, A.; Arcoya, A. Selective hydrogenation of citral over Pt/KL type catalysts doped with Sr, La, Nd and Sm. *Applied Catalysis A: General* **2011**, *401* (1–2), 56–64.

[41] Yi, W.; Tang, G.; Chen, X.; Yang, B.; Liu, X. qvasp: A flexible toolkit for VASP users in materials simulations. *Computer Physics Communications* **2020**, *257*, 107535.

[42] Hafner, J. Ab-initio simulations of materials using VASP: Density-functional theory and beyond. *Journal of computational chemistry* **2008**, *29* (13), 2044–2078.

[43] Seifert, G.; Joswig, J. O. Density-functional tight binding — An approximate density-functional theory method. *Wiley Interdisciplinary Reviews: Computational Molecular Science* **2012**, *2* (3), 456–465.
[44] Geim, A. K. Graphene: Status and prospects. *Science* **2009**, *324* (5934), 1530–1534.
[45] Hafner, J. Materials simulations using VASP — A quantum perspective to materials science. *Computer Physics Communications* **2007**, *177* (1–2), 6–13.
[46] Gradhand, M.; Fedorov, D.; Pientka, F.; Zahn, P.; Mertig, I.; Györffy, B. First-principle calculations of the Berry curvature of Bloch states for charge and spin transport of electrons. *Journal of Physics: Condensed Matter* **2012**, *24* (21), 213202.
[47] Neto, A. C.; Guinea, F.; Peres, N. M.; Novoselov, K. S.; Geim, A. K. The electronic properties of graphene. *Reviews of Modern Physics* **2009**, *81* (1), 109.
[48] Novoselov, K.; *et al.* Electronic properties of graphene. *Physica Status Solidi (b)* **2007**, *244* (11), 4106–4111.
[49] Belardinelli, R.; Pereyra, V. Fast algorithm to calculate density of states. *Physical Review E* **2007**, *75* (4), 046701.
[50] Brihuega, I.; *et al.* Unraveling the intrinsic and robust nature of van Hove singularities in twisted bilayer graphene by scanning tunneling microscopy and theoretical analysis. *Physical Review Letters* **2012**, *109* (19), 196802.
[51] Wu, S.; Zhang, Z.; Watanabe, K.; Taniguchi, T.; Andrei, E. Y. Chern insulators, van Hove singularities and topological flat bands in magic-angle twisted bilayer graphene. *Nature Materials* **2021**, *20* (4), 488–494.
[52] Do, T.-N.; Shih, P.-H.; Gumbs, G.; Huang, D.; Chiu, C.-W.; Lin, M.-F. Diverse magnetic quantization in bilayer silicene. *Physical Review B* **2018**, *97* (12), 125416.
[53] Peng, H.; *et al.* Substrate doping effect and unusually large angle van Hove singularity evolution in twisted bi-and multilayer graphene. *Advanced Materials* **2017**, *29* (27), 1606741.
[54] Zhang, Y.; *et al.* Spin-orbit enhanced superconductivity in Bernal Bilayer graphene. *arXiv preprint arXiv:2205.05087,* **2022**.
[55] Yang, C.-H.; Lin, C.-Y.; Chiu, C.-W.; Liu, C.-T.; Lin, M.-F. Dimension-and lattice-diversified Coulomb excitations in3D, 2D, 1D-nanotube electron gases, graphene and carbon nanotube. *arXiv preprint arXiv:2004.08209,* **2020**.

[56] Yan, W.; *et al.* Angle-dependent van Hove singularities in a slightly twisted graphene bilayer. *Physical Review Letters* **2012**, *109* (12), 126801.

[57] Mori, A.; Sato, M.; Yajima, T.; Konoike, T.; Uchida, K.; Osada, T. Anisotropy of Dirac cones and Van Hove singularity in an organic Dirac fermion system. *Physical Review B* **2019**, *99* (3), 035106.

[58] Xu, S.; *et al.* Tunable van Hove singularities and correlated states in twisted monolayer–bilayer graphene. *Nature Physics* **2021**, *17* (5), 619–626.

[59] Yao, Q.; Chen, X.; van Bremen, R.; Sotthewes, K.; Zandvliet, H. J. Singularities and topologically protected states in twisted bilayer graphene. *Applied physics Letters* **2020**, *116* (1), 011602.

[60] Yuan, N. F.; Isobe, H.; Fu, L. Magic of high-order van Hove singularity. *Nature Communications* **2019**, *10* (1), 1–7.

[61] Ibanescu, M.; Reed, E. J.; Joannopoulos, J. Enhanced photonic band-gap confinement via van hove saddle point singularities. *Physical Review Letters* **2006**, *96* (3), 033904.

[62] Havener, R. W.; Liang, Y.; Brown, L.; Yang, L.; Park, J. Van Hove singularities and excitonic effects in the optical conductivity of twisted bilayer graphene. *Nano Letters* **2014**, *14* (6), 3353–3357.

[63] Sinterhauf, A.; *et al.* Substrate induced nanoscale resistance variation in epitaxial graphene. *Nature Communications, 11* (1), 1–9.

CHAPTER 9

DECORATION OF GRAPHENE NANORIBBONS WITH 5D TRANSITION-METAL ELEMENTS

Wei-Bang Li*, Kuang-I Lin†, Yu-Ming Wang*, Hsien-Ching Chung‡ and Ming-Fa Lin*

*Department of Physics, National Cheng Kung University, Tainan, Taiwan
†Core Facility Center, National Cheng Kung University, Tainan 70101, Taiwan
‡RD Dept., Super Double Power Technology Co., Ltd., Changhua City, Changhua County 500042, Taiwan

Abstract

Graphene is a popular, truly two-dimensional material, possessing a cone-like energy structure near the Fermi level, and is treated as a gapless semiconductor. Its unique properties have motivated researchers to find applications for it. The gapless feature limits the development of graphene nanoelectronics. Making one-dimensional strips of graphene nanoribbons (GNRs) could be one of the more promising routes for modulating the electronic and optical properties of graphene. These properties are highly sensitive to the edge and width of the nanoribbons. The tunability of electronic and optical properties further implies the possibility of applying GNRs. However, the dangling bonds at ribbon edges remain an open issue in GNR systems. Various passivation techniques that could change the physical properties at the ribbon edge are available. This work considers 5d transition-metal elements as guest atoms at the edges. The geometric structure, energy bands, density of states, charge distribution, and optical transitions are discussed.

Keywords: Graphene nanorribons, 5d transition-metal elements, decoration.

9.1. Introduction

A new scientific frontier has arrived since the discovery of graphene. This material has sparked excitement among both fundamental scientists and applied technology engineers for many reasons. Graphene is the first truly two-dimensional material,[1,2] showing a cone-like energy spectrum near the Fermi energy, and it is treated as a gapless semiconductor. The unique electronic structure of graphene results in many fascinating essential properties, such as high carrier mobility at room temperature (>200,000 cm^2/Vs),[3–5] superior thermoconductivity (3,000–5,000 W/mK),[5,6] high transparency to incident light over a wide range of wavelength (97.7%),[7,8] and high modulus (~1 TPa) and tensile strength (~130 GPa).[9] Due to its superior electronic, thermal, mechanical, and optical properties, graphene is perceived as an important potential candidate for synthesizing next-generation electronic and optical devices.

The gapless feature results in a low on/off ratio in graphene-based field-effect transistors (FETs) and limits the development of graphene nanoelectronics. One of the best possible approaches to tuning electronic and optical properties is making one-dimensional (1D) strips of graphene, i.e., graphene nanoribbons (GNRs).[10,11] There is a tremendous variety of fabrication routes for GNR production, including both bottom-up and top-down schemes. From a geometric point of view, graphene cutting seems to be an intuitive route to fabricating GNRs, and the other available routes include the lithographic patterning and etching of graphene,[12–15] sonochemical breaking of graphene,[16–18] oxidation cutting of graphene,[19,20] and metal-catalyzed cutting of graphene.[18,19,21–26] An interesting route is carbon nanotube (CNT) unzipping since a CNT can be treated as a folded or zipped GNR.[27] The available routes for the reverse process include chemical attack,[28,29] laser irradiation,[30] plasma etching,[31,32] metal-catalyzed cutting,[33,34] hydrogen treatment and annealing,[35] intercalation and exfoliation,[36,37] electrochemical unzipping,[38] sonochemical unzipping,[39,40] unzipping functionalized CNTs using scanning tunneling microscope (STM) tips,[41] and electrical unwrapping using transmission electron microscopy (TEM).[42] Other routes include chemical vapor deposition (CVD)[43–45] and chemical synthesis.[46–49] The former involves piecewise linking of molecular precursor monomers, and the latter is much more compatible with the current semiconductor industry.

The electronic and optical properties of GNRs are dominated by the ribbon width and the edge orientation. Zigzag GNRs (ZGNRs) exhibit partial flat subbands near the Fermi level, with peculiar edge states localized at the ribbon edges,[50,51] and the energy gaps of armchair GNRs (AGNRs) scale inversely with the ribbon width.[52,53] The former have been identified from the STM images,[54,55] and the latter have been confirmed by electric conductance[12,16] and tunneling current measurements.[14] Moreover, the edge-dependent absorption selection rules of GNRs have been predicted, i.e., $|\Delta n|$ = odd for ZGNRs and $\Delta n = 0$ for AGNRs, where n is the subband index.[56–58] These fundamental properties can be enriched by external fields, such as magnetic and electric fields.

A uniform static magnetic field perpendicular to the ribbon plane can accumulate the neighboring electronic states, inducing the highly degenerate Landau levels (LLs) with quantized cyclotron orbits.[59] The competition between the lateral confinement and the magnetic confinement enriches the magneto-electronic structures, e.g., partial dispersion-less quasi-Landau levels (QLLs), 1D parabolic subbands, and partial flat subbands.[60,61] Meanwhile, the magneto-optical spectra exhibit many symmetric and asymmetric absorption peaks. The symmetric absorption peaks result from the inter-QLL transitions and obey the magneto-optical selection rule of $|\Delta m| = 1$, where m is an integer. However, the asymmetric absorption peaks originate from the transitions among parabolic subbands and abide by the edge-dependent selection rules.[10] A transverse static electric field generates extra potential energy in GNRs, i.e., the charge carriers experience different site energies.[62,63] The electronic and optical properties are drastically modified. The different potential energies in GNRs restrict the formation of Landau orbits, and the QLLs would tilt, become oscillatory, or exhibit crossings and anticrossings. Moreover, the inter-QLL optical transitions will be severely changed or even destroyed entirely.[10,64,65]

The edge-decorated GNRs, which cover the transition-metal guest adatoms, represent an interesting research strategy due to the availability of highly diversified orbital hybridizations and spin configurations. The 10 active orbitals of 5d can dominate the significant compounds, with special functionalities. However, investigations on the concise quasi-particle mechanisms are almost absent in previous studies. These can be achieved using 1D transition metal-decorated GNRs, including the armchair and zigzag systems. The d orbitals are enriched by the spin-up and spin-down configurations. Their probability distributions are characterized by the 6s and 5d orbitals, which include d_{xy}, d_{yz}, d_{z^2}, d_{xz}, and d_{x^2} and are enriched by the spin-up and spin-down configurations. The spatial electron density, which mainly arises from the linear superposition of 6s and 5d orbitals, is very sensitive to the chemical environment. In this work, the guest adatoms, tantalum (Ta-73) and tungsten (W-74), exhibit strong reactions with dangling carbon bonds at two open edges of ZGNRs and AGNRs. The pristine systems are predicted to display, respectively, ferromagnetic and antiferromagnetic spin configurations on the same sides and across the ribbon center and fully non-magnetic properties. How to modulate and diversify the magnetic behavior through transition-metal decorations is a very interesting research question worth pursuing.

9.2. Geometric Structures for Transition Metal-decorated Graphene Nanoribbon

The atomic numbers of Ta-73 and W-74, which belong to the group of transition-metal elements, are chosen for their unusual quasi-particles. All VASP calculations are performed with delicate analyses to achieve concise pictures of physics, chemistry, and material science. The VASP simulations are well suited for fully exploring the unusual quantum quasi-particles of Ta- and W-decorated 1D armchair and zigzag GNRs. After significant decorations, their critical roles in all essential properties are thoroughly examined and identified from consistent results. We establish an important framework to develop a unified correlation incorporating multi-orbital hybridizations, spin configurations, crystal structures, electronic properties, and optical properties. We find that there are no buckling structures in the Ta- and W-decorated lattices formed via the C–Ta and C–W bonds, as seen in Fig. 9.1

The armchair and zigzag GNRs are good candidates for edge decoration because of their unsaturated bonds. The widths of armchair (N_A) and zigzag (N_Z) GNRs are, respectively, determined by the

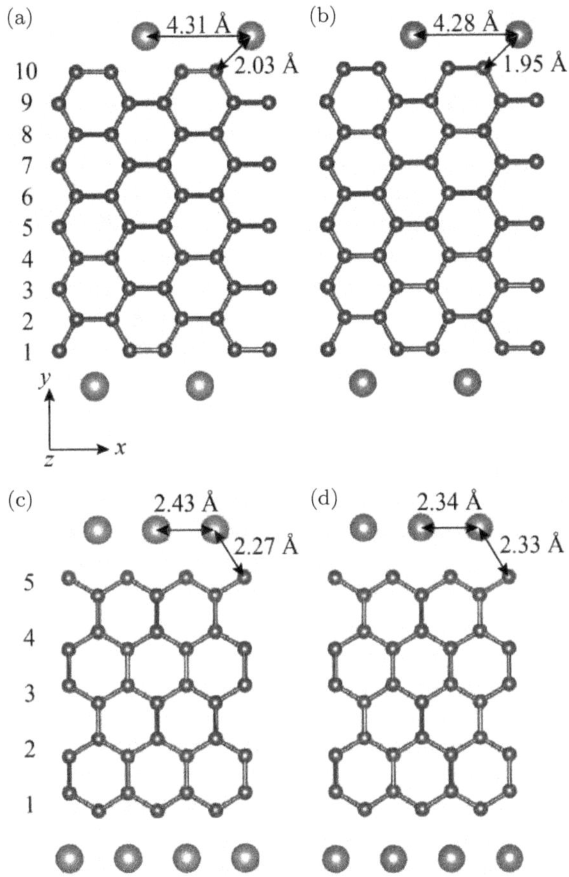

Fig. 9.1. The geometric structure of (a) Ta- and (b) W-decorated armchair systems as well as (c) Ta- and (d) W-decorated zigzag systems. The brown, yellow, and gray balls represent the carbon (C), tantalum (Ta), and tungsten (W) atoms, respectively. The label numbers denote the mth dimer and zigzag lines.

number of dimer lines and zigzag lines along the periodic direction, i.e., the x-direction. In our case, N_A and N_Z are, respectively, 10 and 5. The C–C bond length is 1.42 Å in graphene, but it possesses fluctuations within the 1.42–1.48 Å range in GNRs; the further away the carbon atoms are from the edge, the lesser the fluctuations in the bond lengths.[66] This results in a highly non-uniform physical environment and thus sensitive changes to the $2p_z$–$2p_z$ hopping integrals. The edge structures are totally different before and after the chemical Ta and W decorations. The non-uniform edge structures appear on both armchair and zigzag GNRs. In the former, the Ta–Ta and W–W bond lengths are, respectively, 4.31 and 4.28 Å; the C–Ta and C–W bond lengths are, respectively, 2.03 and 1.95 Å. Moreover, the fluctuations in the C–C bond lengths are reduced to a small range of 1.40–1.43 Å. These reflect the fact that the structures become more stable due to the edge decorations. The C–W bonds seem stronger than the C–Ta ones. For the zigzag cases, the Ta–Ta and W–W bond lengths are, respectively, 2.43 and 2.34 Å. The C–Ta and C–W bond lengths are, respectively, 2.27 and 2.33 Å. The C–Ta bond lengths are shorter than the C–W ones, which indicates that the C–Ta bonds are more stable in the zigzag systems, probably due to interactions between the W atoms.

9.3. Energy Band Structures and Density of States

There are a lot of differences between graphene, few-layered graphene, graphene-related intercalation compounds, and GNR systems. It was known that there is a pair of partial flat valence and conduction bands. Such electronic wave functions are localized near the open zigzag edges. Therefore, they belong to the edge localization states, with almost vanishing group velocities. Bandgaps decline as ribbon widths grow. However, previous studies have shown that the armchair systems possess three different energy bands with respect to different groups of N_A, where $N_A = 3I$, $3I + 1$, or $3I + 2$ and I is a positive integer. In our work, the low-lying energy bands, as clearly illustrated in Fig. 9.2, indicate the dramatic changes after the edge decorations of transition-metal atoms. There are much

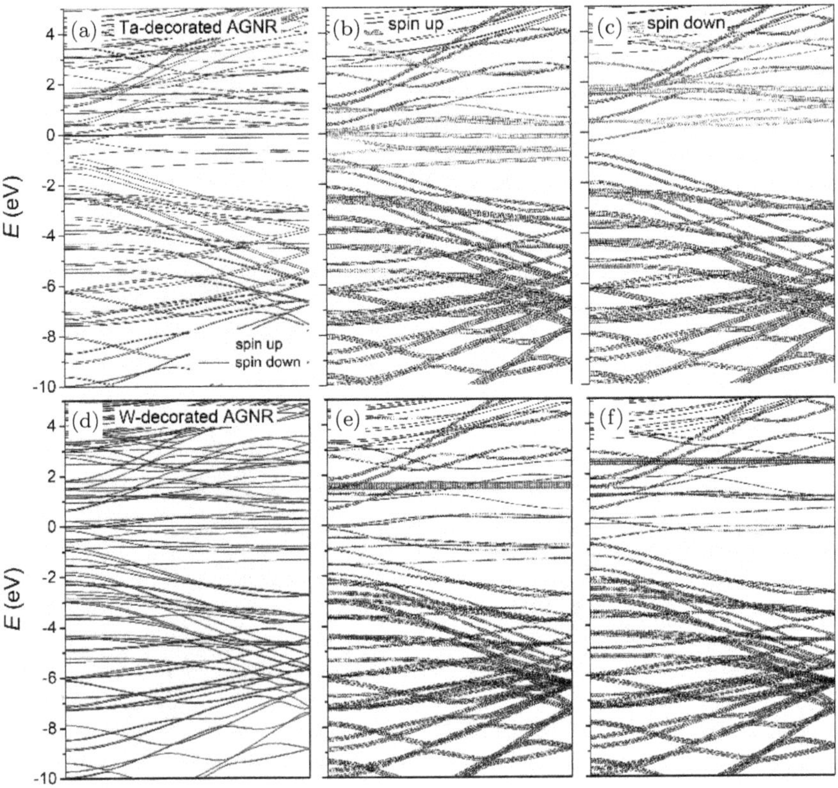

Fig. 9.2. (a) The spin-splitting energy band structures of Ta-decorated AGNRs, where spin up and spin down are indicated by blue and red curves, respectively. (b) The spin-up band structure of Ta-decorated AGNRs, where the red and blue circles indicate the Ta- and C-atom-dominated states, respectively. (c) The spin-down band structure of Ta-decorated AGNRs. (d) The spin-splitting energy band structures of W-decorated AGNRs. (e) The spin-up band structure of W-decorated AGNRs, where the red and blue circles indicate W- and C-atom-dominated states, respectively. (f) The spin-down band structure of W-decorated AGNRs. The corresponding results for (g)–(i) Ta-decorated ZGNRs and (j)–(l) W-decorated ZGNRs.

more subbands, and they are either dominated by C, W, or Ta atoms or co-dominated by C–Ta or C–W. The asymmetric valence and conduction energy subbands are greatly enhanced through the close partnerships of the guest–host and host–host interactions. The band-edge states exhibit apparent changes in their numbers,

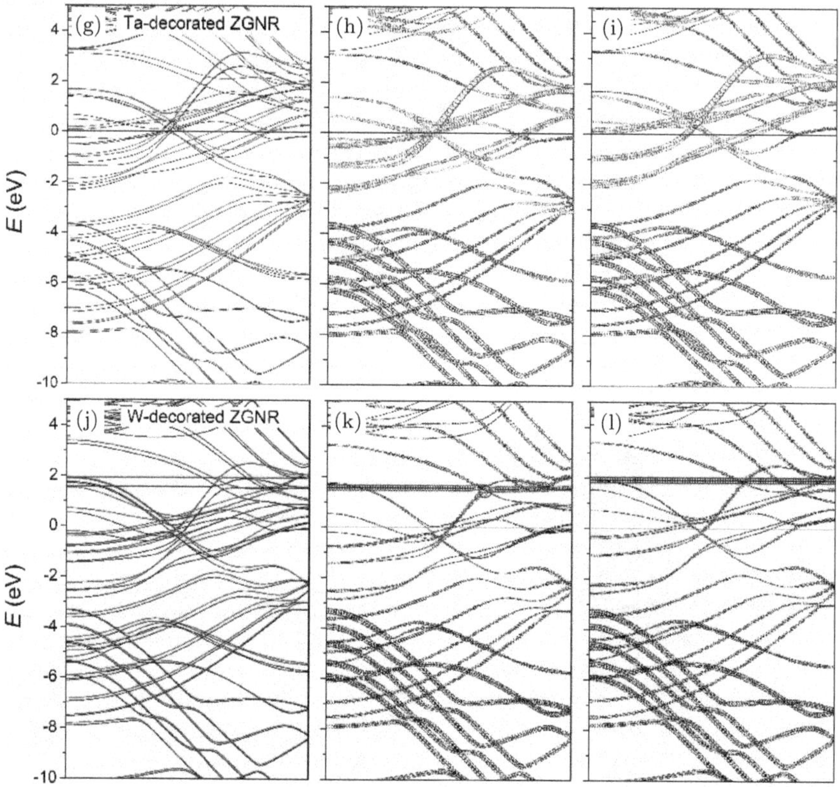

Fig. 9.2. (*Continued*)

energies, critical points, and curvatures. Moreover, the energy band structures can exhibit spin-split behavior, which is mainly attributed to the 5d orbitals of the transition-metal guest atoms, where their splitting energies should be comparable to the on-site spin-dependent Coulomb interactions. Most of the bands which do not cross the Fermi energy (E_F), are either fully occupied or fully unoccupied. So, the net contributions of their magnetic moments are almost zero. However, there are a few bands across the E_F level, and they dominate the magnetic properties. The magnetic momenta of spin-up and spin-down states make distinct contributions to the net magnetic moment. That is to say, the edge-decorated GNRs possess net magnetic momenta not equal to zero when the spin-split energy bands are in different occupied states, e.g., the spin-up energy band is fully occupied but the spin-down one is just partially occupied. In both the armchair and zigzag GNRs, the orbitals of carbon atoms-(2s, $2p_x$, $2p_y$, $2p_z$) are slightly changed, but the spin-split bands of Ta atoms-(6s, $5d_{xy}$, $5d_{yz}$, $5d_{z^2}$, $5d_{xz}$, $5d_{x^2}$) are apparently separated from each other within −2.0 to 2.0 eV in the armchair system. So is the case with the W atoms. The spin-split orbitals of the carbon atoms remain in similar states to each other in the armchair and zigzag systems, but the spin-split bands near E_F of W atoms are very different between the spin-up and spin-down cases. First-principles simulations can be performed to predict the characteristics of the edge-decorated GNRs. To establish a complete quasi-particle framework, the calculated predictions need to be examined by conducting experiments, such as ARPES, which can determine the subband-dependent and spin-dependent Fermi momenta.

There exist highly symmetric points in the energy–wave-vector space, which are called critical points, and they represent the van Hove singularities. For 1D systems, such as GNRs, the density of states can be obtained from the constant energy surface. Figure 9.3 displays the energy-dependent density of states obtained from the atom- and orbital-decomposed spin-split contributions, which include the C and Ta atoms in the armchair system (Figs. 9.3(a) and 9.3(b), respectively) and in the zigzag system (Figs. 9.3(c) and 9.3(d), respectively); the same is shown for the C and W atoms in the armchair system (Figs. 9.4(a) and 9.4(b), respectively) and in the zigzag system (Figs. 9.4(c) and 9.4(d), respectively). In the armchair systems, C-(2s, $2p_x$, $2p_y$) exist at $E < 2.5$ eV, within $-0.8 < E < -0.4$ eV, and within $0.3 < E < 0.4$ eV. There exist spin-split structures mainly on C-$2p_z$ within $-1 < E < 1$ eV; in the zigzag systems, C-(2s, $2p_x$, $2p_y$) appears at $E < -3.5$ eV and within $-2.4 < E < 4.2$ eV, and the splitting of C-$2p_z$ is apparent within $-1 < E < 1$ eV. However, in the armchair systems, the spin-split orbitals of Ta and W atoms exist within a wide range of $-1.8 < E < 4$ eV; in the zigzag systems, the spin-split orbitals of Ta and W atoms exist within $-1.5 < E < 5$ eV. According to the van Hove singularities, C-$2p_z$ and Ta-5d co-dominate weakly within $-5 < E < -4$ eV and -3 eV $< E < -2$ eV and co-dominate

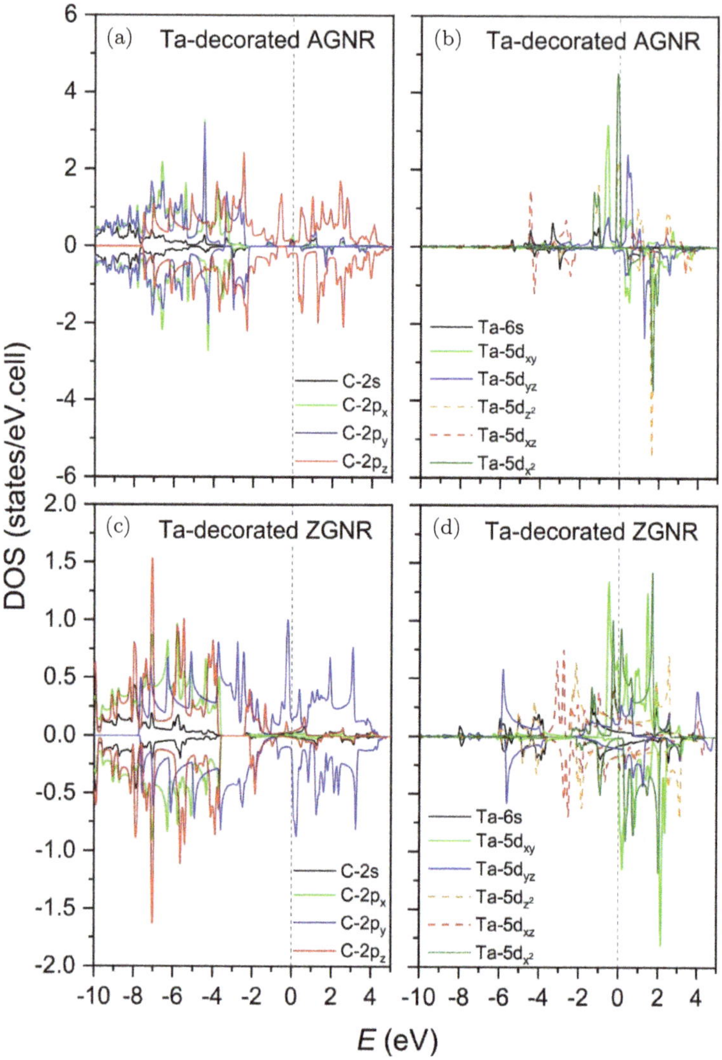

Fig. 9.3. The partial density of states of (a) C and (b) Ta atoms in AGNR systems, and the partial density of states of (c) C and (d) Ta atoms in ZGNR systems. The dashed line indicates $E_F = 0$.

strongly at −0.8, 0.5, and 0.6 eV for the armchair case, and these reveal the existence of C−Ta bonding. Similar to the Ta-decorated case, the C-2p$_z$ and W-5d strongly co-dominate at −0.9, −0.1, 0.3, and −1 eV.

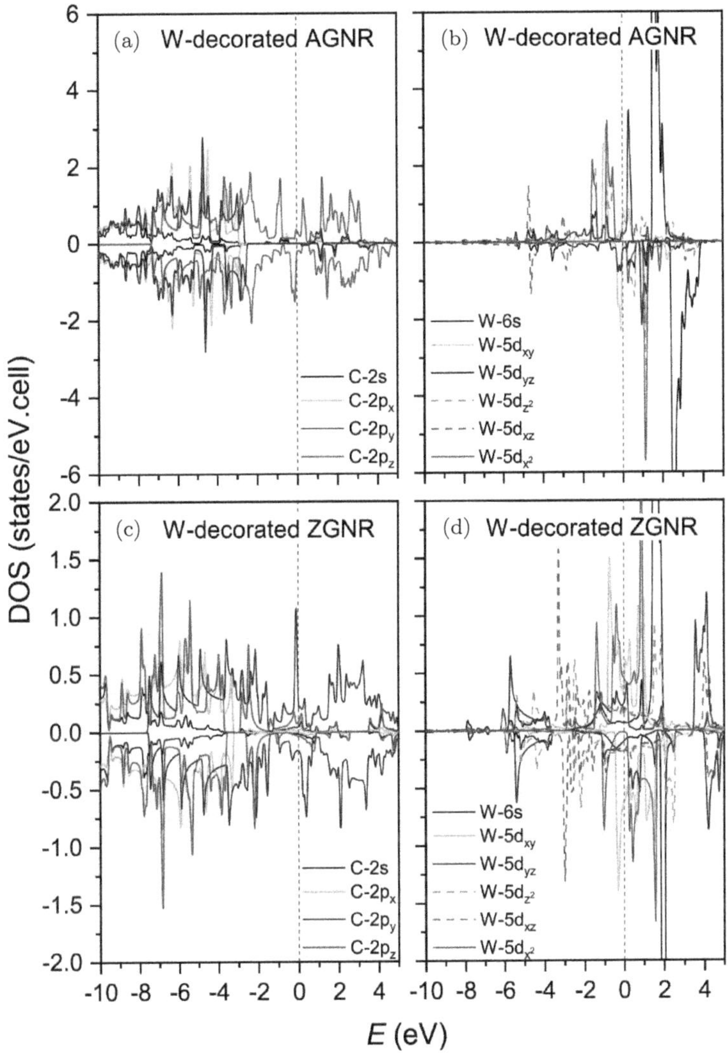

Fig. 9.4. The partial density of states of (a) C and (b) W atoms in AGNR systems, and the partial density of states of (c) C and (d) W atoms in ZGNR systems. The dashed line indicates $E_F = 0$.

In the zigzag system, there are many van Hove singularities near the Fermi level, and the C-2p$_z$ and Ta-5d co-dominate within $-1 < E < 1$ eV and at 2 eV; the Ta-5d dominates at -0.8 and 1.4 eV; and the Ta-6s dominate at 0.2 and 1.7 eV. These reveal that C–C, C–W, or C–Ta and W–W or Ta–Ta must coexist in the edge-decorated GNRs. However, the van Hove singularities of W-6s orbitals appear much strongly within $1 < E < 2$ eV. This might be caused by unsaturated W-6s orbitals, in which one of the electrons in 6s will be transferred to the 5d orbitals to achieve the half-filled constructions of 5d. Also, owing to lattice symmetry, the guest atoms are much closer to each other than in the armchair case. Thus, the W atoms will interact with not only the C atoms but also the nearest W atoms to form metallic bonding. In the W decorations of zigzag systems, the flat bands at 1.6 and 1.9 eV are dominated by W-6s, which is in good agreement with the van Hove singularity density of states; in the armchair case, the flat bands exist at 1.6 and 2.6 eV, which is again in agreement with the van Hove singularities. According to the energy band structures and density of states, we find the coexistence of the multi-orbital hybridizations of C–C, C–W, C–Ta, W–W, and Ta–Ta, which correspond to (2s, $2p_x$, $2p_y$, $2p_z$) – (2s, $2p_x$, $2p_y$, $2p_z$), (2s, $2p_x$, $2p_y$, $2p_z$) – (6s, $5d_{xy}$, $5d_{yz}$, $5d_{z^2}$, $5d_{xz}$, $5d_{x^2}$), (2s, $2p_x$, $2p_y$, $2p_z$) – (6s, $5d_{xy}$, $5d_{yz}$, $5d_{z^2}$, $5d_{xz}$, $5d_{x^2}$), (6s, $5d_{xy}$, $5d_{yz}$, $5d_{z^2}$, $5d_{xz}$, $5d_{x^2}$) – (6s, $5d_{xy}$, $5d_{yz}$, $5d_{z^2}$, $5d_{xz}$, $5d_{x^2}$), and (6s, $5d_{xy}$, $5d_{yz}$, $5d_{z^2}$, $5d_{xz}$, $5d_{x^2}$) – (6s, $5d_{xy}$, $5d_{yz}$, $5d_{z^2}$, $5d_{xz}$, $5d_{x^2}$), respectively. The orbital hybridization of C–C also covers $2p_z - 2p_z$ near the Fermi level.

9.4. Charge Distributions, Charge Variations, and Optical Properties

The edge decorations cause a non-uniform chemical environment in the Ta- and W-decorated GNRs. The spatial charge distributions and variations in the 1D edge-decorated systems are shown in Fig. 9.5. The C, Ta, and W atoms exhibit a spherical probability distribution. The red and green–yellow regions in the spatial charge distributions, respectively, represent the 2s and ($2p_x$, $2p_y$, $2p_z$) orbitals of the C atoms; the red and green–yellow regions also indicate the 6s and ($5d_{xy}$, $5d_{yz}$, $5d_{z^2}$, $5d_{xz}$, $5d_{x^2}$) orbitals of the Ta or W atoms. The calculated results display that the C–C bonding corresponding to the red region is very strong, and there exists weak but significant

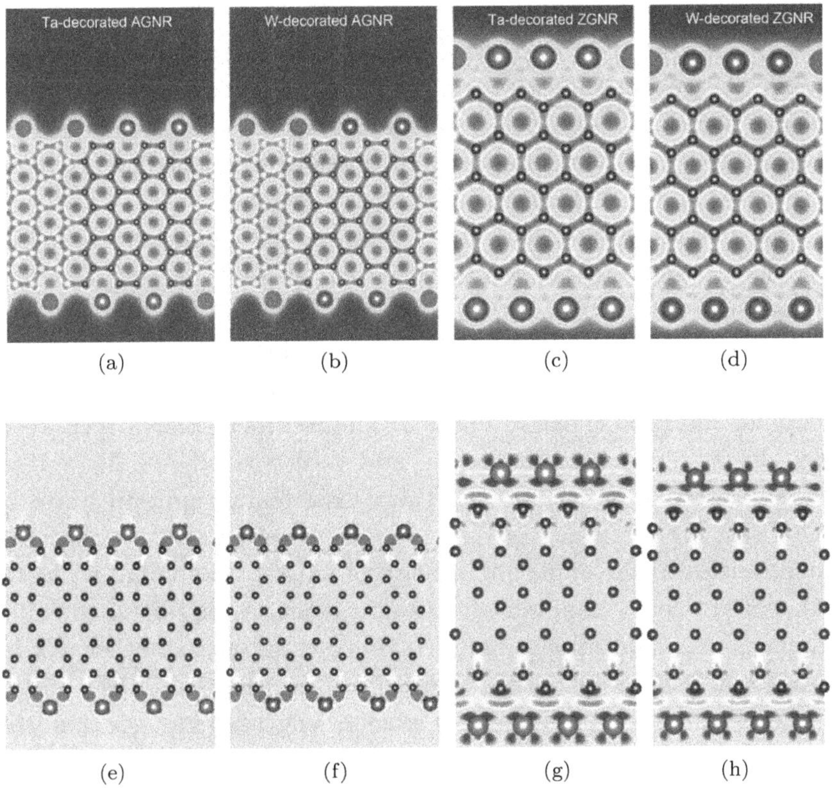

Fig. 9.5. The spatial charge distribution of (a) Ta-decorated AGNRs, (b) W-decorated AGNRs, (c) Ta-decorated ZGNRs, and (d) W-decorated ZGNRs. The charge variations in (e) Ta-decorated AGNRs, (f) W-decorated AGNRs, (g) Ta-decorated ZGNRs, and (h) W-decorated ZGNRs.

bonding on the edged atoms, which are covered by the green region. This reveals that the W and Ta atoms form orbital hybridization mainly with the $2p_z$ of the nearby carbon atoms. Very interestingly, the charge variations show the differences between the armchair and zigzag systems. In the zigzag system, the light-yellow regions exist between the guest atoms after decoration, but seldom in the armchair systems. That is to say, the metallic bonding is more apparent in the zigzag edge in the zigzag systems. However, the red region between the carbon atom and the guest atom is wider in the armchair systems than in the zigzag systems, and this reveals that the carbon–metal bonding is stronger in the armchair systems. These special distributions are caused by the geometric structures; in other words, the guest atoms are far away from each other in the armchair GNRs, but close to each other in the zigzag ones. This complicated mixing of orbital hybridizations is in agreement with the energy band structures and van Hove singularities of the density of states.

We use the orbital-projected net magnetic moments to analyze the distributions of the density of spin, as seen in Fig. 9.6. The transition-metal atoms can make great contributions to the ferromagnetic spin configurations at the edges due to their 5d orbitals. The spin-up and spin-down states are represented, respectively, by the blue and red regions in Fig. 9.6. The results display a very weak spin-down (red region) in the Ta- and W-decorated AGNRs and in W-decorated ZGNRs. However, there exist almost spin-up states in the Ta-decorated zigzag systems, which correspond to the van Hove singularities in the spin-split density of states. Combining with the calculated energy band structures and densities of states, the spin interactions might appear between the guest atoms; in other words, the interactions of ($5d_{xy}$, $5d_{yz}$, $5d_{z^2}$, $5d_{xz}$, $5d_{x^2}$) – ($5d_{xy}$, $5d_{yz}$, $5d_{z^2}$, $5d_{xz}$, $5d_{x^2}$) exist even though their bond lengths are longer than the guest–carbon bond lengths.

Our work shows the optical properties with the imaginary part of the dielectric function, which represent the photon–electron couplings under the requirements of momentum and energy conservations and the spin-dependent Pauli exclusion principle. In the armchair systems of Ta and W decorations, there exists a threshold peak in the imaginary part of the dielectric function, which arises from the part of 5d orbitals about the complicated initial and final states. The pure π-electronic excitations, which possess comparable spin-up and spin-down contributions, come to exist as the second absorption peak at 1.2 and 1 eV for Ta- and W-decorated systems, respectively, as seen in Fig. 9.7. The other prominent absorption peaks belong to the composite quasi-particle pictures, in which they consist of the superposition of C-$2p_z$, C-(2s, $2p_x$, $2p_y$, $2p_z$), C-(2s, $2p_x$, $2p_y$), and parts of the Ta and W orbitals. These unusual atom-, orbital-, and spin-dominated optical quasi-particles clearly

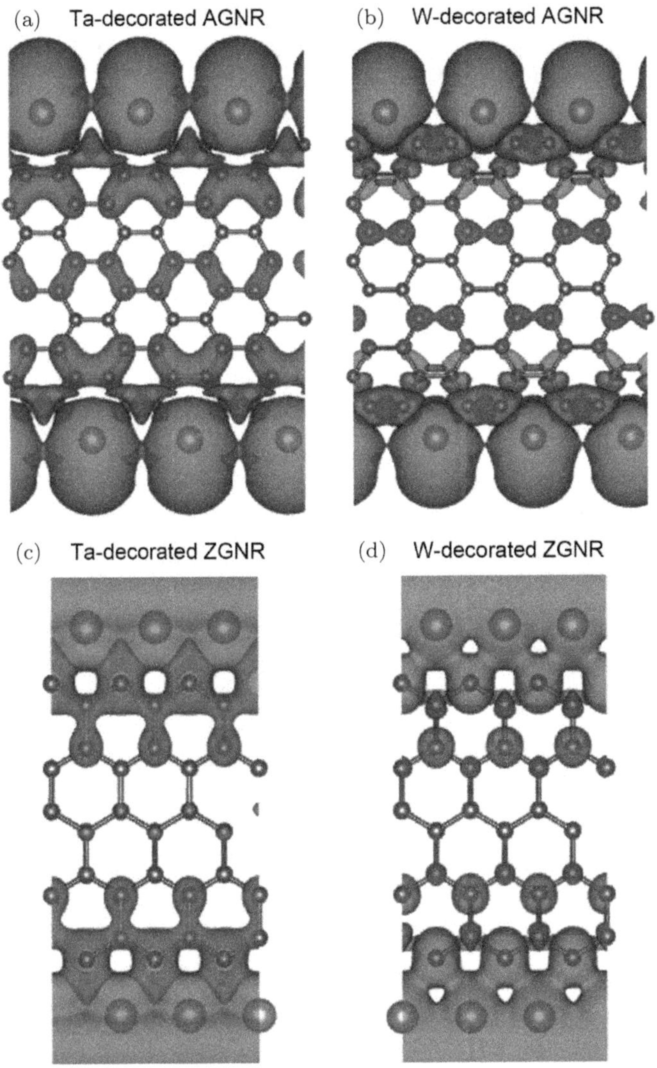

Fig. 9.6. The spatial spin distributions of (a) Ta- and (b) W-decorated AGNRs and (c) Ta- and (b) W-decorated ZGNRs. The spin-up states and spin-down states are indicated by blue and red colors, respectively.

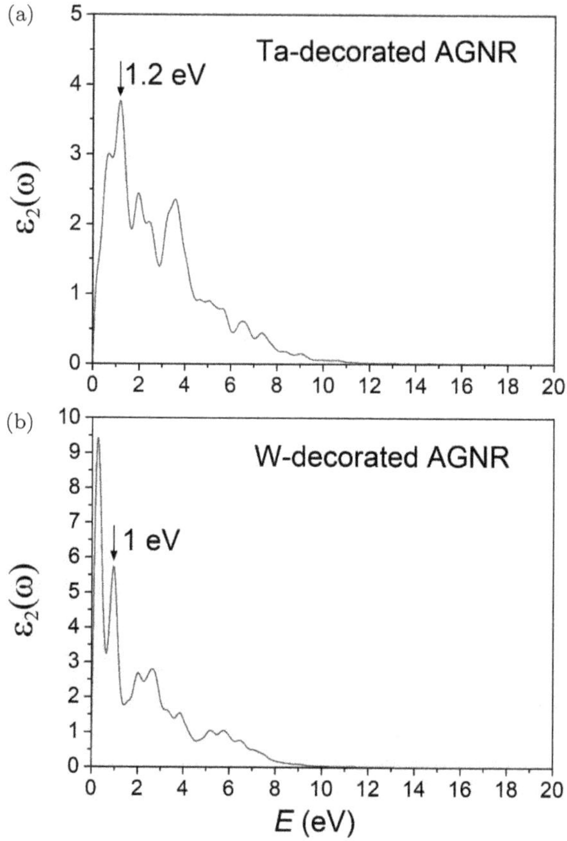

Fig. 9.7. The imaginary part of the dielectric function of (a) Ta- and (b) W-decorated AGNRs.

illustrate diversified excitation phenomena, thus demonstrating the successful development of an enlarged framework. The composite quasi-particles are well characterized by their intrinsic properties, further illustrating an enlarged framework in combination with those described in the other chapters. The high-precision optical measurements of absorption and transmission spectroscopies are available for observing the rich single-particle and collective excitations. It is very interesting and important to investigate these properties through experiments in the future.

9.5. Concluding Remarks

The first-principles calculations with VASP are very useful and successful in investigating the quasi-particles via geometric structures, charge distributions and variations, spin-split energy band structures and density of states, and spatial spin distributions. The transition-metal atoms, which possess 5d orbitals, make great contributions to the compounds in low-lying energies, such as a number of merged structures of the atom-, orbital- and spin-projected van Hove singularities, the atom- and orbital-decomposed magnetic moments, the Ta-/W-induced spin arrangements, and the spin-enriched optical transitions in dielectric functions. The critical properties of the transition metal-edged decorations mainly arise from the orbital hybridizations of host–guest, host–host, and guest–guest atoms. That is to say, the bonding of C–Ta, C–W, Ta–Ta, W–W, and C–C arise from (2s, $2p_x$, $2p_y$, $2p_z$) – (6s, $5d_{xy}$, $5d_{yz}$, $5d_{z^2}$, $5d_{xz}$, $5d_{x^2}$), (2s, $2p_x$, $2p_y$, $2p_z$) – (6s, $5d_{xy}$, $5d_{yz}$, $5d_{z^2}$, $5d_{xz}$, $5d_{x^2}$), (6s, $5d_{xy}$, $5d_{yz}$, $5d_{z^2}$, $5d_{xz}$, $5d_{x^2}$) – (6s, $5d_{xy}$, $5d_{yz}$, $5d_{z^2}$, $5d_{xz}$, $5d_{x^2}$), (6s, $5d_{xy}$, $5d_{yz}$, $5d_{z^2}$, $5d_{xz}$, $5d_{x^2}$) – (6s, $5d_{xy}$, $5d_{yz}$, $5d_{z^2}$, $5d_{xz}$, $5d_{x^2}$), and (2s, $2p_x$, $2p_y$, $2p_z$) – (2s, $2p_x$, $2p_y$, $2p_z$), respectively. In addition, the bonding of C–C covers $2p_z$–$2p_z$. Moreover, according to the spatial spin distributions, the interaction of spin just appears between transition-metal atoms, be it in armchair or zigzag systems. Based on the calculated results, the C-$2p_z$ and five 5d orbitals almost co-dominate the physical and chemical properties of transition metal-decorated GNRs within the low-lying energy. In summary, the quasi-particle properties, including spin-split orbitals and optical properties, have been seldom studied in the past years. It is very difficult to fully and accurately investigate the transition metal-based compounds due to their five complicated 5d orbitals: ($5d_{xy}$, $5d_{yz}$, $5d_{z^2}$, $5d_{xz}$, $5d_{x^2}$). For further exploration and construction of the full framework of quasi-particles, it is worthwhile to combine first-principles simulations with phenomenological models and experimental methods, such as high-resolution angle-resolved photoemission spectroscopy (ARPES) and spin-polarized scanning tunneling spectroscopy (STS), in the future.

The commercialization of electric vehicles (EVs) worldwide[67–69] has resulted in the mass production of lithium-ion (Li-ion) batteries. The retired power batteries of EVs have largely increased, causing resource waste threats. Recycling and utilization of such retired batteries have been suggested,[70–72] since many retired power batteries still possess about 80% of their initial capacity.[73–77] They can be repurposed once again, serving as battery modules in energy storage systems.[78–82] Governments have noticed this serious problem and are prepared to launch policies to deal with the recovery and reuse of repurposed batteries.[82–87] The discovery of graphene and more low-dimensional materials (such as 1D finite-width GNRs discussed in this work)[88,89] has led scientists to believe that these materials can be potential additives in battery materials to improve battery performance and overcome the recycling problems of Li-ion batteries.

Acknowledgments

The author (H. C. Chung) thanks Prof. Ming-Fa Lin for the book chapter invitation and for inspiring him to study this topic. H. C. Chung would like to thank the following contributors to this article for their valuable discussions and recommendations: Jung-Feng Jack Lin, Hsiao-Wen Yang, Yen-Kai Lo, and An-De Andrew Chung. The author (H. C. Chung) thanks Pei-Ju Chien for the discussions on English usage and corrections as well as Ming-Hui Chung, Su-Ming Chen, Lien-Kuei Chien, and Mi-Lee Kao for financial support. This work was supported in part by Super Double Power Technology Co., Ltd., Taiwan, under the project "Development of Cloud-native Energy Management Systems for Medium-scale Energy Storage Systems (https://osf.io/7fr9z/)" (Grant number: SDP-RD-PROJ-001-2020). This work was supported in part by the Ministry of Science and Technology (MOST), Taiwan (grant numbers: MOST 111-2112-M-006-020, MOST 109-2124-M-006-001, and MOST 110-2634-F-006-017). This work was also supported in part by the National Science and Technology Council (NSTC), Taiwan (grant number: NSTC 111-2811-M-006-047).

References

[1] Novoselov, K. S.; Geim, A, K.; Morozov, S. V.; Jiang, D.; Zhang, Y.; Dubonos, S. V.; Grigorieva, I. V.; Firsov, A. A. Electric field effect in atomically thin carbon films. *Science* **2004**, *306*, 666. DOI: 10.1126/science.1102896.

[2] Novoselov, K. S.; Jiang, D.; Schedin, F.; Booth, T. J.; Khotkevich, V. V.; Morozov, S. V.; Geim, A. K. Two-dimensional atomic crystals. *Proceedings of the National Academy of Sciences* **2005**, *102*, 10451. DOI: 10.1073/pnas.0502848102.

[3] Berger, C.; Song, Z. M.; Li, X. B.; Wu, X. S.; Brown, N.; Naud, C.; Mayou, D.; Li, T. B.; Hass, J.; Marchenkov, A. N.; Conrad, E. H.; First, P. N.; de Heer, W. A. Electronic confinement and coherence in patterned epitaxial graphene. *Science* **2006**, *312*, 1191. DOI: 10.1126/science.1125925.

[4] Bolotin, K. I.; Sikes, K. J.; Jiang, Z.; Klima, M.; Fudenberg, G.; Hone, J.; Kim, P.; Stormer, H. L. Ultrahigh electron mobility in suspended graphene. *Solid State Communications* **2008**, *146*, 351. DOI: 10.1016/j.ssc.2008.02.024.

[5] Morozov, S. V.; Novoselov, K. S.; Katsnelson, M. I.; Schedin, F.; Elias, D. C.; Jaszczak, J. A.; Geim, A. K. Giant intrinsic carrier mobilities in graphene and its bilayer. *Physical Review Letters* **2008**, *100*, 016602. DOI: 10.1103/PhysRevLett.100.016602.

[6] Balandin, A. A.; Ghosh, S.; Bao, W. Z.; Calizo, I.; Teweldebrhan, D.; Miao, F.; Lau, C. N. Superior thermal conductivity of single-layer graphene. *Nano Letters* **2008**, *8*, 902. DOI: 10.1021/nl0731872.

[7] Nair, R. R.; Blake, P.; Grigorenko, A. N.; Novoselov, K. S.; Booth, T. J.; Stauber, T.; Peres, N. M. R.; Geim, A. K. Fine structure constant defines visual transparency of graphene. *Science* **2008**, *320*, 1308. DOI: 10.1126/science.1156965.

[8] Bae, S.; Kim, H.; Lee, Y.; Xu, X. F.; Park, J. S.; Zheng, Y.; Balakrishnan, J.; Lei, T.; Kim, H. R.; Song, Y. I.; Kim, Y. J.; Kim, K. S.; Ozyilmaz, B.; Ahn, J. H.; Hong, B. H.; Iijima, S. Roll-to-roll production of 30-inch graphene films for transparent electrodes. *Nature Nanotechnology* **2010**, *5*, 574. DOI: 10.1038/nnano.2010.132.

[9] Lee, C.; Wei, X. D.; Kysar, J. W.; Hone, J. Measurement of the elastic properties and intrinsic strength of monolayer graphene. *Science* **2008**, *321*, 385. DOI: 10.1126/science.1157996.

[10] Chung, H.-C.; Chang, C.-P.; Lin, C.-Y.; Lin, M.-F. Electronic and optical properties of graphene nanoribbons in external fields. *Physical Chemistry Chemical Physics* **2016**, *18*, 7573. DOI: 10.1039/c5cp06533j.

[11] Chung, H. C.; Huang, Y. C.; Lee, M. H.; Chang, C. C.; Lin, M. F. Quasi-Landau levels in bilayer zigzag graphene nanoribbons. *Physica E* **2010**, *42*, 711. DOI: 10.1016/j.physe.2009.11.090.

[12] Han, M. Y.; Ozyilmaz, B.; Zhang, Y. B.; Kim, P. Energy band-gap engineering of graphene nanoribbons. *Physical Review Letters* **2007**, *98*, 206805. DOI: 10.1103/PhysRevLett.98.206805.

[13] Chen, Z. H.; Lin, Y. M.; Rooks, M. J.; Avouris, P. Graphene nano-ribbon electronics *Physica E* **2007**, *40*, 228. DOI: 10.1016/j.physe.2007.06.020.

[14] Tapaszto, L.; Dobrik, G.; Lambin, P.; Biro, L. P. Tailoring the atomic structure of graphene nanoribbons by scanning tunnelling microscope lithography. *Nature Nanotechnology* **2008**, *3*, 397. DOI: 10.1038/nnano.2008.149.

[15] Bai, J. W.; Duan, X. F.; Huang, Y. Rational fabrication of graphene nanoribbons using a nanowire etch mask. *Nano Letters* **2009**, *9*, 2083. DOI: 10.1021/nl900531n.

[16] Li, X. L.; Wang, X. R.; Zhang, L.; Lee, S. W.; Dai. H. J. Chemically derived, ultrasmooth graphene nanoribbon semiconductors. *Science* **2008**, *319*, 1229. DOI: 10.1126/science.1150878.

[17] Wang, X. R.; Ouyang, Y. J.; Li, X. L.; Wang, H. L.; Guo, J.; Dai, H. J. Room-temperature all-semiconducting sub-10-nm graphene nanoribbon field-effect transistors. *Physical Review Letters* **2008**, *100*, 206803. DOI: 10.1103/PhysRevLett.100.206803.

[18] Wu, Z. S.; Ren, W. C.; Gao, L. B.; Liu, B. L.; Zhao, J. P.; Cheng, H. M. Efficient synthesis of graphene nanoribbons sonochemically cut from graphene sheets. *Nano Research* **2010**, *3*, 16. DOI: 10.1007/s12274-010-1003-7.

[19] Ci, L. J.; Song, L.; Jariwala, D.; Elias, A. L.; Gao, W.; Terrones, M.; Ajayan, P. M. Graphene shape control by multistage cutting and transfer. *Advanced Materials* **2009**, *21*, 4487 DOI: 10.1002/adma.200900942.

[20] McAllister, M. J.; Li, J. L.; Adamson, D. H.; Schniepp, H. C.; Abdala, A. A.; Liu, J.; Herrera-Alonso, M.; Milius, D. L.; Car, R.; Prud'homme, R. K.; Aksay, I. A. Single sheet functionalized graphene by oxidation and thermal expansion of graphite. *Chemistry of Materials* **2007**, *19*, 4396. DOI: 10.1021/cm0630800.

[21] Datta, S. S.; Strachan, D. R.; Khamis, S. M.; Johnson, A. T. C. Crystallographic etching of few-layer graphene. *Nano Letters* **2008**, *8*, 1912. DOI: 10.1021/nl080583r.

[22] Ci, L.; Xu, Z. P.; Wang, L. L.; Gao, W.; Ding, F.; Kelly, K. F.; Yakobson, B. I.; Ajayan, P. M. Controlled nanocutting of graphene. *Nano Research* **2008**, *1*, 116 DOI: 10.1007/s12274-008-8020-9.

[23] Severin, N.; Kirstein, S.; Sokolov, I. M.; Rabe, J. P. Rapid trench channeling of graphenes with catalytic silver nanoparticles. *Nano Letters* **2009**, *9*, 457. DOI: 10.1021/nl8034509.

[24] Schaffel, F.; Warner, J. H.; Bachmatiuk, A.; Rellinghaus, B.; Buchner, B.; Schultz, L.; Rummeli, M. H. Shedding light on the crystallographic etching of multi-layer graphene at the atomic scale. *Nano Research* **2009**, *2*, 695. DOI: 10.1007/s12274-009-9073-0.

[25] Schaffel, F.; Warner, J. H.; Bachmatiuk, A.; Rellinghaus, B.; Buchner, B.; Schultz, L.; Rummeli, M. H. On the catalytic hydrogenation of graphite for graphene nanoribbon fabrication. *Physica Status Solidi B* **2009**, *246*, 2540. DOI: 10.1002/pssb.200982293.

[26] Campos, L. C.; Manfrinato, V. R.; Sanchez-Yamagishi, J. D.; Kong, J.; Jarillo-Herrero, P. Anisotropic etching and nanoribbon formation in single-layer graphene. *Nano Letters* **2009**, *9*, 2600–2604. DOI: 10.1021/nl900811r.

[27] Lin, C.-Y.; Yang, C.-H.; Chiu, C.-W.; Chung, H.-C.; Lin, S.-Y.; Lin, M.-F. *Rich Quasiparticle Properties of Low Dimensional Systems.* IOP Publishing, 2021, DOI: 10.1088/978-0-7503-3783-0.

[28] Kosynkin, D. V.; Higginbotham, A. L.; Sinitskii, A.; Lomeda, J. R.; Dimiev, A.; Price, B. K.; Tour, J. M. Longitudinal unzipping of carbon nanotubes to form graphene nanoribbons. *Nature* **2009**, *458*, 872. DOI: 10.1038/nature07872.

[29] Cataldo, F.; Compagnini, G.; Patane, G.; Ursini, O.; Angelini, G.; Ribic, P. R.; Margaritondo, G.; Cricenti, A.; Palleschi, G.; Valentini, F. Graphene nanoribbons produced by the oxidative unzipping of single-wall carbon nanotubes. *Carbon* **2010**, *48*, 2596. DOI: 10.1016/j.carbon.2010.03.063.

[30] Kumar, P.; Panchakarla, L. S.; Rao, C. N. R. Laser-induced unzipping of carbon nanotubes to yield graphene nanoribbons. *Nanoscale* **2011**, *3*, 2127. DOI: 10.1039/c1nr10137d.

[31] Jiao, L.; Zhang, L.; Wang, X.; Diankov, G.; Dai, H. Narrow graphene nanoribbons from carbon nanotubes. *Nature* **2009**, *458*, 877. DOI: 10.1038/nature07919.

[32] Jiao, L.; Zhang, L.; Ding, L.; Liu, J.; Dai, H. Aligned graphene nanoribbons and crossbars from unzipped carbon nanotubes. *Nano Research* **2010**, *3*, 387. DOI: 10.1007/s12274-010-1043-z.

[33] Laura Elias, A.; Botello-Mendez, A. R.; Meneses-Rodriguez, D.; Jehova Gonzalez, V.; Ramirez-Gonzalez, D.; Ci, L.; Munoz-Sandoval, E.; Ajayan, P. M.; Terrones, H.; Terrones, M. Longitudinal cutting of pure and doped carbon nanotubes to form graphitic nanoribbons using metal clusters as nanoscalpels. *Nano Letters* **2010**, *10*, 366. DOI: 10.1021/nl901631z.

[34] Parashar, U. K.; Bhandari, S.; Srivastava, R. K.; Jariwala, D.; Srivastava, A. Single step synthesis of graphene nanoribbons by catalyst particle size dependent cutting of multiwalled carbon nanotubes. *Nanoscale* **2011**, *3*, 3876. DOI: 10.1039/c1nr10483g.

[35] Talyzin. A. V.; Luzan, S.; Anoshkin, I. V.; Nasibulin, A. G.; Jiang, H.; Kauppinen, E. I.; Mikoushkin, V. M.; Shnitov, V. V.; Marchenko, D. E.; Noreus, D. Hydrogenation, purification, and unzipping of carbon nanotubes by reaction with molecular hydrogen: road to graphane nanoribbons. *ACS Nano* **2011**, *5*, 5132. DOI: 10.1021/nn201224k.

[36] Cano-Marquez, A. G.; Rodriguez-Macias, F. J.; Campos-Delgado, J.; Espinosa-Gonzalez, C. G.; Tristan-Lopez, F.; Ramirez-Gonzalez, D.; Cullen, D. A.; Smith, D. J.; Terrones, M.; Vega-Cantu, Y. I. Ex-MWNTs: Graphene sheets and ribbons produced by lithium intercalation and exfoliation of carbon nanotubes. *Nano Letters* **2009**, *9*, 1527. DOI: 10.1021/nl803585s.

[37] Kosynkin, D. V.; Lu, W.; Sinitskii, A.; Pera, G.; Sun, Z. Z.; Tour, J. M. Highly conductive graphene nanoribbons by longitudinal splitting of carbon nanotubes using potassium vapor. *ACS Nano* **2011**, *5*, 968. DOI: 10.1021/nn102326c.

[38] Shinde, D. B.; Debgupta, J.; Kushwaha, A.; Aslam, M.; Pillai, V. K. Electrochemical unzipping of multi-walled carbon nanotubes for facile synthesis of high-quality graphene nanoribbons. *Journal of the American Chemical Society* **2011**, *133*, 4168. DOI: 10.1021/ja1101739.

[39] Jiao, L.; Wang, X.; Diankov, G.; Wang, H.; Dai, H. Facile synthesis of high-quality graphene nanoribbons. *Nature Nanotechnology* **2010**, *5*, 321. DOI: 10.1038/nnano.2010.54.

[40] Xie, L. M.; Wang, H. L.; Jin, C. H.; Wang, X. R.; Jiao, L. Y.; Suenaga, K.; Dai, H. J. Graphene nanoribbons from unzipped carbon nanotubes: Atomic structures, Raman Spectroscopy, and electrical properties. *Journal of the American Chemical Society* **2011**, *133*, 10394. DOI: 10.1021/ja203860a.

[41] Paiva, M. C.; Xu, W.; Proenca, M. F.; Novais, R. M.; Laegsgaard, E.; Besenbacher, F. Unzipping of functionalized multiwall carbon nanotubes induced by STM. *Nano Letters* **2010**, *10*, 1764. DOI: 10.1021/nl100240n.

[42] Kim. K.; Sussman, A.; Zettl, A. Graphene nanoribbons obtained by electrically unwrapping carbon nanotubes. *ACS Nano* **2010**, *4*, 1362. DOI: 10.1021/nn901782g.

[43] Campos-Delgado, J.; Romo-Herrera, J. M.; Jia, X. T.; Cullen, D. A.; Muramatsu, H.; Kim, Y. A.; Hayashi, T.; Ren, Z. F.; Smith, D. J.; Okuno, Y.; Ohba, T.; Kanoh, H.; Kaneko, K.; Endo, M.; Terrones, H.; Dresselhaus, M. S.; Terrones, M. Bulk production of a new form of

sp(2) carbon: Crystalline graphene nanoribbons. *Nano Letters* **2008**, *8*, 2773. DOI: 10.1021/nl801316d.

[44] Wei, D. C.; Liu, Y. Q.; Zhang, H. L.; Huang, L. P.; Wu, B.; Chen, J. Y.; Yu, G. Scalable synthesis of few-layer graphene ribbons with controlled morphologies by a template method and their applications in nanoelectromechanical switches. *Journal of the American Chemical Society* **2009**, *131*, 11147. DOI: 10.1021/ja903092k.

[45] Sprinkle, M.; Ruan, M.; Hu, Y.; Hankinson, J.; Rubio-Roy, M.; Zhang, B.; Wu, X.; Berger, C.; de Heer, W. A. Scalable templated growth of graphene nanoribbons on SiC. *Nature Nanotechnol.* **2010**, *5*, 727. DOI: 10.1038/nnano.2010.192.

[46] Yang, X. Y.; Dou, X.; Rouhanipour, A.; Zhi, L. J.; Rader, H. J.; Mullen, K. Two-dimensional graphene nanoribbons *Journal of the American Chemical Society* **2008**, *130*, 4216. DOI: 10.1021/ja710234t.

[47] Cai. J. M.; Ruffieux, P.; Jaafar, R.; Bieri, M.; Braun, T.; Blankenburg, S.; Muoth, M.; Seitsonen, A. P.; Saleh, M.; Feng, X. L.; Mullen, K.; Fasel, R. Atomically precise bottom-up fabrication of graphene nanoribbons. *Nature* **2010**, *466*, 470. DOI: 10.1038/nature09211.

[48] Blankenburg, S.; Cai, J.; Ruffieux, P.; Jaafar, R.; Passerone, D.; Feng, X.; Muellen, K.; Fasel, R.; Pignedoli, C. A. Intraribbon heterojunction formation in ultranarrow graphene nanoribbons. *ACS Nano* **2012**, *6*, 2020. DOI: 10.1021/nn203129a.

[49] Yi, Z.; Yanfang, Z.; Geng, L.; Jianchen, L.; Xiao, L.; Shixuan, D.; Berger, R.; Xinliang, F.; Mullen, K.; Hong-Jun, G. Direct visualization of atomically precise nitrogen-doped graphene nanoribbons. *Applied Physics Letters* **2014**, *105*, 023101. DOI: 10.1063/1.4884359.

[50] Nakada, K.; Fujita, M.; Dresselhaus, G.; Dresselhaus, M. S. Edge state in graphene ribbons: Nanometer size effect and edge shape dependence. *Physical Review B* **1996**, *54*, 17954. DOI: 10.1103/PhysRevB.54.17954.

[51] Fujita, M.; Wakabayashi, K.; Nakada, K.; Kusakabe, K. Peculiar localized state at zigzag graphite edge. *Journal of the Physical Society of Japan* **1996**, *65*, 1920. DOI: 10.1143/JPSJ.65.1920.

[52] Barone, V.; Hod, O.; Scuseria.; G. E. Electronic structure and stability of semiconducting graphene nanoribbons. *Nano Letters* **2006**, *6*, 2748. DOI: 10.1021/nl0617033.

[53] Son, Y.-W.; Cohen, M. L.; Louie, S. G. Energy gaps in graphene nanoribbons. *Physical Review Letters* **2006**, *97*, 216803. DOI: 10.1103/PhysRevLett.97.216803.

[54] Kobayashi, Y.; Fukui, K.; Enoki, T.; Kusakabe, K.; Kaburagi, Y. Observation of zigzag and armchair edges of graphite using scanning

tunneling microscopy and spectroscopy. *Physical Review B* **2005**, *71*, 193406. DOI: 10.1103/PhysRevB.71.193406.

[55] Kobayashi, Y.; Fukui, K.; Enoki, T.; Kusakabe, K. Edge state on hydrogen-terminated graphite edges investigated by scanning tunneling microscopy. *Physical Review B* **2006**, *73*, 125415. DOI: 10.1103/PhysRevB.73.125415.

[56] Hsu, H.; Reichl, L. E. 2007 Selection rule for the optical absorption of graphene nanoribbons *Physical Review B* **2006**, *76*, 045418. DOI: 10.1103/PhysRevB.76.045418.

[57] Chung, H. C.; Lee, M. H.; Chang, C. P.; Lin, M. F. Exploration of edge-dependent optical selection rules for graphene nanoribbons. *Optics Express* **2011**, *19*, 23350–23363. DOI: 10.1364/OE.19.023350.

[58] Sasaki, K.-I.; Kato, K.; Tokura, Y.; Oguri, K.; Sogawa, T. Theory of optical transitions in graphene nanoribbons *Physical Review B* **2011**, *84*, 085458. DOI: 10.1103/PhysRevB.84.085458.

[59] Landau, L. Diamagnetismus der Metalle *Zeitschrift f\"{u}r Physik* **1930**, *64*, 629. DOI: 10.1007/BF01397213.

[60] Brey, L.; Fertig, H. A. Edge states and the quantized Hall effect in graphene *Physical Review B* **2006**, *73*, 195408. DOI: 10.1103/PhysRevB.73.195408.

[61] Wakabayashi, K.; Fujita, M.; Ajiki, H.; Sigrist, M. Electronic and magnetic properties of nanographite ribbons *Physical Review B* **1999**, *59*, 8271. DOI: 10.1103/PhysRevB.59.8271.

[62] Chung, H.-C.; Yang, P.-H.; Li, T.-S.; Lin, M.-F. Effects of transverse electric fields on Landau subbands in bilayer zigzag graphene nanoribbons. *Philosophical Magazine* **2014**, *94*, 1859. DOI: 10.1080/14786435.2014.897009.

[63] Chung, H.-C.; Lee, M.-H.; Chang, C.-P.; Huang, Y.-C.; Lin, M.-F. Effects of transverse electric fields on quasi-Landau levels in zigzag graphene nanoribbons. *Journal of the Physical Society of Japan* **2011**, *80*, 044602. DOI: 10.1143/jpsj.80.044602.

[64] Chung, H.-C.; Su, W.-P.; Lin, M.-F. Electric-field-induced destruction of quasi-Landau levels in bilayer graphene nanoribbons. *Physical Chemistry Chemical Physics* **2013**, *15*, 868. DOI: 10.1039/C2CP43631K.

[65] Chung, H.-C.; Lin, Y.-T.; Lin, S.-Y.; Ho, C.-H.; Chang, C.-P.; Lin, M.-F. Magnetoelectronic and optical properties of nonuniform graphene nanoribbons. *Carbon* **2016**, *109*, 883. DOI: 10.1016/j.carbon.2016.08.091.

[66] Li, W.-B.; Wang, Y.-M.; Chung, H.-C.; Lin, M.-F. 1D finite-width graphene nanoribbon systems: Alkalization and hydrogenation. *arXiv e-print* arXiv:2206.11162. **2022**. DOI: 10.48550/arXiv.2206.11162.

[67] IEA. *Global EV Outlook 2020.* **2020.**
[68] Chung, H.-C.; Nguyen, T. D. H.; Lin, S.-Y.; Li, W.-B.; Tran, N. T. T.; Thi Han, N.; Liu, H.-Y.; Pham, H. D.; Lin, M.-F. *First-Principles Calculations for Cathode, Electrolyte and Anode Battery Materials.* IOP Publishing, 2021, pp. 16-1 to -43. DOI: 10.1088/978-0-7503-4685-6ch16.
[69] Chung, H.-C. Engineering integrations, potential applications, and outlooks of Li-ion batteries. *engrXiv.* **2020**. DOI: 10.31224/osf.io/swcyg.
[70] Martinez-Laserna, E.; Gandiaga, I.; Sarasketa-Zabala, E.; Badeda, J.; Stroe, D. I.; Swierczynski, M.; Goikoetxea, A. Battery second life: Hype, hope or reality? A critical review of the state of the art. *Renewable and Sustainable Energy Reviews* **2018**, *93*, 701–718. DOI: 10.1016/j.rser.2018.04.035.
[71] Ahmadi, L.; Yip, A.; Fowler, M.; Young, S. B.; Fraser, R. A. Environmental feasibility of re-use of electric vehicle batteries. *Sustainable Energy Technologies and Assessments* **2014**, *6*, 64 DOI: 10.1016/j.seta.2014.01.006.
[72] Zhu, J. E.; Mathews, I.; Ren, D. S.; Li, W.; Cogswell, D.; Xing, B. B.; Sedlatschek, T.; Kantareddy, S. N. R.; Yi, M. C.; Gao, T.; Xia, Y.; Zhou, Q.; Wierzbicki, T.; Bazant, M. Z. End-of-life or second-life options for retired electric vehicle batteries. *Cell Reports Physical Science* **2021**, *2*, 100537. DOI: 10.1016/j.xcrp.2021.100537.
[73] Casals, L. C.; Garcia, B. A.; Canal, C. Second life batteries lifespan: Rest of useful life and environmental analysis. *Journal of Environmental Management* **2019**, *232*, 354–363. DOI: 10.1016/j.jenvman.2018.11.046.
[74] Podias, A.; Pfrang, A.; Di Persio, F.; Kriston, A.; Bobba, S.; Mathieux, F.; Messagie, M.; Boon-Brett, L. Sustainability assessment of second use applications of automotive batteries: Ageing of Li-Ion battery cells in automotive and grid-scale applications. *World Electric Vehicle Journal* **2018**, *9*, 24. DOI: 10.3390/wevj9020024.
[75] Tong, S.; Fung, T.; Klein, M. P.; Weisbach, D. A.; Park, J. W. Demonstration of reusing electric vehicle battery for solar energy storage and demand side management. *Journal of Energy Storage* **2017**, *11*, 200–210. DOI: 10.1016/j.est.2017.03.003.
[76] Wood, E.; Alexander, M.; Bradley, T. H. Investigation of battery end-of-life conditions for plug-in hybrid electric vehicles. *Journal of Power Sources* **2011**, *196*, 5147–5154. DOI: 10.1016/j.jpowsour.2011.02.025.
[77] Chung, H.-C. Failure mode and effects analysis of LFP battery module. *engrXiv.* **2018**. DOI: 10.31224/osf.io/acxsp.

[78] Kamath, D.; Shukla, S.; Arsenault, R.; Kim, H. C.; Anctil, A. Evaluating the cost and carbon footprint of second-life electric vehicle batteries in residential and utility-level applications *Waste Manage.* **2020**, *113*, 497–507. DOI: 10.1016/j.wasman.2020.05.034.

[79] Quinard, H.; Redondo-Iglesias, E.; Pelissier, S.; Venet, P. Fast electrical characterizations of high-energy second life lithium-ion batteries for embedded and stationary applications. *Batteries* **2019**, *5*, 33. DOI: 10.3390/batteries5010033.

[80] Heymans, C.; Walker, S. B.; Young, S. B.; Fowler, M. Economic analysis of second use electric vehicle batteries for residential energy storage and load-levelling. *Energy Policy* **2014**, *71*, 22–30. DOI: 10.1016/j.enpol.2014.04.016.

[81] Casals, L. C.; Garcia, B. A. Second-life batteries on a gas turbine power plant to provide area regulation services. *Batteries* **2017**, *3*, 10. DOI: 10.3390/batteries3010010.

[82] Chung, H.-C.; Nguyen, T. D. H.; Lin, S.-Y.; Li, W.-B.; Tran, N. T. T.; Thi Han, N.; Liu, H.-Y.; Pham, H. D.; Lin, M.-F. *First-Principles Calculations for Cathode, Electrolyte and Anode Battery Materials.* IOP Publishing. 2021, pp. 15-1 to -8. DOI: 10.1088/978-0-7503-4685-6ch15.

[83] Chung, H. C. Charge and discharge profiles of repurposed LiFePO4 batteries based on the UL 1974 standard. *Scientific Data* **2021**, *8*, 165. DOI: 10.1038/s41597-021-00954-3.

[84] Chung, H.-C.; Cheng, Y.-C. Action planning and situation analysis of repurposing battery recovery and application in China. *Journal of Taiwan Energy* **2019**, *6*, 425. DOI: 10.31224/osf.io/nxv7f.

[85] Hossain, E.; Murtaugh, D.; Mody, J.; Faruque, H. M. R.; Sunny, M. S. H.; Mohammad, N. A comprehensive review on second-life batteries: Current state, manufacturing considerations, applications, impacts, barriers and potential solutions, business strategies, and policies. *IEEE Access* **2019**, *7*, 73215–73252. DOI: 10.1109/access.2019.2917859.

[86] Gur, K.; Chatzikyriakou, D.; Baschet, C.; Salomon, M. The reuse of electrified vehicle batteries as a means of integrating renewable energy into the European electricity grid: A policy and market analysis. *Energy Policy* **2018**, *113*, 535–545 DOI: 10.1016/j.enpol.2017.11.002.

[87] Chung, H.-C.; Cheng, Y.-C. Summary of safety standards for repurposing batteries. *Monthly Journal of Taipower's Eng.* **2020**, *860*, 35. DOI: 10.31224/osf.io/d4n3s.

[88] Nguyen, T. D. H.; Lin, S.-Y.; Chung, H.-C.; Tran, N. T. T.; Lin, M.-F. *First-Principles Calculations for Cathode, Electrolyte and Anode Battery Materials.* IOP Publishing. 2021. DOI: 10.1088/978-0-7503-4685-6.

[89] Chung, H.-C.; Chiu, C.-W.; Lin, M.-F. Spin-polarized magneto-electronic properties in buckled monolayer GaAs. *Scientific Reports* **2019**, *9*, 2332. DOI: 10.1038/s41598-018-36516-8.

CHAPTER 10

DECORATION OF GRAPHENE NANORIBBONS WITH 5F RARE-EARTH ELEMENTS

Vo Khuong Dien and Ming-Fa Lin

Department of Physics, National Cheng Kung University, Tainan, Taiwan

Abstract

This study explores the decoration of graphene nanoribbons with 5f rare-earth elements and investigates their impact on the electronic, magnetic, and optical properties. Using first-principles calculations, we demonstrate that the incorporation of 5f elements induces significant changes in the nanoribbons' electronic structure, leading to localized electronic states within the bandgap. The presence of 5f electrons also introduces magnetic moments, while the hybridization between carbon atoms and 5f orbitals gives rise to unique electronic and magnetic phenomena. Additionally, we examine the optical properties, including absorption and emission spectra, providing insights into potential optoelectronic applications. This research advances our understanding of the design and development of graphene nanoribbon-based materials with tailored electronic, magnetic, and optical functionalities.

Keywords: Graphene nanorribons, 5f rare-earth elements, decorations.

The carbon atom, with an atomic number of 6, has four active orbitals of (2s, $2p_x$, $2p_y$, $2p_z$) in any material. Two fully occupied 1s orbitals only present very deep-energy dispersionless states. The current experimental methods can successfully synthesize the following types of pure element crystals: (1) 3D sp^3-diamond with a giant bandgap,[13–15] layered semi-metallic sp^2-graphite in AB (Bernal[16–18]) and ABC stacking (rhombohedral[19–21]), few-layer semi-metallic/narrow-gap graphene systems with various stacking and layer numbers,[22–24] 1D middle-/small-gap graphene nanoribbons,[5,6,25,26] metallic/narrow-/middle-gap carbon nanotubes,[27–29] 0D carbon tori,[30,31] carbon onions,[32,33] C60-related fullerenes,[34–36] carbon chains,[37,38] carbon rings,[39,40] and carbon disks.[41,42] There exist strong relations among the featured crystal symmetries, and so the composite quantum do quasiparticles. For example, a planar and finite-size graphene nanoribbon could be cut from a 2D graphene along the longitudinal direction by chemical methods.[43,44] Its π-electronic states are sampled from those of monolayer graphene under an open boundary condition, which is similar to those of a single-walled carbon nanotube with a periodic one. More interestingly, with two open edges, even with hydrogen passivation, the observable bond-length fluctuations frequently emerge in all graphene nanoribbons.[45] Such intrinsic behaviors are predicted to result in small bandgaps (the metal–semiconductor transitions in armchair and zigzag systems),[46,47] as clearly illustrated by the first-principles calculations[5,26] and the tight-binding model.[48–50] In addition to the planar carbon-honeycomb lattices, there also exist the curved,[51,52] folded,[53] and scrolled structures.[54] A lot of dangling bonds in open edges and surface morphologies can provide an outstanding platform for creating diverse quasi-particle phenomena through chemical modifications (as detailed in Section 2.1) and physical perturbations (Section 2.2).

Strong chemical modifications consist of adsorptions,[11,55] decorations,[5,56] substitutions,[57,58] intercalations/de-intercalations,[12,59] and heterojunctions.[60] The current scientific researches show that 1D graphene nanoribbons are outstanding candidates for fully exploring free conduction electrons due to their interlayer metallic bonds and the close relations among the ferromagnetic, antiferromagnetic, and non-magnetic spin configurations. A planar carbon-honeycomb crystal clearly exhibits perpendicular σ and π bondings, of which the former and the latter are, respectively, responsible for the deep- and low-energy quasi-particle phenomena. The interlayer $2p_z$–$2p_z$ orbital hybridizations and the van der Waals interactions dominate the layered structures with the well-behaved AAA,[61] AB,[62] ABC,[63] and AAB[24] stackings. The guest-adatom adsorptions correspond to the alkalized,[12] hydrogenated,[25] oxidized,[64] and halogenated[9] graphene nanoribbons. According to the 1D metallic model,[65,66] the free carrier density is just proportional to the magnitude of the Fermi momentum. The delicate first-principles simulations can determine the 1D Fermi surfaces very accurately and then identify a linear behavior between the alkali-atom concentration and the transferred electron density. This is valid for any alkali adatoms under the various adsorption cases of adatom concentrations and arrangements.[25,67] However, such examinations cannot be realized in 2D alkalized monolayer graphene systems because of the highly complicated evaluation of conduction electron density.[67] Moreover, the ferromagnetic configurations, which frequently emerge in zigzag graphene nanoribbons, are very sensitive to the alkali positions and concentrations. The predicted magnetic properties could be verified by spin-polarized scanning tunneling spectroscopy (STS).[68,69] Two open-edge boundaries are very suitable for the decorated address based on the delicate first-principles simulations.[5,56] After various guest-atom decorations, the curved graphene nanoribbons clearly present three types of stable crystal structures: (1) a closed carbon nanotube with a hybridized adatom line (detailed in Section 2.1.4), (2) a highly curved system,[51] and (3) a planar one.[5] The third type might have non-closed, hexagonal, pentagonal, or heptagonal structures without/with buckling. The strong partnerships between carbon–carbon and carbon-intercalant bonds should be responsible for all quasi-particle behaviors. Their orbital hybridizations and spin configurations need to be thoroughly examined through the VASP simulations within a unified framework.

The open-edge decorations of graphene nanoribbons, which cover the rare-earth metal guest adatoms, will be an excellent research strategy because of the very complicated orbital hybridizations and spin configurations. This is helpful in developing a broader and wider quasi-particle framework, compared to those described in previously published books.[25,60,67] The 14 active orbitals of 4f/5f can dominate the significant compounds, with special functionalities, e.g., high-temperature superconductors,[70,71] intermetallic systems,[72,73] and the heavy Fermi compounds.[74,75] However, the concise quasi-particle mechanisms are almost absent in the previous studies. They will be achieved using 1D rare-earth metal-decorated graphene nanoribbons. Their probability distributions are characterized by the coordinate-dependent cubic powers: (I) f_{x^3}, (II) f_{xz^2}, (III) f_{yz^2}, (IV) f_{xyz}, (V) $f_{z(x^2-y^2)}$, (VI) $f_{x(x^2-y^2)}$, and (VII) $f_{x(x^2-3y^2)}$, which are enriched by the spin-up and spin-down configurations. The total electron density, which mainly comes from the linear superposition of 4f/5f orbitals, will be very sensitive to the chemical environment. In this work, the guest adatoms exhibit strong reactions with dangling carbon bonds at the two open edges of the zigzag/armchair graphene nanoribbons. The pristine systems are predicted to display, respectively, the ferromagnetic and antiferromagnetic spin configurations on the same sides and the fully non-magnetic properties across the ribbon center.[46,47] How to modulate and diversify the magnetic behaviors through the rare-earth metal decorations should be a very interesting research topic. The atomic numbers of Np-93 and Pu-94, which belong to the group of radiative elements, are chosen for their unusual quasi-particles. All VASP calculations are performed under delicate analyses to achieve the concise pictures of physics, chemistry, and material science.

In this work, the VASP simulations are well suited for fully exploring the unusual quantum quasi-particles of Pd-/Pu-decorated 1D armchair and zigzag graphene nanoribbons (Figs. 10.1(a)–10.1(f)). After significant decorations, their critical roles in all essential properties are thoroughly examined and identified from consistent results. The strong partnerships among the multi-orbital hybridizations, spin configurations, crystal structures, electronic properties, and optical transitions are successfully established under a unified framework. Whether there exist closed polygons with/without bucking is understood from the featured C–Pd/C–Pu bonds. The magnetic configurations might be dependent on the rare-earth metal adatoms, the open zigzag edges, and/or the strong relationships between them. This should be thoroughly explored using the atom-, orbital-, and spin-decomposed magnetic moments, van Hove singularities, and the spatial spin-density distributions. Concise pictures under a highly non-uniform environment are

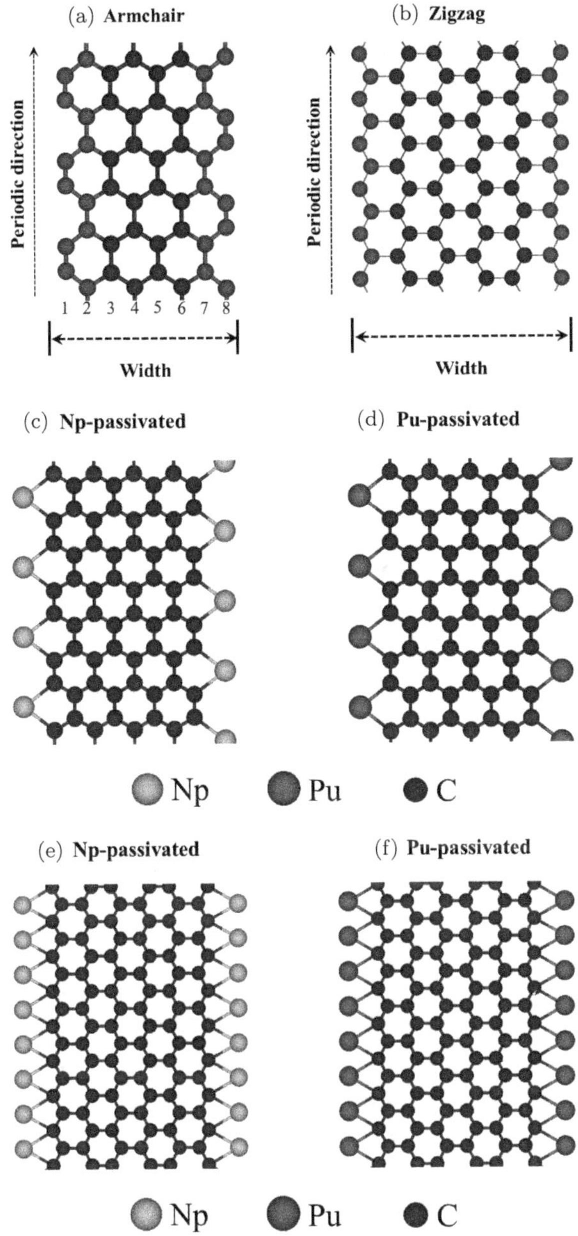

Fig. 10.1. The optimal 1D crystal structures for (a) armchair graphene nanoribbon (Na = 8), (b) zigzag one (Nz = 6), (c)/(d) Np-/Pu-decorated armchair compounds, and (e)/(f) zigzag partners.

obtained by examining the unusual crystal structures, the featured band structures, the very complicated charge densities, and the rich density of states. The drastic orbital deformations of C-[2s, $2p_x$, $2p_y$, $2p_z$] and Pd-/Pu-5f orbitals, and the merged van Hove singularities are available for distinguishing the active orbitals in C–Pd/C–Pu bonds. The calculated optical properties, dielectric functions, and energy loss spectra can provide a useful platform for understanding the single-particle excitations and plasmon modes. The quantum quasi-particles, corresponding to the prominent absorption peaks, will be well characterized by the complicated hybridization orbitals of the initial and final states. Of course, the existence of stable excitons should be checked in detail for the semiconducting systems. The simultaneous progress in the phenomenological models and experimental observations is one of the focuses of the study.

10.1. Np/Pu Decoration of Armchair and Zigzag Graphene Nanoribbons

Achiral armchair and zigzag graphene nanoribbons are outstanding candidates for showing the diversified crystal phases after edge decorations, as clearly illustrated in Figs. 10.1(a)–10.1(f). Na = 8 and Nz = 6 (Figs. 10.1(a) and 10.1(b)), which, respectively, correspond to the dimer and zigzag line numbers along the transverse direction, are chosen for a model study. The C–C bond lengths present the observable fluctuations within the 1.24–1.43 Å or 1.42–1.48 Å range or in pristine Na = 8/Nz = 6 graphene nanoribbons in the absence of edge decorations. The dangling bonds become shorter in an active environment compared with those at the ribbon center. This results in a highly non-uniform physical environment and thus sensitive changes in $2p_z$–$2p_z$ hopping integrals. According to the delicate first-principles simulations[25,47] and the tight-binding model,[76] the Na = $3I + 2$ (where I is an integer) armchair systems are 1D metals/small-gap semiconductors under the uniform/non-uniform cases (Fig. 10.2(a)). Furthermore, all zigzag nanoribbons display the metal–semiconductor transitions near the zone boundary since the partially flat valence and conduction energy subbands change

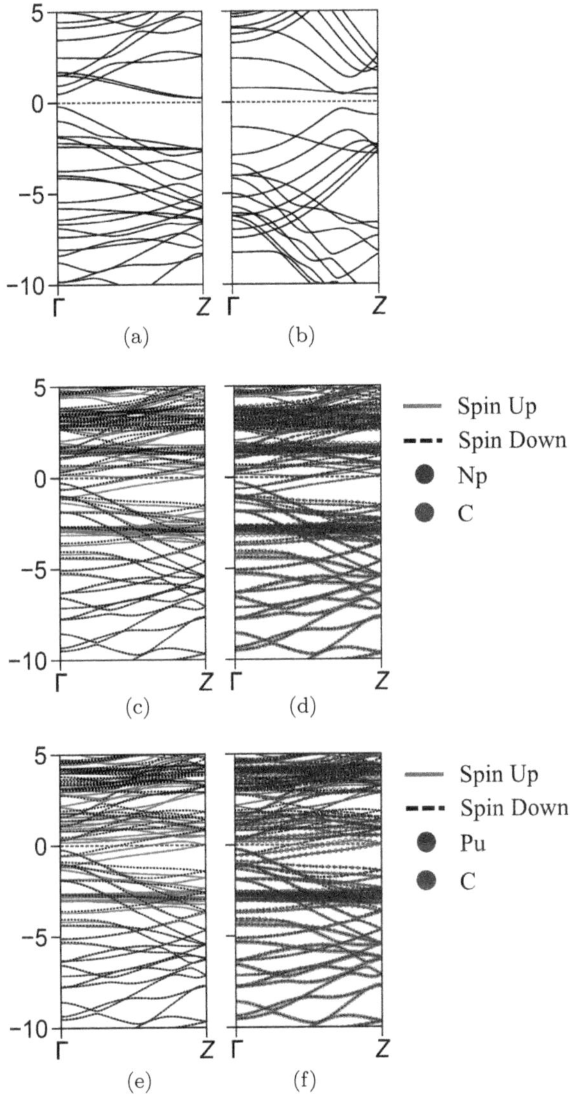

Fig. 10.2. The featured band structures with atom dominances for (a) armchair, (b) zigzag graphene nanoribbons, (c) and (d)/(e) and (f) Np-/Pu-decorated armchair compounds, and (g) and (h)/(i) and (j) zigzag partners, in which the blue and red dots represent the atom-decomposed contributions.

Table 10.1. The C–C, Np–Np/Pu–Pu, and C–Np/C–Pu bond lengths for: armchair graphene nanoribbon, zigzag one, Pd-/Pu-decorated armchair compounds, and zigzag partners.

Armchair nanoribbon					
Np decoration			**Pu decoration**		
C– C (Å)	Np–Np (Å)	C–Np	C–C	Pu–Pu	C–Pu
Zigzag nanoribbon					
C–C	Np–Np	C–Np	C–C	Pu–Pu	C–Pu

from the degenerate state into splitting configurations (the gapless to gapped status). The edge structures are totally different before and after the chemical Np/Pu decorations. A non-uniform pentagon appears in an armchair edge (Figs. 10.1(c) and 10.1(d)), in which the Np–Np/Pu–Pu and C–Np/C–Pu bond lengths (Table 10.1) are, respectively, 4.28 Å/4.29 Å and 2.42 Å/2.39 Å. Furthermore, the non-uniform C–C bond lengths are obviously reduced to 1.41–1.44 Å. These reflect more stable structures due to the edge passivations, in which the C–Pu bond is stronger than the C–Np one. In addition, the hydrogen passivation, which is frequently utilized for stabilizing samples,[5] only shows an open structure because of the very short C–H bonds.

The theoretical predictions of the planar crystal structures are very suitable for examination through high-resolution scanning tunneling microscopy (STM)[77] and reflection low-energy electron diffraction (RLEED).[78,79] However, the X-ray diffraction patterns cannot realize the surface morphology because the responses are too weak. Tunneling electron microscopy (TEM)[80] can observe the side-view geometries, e.g., the interlayer/intertube distance[81–83] and the buckled/folded/scrolled/curved structures.[84] The direction combinations of STM-, RLEED-, and TEM-STS are very powerful in testing the close relationships between the geometric and electronic properties, covering the very successful observations of few-layer graphene systems with the distinct stacking configurations and layer numbers, single-/double-walled carbon nanotubes/carbon nanotube bundles, and armchair/zigzag/chiral graphene nanoribbons. Most of the layered graphenes are identified as being semi-metals sensitive to the weak valence and conduction overlaps. Through the simultaneous progress in experimental and theoretical researches, three types of (m,n) single-walled carbon nanotubes have been identified (details in Ref. [3]): (I) metals with $m = n$, (II) narrow-gap semiconductors with $2m + n = 3I$, and (III) middle-gap ones for others. Moreover, all graphene nanoribbons are semiconductors, and their band-gap magnitudes decline with an increase in width. Three types of variation approaches are clearly revealed in Na $= 3I + 2$, $3I + 1$, and 3I armchair nanoribbons, where the first and second/third types, respectively, belong to narrow- and middle-gap semiconductors.[56] In this work, reliable high-precision measurements are made for exploring the free conduction electrons near the Fermi level (discussed later under the section on the density of states; Fig. 10.5).

10.2. Unusual 1D Band Structures and Wave Functions

Pristine and decorated graphene nanoribbons present the featured band structures and wave functions, compared to those of carbon nanotubes,[27,29] few-layer graphene systems,[23,85] and graphite materials.[86] An armchair graphene nanoribbon, as clearly shown in Fig. 10.2(a), for Na = 8, displays a number of 1D valence and conduction energy subbands, directly reflecting the active carbon-2pz orbitals in a unit cell, The other main features include the indirect gap with the highest/lowest occupied/unoccupied valence/conduction state at the Γ-zone boundary of $E_g^i \approx 0.71\,eV$, the high asymmetry of hole and electron spectra about the Fermi level, the strong energy dispersions in the parabolic, partially flat, and linear forms, the subband anticrossings/crossings/no-crossings, and the many band-edge states with zero-group velocities (van Hove singularities). The different open edges have very strong effects on the low-lying energy spectra. Any zigzag system (Fig 10.2(b) for Nz = 6) exhibits a pair of partially flat valence and conduction subbands, in which their splitting energy determines the bandgaps.

Such electronic wave functions are localized near the open zigzag edges; they belong to the edge localization states with almost vanishing group velocities. Bandgaps decline as ribbon widths grow. Furthermore, there are three types of approaches in armchair systems according to the open boundary condition: (I) narrow gap for Na = $3I + 2$ and (II)/(III) middle-gap semiconductors for $Na = 3I + 1$ or $3I$. Such bandgaps could be examined using direct combinations of STS and STM observations, as illustrated for single-walled carbon nanotubes/carbon nanotube bundles.[87] The theoretical predictions of the occupied valence energy spectra have been successfully verified using high-resolution angle-resolved photoemission spectroscopy (ARPES)[88] because of the well-defined 1D momentum transfer in a planar honeycomb crystal. However, the ARPES measurements on carbon nanotubes have been scarcely performed until now, mainly owing to their prominent deformations along the nanotube axis.

The low-lying band structures, as clearly illustrated in Figs. 10.2(c)–10.2(f), exhibit the dramatic transformations after edge decorations with rare-earth metal adatoms. There are more 1D energy subbands because of the active Np-/Pu–carbon multi-orbital hybridizations (discussed later under the section on the density of states). Each subband is dominated by C or Np/Pu, or co-dominated by (C, Np)/*(C, Pu) (the red and blue circles denoting host and guest atoms, respectively). The asymmetric valence and conduction energy subbands are greatly enhanced through the close partnerships between guest–host and host–host interactions. The drastic changes in the band-edge states include their numbers, energies, critical points, and curvatures. The 1D energy subbands might exhibit spin-split behaviors, mainly owing to the 5f orbitals of the rare-earth metal guest atoms, where their splitting energies should be comparable to the on-site spin-dependent Coulomb interactions (the Hubbard-like Hamiltonians[89]). Most of them belong to the fully occupied/unoccupied valence/conduction states; therefore, their contributions to the net magnetic moment are almost zero. Only a few of the energy subbands, which cross or are close to E_F, dominate the magnetic properties. That is to say, the Fermi momenta of the spin-up and spin-down configurations represent opposite contributions to the net magnetic moment. Furthermore, these dominating free carrier densities would induce an observable spin current under a small bias voltage. The 1D rare-earth metal edge-decorated graphene nanoribbons are expected to create unusual spintronics. More interestingly, the semiconductor–metal transitions emerge in all rare-earth metal-decorated systems, indicating the frequent generations of metallic interaction bonds (discussed in the section dealing with spatial charge density distributions).

The first-principles predictions, which are conducted on pristine and rare-earth metal edge-decorated graphene nanoribbons, behave like narrow- or middle-gap semiconductors and ferromagnetic metals. The former have been successfully verified using the high-resolution ARPES measurements for a few outstanding samples, but in the absence of width-dependent bandgaps.[90] The spin-polarized ARPES observations are available for determining subband- and spin-dependent Fermi momenta (k_F). This is very useful for evaluating the 1D free carrier densities (π/k_F) and thus exploring their quasi-particle transport properties (spintronics). Establishing strong relationships between the tight-binding model and the first-principles simulations becomes very difficult under rare-earth metal-decoration cases because of the complicated orbital hybridization and ferromagnetic spin configurations. For example, reliable parameters for the very rich hopping integrals between the active C-4 and R-7 orbitals would be very difficult to achieve. This open issue is under current investigation. Similar viewpoints could be generalized to transition-metal decoration cases. The diversified quasi-particle phenomena of geometric, electronic, and optical properties are obviously revealed in the modulation strategy. The scientific studies described in Chapters 9 and 10 clearly present a wider quasi-particle framework.

10.3. Highly Anisotropic Charge/Spin-density Distributions

Interestingly, highly non-uniform chemical and physical environments are clearly revealed in 1D pristine and edge-decorated

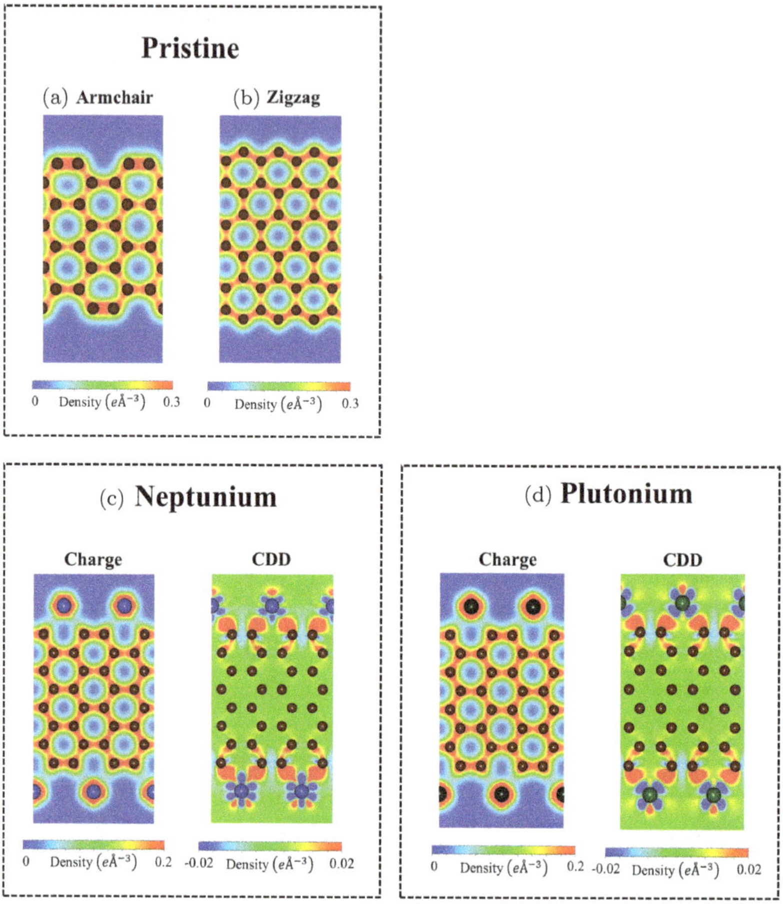

Fig. 10.3. The 2D charge density distributions in (a) armchair graphene nanoribbon, (b) zigzag one, (c)/(d) Nd-/Pu-decorated armchair compounds, and (e)/(f) zigzag partners.

graphene nanoribbons, as shown in Figs. 10.3(a)–10.3(f). The isolated carbon and Np/Pu atoms exhibit spherical probability distributions. Both exhibit radius-dependent red and yellow/green regions, respectively, corresponding to the 2s and ($2p_x$, $2p_y$, $2p_z$) orbitals and the 7s and seven 5f ones. These isotropic charge densities represent the total change after the generation of a planar carbon-honeycomb crystal. Two neighboring carbon atoms are strongly bound together by the σ bondings of (2s, $2p_x$, $2p_y$) orbitals (the dog-bone-like heavy red region). This is responsible for the stable structures of layered graphene systems.[67] Furthermore, the σ and π bondings are perpendicular to each other in the absence of non-planar structures. Of course, the charge densities are different for the edge and other carbon atoms because of the open edges. After prominent edge decorations, the saturated guest–host bonds can create the very rich multi-orbital hybridizations, more strongly in the nearest-or, even, second-nearest-neighbor C–C bonds. The atomic interactions, which are associated with the edge-dependent closed pentagon, play a critical role in the diverse quasi-particle phenomena. Specifically, the edge C–Np/C–Pu bonds clearly illustrate (2s, $2p_x$, $2p_y$, $2p_z$) – (f_{x^3}, f_{xz^2}, f_{yz^2}, (IV) f_{xyz}, (V) $f_{z(x^2-y^2)}$, $f_{x(x^2-y^2)}$, $f_{x(x^2-3y^2)}$), which has not been reported in the previously published works. Such complicated orbital mixings are thoroughly assisted by merged structures of van Hove singularities (as shown in Fig. 10.5).

The spatial spin-density distributions (Figs. 10.4(a)–10.4(e)) are fully assisted by the atom- and orbital-projected net magnetic moments (Table 10.2). The rare-earth metal guest atoms make very large contributions to the ferromagnetic spin configurations at the two edge structures (the very large red regions, e.g., 3.949 μ_B/5.318 μ_B in the Na $=$ 8 decorated armchair graphene nanoribbons, and 3.949 μ_B/5.318 μ_B in the Nz $=$ 6 zigzag ones). The opposite is true for the host carbon atoms, whose region is roughly smaller than 0.2%. All of them display the very weak spin-down distributions under the Pu decoration (the enlarged blue dots). However, both spin up and spin down are mixed together in the Np case. Obviously, the extremely magnetic configurations frequently emerge because of the rare-earth metal edge decorations. The Np–Np/Pu–Pu magnetic interactions should be much stronger than those in the neighboring Np–C/Pu–C bonds, even if their bond-length ratio is roughly equal to two. Such information is very useful in establishing the spin-based ferromagnetic Hamiltonians with reliable parameters and thus expanding the quasi-particle framework. These unusual quasi-particle behaviors could be directly examined by the SQUID/neutron scatterings on the net magnetic moments[91,92] and the spin-polarized STM of the spin-related surface morphologies.[93,94]

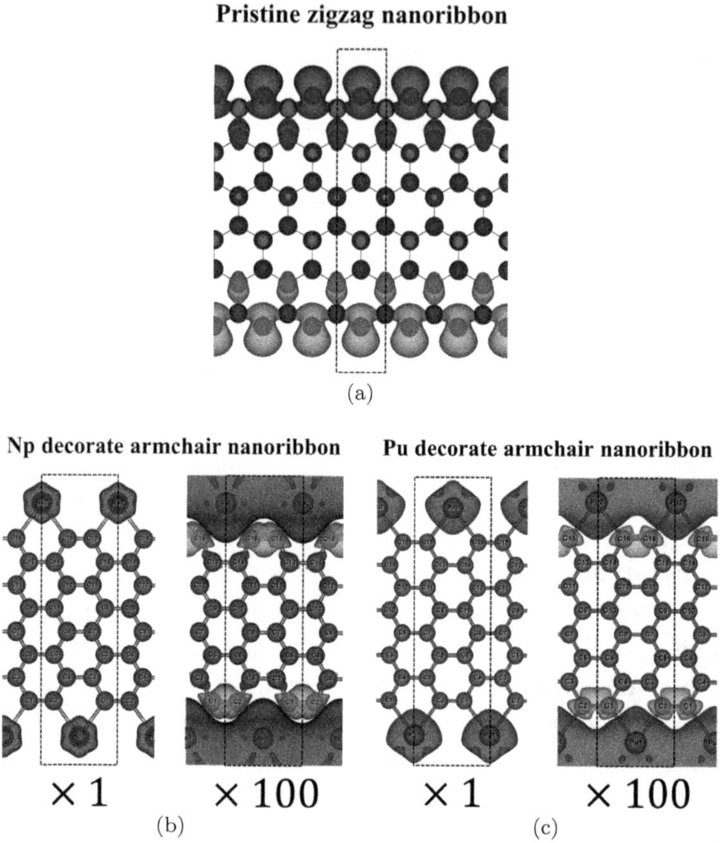

Fig. 10.4. The 2D spin configurations for (a) zigzag graphene nanoribbon, (b)/(c) Nd-/Pu-decorated armchair compounds, and (d)/(e) zigzag partners.

10.4. Rich van Hove Singularities

The critical points originate in the energy–wave-vector space, corresponding to the band-edge states,[95,96] display the irregular group velocities, and thus account for the frequency-dependent van Hove singularities. According to the density-of-states definition, the 3D/2D/1D magnitudes are associated with the constant-energy surface/loop/point integration of the inverse of group velocity, except for the delta-function-like peaks due to the 0D discrete states (e.g., 0D quantum dots). Generally, the 1D graphene-nanoribbon-related

Table 10.2. The atom- and orbital-decomposed magnetic moment (μ_B) for (a) Np or (b) Pu edge-decorated armchair/zigzag graphene nanoribbons.

Atom	**Orbital**								
					(a)				
C1	s	px	py	pz	$5fy(3x^2-y^2)$	$5fxyz$	$5fyz^2$	$5fz^3$	$5fxz^2$
C2	−0.0005	−0.0021	−0.0026	−0.0029					
C3	−0.0005	−0.0021	−0.0026	−0.0029					
C4	0.0005	0.0008	−0.0003	0.0002					
C5	0.0005	0.0008	−0.0003	0.0002					
C6	0.0001	0.0000	0.0000	0.0005					
C7	0.0001	0.0000	0.0000	0.0005					
C8	0.0000	0.0000	0.0000	−0.0001					
C9	0.0000	0.0000	0.0000	−0.0001					
C10	0.0000	0.0000	0.0000	−0.0001					
C11	0.0000	0.0000	0.0000	−0.0001					
C12	0.0001	0.0000	0.0000	0.0005					
C13	0.0001	0.0000	0.0000	0.0005					
C14	0.0005	0.0008	−0.0003	0.0002					
C15	0.0005	0.0008	−0.0003	0.0002					
C16	−0.0005	−0.0021	−0.0026	−0.0029					
Np1					0.9860	0.0234	0.7748	1.0011	0.1996
Np2					0.9860	0.0234	0.7733	1.0011	0.2010
					(b)				
C1	s	px	py	pz	$5fy(3x^2-y^2)$	$5fxyz$	$5fyz^2$	$5fz^3$	$5fxz^2$
C2	0.0000	−0.0008	−0.0017	−0.0027					
C3	0.0000	−0.0008	−0.0017	−0.0027					
C4	0.0004	0.0007	−0.0002	0.0000					
C5	0.0004	0.0007	−0.0002	0.0000					
C6	0.0001	0.0001	0.0000	0.0003					
C7	0.0001	0.0001	0.0000	0.0003					
C8	0.0000	0.0000	0.0000	−0.0001					
C9	0.0000	0.0000	0.0000	−0.0001					
C10	0.0000	0.0000	0.0000	−0.0001					
C11	0.0000	0.0000	0.0000	−0.0001					
C12	0.0001	0.0001	0.0000	0.0003					
C13	0.0001	0.0001	0.0000	0.0003					
C14	0.0004	0.0007	−0.0002	0.0000					
C15	0.0004	0.0007	−0.0002	0.0000					
C16	0.0000	−0.0008	−0.0017	−0.0027					
Pu1					0.7061	0.9662	0.4168	0.9593	0.9778
Pu2					0.7054	0.9662	0.4168	0.9592	0.9778

systems display three kinds of energy dispersions with distinct van Hove singularities: (I) the asymmetric divergent peaks in the inverse of square-root form, with their intensities proportional to the square root of the effective mass for the parabolic energy subbands; (II) the unusual plateau structures through the inverse relations between intensities and group velocities due to the linear valence or conduction energy subbands; and (III) the delta-function-like symmetric peaks arising from the partially flat dispersions. The spin degeneracy and spin split, respectively, were revealed in pristine and rare-earth metal-decorated graphene nanoribbons. The first-principles calculations are, respectively, examined using the non-polarized and spin-polarized STS measurements. This will be very useful in understanding the important differences between the covalent and metallic bonds. The rare-earth metal atoms would be outstanding candidates for chemical modifications and very helpful in establishing a grander quasi-particle framework, mainly owing to their active 14 5f/4f orbitals.

More interestingly, Fig. 10.5 clearly illustrates the energy-dependent density of states through the atom- and orbital-decomposed spin-split contributions for: (a) armchair and (b) zigzag graphene nanoribbons as well as (c) & (d)/(e) & (f) for Np-/Pu-decorated armchair compounds, and (g) & (h)/(i) & (j) zigzag partners. The main features of van Hove singularities are easily modulated by the host and guest atoms. A pristine graphene nanoribbon can display a number of strongly asymmetric peaks across the Fermi level of E, mainly owing to the 1D parabolic energy subbands. Any system with bandgaps is without free carriers at the Fermi level because of the complicated chemical and physical environments (Fig. 10.2(a)). Most importantly, the $2p_z$-(the red curve) and ($2s$, $2p_x$, $2p_y$)-orbital peaks (coffee, green, and blue curves), respectively, appear within the energy ranges of $|E^{c,v}| < 8.0$ eV and $-20.0 < E^{c,v} < -2.1$ eV/0, $3 < E^{c,v} < 2.1$ eV/$6.50 < E^{c,v} < 8.00$ eV. Also, the van Hove singularities of the latter come to exist at the same frequencies but under different intensities, even for $2p_x$ and $2p_y$ orbitals (the destruction of (x, y) plane symmetry). The highest occupied valence states and the lowest unoccupied conduction ones,

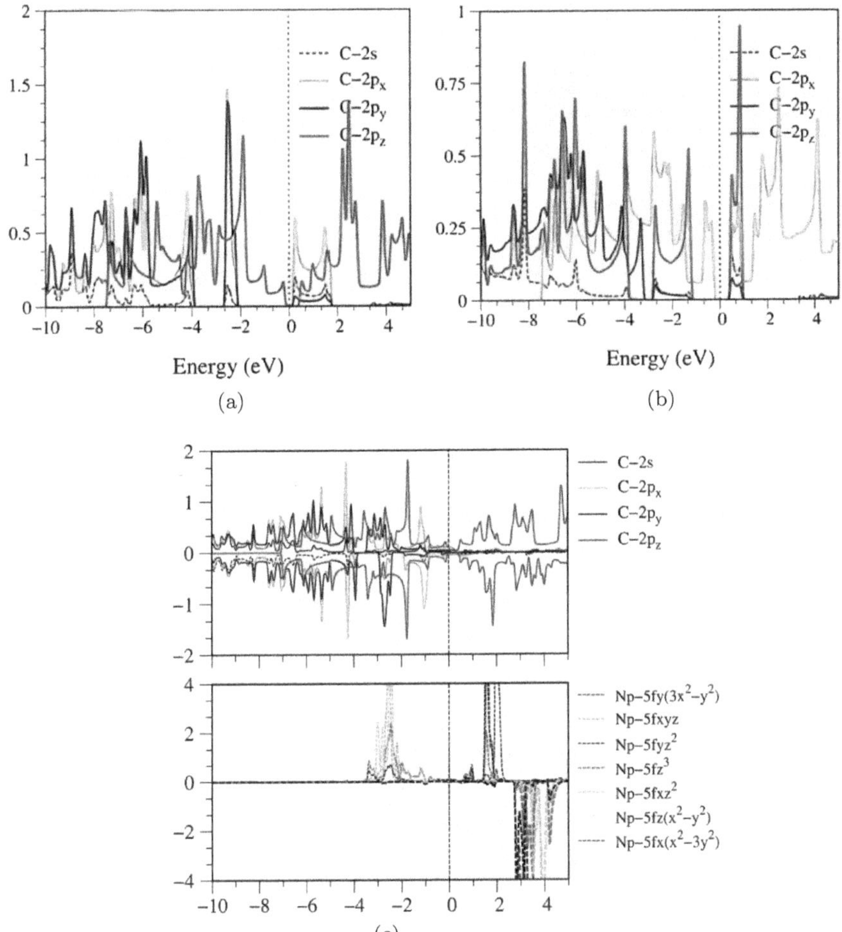

Fig. 10.5. The energy-dependent density of states with the atom- and orbital-decomposed spin-split contributions for (a) armchair graphene nanoribbons, (b) zigzag ones, (c) and (d)/(e) and (f) for Np-/Pu-decorated armchair compounds, and (g) and (h)/(i) and (j) zigzag partners.

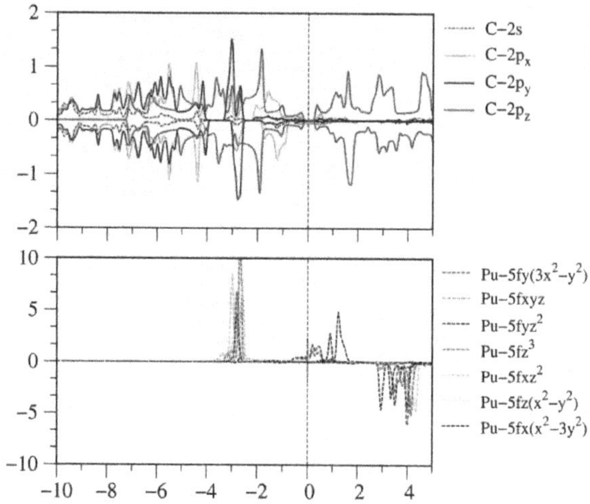

(d)

Fig. 10.5. (*Continued*)

which account for a bandgap, respectively, arise from the $2p_z$ and ($2p_x$, $2p_y$) orbitals, These results are totally different from those obtained for few-layer graphene systems and carbon nanotubes.

Most of the π and σ electronic van Hove singularities do not present the prominent mergences, clearly indicating the perpendicular chemical bondings on a planar honeycomb crystal. The number, frequency, and intensity of van Hove singularities [Refs] show the obvious changes after the rare-earth metal edge decorations, especially for their dominant energy region of $3.50 < E^{c,v} < 4.00$ eV. The significant 14-5f orbitals about ($5f_{x^3}$, $5f_{xz^2}$, $5f_{yz^2}$, $5f_{xyz}$, $5f_{z(x^2-y^2)}$, $5_{fx(x^2-y^2)}$, $5f_{x(x^2-3y^2)}$) reveal the comparable contributions. Furthermore, there exist important differences between the spin-up (the solid curves) and spin-down (the dashed curve) van Hove singularities. As to the spin-split C-(2s, $2p_x$, $2p_y$, $2p_z$), it also displays observable differences, in which the C-$2p_z$ orbitals have the Fermi-surface band properties. This further illustrates the metallic bondings of C–Np/C–Pu and Np–Np/Pu–Pu at the edge boundaries. From the merged van Hove singularities, the multi-orbital hybridizations of C–Np/C–Pu, Np–Np/Pu–Pu and C–C bonds, respectively, exhibit (2s, $2p_x$,

$2p_y$, $2p_z$) – ($5f_{x^3}$, f_{xz^2}, $5f_{yz^2}$, $5f_{xyz}$, $5_{fz(x^2-y^2)}$, $5f_{x(x^2-y^2)}$, $5f_{x(x^2-3y^2)}$), ($5f_{x^3}$, $5f_{xz^2}$, $5f_{yz^2}$, $5f_{xyz}$, $5f_{z(x^2-y^2)}$, $5f_{x(x^2-y^2)}$, $5f_{x(x^2-3y^2)}$) – ($5f_{x^3}$, $5f_{xz^2}$, $5f_{yz^2}$, $5f_{xyz}$, $5f_{z(x^2-y^2)}$, $5f_{x(x^2-y^2)}$, $5f_{x(x^2-3y^2)}$), and (2s, $2p_x$, $2p_y$) – (2s, $2p_x$, $2p_y$) and $2p_z$–$2p_z$. Of course, the ferromagnetic spin configurations are driven by the rare-earth metal atoms. All the concise pictures are very useful in the further development of an expanded quasi-particle framework.

The high-resolution STS measurements[97,98] are rather powerful for observing the van Hove singularities of 2D/1D/0D systems, in which the quantum currents are greatly enhanced by the gate voltages. These observations have been very successful in verifying the geometry-dominated electronic properties in graphene-related materials through direct combinations with STM/TEM/RLEED.[99,100] For example, stacking configurations and layer numbers are responsible for the semi-metallic or small-gap behaviors in layered graphene systems. There are three types of single-walled carbon nanotubes: (I) 1D metals for (m, m) armchair nanotubes, (II) narrow-gap semiconductors for non-armchair (m,n) systems with $2m + n = 3I$ (where I is an integer), and (III) middle-gap semiconductors in the other cases. Armchair graphene nanoribbons possess three types of bang gaps; (I) narrow gaps for Na $= 3I + 2$ and (II)/(III) middle gaps under $3I/3I + 1$. But for zigzag ones, their bandgaps obviously decline with an increase in the ribbon width. The STS/spin-STS examinations of the boundary, quantum-confinement, and decoration effects are mainly focused on the finite bandgap/a 1D gapless metal before/after Np-/Pu-rare-earth metal intercalations, the drastic changes of van Hove singularities across the Fermi level, the important differences between the spin-up and spin-down density of states at the specific energy ranges, and the ferromagnetic metals dominated by the Np/Pu guest atoms. These phenomena are very helpful in reliably establishing the interaction parameters of the critical Hamiltonians. The phenomenological models are very difficult to achieve in an analytic form because of the rather complicated band structures (Fig. 10.2) and density of states (Fig. 10.5). How to induce simultaneous progress among these three approaches remains an open issue (detailed in Chapter 19).

10.5. Unique Optical Transitions

The electromagnetic waves are useful in creating the vertical electronic excitations within the distinct energy regions according to the strengths of chemical bondings, generating the atom-/orbital-/spin-dependent single-particle electron–hole excitations and distinct plasmon modes and thus developing the composite quasi-particles under a grander framework (detailed in several books[3,4,60,101]). Each absorption structure, which appears in the bare response function (the imaginary part of the dielectric function), mainly arises from the joint van Hove singularities due to the initial and final band-edge states and the finite vertical transition probabilities associated with the electric dipole moments. After all the charge screenings, the screened response function (the energy loss functions; Refs) will display the prominent plasmon peaks coexistent with the electron–hole pairs under the obvious Landau dampings. The electromagnetic- and Coulomb-field perturbations are fully assisted by the finite-width quantum confinements and open/decorated edges. These critical factors are responsible for the featured quasi-particle behaviors of optical transitions, covering the threshold absorption frequency (the optical gap or the double of the Fermi-level blueshift), the existence of optical selection rules before and after edge decoration, and three/two kinds of carrier excitations under the decoration/pristine cases (free, π, and σ electronic excitations). 1D graphene nanoribbons only possess the valence/conduction energy subbands with the index numbers as measured from the Fermi level. These cannot serve as good quantum numbers, and they are totally different from the discrete angular momenta in cylindrical carbon nanotubes (the transverse periodic boundary conditions). The edge decoration effects on selection rules will be investigated in detail. Also, whether the stable excitons of the excited valence holes and conduction electrons could exist in narrow- and middle-gap pristine graphene nanoribbons is worthy of thorough studies. The quantum quasi-particles clearly illustrate the diversified excitation phenomena, as shown in Fig. 10.6, for the imaginary-/real-part dielectric functions and energy loss spectra: (a) and (b) armchair graphene nanoribbons, (c) and (d)

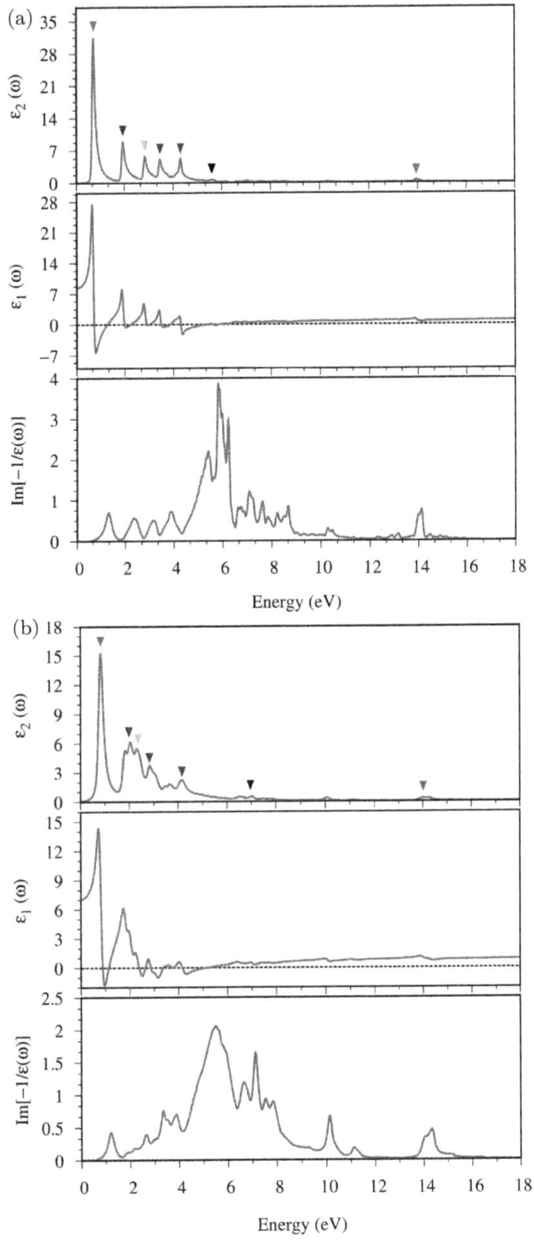

Fig. 10.6. The optical properties, dielectric functions, and energy loss spectra for: (a) armchair graphene nanoribbons, (b) zigzag ones, (c) and (d)/(e) and (f) Np-/Pu-decorated armchair compounds, and (g) and (h)/(i) and (j) zigzag partners.

zigzag ones, (e) and (f)/(g) and (h) Pd-/Pu-decorated armchair compounds, and (i) and (j)/(k) and (l) zigzag counterparts. $\epsilon_2(\omega)$ (the solid red curves in Figs. 10.6(a), 10.6(c), 10.6(e), 10.6(g), 10.6(i), and 10.6(k)) represent the photon–electron couplings under the requirements of momentum and energy conservations, as well as the spin-dependent Pauli exclusion principle. After sufficiently strong dipole scatterings, a number of 1D prominent absorption structures are associated with the joint van Hove singularities. Through delicate analyses using the VASP simulations, the obtained significant excitation frequencies and channels through the orbital hybridization and spin-dependent composite quantum quasi-particles are thoroughly listed in Table 10.3, covering single-particle/collective excitations for (a)/(b) armchair graphene nanoribbon, (c)/(d) zigzag ones, (e)/(f) and (g)/(h) Np- and Pu-decorated armchair compounds and zigzag counterparts. This table and the effective vertical transitions in band structure (the colored arrows in Figs. 10.2(a–b), Figs. 10.2(c–d)/and Figs. 10.2(e–f)) are very helpful for the greatly enhanced development of the composite quasi-particles since the atom-, orbital- and spin-related initial and final states are thoroughly identified from the delicate VASP simulations. As to pristine armchair (Fig. 10.6(a)) and zigzag systems (Fig. 10.6(c)), their optical excitations could be roughly categorized into three excitation types: the (I) π-, (II) composite σ- and π-, and (III) σ-electronic contributions, where the first and third ones survive at, respectively, lower and higher than 3.02 and 6.67 eV. Specifically, the second type is associated with the (2s, $2p_x$, $2p_y$)-initial and $2p_z$-final states, mainly owing to the dangling bonds. Many prominent peaks mainly come from the valence and conduction energy subbands due to finite-width effects. However, it is very difficult to analyze whether the specific selection rules are present in the various cases since good quantum numbers cannot be achieved through them. Such excitation frequency ranges are also revealed in 3D bulk graphite systems, few-layer graphene materials, and carbon nanotubes. This unique quasi-particle behavior only reflects the dominating sp^2 bondings on the planar honeycomb lattices. As to the threshold absorption frequency, the effective absorption excitation channel is indicated by a red-colored vertical arrow, clearly illustrating a sensitive dependence on the low-lying energy subbands near the high-symmetry points. Its value might not be equal to an indirect or direct bandgap. While the charge screening effects are due to all the energy-dependent carriers. The energy loss function, defined as $[-1/\epsilon(\omega)]$, can comprehend the coherent charge oscillations in the different energy ranges (the discrete plasmon modes). The lower- and higher-intensity plasmon peaks, as clearly shown in Figs. 10.6(b), 10.6(d), 10.6(f), 10.6(h), 10.6(j), and 10.6(l). The former, which is accompanied by the strong Landau dampings, mainly arise from a few/or certain valence energy subbands; that is, they are partly due to the π-/σ-electronic states, as observed in discrete π-plasmon modes below ~6.0 eV. Both π and σ plasmon modes due to all the carriers present strong peaks at ~5.78 and 13.91 eV.

The rich quasi-particle behaviors of single-particle and collective excitations are drastically changed by the rare-earth metal edge decorations. All effective excitation channels might be very different, as might the excitation frequency, intensity, number, and form of the prominent absorption peaks. Most of the effective excitation channels are closely related to the spin-up configurations, except for a few. The threshold excitation frequency at 0.09 eV is non-zero even for ferromagnetic metals, mainly owing to the rare-earth metal dominances on the spin-dependent vertical excitations (the red-colored vertical arrows in the band structures shown in Figs. 10.2(c), 10.2(e), 10.2(g), and 10.2(h) and their values in Tables 10.3(e)–10.3(h)). For example, the Np-decorated armchair system presents a threshold peak in the imaginary-prat dielectric function (the red triangle in Fig. 10.6(e)), mainly arising from a part of the 5f-orbitals about the initial and final Np-[$5f_{y(3x^2-y^2)}$, $5f_{x(x^2-3y^2)}$, $5f_{z^3}$, $5f_{yz^2}$] states. The pure π-electronic excitations, which possess comparable spin-up and spin-down contributions, emerge as the second absorption peak at 0.65 eV (the blue triangle). The other prominent absorption peaks belong to the composite quasi-particle pictures, in which they consist of the superposition of C-$2p_z$, C-(2s, $2p_x$, $2p_y$, $2p_z$)/(2s, $2p_x$, $2p_y$), and part of the Np-5f orbitals. These unusual atom-, orbital-, and spin-dominated optical quasi-particles clearly illustrate the diversified excitation phenomena, which have been very successful in developing an enlarged framework. Both π and σ plasmon modes, which are partially assisted by the Np/Pu 5f-orbitals, show obvious changes in their frequencies (6.67 and 14.68 eV for the Np decoration) and intensities, especially for the latter. That the rare-earth metal atoms can make much contribution to the low-lying energy subbands through the metallic bondings and the ferromagnetic spin configurations could be the main reasons. More interestingly, the spin-split multi-orbital hybridizations have been thoroughly examined for each prominent absorption channel. The composite quasi-particles are well characterized by their intrinsic properties, further illustrating an expanded framework in this book chapter. The high-precision optical measurements of absorption and transmission spectroscopies are available for observing the rich single-particle and collective excitations. The measured optical quantity, being assisted by the Kramers–Kronig relations under the physical perturbations and responses (the principle-value integration partnerships between the real- and imaginary-part response functions[102]), can clearly illustrate the frequency-dependent dielectric function and energy loss function. Before and after the rare-earth metal edge decorations, the main features of the π- and σ-electronic vertical excitations are almost totally changed into those of the composite quasi-particles associated with the prominent superposition of the Np-/Pu-5f orbitals. Most of the spin-split absorption peaks arise from the spin-up configurations. The experimental verifications of the first theoretical predictions can provide sufficient information about the charge- and spin-dependent interactions. Notably, the threshold absorption is also independent of the Fermi-level red- or blueshift.

10.6. Concise Conclusions

The delicate VASP simulations have successfully enlarged the composite quantum quasi-particles by the spin-up and spin-down 14-5f orbitals of the rare-earth metal guest atoms in the 1D edge-decorated graphene nanoribbons. The rich and unique multi-orbital hybridizations and ferromagnetic spin configurations are fully examined through the quasi-particle properties: the planar carbon-honeycomb lattices,[103] the closed edges of non-uniform pentagons with two long C–R bonds, the featured band structures with the sensitive atom and spin dominance at distinct energy ranges, more 1D valence/conduction energy subbands and band-edge states, the semiconductor–ferromagnetic–metal transitions and the spin-split Fermi surfaces after the dangling-bond saturations, the disappearance of edge-localized states in zigzag graphene nanoribbons, the dramatic transformations of the spatial charge density distributions under the very strong C–R chemical bondings, the magnetic phase configuration transition between antiferromagnetism and ferromagnetism in the zigzag systems, a number of merged structures of atom-, orbital-, and spin-projected van Hove singularities, the atom- and orbital-decomposed magnetic moments, the Np-/Pu-induced spin arrangements, and the spin-enriched optical transitions in dielectric functions of electron–hole pair excitations and energy loss spectra of plasmon modes.

Most importantly, the critical mechanisms in the C–Pa/C–U, Pa–Pa/U–U, and C–C bonds, respectively, arise from (2s, $2p_x$, $2p_y$, $2p_z$) – ($5f_{x^3}$, $5f_{xz^2}$, $5f_{yz^2}$, $5f_{xyz}$, $5f_{z(x^2-y^2)}$, $5_{fx(x^2-y^2)}$, $5f_{x(x^2-3y^2)}$), ($5f_{x^3}$, $5f_{xz^2}$, $5f_{yz^2}$, $5f_{xyz}$, $5f_{z(x^2-y^2)}$, $5f_{x(x^2-y^2)}$, $5f_{x(x^2-3y^2)}$) – ($5f_{x^3}$, 5_{fxz^2}, $5f_{yz^2}$, $5f_{xyz}$, $5f_{z(x^2-y^2)}$, $5f_{x(x^2-y^2)}$, $5f_{x(x^2-3y^2)}$), and (2s, $2p_x$, $2p_y$) – (2s, $2p_x$, $2p_y$) and $2p_z$–$2p_z$. Furthermore, the ferromagnetic spin interactions are only present in Pa–Pa/U–U bondings because of the extremely localized magnetic moments (Fig. 10.4 and Table 10.2). Such important mechanisms belong to the Heisenberg-like forms,[104] but not to the Hubbard-like forms.[105] Specifically, the threshold absorption frequency is hardly dependent on the magnitude of the Fermi level. Each strong absorption peak can be identified to arise from the initial and final states of the atom- and orbital dominances of C-(2s, $2p_x$, $2p_y$, $2p_z$)/C-$2p_z$ and (2s, $2p_x$, $2p_y$) and a part of Np-/Pu-5f orbitals. Many efficient excitation channels are responsible for the obvious mixing of π and σ electronic electron–hole pair excitations and plasmon modes. The spin-split optical quasi-particles are first revealed in this book chapter by first-principles simulations. The above-mentioned quasi-particle features should be enough to explore the orbital-hybridization-created hopping integrals and the on-site

Coulomb potentials, as well as the spin-related electron–electron interactions (the Heisenberg-like ones+). It seems to be very difficult for rare-earth metal-based compounds because of active 14-5f/ 14-4f orbitals. Whether the phenomenological models are worthy of systematic investigation is one of the focuses of the near-future studies. The chemical decoration effects could be generalized to the folded/curved/nanoscrolled and few-layer graphene, silicene, germanene, tinene, and plumbene nanoribbons for developing a greatly enlarged quasi-particle framework.

The VASP predictions of the geometric, electronic, and optical properties need to be thoroughly examined through different experimental observations. The direct combinations of the high-precision STM/TEM/RLLED and spin-polarized STS are very powerful for testing the close relationships between the optimal geometries and the energy-dependent van Hove singularities, such as bandgaps, semiconductor–metal transitions, strong asymmetric peaks in the inverse of square-root form due to the band-edge states, and spin-split energy spacings. The spin-polarized ARPES is available for making delicate measurements of the quasi-particle energy spectra and lifetimes below the Fermi level [Refs]. where the main features of the former include the high asymmetry about the valence holes and conduction electrons, the spin-split Fermi surfaces, more energy dispersion relations, and band-edge states. The multi-Fermi momenta are expected to exhibit the beating phenomena of the well-known Friedel oscillations in 1D ferromagnetic metals. The non-polarized/spin-polarized optical transmission spectroscopy should be suitable for observing the prominent absorption peaks with the same spin states. Also, the stable/quasi-stable excitons might emerge in pristine small- or middle-gap semiconductors, but not in the ferromagnetic metals. The threshold absorption frequency needs to be examined in detail since it is sensitive to many-body effects and semiconductor–metal transitions. In summary, the VASP simulations, phenomenological models, and experimental observations have shown simultaneous progress through the rare-earth metal decoration of planar graphene nanoribbons and their contributions to a grander quasi-particle framework.

References

[1] Matsunaga, K.; Tanaka, T.; Yamamoto, T.; Ikuhara, Y. First-principles calculations of intrinsic defects in Al_2O_3. *Physical Review B* **2003**, *68*, 085110.

[2] Togo, A.; Tanaka, I. First principles phonon calculations in materials science. *Scripta Materialia* **2015**, *108*, 1–5.

[3] Lin, C.-Y.; Wu, J.-Y.; Chiu, C.-W.; Lin, M.-F. *Coulomb Excitations and Decays in Graphene-related Systems*. CRC Press, 2019.

[4] Lin, C.-Y.; Do, T.-N.; Huang, Y.-K.; Lin, M.-F. Optical properties of graphene in magnetic and electric fields. 2017.

[5] Chang, S.-L.; Lin, S.-Y.; Lin, S.-K.; Lee, C.-H.; Lin, M.-F. Geometric and electronic properties of edge-decorated graphene nanoribbons. *Scientific Reports* **2014**, *4*, 1–8.

[6] Chang, S.-L.; Wu, B.-R.; Wong, J.-H.; Lin, M.-F. Configuration-dependent geometric and electronic properties of bilayer graphene nanoribbons. *Carbon* **2014**, *77*, 1031–1039.

[7] Pak, Y.; Kim, S.-M.; Jeong, H.; Kang, C. G.; Park, J. S.; Song, H.; *et al.* Palladium-decorated hydrogen-gas sensors using periodically aligned graphene nanoribbons. *ACS Applied Materials & Interfaces* **2014**, *6*, 13293–13298.

[8] Zheng, N.; Yang, S.; Xu, H.; Lan, Z.; Wang, Z.; Gu, H. A DFT study of the enhanced hydrogen storage performance of the Li-decorated graphene nanoribbons. *Vacuum* **2020**, *171*, 109011.

[9] Tran, N. T. T.; Nguyen, D. K.; Glukhova, O. E.; Lin, M.-F. Coverage-dependent essential properties of halogenated graphene: A DFT study. *Scientific Reports* **2017**, *7*, 1–13.

[10] Tran, N. T. T.; Gumbs, G.; Nguyen, D. K.; Lin, M.-F. Fundamental properties of metal-adsorbed silicene: A DFT Study. *ACS Omega* **2020**, *5*, 13760–13769.

[11] Nguyen, D. K.; Tran, N. T. T.; Chiu, Y.-H.; Lin, M.-F. Concentration-diversified magnetic and electronic properties of halogen-adsorbed silicene. *Scientific Reports* **2019**, *9*, 1–15.

[12] Li, W.-B.; Lin, S.-Y.; Tran, N. T. T.; Lin, M.-F.; Lin, K.-I. Essential geometric and electronic properties in stage-n graphite alkali-metal-intercalation compounds. *RSC Advances* **2020**, *10*, 23573–23581.

[13] Wei, S.-H.; Zunger, A. Predicted band-gap pressure coefficients of all diamond and zinc-blende semiconductors: Chemical trends. *Physical Review B* **1999**, *60*, 5404.

[14] Surh, M. P.; Louie, S. G.; Cohen, M. L. Band gaps of diamond under anisotropic stress. *Physical Review B* **1992**, *45*, 8239.

[15] Galusha, J. W.; Jorgensen, M. R.; Bartl, M. H. Diamond-structured titania photonic-bandgap crystals from biological templates. *Advanced Materials* **2010**, *22*, 107–110.
[16] Bernal, J. D. The structure of graphite. *Proceedings of the Royal Society of London Series A, Containing Papers of a Mathematical and Physical Character* **1924**, *106*, 749–773.
[17] García, N.; Esquinazi, P.; Barzola-Quiquia, J.; Dusari, S. Evidence for semiconducting behavior with a narrow band gap of Bernal graphite. *New Journal of Physics* **2012**, *14*, 053015.
[18] Charlier, J.-C.; Michenaud, J.-P.; Gonze, X. First-principles study of the electronic properties of simple hexagonal graphite. *Physical Review B* **1992**, *46*, 4531.
[19] Haering, R. Band structure of rhombohedral graphite. *Canadian Journal of Physics* **1958**, *36*, 352–362.
[20] Latychevskaia, T.; Son, S.-K.; Yang, Y.; Chancellor, D.; Brown, M.; Ozdemir, S.; *et al.* Stacking transition in rhombohedral graphite. *Frontiers of Physics* **2019**, *14*, 1–7.
[21] Shi, Y.; Xu, S.; Yang, Y.; Slizovskiy, S.; Morozov, S. V.; Son, S.-K.; *et al.* Electronic phase separation in multilayer rhombohedral graphite. *Nature* **2020**, *584*, 210–214.
[22] Lin, C.-Y.; Wu, J.-Y.; Ou, Y.-J.; Chiu, Y.-H.; Lin, M.-F. Magneto-electronic properties of multilayer graphenes. *Physical Chemistry Chemical Physics* **2015**, *17*, 26008–26035.
[23] Lin, C.-Y.; Wu, J.-Y.; Chiu, Y.-H.; Chang, C.-P.; Lin, M.-F. Stacking-dependent magnetoelectronic properties in multilayer graphene. *Physical Review B* **2014**, *90*, 205434.
[24] Lin, C.-Y.; Huang, B.-L.; Ho, C.-H.; Gumbs, G.; Lin, M.-F. Geometry-diversified Coulomb excitations in trilayer AAB stacking graphene. *Physical Review B* **2018**, *98*, 195442.
[25] Lin, S.-Y.; Tran, N. T. T.; Chang, S.-L.; Su, W.-P.; Lin, M.-F. *Structure-and Adatom-Enriched Essential Properties of Graphene Nanoribbons*. CRC Press, 2018.
[26] Lin, S.-Y.; Tran, N. T. T.; Lin, M.-F. Diversified phenomena in metal- and transition-metal-adsorbed graphene nanoribbons. *Nanomaterials* **2021**, *11*, 630.
[27] Li, T.-S.; Lin, M.-F. Electronic properties of carbon nanotubes under external fields. *Physical Review B* **2006**, *73*, 075432.
[28] Lin, M.-F. Magnetic properties of toroidal carbon nanotubes. *Journal of the Physical Society of Japan* **1998**, *67*, 1094–1097.
[29] Lin, M. F.; Shung, K. W. K. Optical and magneto-optical properties of carbon nanotube bundles. *Journal of the Physical Society of Japan* **1997**, *66*, 3294–3302.

[30] Meunier, V.; Lambin, P.; Lucas, A. Atomic and electronic structures of large and small carbon tori. *Physical Review B* **1998**, *57*, 14886.

[31] Tsai, C.-C.; Shyu, F.-L.; Chiu, C.-W.; Chang, C.-P.; Chen, R.-B.; Lin, M.-F. Magnetization of armchair carbon tori. *Physical Review B* **2004**, *70*, 075411.

[32] Tomita, S.; Sakurai, T.; Ohta, H.; Fujii, M.; Hayashi, S. Structure and electronic properties of carbon onions. *The Journal of Chemical Physics* **2001**, *114*, 7477–7482.

[33] Zeiger, M.; Jäckel, N.; Mochalin, V. N.; Presser, V. Carbon onions for electrochemical energy storage. *Journal of Materials Chemistry A* **2016**, *4*, 3172–3196.

[34] Deguchi, S.; Alargova, R. G.; Tsujii, K. Stable dispersions of fullerenes, C60 and C70, in water. Preparation and characterization. *Langmuir* **2001**, *17*, 6013–6017.

[35] Prilutski, Y.; Durov, S.; Bulavin, L.; Pogorelov, V.; Astashkin, Y.; Yashchuk, V.; *et al.* Study of structure of colloidal particles of fullerenes in water solution. *Molecular Crystals and Liquid Crystals Science and Technology Section A Molecular Crystals and Liquid Crystals* **1998**, *324*, 65–70.

[36] Prylutskyy, Y. I.; Petrenko, V.; Ivankov, O.; Kyzyma, O.; Bulavin, L.; Litsis, O.; *et al.* On the origin of C60 fullerene solubility in aqueous solution. *Langmuir* **2014**, *30*, 3967–3970.

[37] Zhang, K.; Zhang, Y.; Shi. L. A review of linear carbon chains. *Chinese Chemical Letters* **2020**, *31*, 1746–1756.

[38] Jochnowitz, E. B.; Maier, J. P. Electronic spectroscopy of carbon chains. *Annual Review of Physical Chemistry* **2008**, *59*, 519–544.

[39] Hunter, J.; Fye, J.; Jarrold, M. F. Carbon rings. *The Journal of Physical Chemistry* **1993**, *97*, 3460–3462.

[40] McCarthy, M. C.; Thaddeus, P. Microwave and laser spectroscopy of carbon chains and rings. *Chemical Society Reviews* **2001**, *30*, 177–185.

[41] Ferrari, A. C. Diamond-like carbon for magnetic storage disks. *Surface and Coatings Technology* **2004**, *180*, 190–206.

[42] Jiménez-Soto, J. M.; Cárdenas, S.; Valcárcel, M. Evaluation of carbon nanocones/disks as sorbent material for solid-phase extraction. *Journal of Chromatography A* **2009**, *1216*, 5626–5633.

[43] Cai, J.; Pignedoli, C. A.; Talirz, L.; Ruffieux, P.; Söde, H.; Liang, L.; *et al.* Graphene nanoribbon heterojunctions. *Nature Nanotechnology* **2014**, *9*, 896–900.

[44] Senese, A. D.; Chalifoux, W. A. Nanographene and graphene nanoribbon synthesis via alkyne benzannulations. *Molecules* **2018**, *24*, 118.

[45] Poljak, M.; Song, E. B.; Wang, M.; Suligoj, T.; Wang, K. L. Influence of edge defects, vacancies, and potential fluctuations on transport properties of extremely scaled graphene nanoribbons. *IEEE Transactions on Electron Devices* **2012**, *59*, 3231–3238.
[46] Yang, L.; Park, C.-H.; Son, Y.-W.; Cohen, M. L.; Louie, S. G. Quasiparticle energies and band gaps in graphene nanoribbons. *Physical Review Letters* **2007**, *99*, 186801.
[47] Han, M. Y.; Özyilmaz, B.; Zhang, Y.; Kim, P. Energy band-gap engineering of graphene nanoribbons. *Physical Review Letters* **2007**, *98*, 206805.
[48] Chung, H.-C.; Lee, M.-H.; Chang, C.-P.; Huang, Y.-C.; Lin, M.-F. Effects of transverse electric fields on quasi-landau levels in zigzag graphene nanoribbons. *Journal of the Physical Society of Japan* **2011**, *80*, 044602.
[49] Chung, H.-C.; Yang, P.-H.; Li, T.-S.; Lin, M.-F. Effects of transverse electric fields on Landau subbands in bilayer zigzag graphene nanoribbons. *Philosophical Magazine* **2014**, *94*, 1859–1872.
[50] Wu, J.-Y.; Chen, L.-H.; Li, T.-S.; Lin, M.-F. Optical properties of graphene nanoribbon in a spatially modulated magnetic field. *Applied Physics Letters* **2010**, *97*, 031114.
[51] Chang, S.-L.; Wu, B.-R.; Yang, P.-H.; Lin, M.-F. Curvature effects on electronic properties of armchair graphene nanoribbons without passivation. *Physical Chemistry Chemical Physics* **2012**, *14*, 16409–16414.
[52] Lin, C.-Y.; Wu, J.-Y.; Chang, C.-P.; Lin, M.-F. Magneto-optical selection rules of curved graphene nanoribbons and carbon nanotubes. *Carbon* **2014**, *69*, 151–161.
[53] Chang, S.-L.; Wu, B.-R.; Yang, P.-H.; Lin, M.-F. Geometric, magnetic and electronic properties of folded graphene nanoribbons. *RSC Advances* **2016**, *6*, 64852–64860.
[54] Savin, A.; Korznikova, E.; Lobzenko, I.; Baimova, Y. A.; Dmitriev, S. Symmetric scrolled packings of multilayered carbon nanoribbons. *Physics of the Solid State* **2016**, *58*, 1278–1284.
[55] Lin, S.-Y.; Chang, S.-L.; Chen, H.-H.; Su, S.-H.; Huang, J.-C.; Lin, M.-F. Substrate-induced structures of bismuth adsorption on graphene: A first principles study. *Physical Chemistry Chemical Physics* **2016**, *18*, 18978–18984.
[56] Lin, Y.-T.; Chung, H.-C.; Yang, P.-H.; Lin, S.-Y.; Lin, M.-F. Adatom bond-induced geometric and electronic properties of passivated armchair graphene nanoribbons. *Physical Chemistry Chemical Physics* **2015**, *17*, 16545–16552.

[57] Nguyen, D. K.; Lin, S.-Y.; Tran, N. T. T.; Liu, H.-Y.; Lin, M.-F. Rich essential properties of Si-doped graphene. In *Green Energy Materials Handbook*. CRC Press, 2019, pp. 83–107.
[58] Pham, H. D.; Gumbs, G.; Su, W.-P.; Tran, N. T. T.; Lin, M.-F. Unusual features of nitrogen substitutions in silicene. *RSC Advances* **2020**, *10*, 32193–32201.
[59] Li, W.-B.; Lin, S.-Y.; Lin, M.-F.; Lin, K.-I. Essential electronic properties of stage-1 Li/Li+-graphite-intercalation compounds for different concentrations. *Condensed Matter* **2022**, *7*, 35.
[60] Lin, S.-Y.; Liu, H.-Y.; Nguyen, D. K.; Tran, N. T. T.; Pham, H. D.; Chang, S.-L.; *et al. Silicene-Based Layered Materials*. IOP Publishing Ltd. 2020.
[61] Mohammadi, Y.; Moradian, R.; Tabar, F. S. Effects of doping and bias voltage on the screening in AAA-stacked trilayer graphene. *Solid State Communications* **2014**, *193*, 1–5.
[62] Mattausch, A.; Pankratov, O. Ab initio study of graphene on SiC. *Physical Review Letters* **2007**, *99*, 076802.
[63] Koshino, M. Interlayer screening effect in graphene multilayers with A B A and A B C stacking. *Physical Review B* **2010**, *81*, 125304.
[64] Valentini, F.; Carbone, M.; Palleschi, G. Graphene oxide nanoribbons (GNO), reduced graphene nanoribbons (GNR), and multi-layers of oxidized graphene functionalized with ionic liquids (GO–IL) for assembly of miniaturized electrochemical devices. *Analytical and Bioanalytical Chemistry* **2013**, *405*, 3449–3474.
[65] Delgado, V.; Marques, R. Surface impedance model for extraordinary transmission in 1D metallic and dielectric screens. *Optics Express* **2011**, *19*, 25290–25297.
[66] Menyhard, N. On the critical behaviour in quasi-1D metallic systems. *Journal of Physics C: Solid State Physics* **1979**, *12*, 1297.
[67] Tran, N. T. T.; Lin, S.-Y.; Lin, C.-Y.; Lin, M.-F. *Geometric and Electronic Properties of Graphene-related Systems: Chemical Bonding Schemes*. CRC Press, 2017.
[68] Pascal, R.; Zarnitz, C.; Bode, M.; Wiesendanger, R. Atomic and local electronic structure of Gd thin films studied by STM and STS. *Physical Review B* **1997**, *56*, 3636.
[69] Pascal, R.; Zarnitz, C.; Tödter, H.; Bode, M.; Getzlaff, M.; Wiesendanger, R. Electronic structure of Gd and Tb on W (110) in the submonolayer coverage regime studied by STM and STS. *Applied Physics A: Materials Science & Processing* **1998**, *66*.
[70] Dagotto, E. Correlated electrons in high-temperature superconductors. *Reviews of Modern Physics* **1994**, *66*, 763.

[71] Ginsberg, D. M. *Physical Properties of High Temperature Superconductors I.* World Scientific, 1998.

[72] Pop, I.; Dihoiu, N.; Coldea, M. Magnetic behaviour of the inter metallic system $Gd_2Ni_{17-x}Al_r$. *Philosophical Magazine B* **1979**, *39*, 245–252.

[73] Prasad, L.; Mikula, A. Effect of temperature on inter-metallic associations in Sb–Zn liquid alloys. *Journal of Alloys and Compounds* **2000**, *299*, 175–182.

[74] Kong, Z.; Duan, Y.; Qu, D.; Bao, L.; Peng, M. First-principles study on the elasticity and thermal properties of heavy Fermi compounds CePtSix (X = 1, 2, 3). *Physica B: Condensed Matter* **2022**, *624*, 413371.

[75] Pépin, C. Fractionalization and fermi-surface volume in heavy-fermion compounds: The case of $YbRh_2Si_2$. *Physical Review Letters* **2005**, *94*, 066402.

[76] Hancock, Y.; Uppstu, A.; Saloriutta, K.; Harju, A.; Puska M. J. Generalized tight-binding transport model for graphene nanoribbon-based systems. *Physical Review B* **2010**, *81*, 245402.

[77] Binnig, G.; Rohrer, H.; Gerber, C.; Weibel, E. Surface studies by scanning tunneling microscopy. *Physical Review Letters* **1982**, *49*, 57.

[78] Larsen, P.; Dobson, P. *Reflection High-energy Electron Diffraction and Reflection Electron Imaging of Surfaces.* Springer Science & Business Media, 2012.

[79] Stern, R.; Perry, J.; Boudreaux, D. Low-energy electron-diffraction dispersion surfaces and band structure in three-dimensional mixed Laue and Bragg reflections. *Reviews of Modern Physics* **1969**, *41*, 275.

[80] Hershfinkel, M.; Gheber, L.; Volterra, V.; Hutchison, J.; Margulis, L.; Tenne, R. Nested polyhedra of MX_2 (M = W, Mo; X = S, Se) probed by high-resolution electron microscopy and scanning tunneling microscopy. *Journal of the American Chemical Society* **1994**, *116*, 1914–1917.

[81] Brihuega, I.; Mallet, P.; González-Herrero, H.; De Laissardière, G. T.; Ugeda, M.; Magaud, L.; *et al.* Unraveling the intrinsic and robust nature of van Hove singularities in twisted bilayer graphene by scanning tunneling microscopy and theoretical analysis. *Physical Review Letters* **2012**, *109*, 196802.

[82] Marchini, S.; Günther, S.; Wintterlin, J. Scanning tunneling microscopy of graphene on Ru (0001). *Physical Review B* **2007**, *76*, 075429.

[83] Giusca, C. E.; Tison, Y.; Stolojan, V.; Borowiak-Palen, E.; Silva, S. R. P. Inner-tube chirality determination for double-walled carbon nanotubes by scanning tunneling microscopy. *Nano Letters* **2007**, *7*, 1232–1239.
[84] Sato, T.; Iwatsuki, M.; Tochihara, H. Detection of the flip-flop motion of buckled dimers on a Ge (001) surface by STM. *Microscopy* **1999**, *48*, 1–7.
[85] Lin, M.-F.; Chuang, Y.-C.; Wu, J.-Y. Electrically tunable plasma excitations in AA-stacked multilayer graphene. *Physical Review B* **2012**, *86*, 125434.
[86] Lin, C.-Y.; Chen, R.-B.; Ho, Y.-H.; Lin, M.-F. *Electronic and Optical Properties of Graphite-related Systems.* CRC Press, 2017.
[87] Olk, C. H.; Heremans, J. P. Scanning tunneling spectroscopy of carbon nanotubes. *Journal of Materials Research* **1994**, *9*, 259–262.
[88] Kumigashira, H.; Yang, S.-H.; Yokoya, T.; Chainani, A.; Takahashi, T.; Uesawa, A.; *et al.* High-resolution angle-resolved photoemission spectroscopy of CeBi. *Physical Review B* **1996**, *54*, 9341.
[89] Martins, M.; Ramos, P. The quantum inverse scattering method for Hubbard-like models. *Nuclear Physics B* **1998**, *522*, 413–470.
[90] Lv, B.; Qian, T.; Ding, H. Angle-resolved photoemission spectroscopy and its application to topological materials. *Nature Reviews Physics* **2019**, *1*, 609–626.
[91] Su, K.; Chi, P.; Paul, T.; Chung, C.; Chen, W.; Su, Y.; *et al.* Lithiation and delithiation induced magnetic switching and electrochemical studies in α-$LiFeO_2$ based Li ion battery. *Materials Today Physics* **2021**, *18*, 100373.
[92] Rossat-Mignod, J.; Regnault, L.; Vettier, C.; Bourges, P.; Burlet, P.; Bossy, J.; *et al.* Neutron scattering study of the $YBa_2Cu_3O_{6+x}$ system. *Physica C: Superconductivity* **1991**, *185*, 86–92.
[93] Bode, M.; Getzlaff, M.; Wiesendanger, R. Spin-polarized vacuum tunneling into the exchange-split surface state of Gd (0001). *Physical Review Letters* **1998**, *81*, 4256.
[94] Kubetzka, A.; Bode, M.; Pietzsch, O.; Wiesendanger, R. Spin-polarized scanning tunneling microscopy with antiferromagnetic probe tips. *Physical Review Letters* **2002**, *88*, 057201.
[95] Kittel, C. *Introduction to Solid State Physics*, 8th edn. 2021.
[96] Cohen, M. L.; Louie, S. G. *Fundamentals of Condensed Matter Physics.* Cambridge University Press, 2016.
[97] Bonnell, D. A. Scanning tunneling microscopy and spectroscopy of oxide surfaces. *Progress in Surface Science* **1998**, *57*, 187–252.

[98] Tsukada, M.; Kobayashi, K.; Isshiki, N.; Kageshima, H. First-principles theory of scanning tunneling microscopy. *Surface Science Reports* **1991**, *13*, 267–304.

[99] Hansma, P.; Elings, V.; Marti, O.; Bracker, C. Scanning tunneling microscopy and atomic force microscopy: Application to biology and technology. *Science* **1988**, *242*, 209–216.

[100] Garnaes, J.; Kragh, F.; Mo/rch, K.; Thölén, A. Transmission electron microscopy of scanning tunneling tips. *Journal of Vacuum Science & Technology A: Vacuum, Surfaces, and Films* **1990**, *8*, 441–444.

[101] Lin, C.-Y.; Ho, C.-H.; Wu, J.-Y.; Do, T.-N.; Shih, P.-H.; Lin, S.-Y.; *et al. Diverse Quantization Phenomena in Layered Materials.* CRC Press, 2019.

[102] Fox, M. *American Association of Physics Teachers.* 2002.

[103] Lui, C. H.; Liu, L.; Mak, K. F.; Flynn, G. W.; Heinz, T. F. Ultraflat graphene. *Nature* **2009**, *462*, 339–341.

[104] Mermin, N. D.; Wagner, H. Absence of ferromagnetism or antiferromagnetism in one-or two-dimensional isotropic Heisenberg models. *Physical Review Letters* **1966**, *17*, 1133.

[105] Coury, M.; Dudarev, S.; Foulkes, W.; Horsfield, A.; Ma, P.-W.; Spencer, J. Hubbard-like Hamiltonians for interacting electrons in s, p, and d orbitals. *Physical Review B* **2016**, *93*, 075101.

[106] Prange, R.; Korenman, V. Local-band theory of itinerant ferromagnetism. IV. Equivalent Heisenberg model. *Physical Review B* **1979**, *19*, 4691.

[107] Oka, H.; Brovko, O. O.; Corbetta, M.; Stepanyuk, V. S.; Sander, D.; Kirschner, J. Spin-polarized quantum confinement in nanostructures: Scanning tunneling microscopy. *Reviews of Modern Physics* **2014**, *86*, 1127.

[108] Kleiber, M.; Bode, M.; Ravlić, R.; Tezuka, N.; Wiesendanger, R. Magnetic properties of the Cr (001) surface studied by spin-polarized scanning tunneling spectroscopy. *Journal of Magnetism and Magnetic Materials* **2002**, *240*, 64–69.

[109] Osterwalder, J. Spin-polarized photoemission. In *Magnetism: A Synchrotron Radiation Approach.* Springer, 2006, pp. 95–120.

CHAPTER 11

HETEROJUNCTIONS OF MONO-/BILAYER GRAPHENE ON TRANSITION-METAL SUBSTRATES

Vo Khuong Dien and Ming-Fa Lin

Department of Physics, National Cheng Kung University, Tainan, Taiwan

Abstract

This study explores the electronic and structural properties of heterojunctions formed by mono-/bi-layers of graphene on a silver (Ag) (111) substrate. Using first-principles calculations, we investigate the influence of the Ag(111) substrate on graphene's electronic structure and interfacial properties. Our findings highlight the significant modifications in the band structure, including the emergence of interface states and a bandgap opening. We also examine the charge transfer phenomena and its impact on electrical conductivity and carrier mobility. Additionally, we investigate the optical properties, such as absorption spectra and excitonic behavior, offering insights into potential optoelectronic applications. This research advances our understanding of graphene-silver heterostructures for the design and development of novel electronic and optoelectronic devices.

Keywords: Heterojunctions, mono and bilayer graphene, Ag(111).

Each carbon atom has four active orbitals in any condensed-matter system, including 3D diamonds,[18] bulk graphite systems,[19] layered graphene materials,[20] 1D coaxial carbon nanotubes,[21] graphene nanoribbons,[15] 0D carbon onions,[22] quantum dots,[23] fullerenes,[24] rings,[25] and chains.[26]The distinct dimensionalities, $sp^3/sp^2/sp$

chemical bondings, stacking configurations, layer numbers, coaxial cylindrical symmetries, planar/curved/folded/scrolled carbon-based lattices, open achiral/chiral edges, and finite-size quantum confinements play critical roles in determining the application of these materials in the basic and applied sciences. Several graphene-related books have reviewed the successful developments with simultaneous progress in phenomenological models,[27,28] first-principles simulations,[14,20] and experimental observations.[29,30] These outstanding works clearly present the rich chemical environments and the unique crystal symmetries. The close relationships with optimal geometries obviously lead to diverse quasi-particle phenomena. A 1D single-walled carbon nanotube[31,32] or a graphene nanoribbon[33,34] could be regarded as a 2D rolled-up graphene in the cylindrical form or a cut graphitic sheet along the longitudinal direction. The periodic boundary conditions, which are fully assisted by the curvature effects of π bondings, can account for three types of carbon nanotubes: metallic, narrow-, and middle-gap semiconductors. Of course, all carbon-related systems possess varying active surfaces and open boundaries, where a lot of dangling bonds exist, arising mainly from the carbon-$2p_z$ orbitals perpendicular to the honeycomb lattice. The chemical modifications are expected to generate a lot of emergent binary, ternary, or quaternary compounds (detailed in Sections 2.1–2.5), e.g., heterojunctions (the subsystem composites, adsorptions, substitutions, intercalations/de-intercalations, and decorations). Their systematic investigations will greatly diversify quantum quasi-particles through developed theoretical models and experimental measurements.

Few-layer graphene systems clearly present diverse quasi-particle behaviors, as clearly described in the previously published books.[14,15] Their hexagonal honeycomb symmetries, highly stable planes, weak, but important van der Waals interactions, stacking configurations, layer numbers, and non-magnetic features play critical roles in various Hamiltonians. The position-dependent carbon–carbon interactions have been thoroughly explored for more than 50 years, which has resulted in well-characterized formulas that are suitable for layered graphene materials[35,36] and 1D coaxial carbon nanotubes.[37] The very successful theoretical predictions on the low-energy physical phenomena have been verified through high-resolution measurements using STS-STM,[29] ARPES,[30] EELS,[39] optical transmission spectroscopy,[39] and quantum transport,[40] respectively, corresponding to geometry–gap relationships, occupied energy bands, electron–hole pair excitations and plasmon modes, prominent absorption structures, and regular/irregular Hall conductivities. Monolayer graphene displays a linear and isotropic Dirac cone structure, in which the Fermi momentum is just the Dirac points. It belongs to a zero-gap semiconductor because it has no free carriers at E_F. With significant interlayer $2p_z$–$2p_z$ orbital hybridizations, bilayer and trilayer graphene systems exhibit diversified band structures and wave functions, e.g., the partially flat, linear, parabolic, oscillatory, sombrero-shaped, or 1D constant-energy-loop energy dispersions in the sliding/rotational bilayer materials[2,41,42] as well as the trilayer AAA,[43] ABA,[44] ABC,[45] or AAB[44] stackings. The above-mentioned band structures and wave functions will become more complicated after chemical modifications so that the π- and σ-electronic Hamiltons cannot be well characterized through the concise pictures. This is thoroughly discussed under heterojunctions in Section 11.2.

The unusual heterojunctions could be regarded as composite compounds with rich modulations of multi-orbital hybridizations and ferromagnetic/antiferromagnetic spin configurations. Such outstanding candidates can provide diverse quasi-particle phenomena and thus greatly contribute to a grand framework. The current experimental syntheses clearly show that the 2D Group-IV materials[46–48] and the binary and ternary semiconductor compounds of III–VI,[49] III–V,[50] II–VI,[51] II–V,[52] and IV–IV[52] have been very successfully grown on various distinct substrates. These mainstream materials play critical roles in the industrial development of functional chips,[53] which dominate the world economy.[54] Of course, there also exist plenty of semiconductor–metal[55] and metal–metal[56] compounds. Moreover, the composite cathode–electrolyte–anode materials[57] clearly exhibit high-performance ion transport through the composite boundaries of ion-based batteries during charging and discharging reactions. Monolayer graphene and bilayer AB stacking on platinum[111] are chosen for a model study. The latter is very rarely found in nature.[58] Pt, with its five isomers of Pt^{190}, Pt^{192}, Pt^{194},Pt^{196}, and Pt^{198}, has an atomic number of 78, so that the outermost ten 5d orbitals are fully occupied under an isolated condition. To realize the sufficiently strong chemical bindings in condensed-matter systems, its one or two electrons should be enhanced to 7s and/or 6d/7p for greatly reduced ground-state energies. Which kinds of orbitals take part in the real multi-orbital hybridizations and ferromagnetic spin configurations need to be thoroughly examined from the geometric, electronic, magnetic, and optical properties.

First-principles calculations, rather than phenomenological models, are very useful for thoroughly exploring the unusual hetero-junctions corresponding to graphene–platinum composites, mainly owing to the rich and unique multi-orbital hybridizations and ferromagnetic spin configurations associated with transition-metal substrates. The Moire superlattices, which correspond to the commensurate unit cells of two composite subsystems, are chosen from the optimal VASP simulations. The crystal structures, covering the C–C, C–Pt, and Pt–Pt bonds, are clearly visible along the longitudinal and transverse directions, especially for the observable bond-length fluctuations close to the boundary. To thoroughly illustrate the critical mechanisms, the strong relationships are clearly identified among the unusual chemical/physical environments, the ferromagnetic/metallic band structures with the sensitive atom and spin dependences, three specific regions of three different chemical bondings,[Refs] the strongly spatial modulations of spin-density arrangements,[Refs] a plenty of merged structures from the atom-/orbital-/spin-decomposed van Hobe singularities, and the single-particle excitations and plasmon modes associated with the significant orbitals and spins of the initial and final states. These quasi-particle behaviors are consistent with one another in determining which kinds of four-(s, p_x, p_y, p_z), and ten-5d are the dominant ones. The decisive pictures are very helpful for the development of unique heterojunction Hamiltonians. Examining the VASP predictions through high-resolution experimental measurements can greatly contribute to a grand quasi-particle framework, especially those from X-ray diffractions/RLEED,[59] spin-polarized ARPES,[60] and optical reflectance/absorption/transmission spectroscopy.[40]

11.1. Unique Heterojunction Crystal Structures

The top and side views of six kinds of heterojunctions, which correspond to graphene/platinum composites,[111] consisting of single-layer graphene and tri-, four-, or five-layer substrate, are shown in Figs. 11.1(a) and 11.1(b), Figs. 11.1(c) and 11.1(d), and Figs. 11.1(e) and 11.1(f), respectively, and their related bilayer AB stacking composites are shown in Figs.11.1(g) and 11.1(h), Figs. 11.1(i) and 11.1(j), and Figs. 11.1(k) and 11.1(l), respectively. These crystal structures can provide sufficient physical, chemical, and material environments for developing diverse quasi-particle phenomena. There exists the intralayer C–C network. The very strong σ bondings with

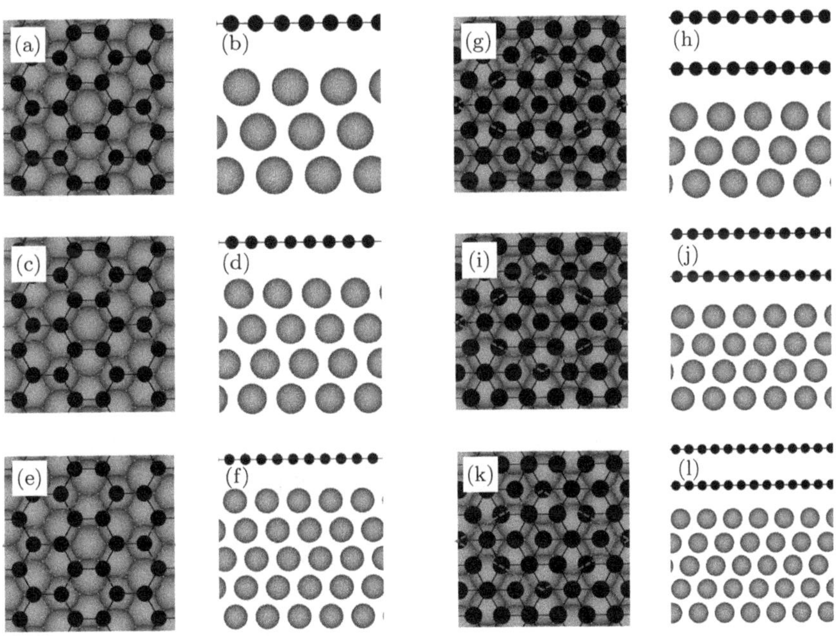

Fig. 11.1. The optimal top and side views of graphene/Pt(111) heterojunctions: (a)/(b) monolayer/trilayer, (c)/(d) monolayer/four-layer, (e)/(f) monolayer/five-layer, (g)/(h) bilayer AB stacking/trilayer, (i)/(j) bilayer/four-layer, and (k)/(l) trilayer/five-layer.

Table 11.1. The observable bond-length fluctuations near the heterojunctions of graphene–platinum composites: single-layer-/bilayer AB stacking/trilayer/four-layer/five-layer or AB stacking/trilayer/four-layer/penta-layer.

System	**C–C bonding**	**Pt–Pt bond in the first Pt layer**	**Pt–Pt bond in the second Pt layer**	**Pt–Pt bond in third Pt layer**	**Pt–Pt bond in the fourth Pt layer**	**Pt–Pt bond in the fifth Pt layer**
AB graphene/ three-layer Pt	1.413–1.416	2.840	2.829	2.829	—	—
AB graphene/ four-layer Pt	1.414–1.417	2.836	2.829	2.829	2.827	—
AB graphene/ five-layer Pt	1.414–1.417	2.836	2.829	2.829	2.828	2.829
System	**C–C bonding**	**Pt-Pt bond in 1st Pt layer**	**Pt-Pt bond in 2nd Pt layer**	**Pt-Pt bond in 3rd Pt layer**	**Pt-Pt bond in 4th Pt layer**	**Pt-Pt bond in 5th Pt layer**
Monolayer C/ three-layer Pt	1.413– 1.416	2.840	2.829	2.829	—	—
Monolayer C/ four-layer Pt	1.414–1.417	2.836	2.829	2.829	2.827	—
Monolayer C/ five-layer Pt	1.414–1.417	2.836	2.829	2.829	2.828	2.829

short C–C bond lengths, listed in Table 11.1 (~1.413–1.416 Å), are responsible for the planar honeycomb structures even close to the C–Pt heterojunction. The intralayer and interlayer C-2p$_z$ orbitals, respectively, generate the π bondings and van der Waals interactions (bilayer AB stacking). The C–Pt bonds, which are associated with the active orbitals of (2s, 2p$_x$, 2p$_y$, 2p$_z$, 5d$_{x^2-y^2}$, 5d$_{yz}$, 5d$_{zx}$, 5d$_{xy}$) (thoroughly examined later, detailed in Section 11.6), are longer than 3.1 Å. They play critical roles in mediating the strong spatial modulations of metallic bonding and magnetic configurations. The platinum substrate, with Pt–Pt bond lengths in the range of 2.827–2.840 Åcan initiate the free carrier density distribution and the ferromagnetic spin arrangements. The detailed results are discussed in Section 11.3. A primitive unit cell, as clearly shown by a red dashed curve in Fig. 11.1(a)), consists of eight carbon and Pt atoms in each layer. There are a number of atoms/active orbitals in the delicate VASP simulations. Similar studies could be generalized to the composite heterojunctions of buckled silicene, germanene, tinene, and plumbene systems, which are frequently revealed to be successful in sample substrate syntheses.[61–63] The important difference among them would become an effective modulation strategy.

The high-resolution measurements of STM/TEM/RLEED[59,64,65] and RLEED/X-ray[59] can, respectively, observe the 2D/1D/0D and 3D crystal structures. Scanning tunneling microscopy consists of a gate-voltage-dependent nanoscale perturbation needle and a planar or curved surface network. The quantum tunneling current, which is very weak according to the model predictions, remains constant at any position during the needle scanning of the whole morphology. The different heights of a nanoscale device are responsible for its atomic profiles, e.g., the very successful STM observations of achiral/chiral hexagons in single-walled carbon nanotubes made in 1998[66] and those of open edges in finite-size graphene nanoribbons.[67] Additionally, the STS measurements can determine the geometry-diversified electronic properties (detailed in Section 11.4). Concerning tunneling electron microscopy, the elastic scatterings of an incident electron beam are responsible for the diffraction pattern, as observed using X-rays. Their measurements are very useful for the side-view geometric parameters, e.g., the interlayer/intertube distances in layered graphene systems/coaxial carbon nanotubes, and the height difference between two buckled sublattices in silicene, germanene, tinene, and plumbene.

The very high-energy electrons are found to be likely to induce inelastic scattering events and provide the screened energy loss spectra (the Coulomb-field excitation spectra). The delicate verifications of the crystal structures of mono- and bilayer AB stackings as well as the tri-, four-, and five-layer Pt composites clearly illustrate the position-dependent chemical bonds, reveal the rather active chemical, physical, and material environments, and thus generate the diverse quantum quasi-particles.

11.2. Rich Band Structures and Wave Functions

The graphene–platinum composites, as clearly illustrated in Figs. 11.2(a)–11.2(n), can exhibit diversified quasi-particle energy spectra, mainly owing to the hexagonal C-honeycomb crystals, the zone-folding effects, and the σ- and π-covalent bondings/van der Waals semi-metallic interactions/metallic C–Pt and Pt–Pt charge distributions. The main features of monolayer graphene, bilayer AB stacking and compounds are totally different from one another, e.g., the perfectly defined/well-behaved/vanishing π-electronic energy subbands, the transitions among a zero-gap semiconductor, semi-metal, and metal, the vanishing/band-overlap, the very sensitive atom and orbital dependences for heterojunctions, and the almost identical σ-electronic energy spectra and wave functions. These significant properties represent an enlarged quasi-particle framework and require the systematic investigation of the various heterojunctions.

The pristine monolayer graphene and bilayer AB stacking, as shown in Figs. 11.2(a)–11.2(h), respectively, belong to a zero-gap semiconductor and a semi-metal. The concise mechanisms mainly arise from the three uniform nearest-neighboring $2p_z$-hopping integrals and the position-dependent interlayer van der Waals interactions. Their electronic states can be well characterized by the π and σ chemical bondings perpendicular to each other. A single-layer graphene clearly displays the π valence subband along the KMΓ path with a $\sim$ 7 eV width and the σ ones of ($2p_x$, $2p_y$) orbitals along the ΓMK high-symmetry points in the presence of a $\sim$ 7 eV energy difference. The Dirac point, which is initiated from the K/K' valley (the corner of the first Brillouin zone[Refs]), just crosses the Fermi level and serves as a unique Fermi momentum. As a result, its density of states vanishes there. Interestingly, their anticrossing behaviors are absent because of a planar honeycomb crystal. The small but observable valence and conduction band overlaps emerge in the bilayer AB stacking, mainly owing to the interlayer van der Waals interactions. The low-density free valence holes and conduction electrons, which simultaneously appear in a semi-metallic system,

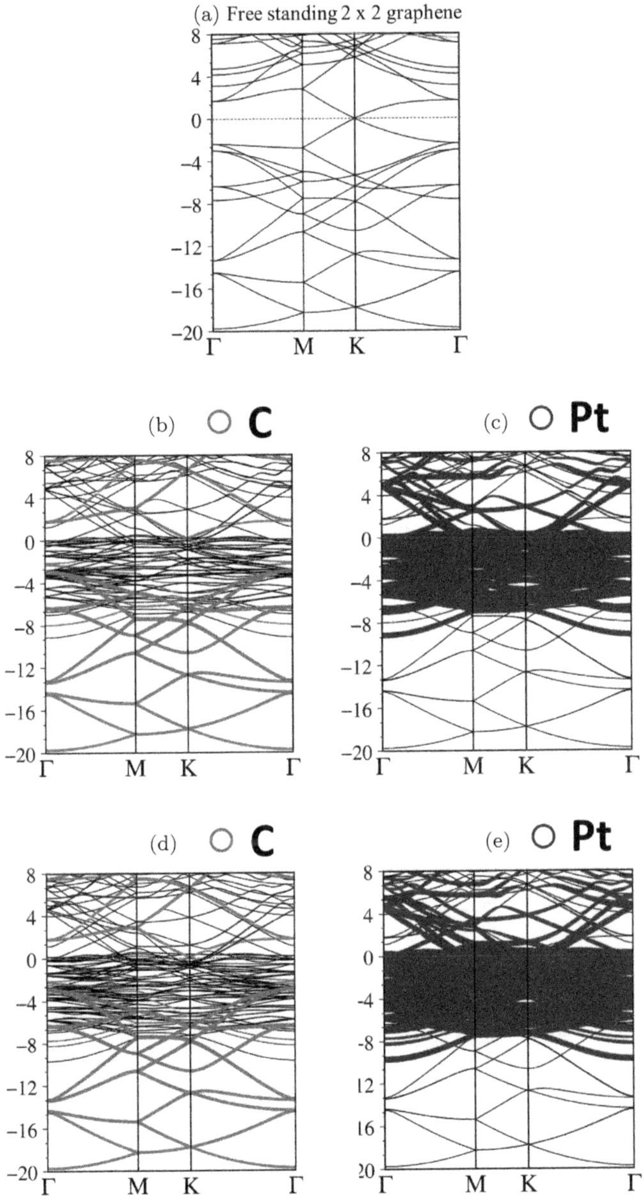

Fig. 11.2. The featured band structures with C-/Pt-atom dominances for graphene–platinum composites: (a) monolayer graphene, (b)/(c) monolayer graphene/trilayer platinum, (d)/(e) monolayer/four-layer, (f)/(g) monolayer/fiver-layer, (h) bilayer AB stacking, (i)/(j) bilayer/trilayer, (k)/(l) bilayer/four-layer, and (m)/(n) bilayer/five-layer.

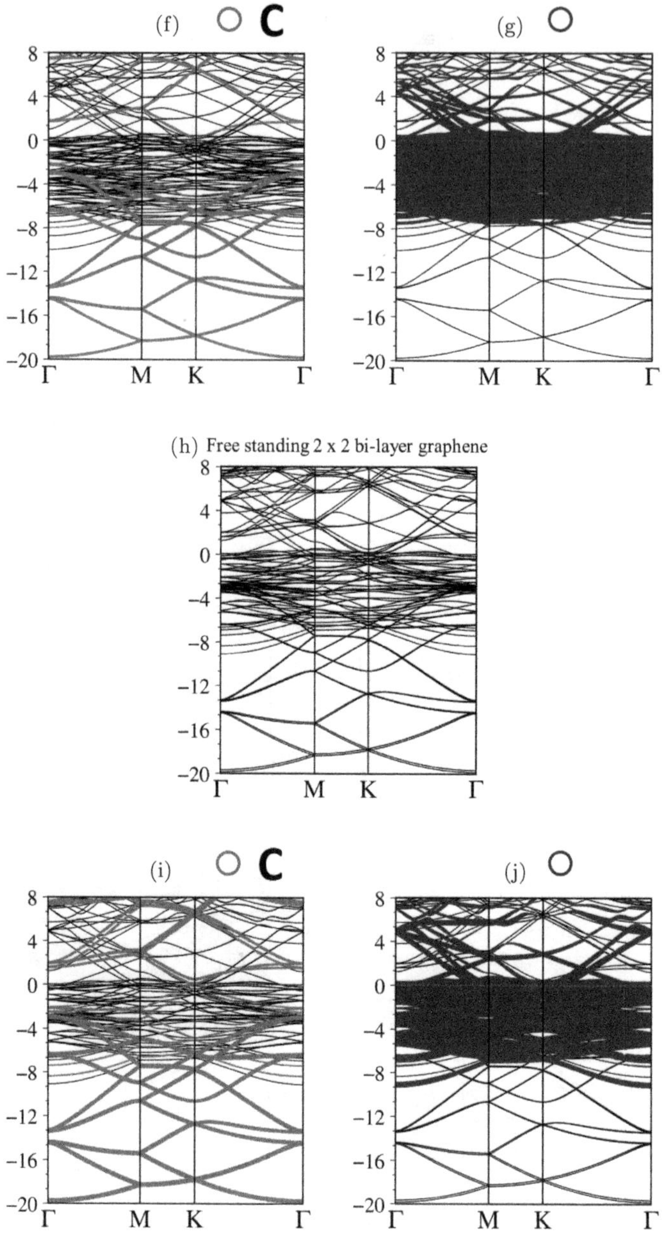

Fig. 11.2. (*Continued*)

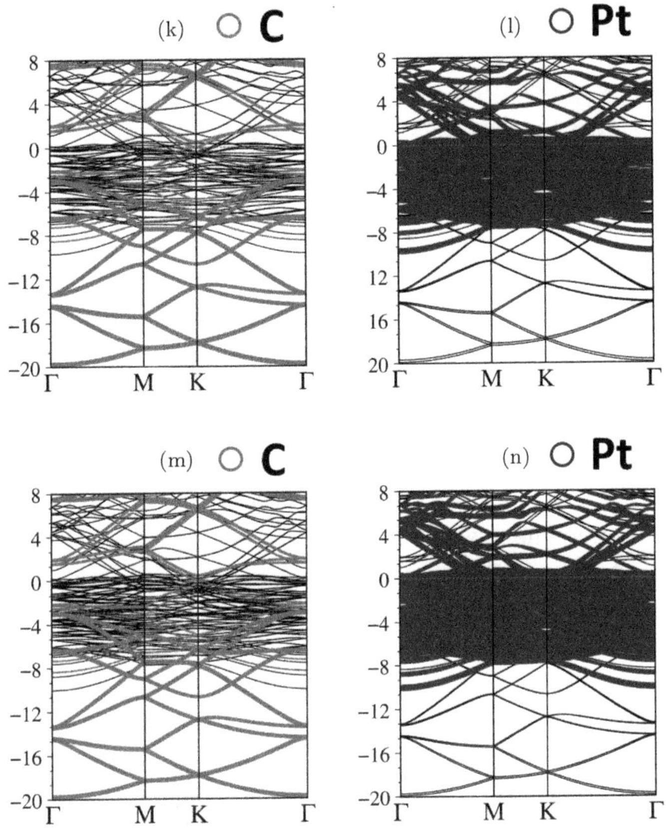

Fig. 11.2. (*Continued*)

make identical contributions to its low-energy physics, such as the temperature-dependent Coulomb-field electron–hole pair excitations and plasmon modes and the geometry-enriched optical absorption structures. Moreover, there exist two pairs of low-lying parabolic energy dispersions. Stacking configurations and layer numbers play critical roles in determining the featured band structure, as shown by systematic investigations of layered graphenes, carbon nanotubes, and graphene nanoribbons.

Of course, the chemical modifications of heterojunctions are very powerful in diversifying their electronic properties. There are more Pt and C atoms in an enlarged unit cell, where the intra- and interlayer

C–C bonds, heterojunction C–Pt bonds, metallic Pt–Pt bonds, and Pt-created ferromagnetic spin configurations play critical roles in determining the featured band structures. The drastic changes in energy spectra and wave functions (Figs. 11.2(b)–11.2(g) and Figs. 11.2(i)–11.2(n), respectively, for monolayer graphene and bilayer AB stacking) encompass: (1) more rich and complicated energy subbands in the increment platinum layer, (2) the greatly enhanced asymmetry of valence and conduction states about the Fermi level ($E_F = 0$), (3) the rather sensitive atom dependences at different energy ranges, (4) a lot of subbands across $E_F = 0$ under the well-characterized Fermi momenta (the multi-Fermi surfaces[Refs]), (5) the almost disappearance of low-lying π-electronic states, and (6) mission impossible in well observing the σ-electronic valence subbands. More interestingly, the multi-Fermi momenta are able to generate the beating phenomena associated with the clear superposition of long-range Friedel oscillations under the charged impurity screenings of free carriers. Obviously, it is very difficult to obtain a good fit of such complicated band structures from the tight-binding model. This directly reflects the unusual superposition of covalent, semi-metallic, and metallic bondings. All charge-dependent intrinsic interactions are simultaneously required in achieving the concise pictures of chemical, physical, and material science. The current and near-future research needs to evaluate whether the phenomenological models in graphene–platinum heterojunctions can be effectively developed under an enlarged quasi-particle framework.

The angle-resolved photoemission spectroscopy (SP-ARPES.[16]) is available for observing the occupied quasi-particle energy spectra and lifetimes. When a photon beam with sufficiently high momentum and kinetic energy is incident on a standard sample, its coupling with the well-behaved electronic states is characterized under the conservation of the transferred momentum and energy as well as the Pauli exclusion principle.[Refs] In general, such bi-particle scatterings can simultaneously generate excited valence holes and conductive electrons. The intermediate quasi-particles will further experience a lot of elastic and inelastic scatterings before their arrival at the sample surface. These quantum particles are capable of entering an outside free space if their energies are sufficient to overcome specific work functions. Obviously, the substrate thickness plays a critical role in ensuring meaningful ARPES measurements.[68] Successful examinations have been frequently done for the high-symmetry 2D layered crystals and 1D planar finite-size nanoribbons, e.g., an isotropic Dirac cone structure in monolayer graphene,[30] the blueshifted Fermi level of potassium-adsorbed graphene systems,[Refs] the strongly modified energy dispersions due to stacking- and layer-number-enriched van der Waals interactions, the prominent quantum confinements on the enlarged bandgaps, and the drastic changes in group velocities. Of course, the high-precision spin-polarized ARPES measurements are well-suited for observing the 2D graphene–platinum heterojunctions in terms of the spin-split multi-separated Femi surfaces, the absence of the Dirac cone blueshift,[Refs] and the sensitive C and Pt dominances at different energy ranges. The strong cooperation with the delicate VASP simulations would be very useful in providing more information about the intralayer covalent C–C, interlayer semi-metallic C–C, and metallic C–Pt/Pt–Pt bonds.

11.3. Spatial Modulations of Charge Density Distributions

The spatial charge distributions directly reflect the highly non-uniform and anisotropic environments, which can be used to establish the unique Hamiltonians of graphene–platinum composite heterojunctions. These are clearly illustrated along the longitudinal and transverse directions for (a)/(b) monolayer graphene, (c)/(d) monolayer graphene/trilayer platinum, (e)/(f) monolayer/four-layer, (g)/(h) monolayer/five-layer, (i)/(j) bilayer AB stacking, (k)/(l) bilayer/trilayer, (m)/(n) bilayer/four-layer, as well as, (o)/p) bilayer/five-layer. The isolated carbon and platinum atoms only present the spherical probability distributions, in which the radius-dependent red and yellow/green regions, respectively, correspond to 2s and ($2p_x$, $2p_y$, $2p_z$) orbitals as well as 6s/(5s, $5p_x$, $5p_y$, $5p_z$) and eight-5d ones. Also, note that both isolated and bonding platinum atoms possess the same atomic configurations in the two occupied halves of ten-5f orbitals, that is, the fully occupied 6s orbitals do not play a critical role in the active multi-orbital hybridizations. The following four kinds of chemical bondings exist in graphene–platinum heterojunctions.[Refs] (I) A planar carbon-honeycomb crystal (Fig. 11.3(a)) clearly presents two neighboring carbon atoms under the very strong σ bondings of (2s, $2p_x$, $2p_y$) orbitals (the dog-bone-like dark red region). This accounts for the stable structures in the layered graphene systems.[Refs] Furthermore, the σ and π bondings are

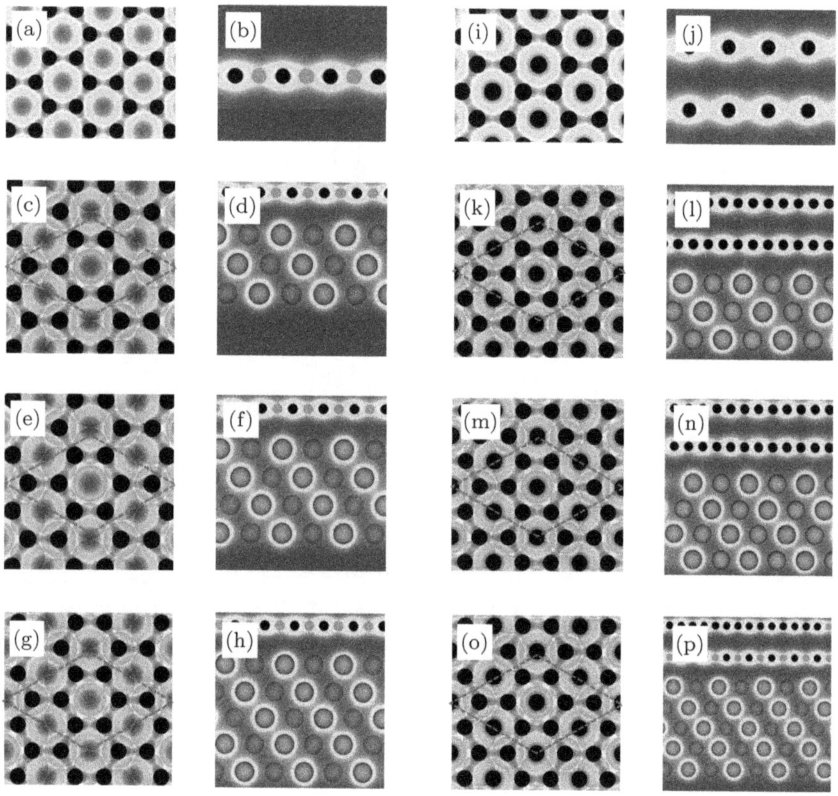

Fig. 11.3. The spatial charge density distributions of graphene–platinum heterojunctions along the longitudinal and transverse directions: (a)/(b) single-layer graphene, (c)/(d) monolayer graphene/trilayer platinum, (e)/(f) monolayer/four-layer, (g)/(h) monolayer/fiver-layer, (i)/(j) bilayer AB stacking, (k)/(l) bilayer/trilayer, (m)/(n) bilayer/four-layer, and (o)/p) bilayer/five-layer.

perpendicular to each other in the absence of deformed structures. (II) The interlayer $2p_z$–$2p_z$ hybridizations in neighboring graphitic sheets (Fig. 11.3(j))[Refs] are capable of exhibiting semi-metallic behavior through light but significant charge density extension along the perpendicular directions. Such interactions, which strongly depend on stacking configurations and layer numbers, create diverse quasi-particle phenomena, as described in the previously published books.[Refs] (III) The mediated C–Pt bonds strongly modulate the density distributions of free carriers close to the specific heterojunction. (IV) The metallic Pt–Pt bonds are initiated from the substrate. The concise pictures of multi-orbital hybridizations cover (i) (2s, $2p_x$, $2p_y$) – (2s, $2p_x$, $2p_y$) and $2p_z$–$2p_z$ perpendicular to each other on a planar carbon-honeycomb crystal,[Refs] (ii) van der Waals interactions of interlayer $2p_z$–$2p_z$ hybridizations, (iii) $2p_z$ – ($5d_{z^2}$, $5d_{x^2-y^2}$, $5d_{yz}$, $5d_{zx}$, $5d_{xy}$) at al heterojunction, and (iv) ($5d_{z^2}$, $5d_{x^2-y^2}$, $5d_{yz}$, $5d_{zx}$, $5d_{xy}$) – ($5d_{z^2}$, $5d_{x^2-y^2}$, $5d_{yz}$, $5d_{zx}$, $5d_{xy}$) within the metallic substrate.[Refs] The multi-orbital hybridizations can determine the hopping integrals and on-site ionization energies in the tight-binding model. These features are fully assisted by the ferromagnetic spin configurations (as noted in the following section), leading to the diversified quasi-particle phenomena.

11.4. Atom- and Orbital-decomposed Van Hove Singularities

The energy-dependent density of states (DOS(E)), which cover the occupied and unoccupied energy spectra across the Fermi level, exhibit a lot of van Hove singularities corresponding to the critical points with the irregular/zero-group velocities in the energy–wave-vector space. Their prominent structures directly reflect the strong response abilities/charge screenings, being characterized by the longitudinal-/transverse-field dielectric functions. Their forms are roughly analytical for 2D/1D/0D condensed-matter systems, but not for 3D ones. Through delicate calculations and analyses, the 2D density of states is expressed as the 1D constant-E-loop integration of the inverse of energy-dependent group velocity. Similar definitions

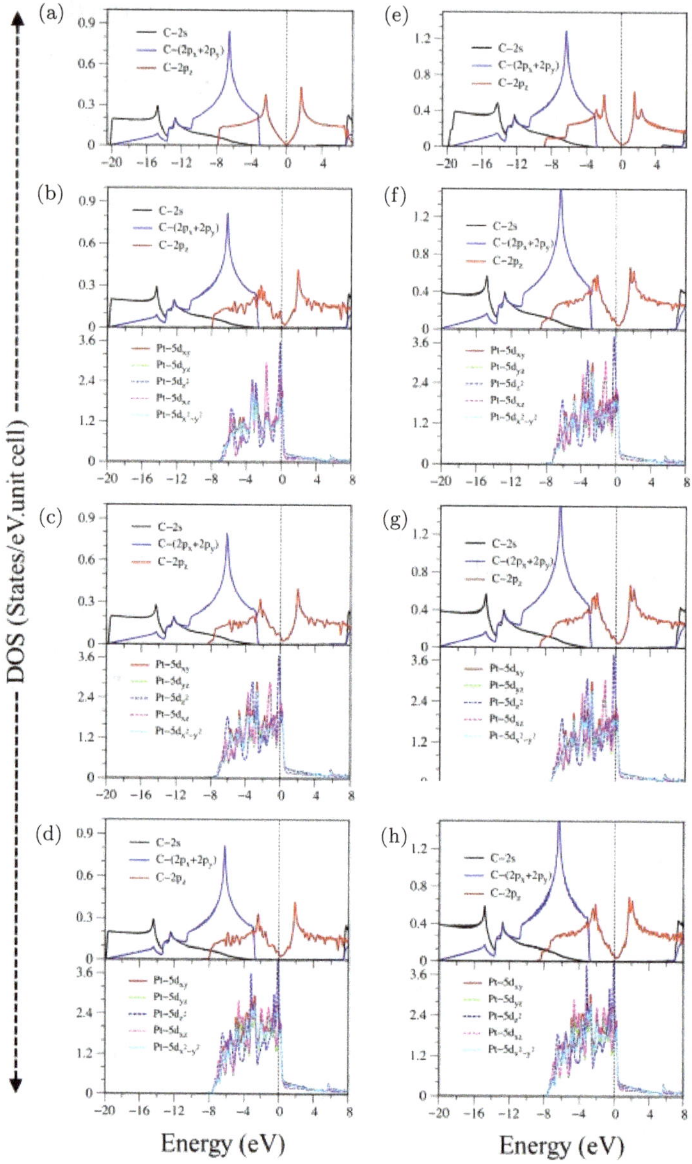

Fig. 11.4. The atom- and orbital-decomposed density of states for graphene–platinum composites: (a) pristine monolayer graphene, (b) monolayer graphene/trilayer platinum, (c) monolayer/four-layer, (d) monolayer/fiver-layer, (e) bilayer AB stacking, (f) bilayer/trilayer, (g) bilayer/four-layer, and (h) bilayer/five-layer.

are presented for other dimensionalities. There are five kinds of van Hove singularities: (I) a V-shaped structure through the linear and isotropic Dirac cone, (II) the broadening shoulders under the parabolic energy dispersions, (III) the logarithmically divergent peals associated with the saddle points, (IV) the delta-function-like prominent peaks due to the partially flat energy subbands,[Refs] and (V) the asymmetric peaks of the inverse of square-root form from the subband anticrossing behaviors (1D constant-energy loops), A lot of merged van Hove singularities frequently emerge in few-layer graphene–platinum composite compounds because of very complicated multi-orbital hybridizations within the spin-dependent Pt–Pt, C–Pt, and C–C bonds. Such behaviors, which are fully assisted by other featured essential properties, are very useful in determining the concise pictures, thus greatly contributing to a quasi-particle framework.

The VASP predictions on van Hove singularities could be directly verified from the high-resolution STS measurements,[29] as successfully done for 2D layered graphene, 1D graphene nanoribbons, coaxial carbon nanotubes, and 0D carbon quantum dots, but not for the 3D materials. Scanning tunneling spectroscopy (STS), an extension of STM, can measure the energy-dependent density of states through weak but significant quantum tunneling currents between a nanoscale probing needle and the sample surface, The position-dependent quantum currents are collected together under the same gate voltage, whose first derivative is the differential conductance,[Refs] which is roughly proportional to the average density of states. Then, similar data are taken at different gates. The resulting sensitive relations between dI/dV and V correspond to those of $D(E)$. Their prominent current responses mainly come from van Hove singularities of band-edge states (Section 10.2). The STS examinations are able to provide useful information about the metallic, semi-metallic, or semiconducting property across the Fermi level,[Refs] as well as the numbers energies, irregular forms, and intensities of valence and conduction van Hove singularities. However, the atom-, orbital-, and spin-decomposed results are meaningless; that is, the experimental analyses cannot properly distinguish the separate contributions.

The concise orbital hybridizations and spin configurations, which are associated with the C–C, C–Pt, and Pt–Pt bonds, are roughly examined through the VASP simulations and experimental observations. Their very strong relationships will be very useful in establishing the theoretical models from phenomenological viewpoints. This is under current investigation, though it is difficult to achieve.

11.5. Quantum Quasi-particles in Optical Excitations

Pristine graphene and bilayer graphene clearly exhibit the unusual optical excitation phenomena, as indicated in Figs. 11.5(a) and 11.5(e). The imaginary part of the frequency-dependent bare response function $Im\left[\epsilon\left(\omega\right)\right]$ shows two prominent absorption structures at 5.51 and 12.90 eV. The featured results reflect the different orbital hybridizations of the 2D carbon-honeycomb crystal, with the first and second peaks, respectively, arising from the inter-π-band transitions associated with the saddle M-point and the excitations of the σ band from the M point. Such identifications are delicately made from the orbital-decomposed van Hove singularities; that is, the initial and/or final states must possess the singular densities of states. Also, note that the transitions between the π and σ states are totally absent, even though their joint density of states is significant. This is due to the fact that the π and σ wave functions are almost perpendicular to each other in the planar lattice, and thus, their electric dipole moments vanish under the quantum averages of the gradient operator. Following the charge screenings of all valence electrons, the energy loss functions of graphene clearly reveal the coherent features of the charge oscillations. The π plasmon and the $\pi + \sigma$ plasmon, respectively, survive at 5.810 and 16.80 eV. Apparently, such energy-dependent plasmons belong to the quantization modes of the collective excitations of π and all valence electrons under the long-wavelength limit. Such collective excitations could coexist with the prominent Landau damping of the excited electron–hole pairs. These two kinds of quasi-particle excitations cooperate with each other to present the prominent absorption coefficient within the frequency range of $\omega < 20$ eV.

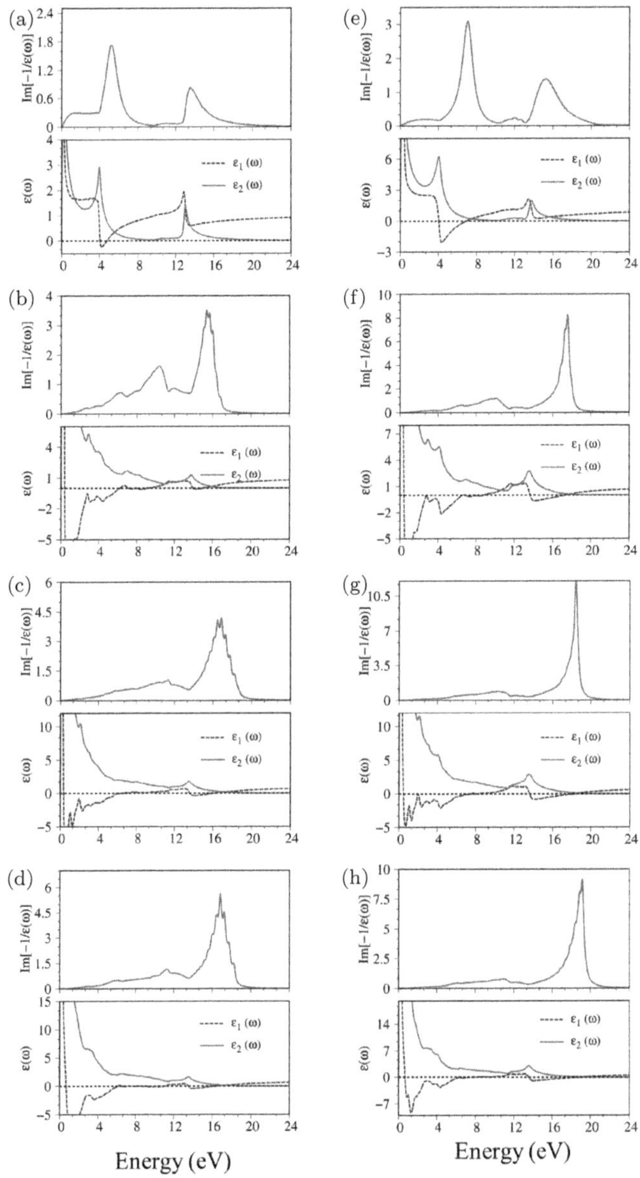

Fig. 11.5. The real and imaginary parts of the dielectric functions and energy loss spectra of graphene–platinum composites: (a) single-layer graphene, (b) monolayer graphene/trilayer platinum, (c) monolayer/four-layer, (d) monolayer/penta-layer, (e) bilayer AB stacking, (f) bilayer/trilayer, (g) bilayer/four-layer, and (h) bilayer/penta-layer.

Concerning the substrate effects, the low-frequency absorption spectra present a significant change. For example, with the enhancement of the low-frequency absorption spectra due to the intraband transition of the platinum states, the prominent peaks belonging to the $\pi \rightarrow \pi^*$interband transitions are changed. More interestingly, the higher-frequency peak that belong to the $\sigma \rightarrow \sigma^*$interband transitions is almost unchanged since these states remain the same. Following the charge screening of all carriers, the energy loss function of graphene on platinum displays two obvious π and $\pi+\sigma$ plasmons. The former is gradually reduced.

11.6. Concise Pictures of Quantum Quasi-particles

The monolayer graphene and bilayer AB stacking/platinum substrate heterojunctions clearly illustrate the diversified quasi-particle phenomena, making them outstanding candidates for developing a grander framework. The first-principles calculations, phenomenological models, and experiment observations also display the strong relationships of simultaneous progress. The concise mechanisms are achieved from (I) the very active crystal environments of the intra-/interlayer C–C and Pt–Pt bonds as well as the C–Pt bonds, (II) the atom- and orbital-dominated electronic wave functions, (III) the unique spatial modulations of charge and spin densities along the longitudinal and transverse directions, (IV) a plenty of merged van Hove singularities due to multi-orbital hybridizations in five kinds of chemical bonds, and (VI) the prominent absorption structures of the dielectric functions due to the specific orbitals within the initial and final states as well as the plasmon peaks in the energy loss spectra. Specifically, the redshift at the Fermi level could be estimated from the minimum density states close below E_F and the optical gap. The distinct multi-orbital hybridizations consist of (1) (2s, $2p_x$, $2p_y$) – (2s, $2p_x$, $2p_y$) and $2p_z$–$2p_z$ perpendicular to each other on a planar carbon-honeycomb crystal,[Refs] (2) the interlayer van der Waals interactions of the $2p_z$–$2p_z$ mixings, (3) $2p_z$ – ($5d_{z^2}$, $5d_{x^2-y^2}$, $5d_{yz}$, $5d_{zx}$, $5d_{xy}$) for the carbon–transition-metal heterojunction, and ($5d_{z^2}$, $5d_{x^2-y^2}$, $5d_{yz}$,

$5d_{zx}$, $5d_{xy}$) – ($5d_{z^2}$, $5d_{x^2-y^2}$, $5d_{yz}$, $5d_{zx}$, $5d_{xy}$) in the metallic substrate.[Refs] The metallic bondings mainly arise from the neighboring platinum atoms, exhibit large fluctuations near the unusual boundary, and then reveal a great decline within the very strong covalent σ bondings of honeycomb lattices. In addition, carbon and platinum atoms in their heterojunctions, respectively, display (2s, $2p_x$, $2p_y$, $2p_z$) and ($5d_{x^2-y^2}$, $5d_{yz}$, $5d_{zx}$, $5d_{xy}$). Obviously, the above-mentioned quasi-particle phenomena are very suitable for expanding a quantum quasi-particle framework since these results have not appeared in the previous predictions[Refs] and measurements. The delicate first-principles studies could be generalized to the various heterojunctions, covering semiconductor–semiconductor/semiconductor–metal/metal–metal composites and the cathode/electrolyte/anode combined materials of ion-based batteries.[Refs/Refs/Refs] In addition, the stationary ion transports require a unified theory during the high-performance operations, where the time-dependent chemical reactions, the modulations of crystal symmetries, and the well-characterized elastic/inelastic scatterings[Refs] should display strong relationships. (1) The first-principles calculations require further high-resolution experimental examinations. Few-layer graphenes on platinum[111] have been successfully grown in laboratories. STM, TEM, RLEED, or X-ray[Refs/Refs/Refs/Refs] could be utilized to observe the position-dependent chemical bonds in a primitive unit cell, where the first, second, or third technique and the third or fourth technique are, respectively, suitable for layered and bulk samples because of the response strength. The ARPES observations could be made through a small enough depth since the photon-excited electronic states experience many inelastic scatterings and should have sufficient kinetic energy to overcome the surface work function. Their meaningful measurements of the occupied quasi-particle energy spectra and lifetimes are applicable only for thin heterojunctions. The direct combination of STM and STS is very powerful in examining the prominent relationships between crystal symmetries and band properties as well as the many strong van Hove singularities across the Fermi level. Their examinations cannot verify the atom-/orbital-decomposed van Hove singularities. Moreover, optical reflectance, absorption, or transmission spectroscopy can measure the single-particle dielectric functions and many-body energy loss spectra (electron–hole pairs and plasmons[Refs]) and then identify their strong relationships with the multi-orbitals of the initial and final states through a strong cooperation with first-principles calculations. However, for the phenomenological models,[Refs] it would be very difficult to establish reliable parameters pertaining to the complicated orbital-dependent hopping integrals[Refs] and spin-induced modulation interactions.[Refs] The applicability of empirical formulas to atomic interactions is excellent only for the C–C bonds,[Refs] but not for the C–Pt and Pt–Pt ones.

References

[1] Sharma, B.; Purohit, R. *Semiconductor Heterojunctions.* Elsevier. 2015.

[2] Geim, A. K.; Novoselov, K. S. The rise of graphene. In *Nanoscience and Technology: A Collection of Reviews from Nature Journals.* World Scientific, 2010, pp. 11–19.

[3] Arafune, R.; Lin, C.-L.; Kawahara, K.; Tsukahara, N.; Minamitani, E.; Kim, Y.; *et al.* Structural transition of silicene on Ag (111). *Surface Science* **2013**, *608*, 297–300.

[4] Acun, A.; Zhang, L.; Bampoulis, P.; Farmanbar, M. V.; van Houselt, A.; Rudenko, A.; *et al.* Germanene: The germanium analogue of graphene. *Journal of Physics: Condensed Matter* **2015**, *27*, 443002.

[5] Zhu, F.-F.; Chen, W.-J.; Xu, Y.; Gao, C.-l.; Guan, D.-D.; Liu, C.-h.; *et al.* Epitaxial growth of two-dimensional stanene. *Nature Materials* **2015**, *14*, 1020–1025.

[6] Yuhara, J.; He, B.; Matsunami, N.; Nakatake, M.; Le Lay, G. Graphene's latest cousin: Plumbene epitaxial growth on a "nano WaterCube". *Advanced Materials* **2019**, *31*, 1901017.

[7] Bahuguna, B. P.; Saini, L.; Sharma, R. O.; Tiwari, B. Hybrid functional calculations of electronic and thermoelectric properties of GaS, GaSe, and GaTe monolayers. *Physical Chemistry Chemical Physics* **2018**, *20*, 28575–28582.

[8] Mohammad, S. N.; Morkoç, H. Progress and prospects of group-III nitride semiconductors. *Progress in Quantum Electronics.* **1996**, *20*, 361–525.

[9] M. Afzaal and P. O'Brien. Recent developments in II–VI and III–VI semiconductors and their applications in solar cells. *Journal of Materials Chemistry* **2006**, *16*, 1597–1602.

[10] Lazarev, V.; Shevchenko, V. I.; Grinberg, I. K.; Sobolev, V. *Group II-V Semiconductor Compounds*. Moscow Izdatel Nauka, 1978.

[11] Hu, Z.; Ding, Y.; Hu, X.; Zhou, W.; Yu, X.; Zhang, S. Recent progress in 2D group IV–IV monochalcogenides: Synthesis, properties and applications. *Nanotechnology* **2019**, *30*, 252001.

[12] Lin, S.-Y.; Liu, H.-Y.; Nguyen, D. K.; Tran, N. T. T.; Pham, H. D.; Chang, S.-L.; *et al. Silicene-Based Layered Materials*. IOP Publishing Ltd., 2020.

[13] Wang, W.; Chen, S.; Yang, P.-X.; Duan, C.-G.; Wang, L.-W. Si: WO 3 heterostructure for Z-scheme water splitting: An AB initio study. *Journal of Materials Chemistry A* **2013**, *1*, 1078–1085.

[14] Tran, N. T. T.; Lin, S.-Y.; Lin, C.-Y.; Lin, M.-F. *Geometric and Electronic Properties of Graphene-related Systems: Chemical Bonding Schemes*. CRC Press. 2017.

[15] Lin, S.-Y.; Tran, N. T. T.; Chang, S.-L.; Su, W.-P.; Lin, M.-F. *Structure and Adatom-Enriched Essential Properties of Graphene Nanoribbons*. CRC Press, 2018.

[16] Lv, B.; Qian, T.; Ding, H. Angle-resolved photoemission spectroscopy and its application to topological materials. *Nature Reviews Physics* **2019**, *1*, 609–626.

[17] Jin, W.; Yeh, P.-C.; Zaki, N.; Zhang, D.; Sadowski, J. T.; Al-Mahboob, A.; *et al.* Direct measurement of the thickness-dependent electronic band structure of MoS 2 using angle-resolved photoemission spectroscopy. *Physical Review Letters* **2013**, *111*, 106801.

[18] Gabrysch, M. *Electronic Properties of Diamond*. Institutionen för teknikvetenskaper, 2008.

[19] Lin, C.-Y.; Chen, R.-B.; Ho, Y.-H.; Lin, M.-F. *Electronic and Optical Properties of Graphite-related Systems*. CRC Press, 2017.

[20] Wong, J.-H.; Wu, B.-R.; Lin, M.-F. Strain effect on the electronic properties of single layer and bilayer graphene. *The Journal of Physical Chemistry C*. **2012**, *116*, 8271–8277.

[21] Lin, C.-Y.; Wu, J.-Y.; Chang, C.-P.; Lin, M.-F. Magneto-optical selection rules of curved graphene nanoribbons and carbon nanotubes. *Carbon*, **2014**, *69*, 151–161.

[22] Tomita, S.; Sakurai, T.; Ohta, H.; Fujii, M.; Hayashi, S. Structure and electronic properties of carbon onions. *The Journal of Chemical Physics* **2001**, *114*, 7477–7482.

[23] Lim, S. Y.; Shen, W.; Gao, Z. Carbon quantum dots and their applications. *Chemical Society Reviews* **2015**, *44*, 362–381.

[24] Dresselhaus, M.; Dresselhaus, G.; Eklund, P. Fullerenes. *Journal of Materials Research* **1993**, *8*, 2054–2097.

[25] Hunter, J.; Fye, J.; Jarrold, M. F. Carbon rings. *The Journal of Physical Chemistry* **1993**, *97*, 3460–3462.

[26] Zhang, K.; Zhang, Y.; Shi, L. A review of linear carbon chains. *Chinese Chemical Letters* **2020**, *31*, 1746–1756.

[27] Chung, H.-C.; Chang, C.-P.; Lin, C.-Y.; Lin, M.-F. Electronic and optical properties of graphene nanoribbons in external fields. *Physical Chemistry Chemical Physics* **2016**, *18*, 7573–7616.

[28] Lin, M.-F.; Shyu, F.-L. Optical properties of nanographite ribbons. *Journal of the Physical Society of Japan* **2000**, *69*, 3529–3532.

[29] Ridene, M.; Girard, J.; Travers, L.; David, C.; Ouerghi, A. STM/STS investigation of edge structure in epitaxial graphene. *Surface Science* **2012**, *606*, 1289–1292.

[30] Sprinkle, M.; Siegel, D.; Hu, Y.; Hicks, J.; Tejeda, A.; Taleb-Ibrahimi, A.; *et al.* First direct observation of a nearly ideal graphene band structure. *Physical Review Letters* **2009**, *103*, 226803.

[31] Liu, H.-Y.; Lin, M.-F.; Wu, J.-Y. Essential electronic properties of silicon nanotubes. *Nanomaterials* **2021**, *11*, 2475.

[32] Li, T.-S.; Lin, M.-F. Electronic properties of carbon nanotubes under external fields. *Physical Review B* **2006**, *73*, 075432.

[33] Wu, J.-Y.; Chen, L.-H.; Li, T.-S.; Lin, M.-F. Optical properties of graphene nanoribbon in a spatially modulated magnetic field. *Applied Physics Letters* **2010**, *97*, 031114.

[34] Chang, S.-L.; Lin, S.-Y.; Lin, S.-K.; Lee, C.-H.; Lin, M.-F.. Geometric and electronic properties of edge-decorated graphene nanoribbons. *Scientific Reports* **2014**, *4*, 1–8.

[35] Wang, X.; You, H.; Liu, F.; Li, M.; Wan, L.; Li, S.; *et al.* Large-scale synthesis of few-layered graphene using CVD. *Chemical Vapor Deposition* **2009**, *15*, 53–56.

[36] Lv, R.; Terrones, M. Towards new graphene materials: Doped graphene sheets and nanoribbons. *Materials Letters* **2012**, *78*, 209–218.

[37] Salunkhe, R. R.; Lin, J.; Malgras, V.; Dou, S. X.; Kim, J. H.; Yamauchi, Y. Large-scale synthesis of coaxial carbon nanotube/Ni (OH) 2 composites for asymmetric supercapacitor application. *Nano Energy* **2015**, *11*, 211–218.

[38] Bertoni, G.; Calmels, L.; Altibelli, A.; Serin, V. First-principles calculation of the electronic structure and EELS spectra at the graphene/Ni (111) interface. *Physical Review B* **2005**, *71*, 075402.

[39] Generalov, A.; Dedkov, Y. S. EELS study of the epitaxial graphene/Ni (1 1 1) and graphene/Au/Ni (1 1 1) systems. *Carbon* **2012**, *50*, 183–191.

[40] Taft, E.;Philipp, H. Optical properties of graphite. *Physical Review* **1965**, *138*, A197.

[41] Wakabayashi, K.; Okada, S.; Tomita, R.; Fujimoto, S.; Natsume, Y. Edge states and flat bands of graphene nanoribbons with edge modification. *Journal of the Physical Society of Japan* **2010**, *79*,034706.

[42] McCann, E.; Koshino, M. The electronic properties of bilayer graphene. *Reports on Progress in Physics* **2013**, *76*, 056503.

[43] Mohammadi, Y.; Moradian, R.; Tabar, F. S. Effects of doping and bias voltage on the screening in AAA-stacked trilayer graphene. *Solid State Communications* **2014**, *193*, 1–5.

[44] Cong, C.; Yu, T.; Sato, K.; Shang, J.; Saito, R.; Dresselhaus, G. F.; *et al.* Raman characterization of ABA-and ABC-stacked trilayer graphene. *ACS Nano* **2011**, *5*, 8760–8768.

[45] Koshino, M. Interlayer screening effect in graphene multilayers with A B A and A B C stacking. *Physical Review B* **2010**, *81*, 125304.

[46] Mishra, N.; Boeckl, J.; Motta, N.; Iacopi, F. Graphene growth on silicon carbide: A review. *Physica Status Solidi (a)* **2016**, *213*, 2277–2289.

[47] Lalmi, B.; Oughaddou, H.; Enriquez, H.; Kara, A.; Vizzini, S.; Ealet, B.;*et al.* Epitaxial growth of a silicene sheet. *Applied Physics Letters* **2010**, *97*, 223109.

[48] d'Acapito, F.; Torrengo, S.; Xenogiannopoulou, E.; Tsipas, P.; Velasco, J. M.; Tsoutsou, D.; *et al.* Evidence for Germanene growth on epitaxial hexagonal (h)-AlN on Ag (1 1 1). *Journal of Physics: Condensed Matter* **2016**, *28*, 045002.

[49] Kokh, K.; Andreev, Y. M.; Svetlichnyi, V.; Lanskii, G.; Kokh, A. Growth of GaSe and GaS single crystals. *Crystal Research and Technology* **2011**, *46*, 327–330.

[50] Nakamura, S. N. S. GaN growth using GaN buffer layer. *Japanese Journal of Applied Physics* **1991**, *30*, L1705.

[51] Wang, J.; Isshiki, M. Wide-bandgap II-VI semiconductors: Growth and properties. Springer handbook of electronic and photonic materials. **2007**, 325.

[52] Chelluri, B.; Chang, T.; Ourmazd, A.; Dayem, A.; Zyskind, J.; Srivastava, A. Molecular beam epitaxial growth of the II-V semiconductor compound Zn3As2. *Applied Physics Letters.* **1986**, *49*, 1665–1667.

[53] Mukherjee, M.; Banerjee, S.; Banerjee, J. Dynamic characteristics of III-V and IV-IV semiconductor based transit time devices in the terahertz regime: A comparative analysis. *Terahertz Science and Technology* **2010**, *3*, 97–109.

[54] Malerba, F. *The Semiconductor Business: The Economics of Rapid Growth and Decline.* University of Wisconsin Press, 1985.

[55] Hasegawa, H.; Sawada, T. On the electrical properties of compound semiconductor interfaces in metal/insulator/semiconductor structures and the possible origin of interface states. *Thin Solid Films* **1983**, *103*, 119–140.

[56] Desoize, B. Metals and metal compounds in cancer treatment. *Anticancer Research* **2004**, *24*, 1529–1544.

[57] Whittingham, M. S. Lithium batteries and cathode materials. *Chemical Reviews* **2004**, *104*, 4271–4302.

[58] Chen, A.; Holt-Hindle, P. Platinum-based nanostructured materials: Synthesis, properties, and applications. *Chemical Reviews* **2010**, *110*, 3767–3804.

[59] Odahara, G.; Otani, S.; Oshima, C.; Suzuki, M.; Yasue, T.; Koshikawa, T. In-situ observation of graphene growth on Ni (111). *Surface Science* **2011**, *605*, 1095–1098.

[60] Ghosh, B.; Mondal, D.; Kuo, C.-N.; Lue, C. S.; Nayak, J.; Fujii, J.; *et al.* Observation of bulk states and spin-polarized topological surface states in transition metal dichalcogenide Dirac semimetal candidate NiTe 2. *Physical Review B* **2019**, *100*, 195134.

[61] Takagi, N.; Lin, C.-L.; Kawahara, K.; Minamitani, E.; Tsukahara, N.; Kawai, M.; *et al.* Silicene on Ag (1 1 1): Geometric and electronic structures of a new honeycomb material of Si. *Progress in Surface Science* **2015**, *90*, 1–20.

[62] Martínez, E. A.; Fuhr, J. D.; Grizzi, O.; Sánchez, E. A.; Cantero, E. D. Growth of germanene on Al (111) hindered by surface alloy formation. *The Journal of Physical Chemistry C* **2019**, *123*, 12910–12918.

[63] Eltinge, S.; Ismail-Beigi, S. Epitaxial binding and strain effects of monolayer stanene on the Al 2 O 3 (0001) surface. *Physical Review Materials* **2022**, *6*, 014007.

[64] Tang, S.; Zhang, Y.; Tian, Y.; Jin, S.; Zhao, P.; Liu, F.; *et al.* A two-dimensional structure graphene STM tips fabricated by microwave plasma enhanced chemical vapor deposition. *Carbon* **2017**, *121*, 337–342.

[65] Xu, Z.; Bando, Y.; Liu, L.; Wang, W.; Bai, X.; Golberg, D. Electrical conductivity, chemistry, and bonding alternations under graphene oxide to graphene transition as revealed by in situ TEM. *ACS Nano* **2011**, *5*, 4401–4406.

[66] Odom, T. W.; Huang, J.-L.; Kim, P.; Lieber, C. M. Atomic structure and electronic properties of single-walled carbon nanotubes. *Nature* **1998**, *391*, 62–64.

[67] Xu, C.; Zou, R.; Peng, Y.; Liu, Q.; Ruan, S.; Hu, J. In situ transmission electron microscope studies on one-dimensional nanomaterials: Manipulation, properties and applications. *Progress in Materials Science* **2020**, *113*, 100674.

[68] Lu, D.; Yi, M.; Mo, S.-K.; Analytis, J.; Chu, J.-H.; Erickson, A.; *et al.* ARPES studies of the electronic structure of LaOFe (P, As). *Physica C: Superconductivity* **2009**, *469*, 452–458.

CHAPTER 12

HETEROJUNCTIONS OF MONO-/BILAYER GRAPHENE ON RARE-EARTH METAL SUBSTRATES

Shih-Yang Lin and Ming-Fa Lin

Department of Physics, National Cheng Kung University, Tainan, Taiwan

Abstract

This study investigates the electronic and structural properties of heterojunctions formed by mono-/bi-layers of graphene on a cerium oxide (CeO_2) substrate. By utilizing first-principles calculations, we analyze the effects of the CeO_2 substrate on graphene's electronic structure and interfacial properties. The presence of the CeO_2 substrate induces significant modifications in the band structure, including the formation of interface states and changes in the bandgap. The interfacial bonding and lattice matching play crucial roles in determining the heterojunction's electronic behavior. Furthermore, we examine the charge transfer phenomena and its influence on electrical conductivity and carrier mobility. This research provides insights into the design and optimization of graphene-based devices on CeO_2 substrates for potential electronic and optoelectronic applications.

Keywords: Heterojuntions, monolayer and bilayers graphene, CeO_2.

12.1. Monolayer graphene on CeO_2 substrate

Recently, the unique material cerium oxide (CeO_2) has become an important material because of its applications in the chemical and material fields,[1–9] including as catalysts for the elimination of toxic gases,[10–13] oxygen sensors,[12–17] and green energy materials in fuel cells. The feature-rich storage and transport properties of oxygen in CeO_2 are related to these applications. Two-dimensional (2D) nanomaterials have significant advantages for their kinetics and reactivity, which are potential elements for future nanodevices. Interestingly, the magnitude of the electronic conductivity of CeO_2 largely increases when it becomes a nanocrystalline material.[23–26] CeO_2 has been a mainstream material of investigation in recent years, mainly owing to its wide range of applications. The 5d and 4f orbitals are critical factors in exploring the multi-hybridization of orbitals between them and carbon atoms.

Graphene possesses nanoscale thickness and a honeycomb lattice, which have attracted a lot of theoretical and experimental studies.[27–36] This unique material exhibits many unusual properties, such as tunable electronic densities, ultrafast carrier mobility, and Dirac cone-shaped electronic structures. Graphene-related nanomaterials promise great potential in next-generation devices, such as high-frequency FET transistors, ultrafast rechargeable solid batteries, and large reversible green energy storages.[37–44] Monolayer graphene, with vanishing density of states at the Fermi level, is a zero-gap semiconductor, but few-layer graphenes are semi-metals that exhibit a low free carrier density. Its largely enhanced free carriers and the opening of its bandgap are critical factors for its potential applications. The electronic properties of graphene can be modulated easily through dopants, guest-atom substitutions, heterojunction interactions, the effect of substrates, layer stacking configurations, and external electric and magnetic fields. Among these methods, the substrate effects are the most effective way of dramatically changing the energy gaps or inducing metal–semiconductor transitions. Through detailed calculations, the critical orbital hybridizations are proposed to comprehend the essential physical properties.

A monolayer graphene on a CeO_2 (100) surface presents rich geometric and electronic properties, including non-uniform chemical bonding environments, the enhanced buckling of single-layer graphene, and the significant Ce–C multi-orbital hybridizations. The substrate effects are fully taken into account by applying the sufficiently wide bandgap of CeO_2. The four-layer substrate of Ce and O atoms, as clearly illustrated in Fig. 12.1, is utilized in

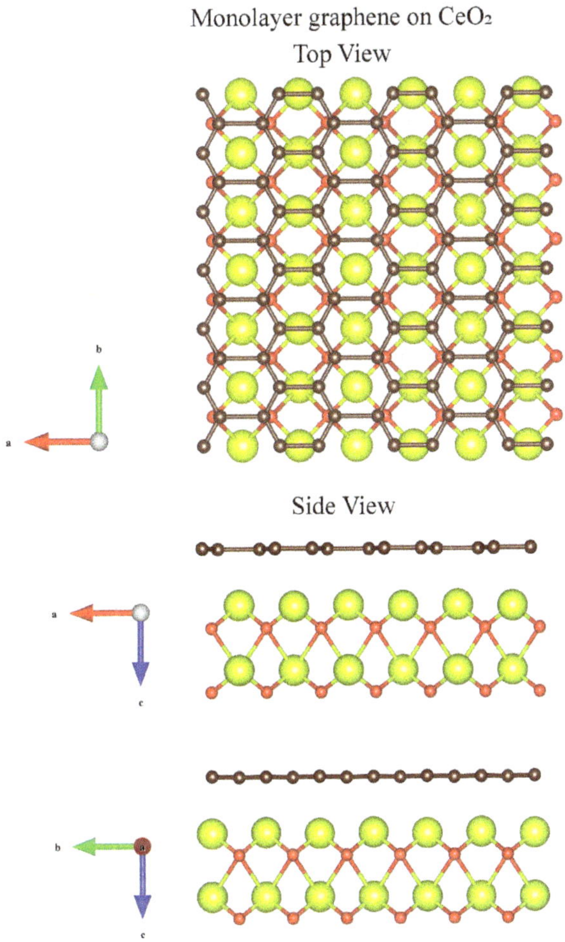

Fig. 12.1. The optimal geometric structure of graphene–CeO_2. The brown, yellow, and red balls, respectively, correspond to the carbon, cerium, and oxygen atoms.

first-principles calculations and reduced from an eight-layer substrate. The optimal geometry obviously shows that the four-layer and eight-layer substrates are almost identical in their bond lengths and bond angles. There exist four, eight, and eight atoms of Ce, O, and C, respectively, enclosed within an enlarged unit cell, as shown in Fig. 12.1. The lattice constant in the x- and y-directions is $\sim$5.19 Å. The variation in the C–C bond lengths, height differences between two interlayer sublattices, and interlayer distances between two adjacent layers directly come from the above-mentioned mechanisms. The most significant C–C/Ce–C/Ce–O bond lengths are different within an enlarged unit cell, and they are in the ranges of 1.494–1.835/2.727–2.844/2.107–2.678 Å. The height difference in the first-layer carbon atoms is approximately 0.0422 Å, where the C atoms which are close to the top Ce atoms possess the higher interlayer distance. This indicates the existence of non-orthogonality in π and σ bondings for graphene, i.e., the sp^3 bondings. In broad terms, the average C–C bonds are larger than those in pristine graphene systems. A significant fluctuation is revealed in the heterojunction so that the C–Ce chemical bonds are expected to result in multi-orbital hybridizations (discussed later for charge density distribution in Fig. 12.3). Apparently, the feature-rich chemical bondings (the non-uniform geometric structures) merge in the intralayer and interlayer atomic interactions related to the specific Moire superlattices. Hence, this will lead to high barriers for the simulation of this compound using tight-binding methods. The theoretical predictions on the main geometric features could be further tested through high-resolution STM, TEM, and LEED measurements.

The graphene–CeO_2 compound exhibits unusual band structures, as clearly shown in Figs. 12.2(a)–12.2(f). Within a pure zone-folding effect of the specific 1×2 Moire superlattice, graphene–CeO_2 (Fig. 12.2(a)) possesses more energy subbands for a greatly reduced first Brillouin zone, as shown in Fig. 12.2(a). The black and red colors represent the spin-up and spin-down states, respectively. It is clear that the deep energy region exhibits very few spin-splitting states, while the low-energy region reveals much more significant spin-splitting bands, especially near the S point. The carbon atom-dominated states are indicated by the green circles. We can identify the slightly separated linearly intersecting structures purely due to the $2p_z$ orbitals of carbon atoms at about $-1.0\,\text{eV}$ along the ΓY direction, as indicated in Fig. 12.2(b). The deeper σ valence subbands are initiated at about $E_v \leq -1.60\,\text{eV}$ near the X point. With a decrease in state energy, the $2p_z$–π and $[2p_x, 2p_y]$–σ valence bands might be mixed together through the anti-crossing behavior, where one-third and two-thirds of the electronic valence states are dominated by the former and the latter, respectively. In particular,

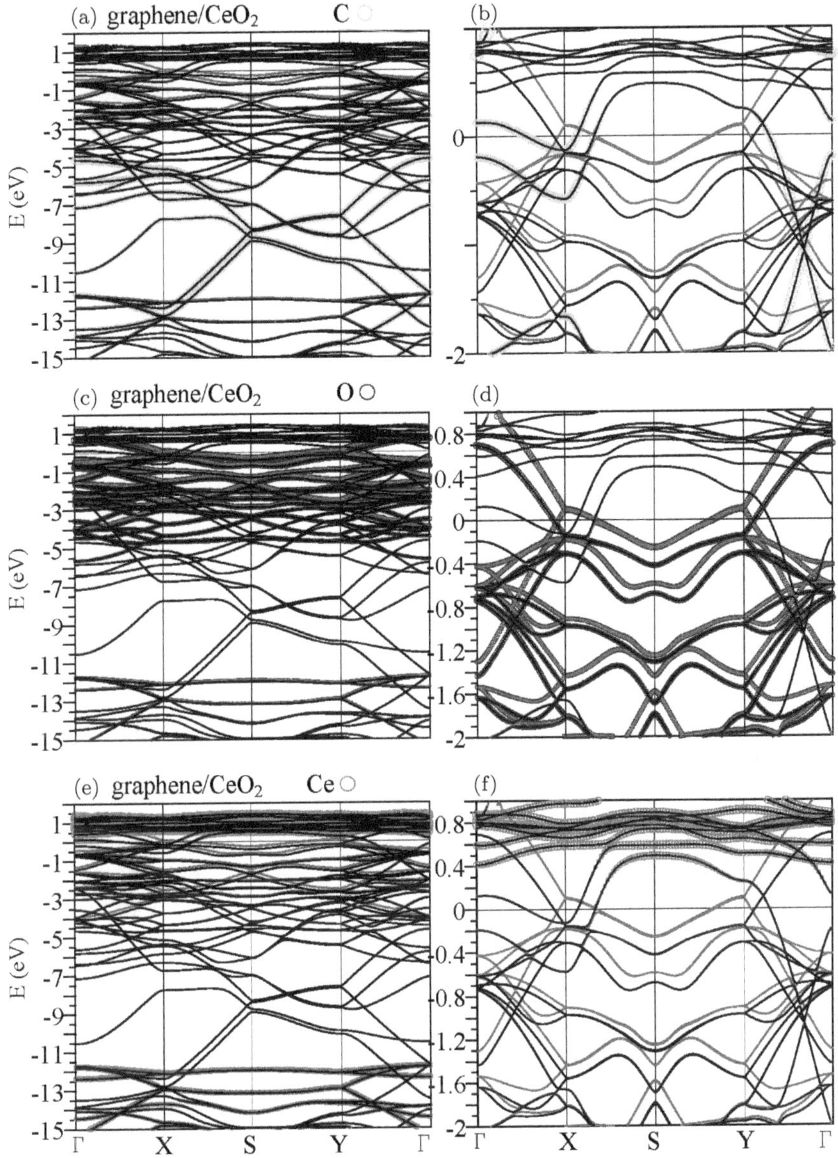

Fig. 12.2. Band structures of graphene–CeO_2 with: (a)/(b) carbon-dominated bands, (c)/(d) oxygen-dominated bands, and (e)/(f) cerium-dominated bands in wide-/low-energy region.

the deeper valence bands ranging from −6 to −11 eV purely come from the σ-band state of carbon (green circle in Fig. 12.2(a)).

The oxygen atom-dominated states present a significant difference through the rather strong hybridization effects, as clearly displayed in Figs. 12.2(c) and 12.2(d). A plenty of band-edge states, such as the extreme and saddle points of the energy–wave-vector space, are mainly dominated by the oxygen atoms in the range of −5.0 to 1.0 eV. The asymmetric spectra of electron and hole energy subbands about the Fermi level are mainly originated from the interlayer interactions between the neighboring Ce–O bonds. On the other hand, Ce atoms have their atomic contributions mainly focused in the ranges of 0.4 to 1.5 eV and −11.5 to −15.0 eV, which are very different from the previously discussed ones, as shown in Figs. 12.2(e) and 12.2(f). For the low-lying energy region, the Ce-dominated bands are located in the range of 0.4 to 1.5 eV. Such bands have much weaker dispersion compared with the previously discussed C- and O-dominated bands. Similar characteristics could also be found in the deep energy region of 11.5 to −15.0 eV. Apparently, the original Dirac cone structures of graphene across the Fermi level are thoroughly destroyed under the C–Ce interlayer interactions at the heterojunctions. The π chemical bondings, which are continuously extended to about −6.0 eV, are largely destroyed by the significant multi-orbital hybridizations in the C−Ce chemical bonds. In addition, the initial σ valence bands behave similarly; hence, it is very difficult to redefine the π- and σ-band widths under such band modifications.

The above-mentioned band features predicted through first principles are hard to simulate using the tight-binding method. As for the atom- and orbital-decomposed electronic states, they will be further examined through the important van Hove singularities, as shown in Fig. 12.5. Similar atom dominances could be found in the bilayer case (which is discussed later). There exist close relations between the complicated band structures and Moire superlattices. Several stable geometries of CeO_2 compounds are tested through experimental examinations. Apparently, the significant zone-folding and strong substrate effects are directly combined under any conditions, thus leading to the geometry-diversified electronic energy spectra. The ARPES measurements of the whole valence bands might become very difficult as a result of certain valence subbands having non-monotonous k-dependences across E_F simultaneously, especially for those of Ce and O due to their large atomic weights.

The spatial charge density presents the changes after the formation of a heterojunction of graphene–CeO_2 in Fig. 12.3, enabling

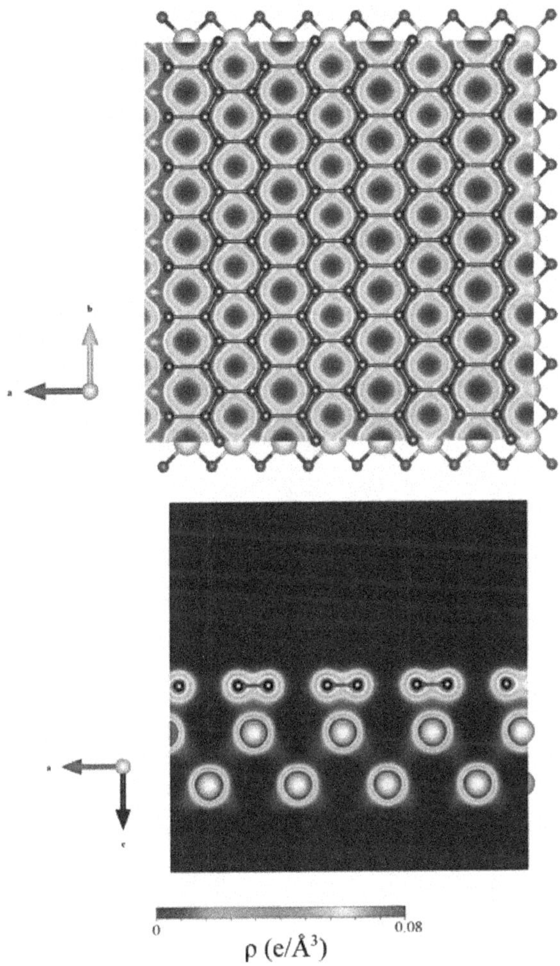

Fig. 12.3. The spatial charge density of graphene–CeO_2.

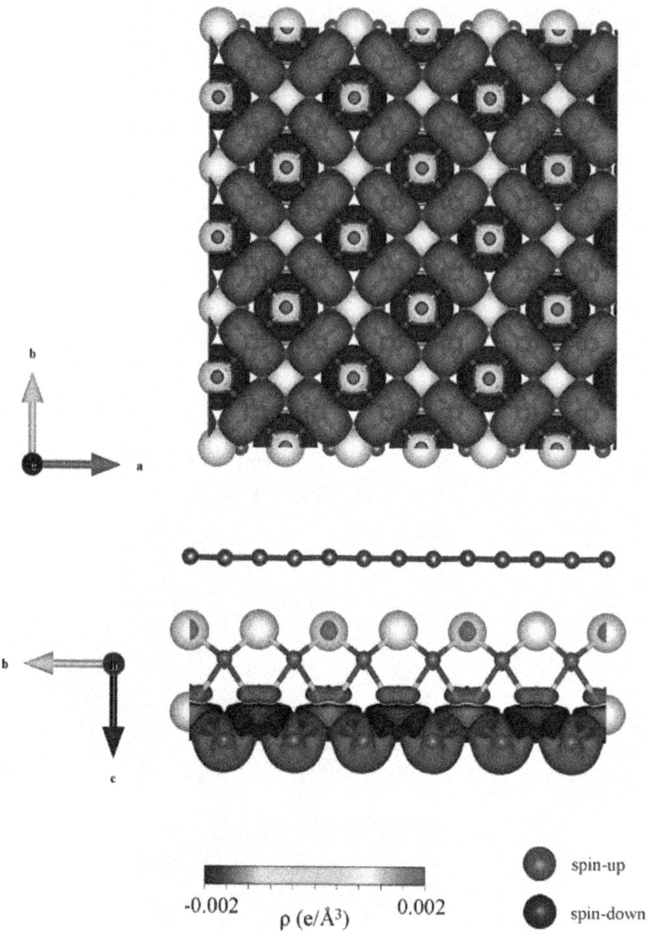

Fig. 12.4. The spatial spin density of graphene–CeO_2.

us to effectively examine the critical multi-orbital hybridizations in the C–C, C–Ce, and Ce–Ce chemical bonds. The intralayer and interlayer atomic interactions should be the critical factors. From the graphene plane projection, carbon atoms exhibit significant orbital hybridizations with their nearest-neighboring ones, as clearly illustrated by the red and yellow regions in the upper panel of Fig. 12.3. The C–Ce chemical bonds, which cause the buckling of the graphene honeycomb lattice and creates major deformation of the graphene layer, play an important role in the unusual carrier density, as indicated by the light-blue region between the C and Ce atoms (bottom panel of Fig. 12.3). Interestingly, the very complicated multi-orbital hybridizations of C–Ce bonding are further supported by the (C, Ce)-co-dominant band structure under certain energy ranges (e.g., from 0.4 to 1.5 eV and from 11.5 eV to −15.0 eV in Fig. 12.2). Both π and σ chemical bondings in the Si-honeycomb lattice are affected/distorted by the substrate effects. Hence, the well-defined π and σ band behaviors are completely absent in the electronic properties. The well-defined Dirac cone viewpoints would become meaningless when accounting for the CeO_2-substrate-related graphene systems. Specifically, the charge density around the Ce atom center is quite high, as indicated by the wide and heavy red region, originating from 12 valence electrons in the first-principles calculations.

The main characteristics of the complicated band structure are directly reflected in the 2D van Hove singularities, as indicated in Fig. 12.5. In general, the band-edge states cover the saddle point, extreme points, and partially flat dispersions. These band-edge states are able to generate the shoulders, logarithmically symmetric peaks, square-root asymmetric peaks, and delta-function-like peaks. There are a huge number of van Hove singularities that describe well the neighboring special structures, but only the merged and broadening ones could be observed in the calculated results. Such results include the contributions of C, Ce, and O atoms, which are indicated, respectively, by the red, blue, and orange curves in Fig. 12.5(a). The above-mentioned four kinds of van Hove singularities might become unique because of the strong zone-folding effects on the great enhancement of the critical points in the density of states.

Significantly, the density of states is relatively small but possesses a finite value at the Fermi level, directly indicating the metallic property. This behavior is similar to those in metal atom-doped graphene and AB-stacked graphite. In particular, the prominent two-peak structure near the Fermi level, which corresponds to the weak dispersion structures, is mainly dominated by cerium atoms (blue curve in Fig. 12.5(d), ranging from 0 to 2 eV). Moreover, the distribution range of C-$2p_z$ exhibits a dip structure at about

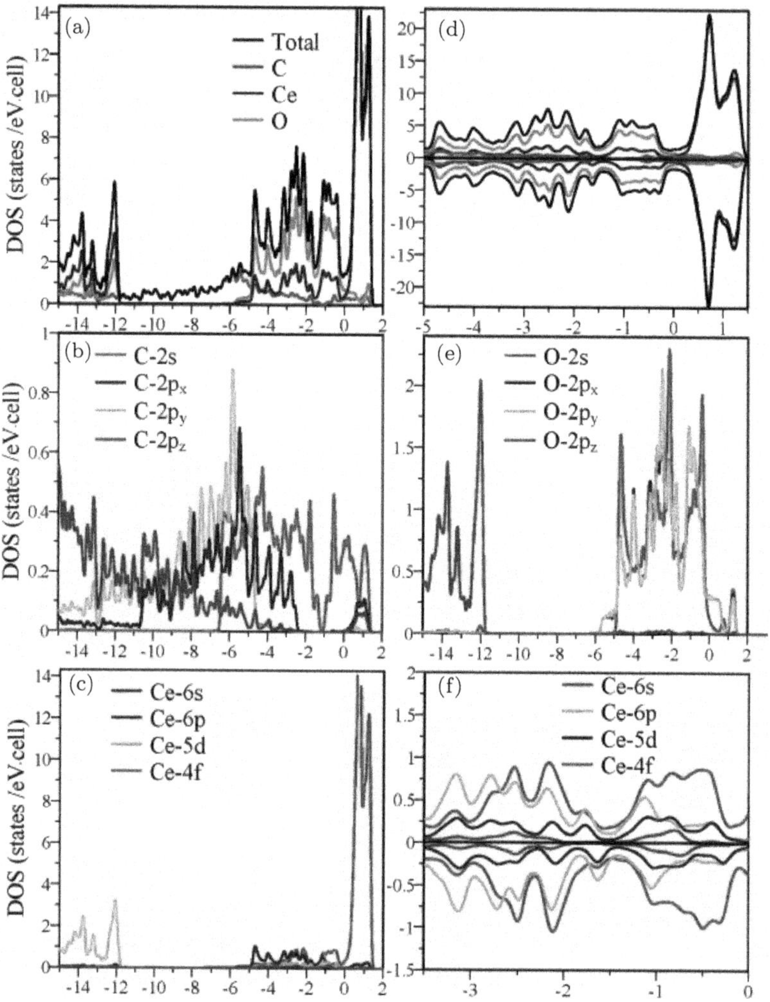

Fig. 12.5. The atom- and orbital-projected density of states of the graphene–CeO_2 system.

−1.1 eV (the red curve), and the whole energy range of C-$2p_z$ is much larger than that (∼7 eV) in monolayer graphene. These properties are consistent with the linearly intersecting bands and the absence of the well-defined π-electronic energy bands and chemical bonding (Fig. 12.2). Also, a similar phenomenon is revealed in the σ-electronic orbitals, where the 2s and $2p_x + 2p_y$ orbitals are represented by the purple, blue, and green curves. That is to say, the (2s, $2p_x$, $2p_y$, $2p_z$)-orbitals of carbon atoms make non-negligible contributions in the energy region of −2.0 to −6.0 eV, which need to be investigated simultaneously. For example, the significant chemical bondings of C–C, Ce–C, and Ce–O co-dominate the low-energy multi-peak structures. In particular, at very deep energies, the 5d orbitals of cerium, the 2s orbitals of oxygen, and the (2s, $2p_x$, $2p_y$) orbitals of carbon dominate the energy spectrum from −12 to −15 eV; therefore, the σ-electronic bonding of C–C could survive under the strong CeO_2–substrate interactions, as could the Ce–O bonds. On the other hand, the spin-split states are more significant in the low-energy region of −3.5 to 0 eV, as indicated in Figs. 12.5(d) and 12.5(f), especially for the 6s, 6p, 5d, and 4f orbitals of cerium. The above-mentioned van Hove singularities could be further verified by future experiments.

12.2. AB-stacked bilayer graphene/CeO_2

A bilayer graphene subsystem on CeO_2 could present plenty of quasi-stable stacking, where the most stable configuration would be the AB stacking in the pristine cases, which is chosen as a model for this study. Its Moire superlattice and lattice constant are almost identical to those in the monolayer graphene in the CeO_2-substrate compound, as clearly illustrated in Fig. 12.6. In particular, the buckling degree quickly declines for the second-layer graphene and the others. This result directly reflects the fact that the second layer of graphene is very difficult to generate rather strong multi-orbital hybridizations with a CeO_2 substrate. There exist the non-uniform C–C/C–Ce/Ce–Ce bonds, a feature which is similar to that of monolayer graphene–CeO_2 systems. Compared with a monolayer compound (Fig. 12.1), the buckling degree is slightly decreased in the first graphene layer. In addition to the optimal geometric structure, the layer number, stacking configuration, and substrate play important roles in the electronic properties.

A bilayer AB-stacked graphene–CeO_2 compound presents spin-split valence and conduction bands, as shown in Fig. 12.7. However,

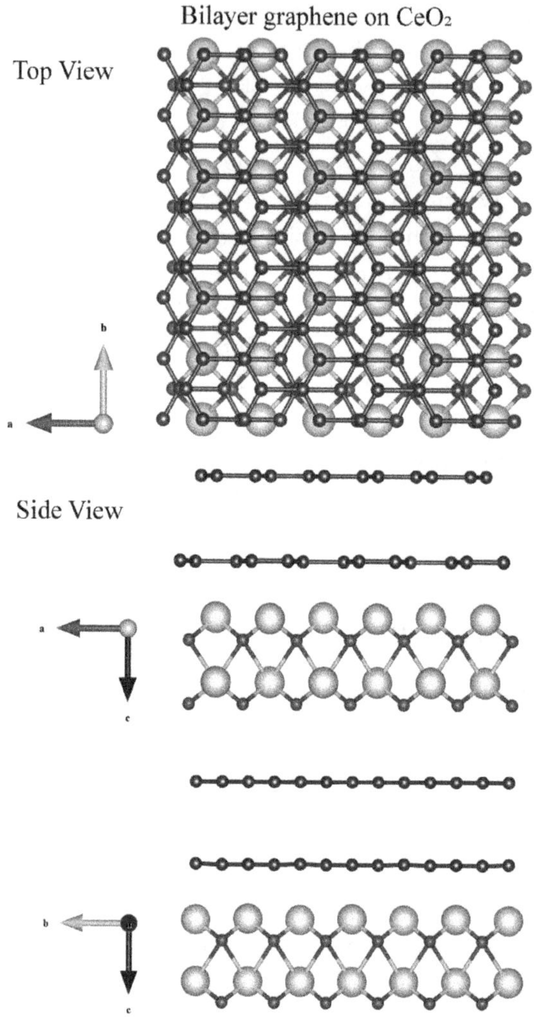

Fig. 12.6. The optimal geometric structure of AB-stacking bilayer graphene–CeO_2. The brown, yellow, and red balls, respectively, correspond to the carbon, cerium, and oxygen atoms.

this system is slightly different from the monolayer graphene–substrate system (Fig. 12.2(b)) in terms of the split spacing, i.e., a much smaller spin splitting is found. The spin splitting could be observed in the low-energy range, in which the largest energy spacing might reach $\sim$0.01 eV, but there are no split bands in the deeper-energy region below −5.0 eV. It is clear that the deep energy region exhibits very few spin-splitting states, which is similar to that of monolayer graphene–CeO_2 systems. The carbon atom-dominated states are indicated by the green circles, as shown in Fig. 12.7(a). Such dominances are extended from −15 to 1 eV. Specifically, in the energy region ranging from −6 to −11.0 eV, the Ce- and O-dominated bands are absent, as indicated in Figs. 12.7(c) and 12.7(e). This means that the middle-energy bands are purely dominated by the carbon atoms. On the other hand, the low-lying valence and conduction bands within the range of $-2 \leq E^{c,v} \leq 1$ eV, which cross the Fermi levels, are capable of creating the carbon-dominated occupied and unoccupied electronic states with linearly intersecting dispersions. It is well known that the bilayer graphene possesses two pairs of linearly intersecting bands, and this feature persists under the substrate effects, as indicated by the green circles in Fig. 12.7(b). However, it exhibits anisotropic Fermi momenta and anti-crossing behavior, which are different from the pristine graphene bilayer. As for the oxygen-dominated bands, they mainly exhibit in the range of −5.5 to 1.0 eV, a feature similar to that of previously discussed monolayer systems. The high-resolution ARPES/SP-ARPES measurements are available for examining the theoretical predictions of the significant substrate effects on the electronic energy spectra. To fully understand the substrate effects on the bilayer AB-stacking system, the multi-orbital hybridizations in complicated chemical bonds are directly studied from the spatial charge distributions, as shown in Fig. 12.8. The significant in-plane orbital hybridizations of carbon, as clearly indicated by the green and red/yellow colors in the upper panel of Fig. 12.8, are responsible for the undefined π and σ states. Also, complicated multi-orbital hybridizations of Ce and C atoms are revealed by the significant light-blue region between the carbon and cerium atoms (bottom panel of Fig. 12.8), leading to a thorough absence of a modified Dirac cone in the band structure and a V-shaped van Hove singularity. Their critical multi-orbital hybridizations are delicately identified from the atom- and orbital-decomposed density of states or the atom-dominated band structure.

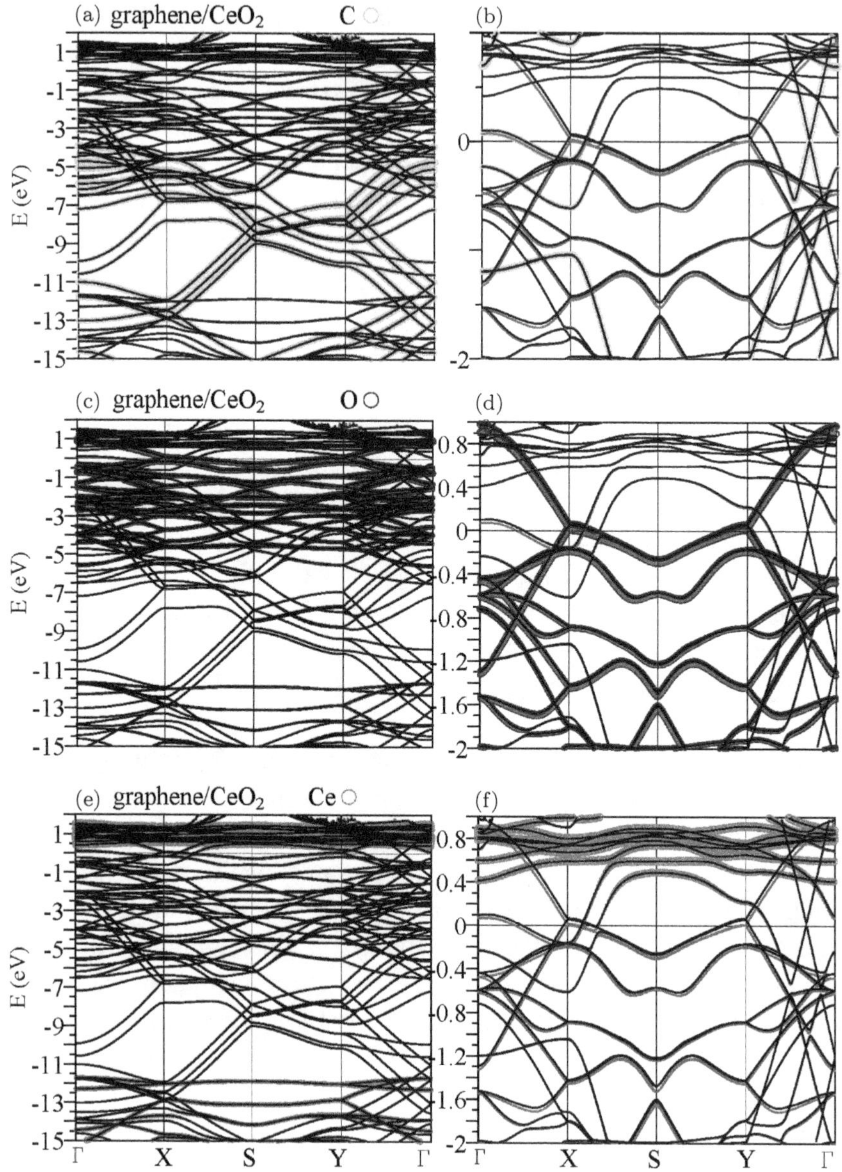

Fig. 12.7. Band structures of AB-stacking bilayer graphene–CeO_2 with (a)/(b) carbon-dominated bands (c)(d) oxygen-dominated bands, and (e)/(f) cerium-dominated bands in the wide-/low-energy region.

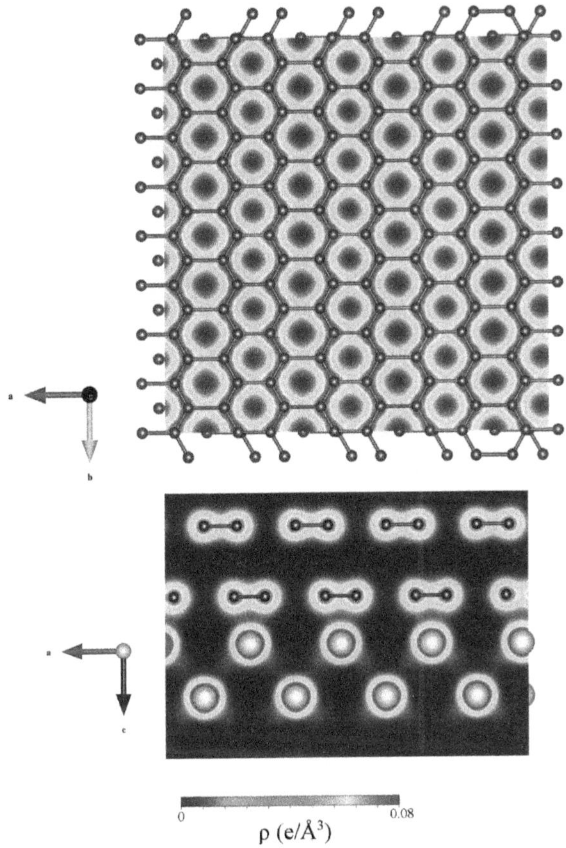

Fig. 12.8. The spatial charge density of AB-stacking bilayer graphene–CeO_2.

The ferromagnetic configuration of the bilayer AB graphene–CeO_2 compound is further analyzed based on the spatial spin-density distribution, as indicated by Fig. 12.9. Significantly, there exists a spin arrangement around the bottom layer of Ce atoms and the first layer of C atoms. The spin-up electronic states mainly originate from the C atoms away from the Ce atoms of the nearest layer, while the C atoms closest to the Ce atoms exhibit spin-down states of a larger magnitude. The spin-down magnetic moment also revealed a much larger value for the bottom layer of Ce. Their 4f orbitals make prominent contributions to the magnetic phenomena. The above-mentioned spin arrangement is consistent with the atom- and

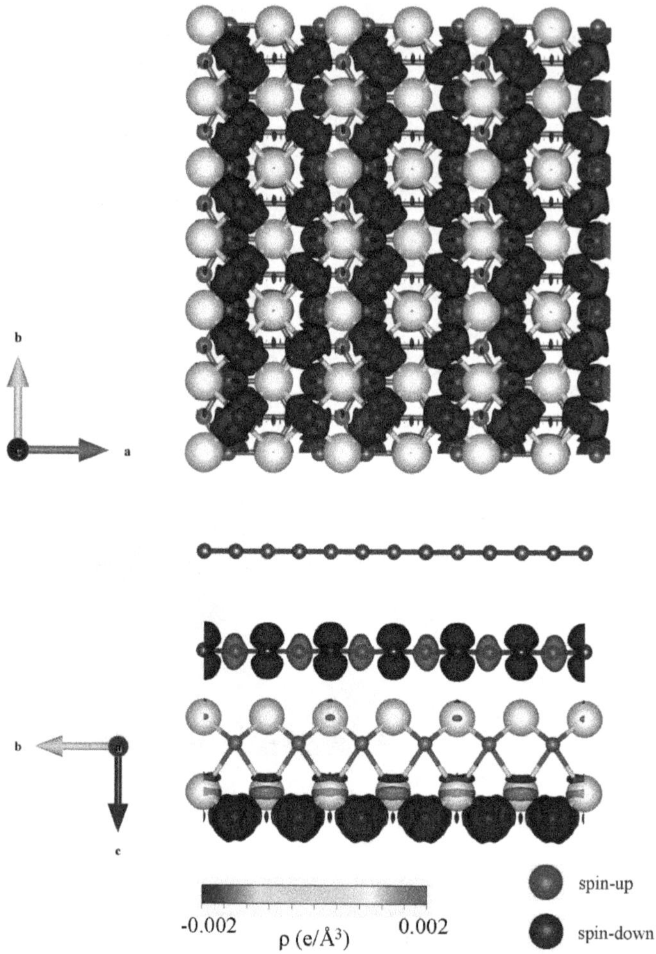

Fig. 12.9. The spatial spin density of AB-stacking bilayer graphene–CeO_2.

spin-dominated low-lying occupied and unoccupied states across the Fermi level (the spin-split energy bands in Fig. 12.7).

The atom-dominated and spin-split band structure directly reflects the atom-, orbital-, and spin-projected densities of states, as shown in Fig. 12.10. It is obtained by delicately determining the critical orbital hybridizations in C–C, C–Ce, and Ce–O bonds, as well as the ferromagnetic spin configuration. The (2s, $2p_x + 2p_y$, $2p_z$) and (6s, 6p, 5d, 4f) orbitals arise from the C and Ce atoms,

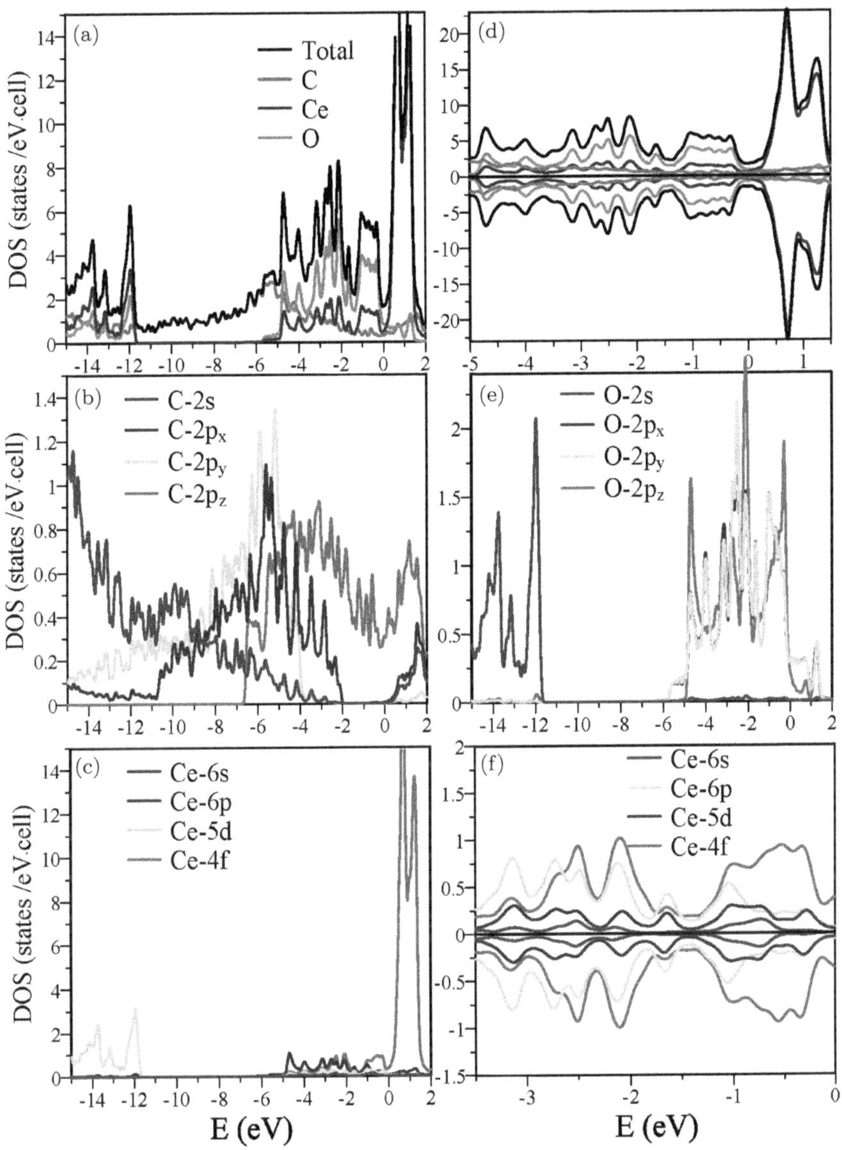

Fig. 12.10. The atom- and orbital-projected density of states of the AB-stacking bilayer graphene–CeO_2 system.

respectively, making significant contributions to a lot of van Hobe singularities. In general, the important difference from the monolayer system lies in the intensities of carbon-dominated states, which are almost twice as large. The $2p_z$ states, being another difference, do not show a V shape across the all-energy region. The spin-split states are much less significant in the bilayer system, as indicated by Fig. 12.10(f). Moreover, the 4f orbitals of Ce contribute mostly to the low-energy prominent peaks, as shown by the red curve in Fig. 12.10(c), which is a feature similar to that of the monolayer systems.

References

[1] Spezzati, G.; Benavidez, A. D.; DeLaRiva, A. T.; Su, Y.; Hofmann, J. P.; Asahina, S.; Olivier, E. J.; Neethling, J. H.; Miller, J. T.; Datye, A. K.; Hensen, E. J. M. CO oxidation by Pd supported on CeO2(100) and CeO2(111) facets. *Applied Catalysis B: Environmental* **2019**, *243*, 36–46.

[2] Jomjaree, T.; Sintuya, P.; Srifa, A.; Koo-amornpattana, W.; Kiatphuengporn, S.; Assabumrungrat, S.; Sudoh, M.; Watanabe, R.; Fukuhara, C.; Ratchahat, S. Catalytic performance of Ni catalysts supported on CeO2 with different morphologies for low-temperature CO2 methanation. *Catalysis Today* 2021, *375*, 234–244.

[3] Gao, Y.; Zhang, Z.; Li, Z.; Huang, W. Understanding morphology-dependent CuO-CeO2 interactions from the very beginning. *Chinese Journal of Catalysis* 2020, *41*, 1006–1016.

[4] Zhang, J.; Yang, Y.; Liu, J.; Xiong, B. Mechanistic understanding of CO2 hydrogenation to methane over Ni/CeO2 catalyst. *Applied Surface Science* **2021**, *558*, 149866.

[5] Choi, E. J.; Lee, Y. H.; Lee, D.-W.; Moon, D.-J.; Lee, K.-Y. Hydrogenation of CO2 to methanol over Pd–Cu/CeO2 catalysts. *Molecular Catalysis* **2017**, *434*, 146–153.

[6] Ratchahat, S.; Surathitimethakul, S.; Thamungkit, A.; Mala, P.; Sudoh, M.; Watanabe, R.; Fukuhara, C.; Chen, S. S.; Wu, K. C. W.; Charinpanitkul, T. Catalytic performance of Ni/CeO2 catalysts prepared from different routes for CO2 methanation. *Journal of the Taiwan Institute of Chemical Engineers* **2021**, *121*, 184–196.

[7] Kourtelesis, M.; Moraes, T. S.; Mattos, L. V.; Niakolas, D. K.; Noronha, F. B.; Verykios, X. The effects of support morphology on the performance of Pt/CeO2 catalysts for the low temperature steam reforming of ethanol. *Applied Catalysis B: Environmental* **2021**, *284*, 119757.

[8] Ren, Z.; Peng, F.; Li, J.; Liang, X.; Chen, B. Morphology-dependent properties of Cu/CeO2 catalysts for the water-gas shift reaction. *Catalysts* **2017**, *7*, 48.
[9] Rashed, M. N.; Touchy, A. S.; Chaudhari, C.; Jeon, J.; Siddiki, S. M. A. H.; Toyao, T.; Shimizu, K. Selective C3-alkenylation of oxindole with aldehydes using heterogeneous CeO2 catalyst. *Chinese Journal of Catalysis* **2020**, *41*, 970–976.
[10] Ohno, H.; Ikhlayel, M.; Tamura, M.; Nakao, K.; Suzuki, K.; Morita, K.; Kato, Y.; Tomishige, K.; Fukushima, Y. Direct dimethyl carbonate synthesis from CO2 and methanol catalyzed by CeO2 and assisted by 2-cyanopyridine: A cradle-to-gate greenhouse gas emission study. *Green Chemistry* **2021**, *23*, 457–469.
[11] Renganathan, B.; Rao, S. K.; Ganesan, A. R.; Deepak, A. High proficient sensing response in clad modified ceria doped tin oxide fiber optic toxic gas sensor application. *Sensors and Actuators A: Physical* **2021**, *332*, 113114.
[12] Motaung, D. E.; Mhlongo, G. H.; Makgwane, P. R.; Dhonge, B. P.; Cummings, F. R.; Swart, H. C.; Ray, S. S. Ultra-high sensitive and selective H2 gas sensor manifested by interface of n–n heterostructure of CeO2-SnO2 nanoparticles. *Sensors and Actuators B: Chemical* **2018**, *254*, 984–995.
[13] Naganaboina, V. R.; Singh, S. G. Graphene-CeO2 based flexible gas sensor: Monitoring of low ppm CO gas with high selectivity at room temperature. *Applied Surface Science* **2021**, *563*, 150272.
[14] Ramshanker, N.; Ganapathi, K. L.; Bhat, M. S.; Mohan, S. RF sputtered CeO2 thin films-based oxygen sensors. *IEEE Sensors Journal* **2019**, *19*, 10821–10828.
[15] Ramshanker, N.; Ganapathi, K. L.; Varun, N.; Bhat, M. S.; Mohan, S. Development of CeO 2 -HfO 2 mixed oxide thin films for high performance oxygen sensors. *IEEE Sensors Journal* **2021**, *21*, 18326–18333.
[16] Beie, H.-J.; Gnörich, A. Oxygen gas sensors based on CeO2 thick and thin films. *Sensors and Actuators B: Chemical* **1991**, *4*, 393–399.
[17] Izu, N.; Shin, W.; Murayama, N.; Kanzaki, S. Resistive oxygen gas sensors based on CeO2 fine powder prepared using mist pyrolysis. *Sensors and Actuators B: Chemical* **2002**, *87*, 95–98.
[18] Wang, B.; Zhu, B.; Yun, S.; Zhang, W.; Xia, C.; Afzal, M.; Cai, Y.; Liu, Y.; Wang, Y.; Wang, H. Fast ionic conduction in semiconductor CeO2-δ electrolyte fuel cells. *NPG Asia Materials* **2019**, *11*, 51.
[19] Akbar, M.; Tu, Z.; Jin, B.; Mushtaq, N.; He, Z.; Dong, W.; Wang, B.; Wang, X.; Xia, C. Demonstrating the dual functionalities of CeO2–CuO composites in solid oxide fuel cells. *International Journal of Hydrogen Energy* **2021**, *46*, 9938–9947.

[20] Li, L.; Zhu, B.; Zhang, J.; Yan, C.; Wu, Y. Electrical properties of nanocube CeO2 in advanced solid oxide fuel cells. *International Journal of Hydrogen Energy* **2018**, *43*, 12909–12916.
[21] Li, J.; Xie, J.; Li, D.; Yu, L.; Xu, C.; Yan, S.; Lu, Y. An interface heterostructure of NiO and CeO2 for using electrolytes of low-temperature solid oxide fuel cells. *Nanomaterials* **2021**, *11*, 2004.
[22] Xing, Y.; Akbar, M.; Yousaf, M.; Shah, M. Y.; Xia, C.; Gao, J.; Wang, X. CeO2 coated NaFeO2 proton-conducting electrolyte for solid oxide fuel cell. *International Journal of Hydrogen Energy* **2021**, *46*, 9855–9860.
[23] Melchionna, M.; Bevilacqua, M.; Fornasiero, P. The electrifying effects of carbon-CeO2 interfaces in (electro) catalysis. *Materials Today Advances* **2020**, *6*, 100050.
[24] Lucid, A. K.; Keating, P. R. L.; Allen, J. P.; Watson, G. W. Structure and reducibility of CeO2 doped with trivalent cations. *The Journal of Physical Chemistry C* **2016**, *120*, 23430–23440.
[25] Kogut, I.; Wollbrink, A.; Steiner, C.; Wulfmeier, H.; El Azzouzi, F.-E.; Moos, R.; Fritze, H. Linking the electrical conductivity and non-stoichiometry of thin film Ce1−xZrxO2−δ by a resonant nanobalance approach. *Materials (Basel).* **2021**, *14*, 748.
[26] Sun, L.; Zhou, L.; Yang, C.; Yuan, Y. CeO2 nanoparticle-decorated reduced graphene oxide as an efficient bifunctional electrocatalyst for oxygen reduction and evolution reactions. *International Journal of Hydrogen Energy* **2017**, *42*, 15140–15148.
[27] Lai, L.; Lu, J.; Wang, L.; Luo, G.; Zhou, J.; Qin, R.; Chen, Y.; Li, H.; Gao, Z.; Li, G.; Mei, W. N.; Maeda, Y.; Akasaka, T.; Sanvito, S. Magnetism in carbon nanoscrolls: Quasi-half-metals and half-metals in pristine hydrocarbons. *Nano Research* **2009**, *2*, 844–850.
[28] Nishitani, S.; Sekiya, R.; Haino, T. Chirality-embedded nanographenes *Angewandte Chemie International Edition* **2020**, *59*, 669–673.
[29] Liu, N.; Bo, G.; Liu, Y.; Xu, X.; Du, Y.; Dou, S. X. Recent progress on Germanene and functionalized Germanene: Preparation, characterizations, applications, and challenges. *Small* **2019**, *15*, 1805147.
[30] Goodenough, J. B.; Park, K. S. The Li-ion rechargeable battery: A perspective. *Journal of the American Chemical Society* **2013**, *135*, 1167–1176.
[31] Nguyen, C. V.; Ngoc Hieu, N.; Duque, C. A.; Quoc Khoa, D.; Van Hieu, N.; Van Tung, L.; Vinh Phuc, H. Linear and nonlinear magneto-optical properties of monolayer phosphorene. *Journal of Applied Physics* **2017**, *121*, 045107.
[32] Wu, M. S.; Xu, B.; Chen, L. Q.; Ouyang, C. Y. Geometry and fast diffusion of AlCl4 cluster intercalated in graphite. *Electrochimica Acta* **2016**, *195*, 158–165.

[33] Siebentritt, S.; Pues, R.; Rieder, K. H.; Shikin, A. M. Surface phonon dispersion in graphite and in a lanthanum graphite intercalation compound. *Physical Review B – Condensed Matter and Materials Physics* **1997**, *55*, 7927–7934.

[34] Lee, C. H.; Chen, S. C.; Yang, C. K.; Su, W. S.; Lin, M. F. Low-energy electronic structures of nanotube-nanoribbon hybrid systems. *Computer Physics Communications* **2011**, *182*, 68–70.

[35] Okamoto, Y. Density functional theory calculations of alkali metal (Li, Na, and K) graphite intercalation compounds. *The Journal of Physical Chemistry C 2014*, *118*, 16–19.

[36] Thess, A.; Lee, R.; Nikolaev, P.; Dai, H.; Petit, P.; Robert, J.; Xu, C.; Lee, Y. H.; Kim, S. G.; Rinzler, A. G.; Colbert, D. T.; Scuseria, G. E.; Tománek, D.; Fischer, J. E.; Smalley, R. E. Crystalline ropes of metallic carbon nanotubes. *Science (80-)* **1996**, *273*, 483–487.

[37] Cai, B.; Zhang, S.; Hu, Z.; Hu, Y.; Zou, Y.; Zeng, H. Tinene: A two-dimensional Dirac material with a 72 meV band gap. *Physical Chemistry Chemical Physics* **2015**, *17*, 12634–12638.

[38] Koh, Y. K.; Bae, M. H.; Cahill, D. G.; Pop, E. Heat conduction across monolayer and few-layer graphenes. *Nano Letters* **2010**, *10*, 4363–4368.

[39] Pang, Q.; Zhang, C. L.; Li, L.; Fu, Z. Q.; Wei, X. M.; Song, Y. L. Adsorption of alkali metal atoms on germanene: A first-principles study. *Applied Surface Science* **2014**, *314*, 15–20

[40] Monshi, M. M.; Aghaei, S. M.; Calizo, I. Edge functionalized germanene nanoribbons: Impact on electronic and magnetic properties. *RSC Advces* **2017**, *7*, 18900–18908.

[41] Molle, A.; Grazianetti, C.; Tao, L.; Taneja, D.; Alam, M. H. Akinwande D 2018 Silicene, silicene derivatives, and their device applications. *Chemical Society Reviews 47*, 6370–6387.

[42] Liu, Y.; Weiss, N. O.; Duan, X.; Cheng, H. C.; Huang, Y.; Duan, X. Van der Waals heterostructures and devices. *Nature Reviews Materials* **2016**, *1*, 16042.

[43] Zhou, S. Y.; Gweon, G. H.; Fedorov, A. V.; First, P. N.; De Heer, W. A.; Lee, D. H.; Guinea, F.; Castro Neto, A. H.; Lanzara, A. Substrate-induced bandgap opening in epitaxial graphene. *Nature Reviews Materials* **2007**, *6*, 770–775.

[44] Sprinkle, M.; Ruan, M.; Hu, Y.; Hankinson, J.; Rubio-Roy, M.; Zhang, B.; Wu, X.; Berger, C.; De Heer, W. A. Scalable templated growth of graphene nanoribbons on SiC. *Nature Nanotechnology* **2010**, *5*, 727–731.

CHAPTER 13

STRUCTURAL DIVERSITY AND OPTOELECTRONIC PROPERTIES OF CHEMICALLY MODIFIED PENTAGONAL QUANTUM DOTS

Nguyen Thanh Tien*, Pham Thi Bich Thao* and Ming-Fa Lin†

**College of Natural Science, Can Tho University, Vietnam*
†Department of Physics, National Cheng Kung University, Taiwan

Abstract

A first-principles study of the structural diversity and optoelectronic properties of the small penta-graphene (PG) quantum dots (PGQDs) of various sizes is performed. The stability and optoelectronic properties of the PGQDs are investigated under the effects of chemical modifications. PGQDs are edge-functionalized by Si, P, O, and F atoms. Furthermore, PGQDs are also doped with B and P atoms. An increase in the size of the PGQDs leads to a decrease in their bandgaps. All edge-functionalized structures are stable with strong electronic quantization and exhibit semiconducting properties, except for the P-edge functional structure (metallic nature). There is a semiconductor–metal transition with some doped structures. We did not observe absorption peaks in the visible region for hydrogen-passivated PGQDs. However, some absorption peaks appear in this region for edge-passivated or doped PGQDs. There are changes in the electronic properties of PGQD samples containing the impurities B, P, or BP, which also causes a shift in the peak of the spectrum to the visible region from the ultraviolet region of the corresponding pure sample because of various hybridization effects in PG (sp^2 and sp^3) and PGQDs with edge passivation and atom doping. A solid optical polarization effect occurs with some structures, e.g., H-36-B1P2 and F-ZZ-36. This represents its significant asymmetry and anisotropy compared to the other structures. The enhanced reactivity and the controllable electronic properties by size change, edge passivation, and doping make PGQDs ideal for new nanodevice applications.

Keywords: Pentagonal quantum dots, Structural and optoelectronic properties.

13.1. Introduction

Graphene quantum dots (GQDs) have attracted much attention due to their unique optoelectronic properties and relevant application.[1–3] The significant properties of GQDs arise from the electron confinement in the finite-size graphene structures (clusters, fullerenes, etc.) that lead to the opening of the bandgap and the quantization of electronic energy. Cutting graphene sheet into small clusters can create GQDs with various shapes and edges. The bandgap strongly depends on the size, shape, and edges of the GQDs. Moreover, magnetic properties are also governed by the edge electron states.[4]

Recently, two-dimensional (2D) penta-graphene (PG) and its one-dimensional ribbons (PGNRs) have received considerable attention because they are semiconductors that can be optimally combined with graphene and 2D-like materials.[5,6] Its potential applications have been recently explored, as have the possibilities of functionalization, adsorption, and atomic substitution to modify its properties.[7] Some of these modifications may change the semiconducting character of PGs and PGNRs.[8–11] Another PG derivative is penta-graphene quantum dots (PGQDs), in which the quantum confinement and edge morphology can offer additional functionality to PG.[12] Moreover, the surface or the edges of PGQDs can be chemically functionalized to increase their interactivity, tune the energy gap, enhance physicochemical capability, and exhibit new physical properties. These structures have a valuable asymmetry property.[13] Therefore, the study of the structural stability as well as the electronic and optical properties of PGQDs is essential.

In this chapter, we present the electronic and optical properties of PGQDs using density functional theory (DFT). Our calculations reveal that the electronic and optical properties of PGQD structures depend strongly on their size, morphology, the type of edge functional atom, and the type of doping atom. The cause of that effect can be traced back to a variety of atomic orbital hybridizations.

This chapter is organized as follows. We first discuss the geometric structures of PGQDs and evaluate their stability by calculating the formation energy. Next, we study their electronic properties by calculating their electronic band structure and electronic state density. Finally, we investigate their optical properties using the complex dielectric functions in the Kubo–Greenwood form.

13.2. Methodology

From the 2D PG structure, PGQDs with three different sizes (H-12, H-21, and H-36) and four different morphologies (H-AA-31, H-SS-35, H-ZA-37, and H-ZZ-36) are generated. Next, H-ZZ-36 is passivated by silicon (Si), phosphorus (P), oxygen (O), and fluorine (F). In addition, the structures of ZZ-36 passivated by alternate edge termination (a combination of two atoms) are also established. Due to its stability and high symmetry, H-ZZ-36 is further doped or co-doped with boron (B) and phosphorous (P). The structures were optimized by CASTEP code[14] within the DFT method using the generalized gradient approximation (GGA) for the Perdew–Burker–Ernzerhof (PBE) exchange-correlation functional,[15] with the following conditions: a k-point mesh of $1 \times 1 \times 1$ and a cutoff energy of 500–1000 eV. To avoid interaction between periodic images of the structure, a 15 Å vacuum region was applied for the x-, y-, and z-directions of the studied samples. The convergence precision of energy for the maximum energy change, the maximum residual force on each atom, the maximum stress, and the maximum displacement were 10^{-6} eV/atom, 0.01 eV/Å, 0.1 GPa, and 0.001 Å, respectively.

The electronic and optical properties of the optimized structures were performed in the Atomistix ToolKit (ATK) software package (version 2018) by DFT 1/2 computing with the GGA-PBE function and a double-zeta polarized basis set.[16,17] Such a computational combination was successfully employed in our recent reports.[18]

In all calculations, we use one k-point (G point), a mesh cutoff energy of 1000 eV and set the Fermi level to zero ($E_F = 0$).

The optical properties of the investigated structures are determined by the complex dielectric function:

$$\varepsilon(\omega) = \varepsilon_1(\omega) + i\varepsilon_2(\omega), \tag{13.1}$$

where ε_1 and ε_2 are the real and imaginary parts of the dielectric function, respectively, and ω indicates the angular frequency of the incident photon. The dielectric function is related to the susceptibility as $\varepsilon(\omega) = 1 + \chi(\omega)$. The susceptibility, $\chi(\omega)$, is calculated using the Kubo–Greenwood formula:[19]

$$\chi_{ij}(\omega) = -\frac{e^2\hbar^4}{m_e^2\varepsilon_0 V\omega^2}\sum_{mn}\frac{f(E_m) - f(E_n)}{E_{nm} - \hbar\omega - i\hbar\Gamma} \times \pi_{nm}^i \pi_{mn}^j, \tag{13.2}$$

where $\pi_{mn}^{i,j}$ means the i, j components of the dipole matrix element between the n and m states, Γ is the broadening, V is the volume, $f(E_m/E_n)$ is the Fermi function, and $E_m(E_n)$ corresponds to the eigenvalues of the $m(n)$ states. The extinction coefficient κ is calculated as

$$\kappa = \sqrt{\frac{\sqrt{\varepsilon_1^2 + \varepsilon_2^2}}{2} - \frac{\varepsilon_1}{2}}. \tag{13.3}$$

From the extinction coefficient κ, we can calculate the optical absorption coefficient α of the materials:

$$\alpha = \frac{2\omega\kappa}{c}. \tag{13.4}$$

Here, c the velocity of light in vacuum.

13.3. Results and Discussion

13.3.1. *Effect of size on electronic and optical properties of PGQD*

Figure 13.1 displays the structures after relaxation. It is obvious that the structural form is almost identical at three different sizes, and

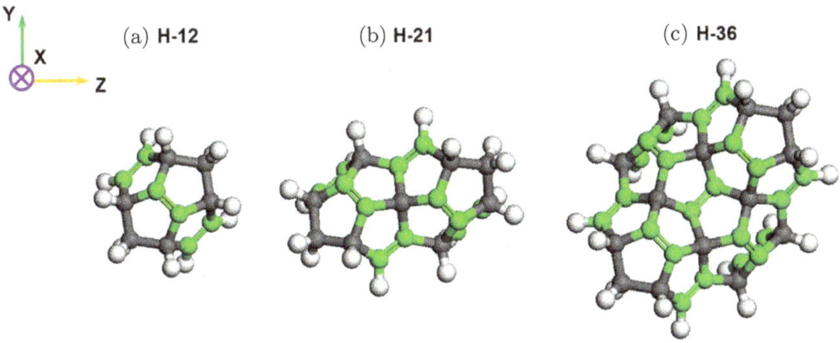

Fig. 13.1. Configurations of PGQDs after optimization: The gray- and green-colored balls represent the sp^3- and sp^2-hybridized carbon atoms, respectively.

significant variation is observed at the edges. To evaluate the stability of the structure, the formation energy is determined:

$$E_{\text{form}} = \frac{(E_{\text{total}} - n_C E_C - n_H E_H)}{n_C + n_H}, \tag{13.5}$$

with the total energy for one unit cell being E_{total}. The isolated energy for one C/H atom is $E_{C/H}$, and the number of C and H atoms is denoted by n_C and n_H, respectively.

The formation energies of the studied PGQDs are shown in Table 13.1. In all structures, E_{form} values are negative and vary between −6.48 and −7.22 eV, indicating that our PGQDs are stabilized. As the quantum dot size increases, the formation energy decreases. Compared with the others, H-36 shows more stability.

Table 13.1 also provides information about the average bond length of each type of bond in the sample. Specifically, the average

Table 13.1. Formation energy, energy gap, and average bond length of PGQDs.

System	Formation energy/ atom (eV)	Energy gap (eV)	Bond lengths (Å)				
			sp^2–sp^2	sp^2–sp^3	sp^3–sp^3	sp^2–edge	sp^3–edge
H-12	−6,48	4.42	1.342	1.503	1.569	1.092	1.102
H-21	−6,83	3.94	1.345	1.527	1.566	1.092	1.103
H-36	−7,22	3.60	1.346	1.500	1.583	1.091	1.103

length of each type of bond is determined as the average of all such bonds in the sample. The average bond length between two sp^3-hybridized carbon atoms, two sp^2-hybridized carbon atoms, between one sp^2-hybridized carbon atom and one sp^3-hybridized carbon atom, between one sp^2-hybridized carbon atom and an edge hydrogen atom, and between one sp^3-hybridized carbon atom and an edge hydrogen atom (before and after optimization) are denoted by sp^2–sp^2, sp^3–sp^3, sp^2–sp^3, sp^2–edge, and sp^3–edge, respectively. From the optimization parameters in Table 13.1 and Fig. 13.1, we note that the difference in the bond lengths among the surveyed samples is not significant. This implies that the quantum dot size does not affect the average bond lengths of the atoms in the sample.

The band structure of PGQDs with increasing size is presented in Fig. 13.2. It can be clearly seen that the bandgap gradually decreases when the sample size increases. In addition, the number of subbands increases as the sample size increases. This result is further clarified through the total state density (DOS) and specific state density (PDOS) graphs depicted in Fig. 13.2. The sample size increases, and the number of atoms in the system increases; from the PDOS plots, the contribution of each of the atom types also increases. This leads to a gradual narrowing of the bandgap in the band structure.

The static dielectric constant of these structures ($\varepsilon_1(0)$) of PGQDs varies from 1.01 to 1.03. This range of values is smaller than that of the monolayer PG and PGNR.[18,22]

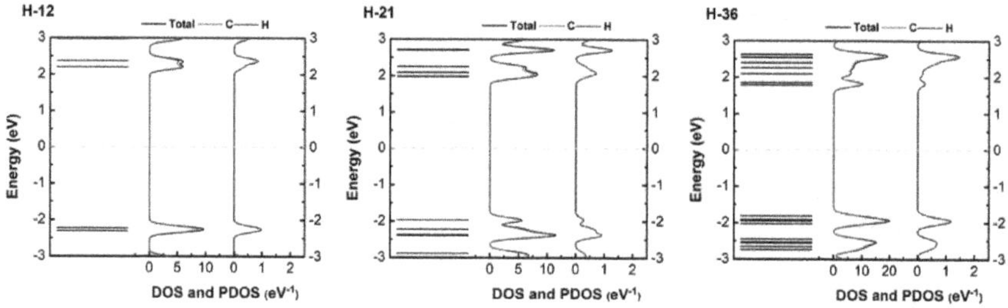

(a) (b) (c)

Fig. 13.2. Band structure, the density of state (DOS), and the partial density of state (PDOS) of hydrogen-passivated PGQDs.

In Fig. 13.3, the first peak of the real parts $\varepsilon_1(\omega)$ of the hydrogen-passivated PGQDs shifts toward the high-energy region, which is called the blueshift. As the size of the PGQDs increases, this peak shifts to a lower-energy region. We observe a notable spatial anisotropy in all three directions for H-21. The anisotropy is almost negligible in the x- and y- directions for other structures. In Fig. 13.3, the first peak $\varepsilon_2(\omega)$ of hydrogen-passivated PGQDs occurs at the largest energy of 4.5 eV. In addition, as the sample size increases, the first peak $\varepsilon_2(\omega)$ shifts toward the lower-energy region. The energy values of the first peak $\varepsilon_2(\omega)$ of the investigated samples correspond to the bandgap widths of these samples in Table 13.1.

The absorption coefficient of the different-sized PGQDs is calculated from the real and imaginary parts of the dielectric function, which is depicted in Fig. 13.4. In general, the absorption intensity decreases, and the peak of the spectrum shifts to a lower-energy region as the quantum dot size increases for all the investigated

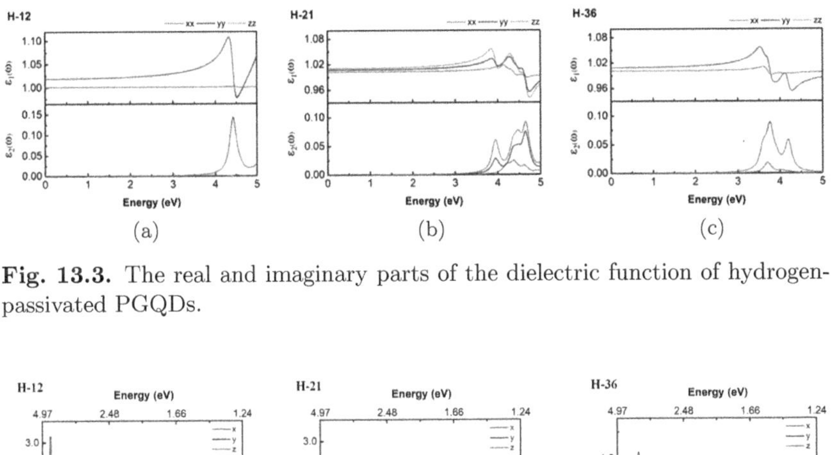

Fig. 13.3. The real and imaginary parts of the dielectric function of hydrogen-passivated PGQDs.

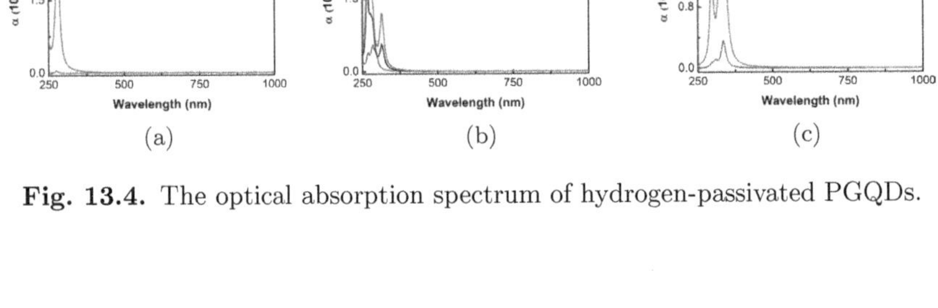

Fig. 13.4. The optical absorption spectrum of hydrogen-passivated PGQDs.

samples. Absorption peaks in the visible region were not observed for the hydrogen-passivated PGQDs.

13.3.2. *Effect of doping on electronic and optical properties of PGQD*

Among the investigated structures, H-36 exhibits the highest stability due to its lowest formation energy. However, the bandgap of this structure is still quite large, so the absorption peaks are in the ultraviolet region. To explore the novel electronic and optical properties of this structure, H-36 was doped with B and P at the sp^3- and sp^2-hybridized carbon atoms, respectively (H-36-B1, H-36-B2, H-36-P1, and H-36-P2). The above structure is also co-doped with B and P at the two characteristic carbon positions of the pentagonal structure (H-36-B1P2 and H-36-B2P1).

Figure 13.5 shows the formation energy of six doped PGQDs. For the single-doped PGQD, the doped configuration at the sp^2-hybridized carbon atom position exhibits greater stability for B and P doping. This result is similar to that obtained for the post-optimal structural morphology. In Figs. 13.5(a) and 13.5(c), it can be clearly seen that the five penta-ring structure is significantly deformed, while the stable structure is observed in Figs. 13.5(b) and 13.5(d).

The co-doping for H-36-B1P2 and H-36-B2P1 shows more stability than single doping when considering the formation energy and optimized structural morphology. In this case, H-36-B2P1 has a lower formation energy and a more stable penta-ring structure. Meanwhile, in H-36-B2P1, the bond lengths around the P atom change dramatically and increase the buckling of the penta-ring.

In this section, we consider the effect of impurities on the electronic properties of PGQDs. The band structures of the six mentioned structures are shown in Fig. 13.6. For single-doped PGQDs (H-36-B1, H-36-B2, H-36-P1, and H-36-P2), there appears to be a subband that coincides with the Fermi level. This result implies a transition from semiconductor to metal, compared to the pristine structure (H-36). The change is homologous to that in the B- or N-doped penta-nanoribbon structure. In contrast,

B and P co-doped PGQDs (H-36-B1P2 and H-36-B2P1) exhibit semiconductor properties. Specifically, their bandgaps are 1.50 and 1.14 eV, respectively.

The PDOS graph in Fig. 13.6 and the electron localization function (ELF) in Fig. 13.7 can further clarify the influence of

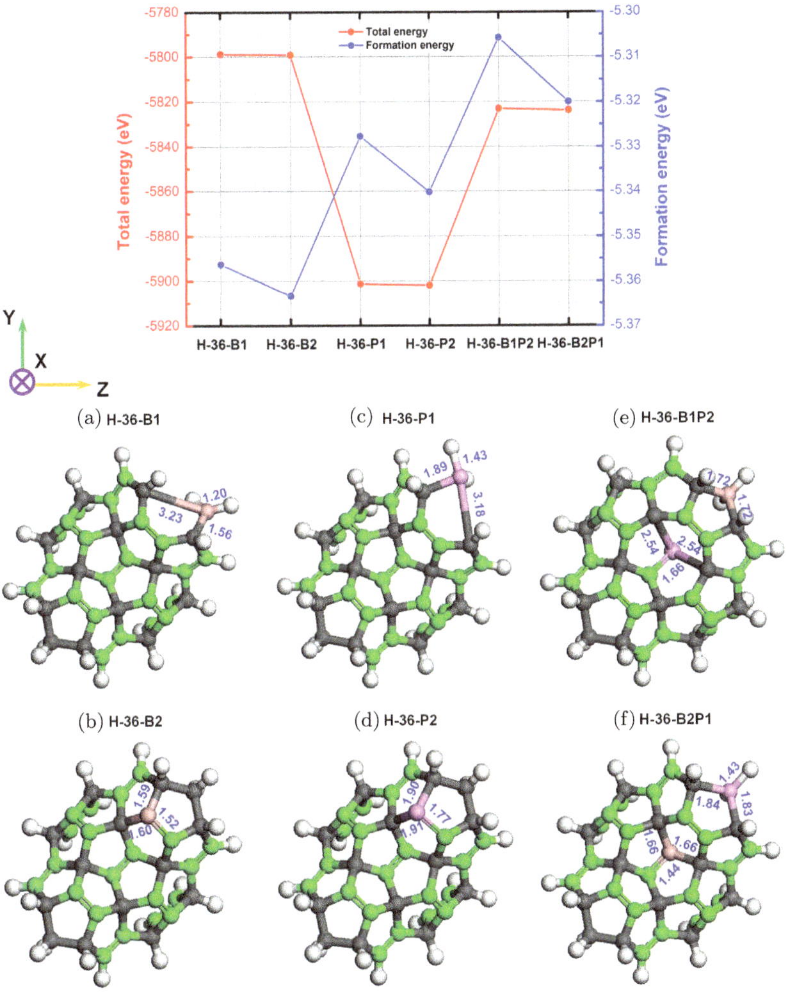

Fig. 13.5. (top) Total and formation energy graphs of PGQDs doped with B, P, or BP. (bottom) Optimized structures of six investigated PGQD dopings. The pink- and violet-colored balls represent the B and P atoms, respectively.

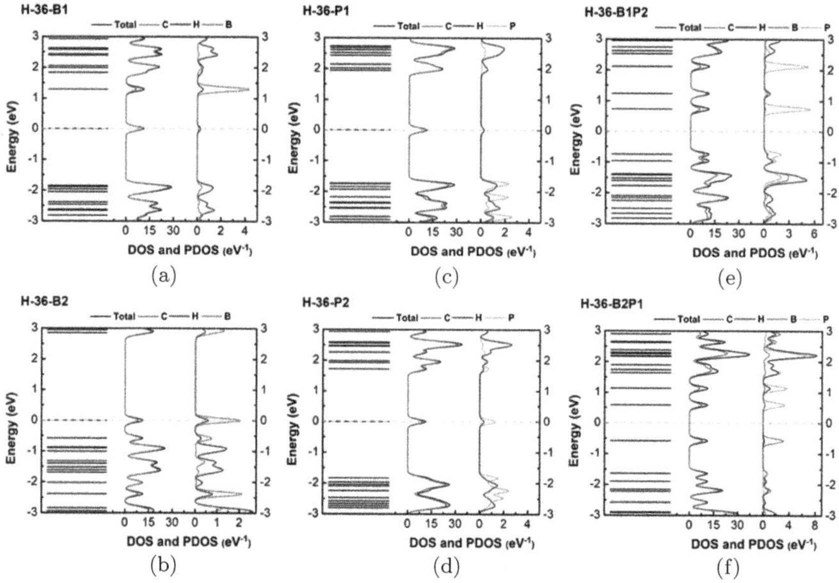

Fig. 13.6. Band structure, the density of state (DOS), and the partial density of state (PDOS): (a) H-36-B1, (b) H-36-B2, (c) H-36-P1, (d) H-36-P2, (e) H-36-B1P2, and (f) H-36-B2P1.

B and P atoms on the electronic properties of PGQDs. For H-36-B1, H-36-B2, H-36-P1, and H-36-P2, the doping atom creates a state at the Fermi level that causes the corresponding subband to appear in the band structure. Conversely, in co-doped PGQD, the states of the atoms are not found at the Fermi level. In addition, the contribution from P is more significant than that from B at subbands near the Fermi level. In Fig. 13.7, we can also observe increasing electron localization around the B and P atoms.

The optical properties of doped PGQDs are presented in this section. The absorption coefficient explores the absorption capacity of the studied structures. To obtain this result, we first determine the real and imaginary parts of the dielectric function.

In Fig. 13.8, the real and imaginary parts of the dielectric function of all doped samples are shown. For the single-doped PGQDs, we observe a significant difference in all three polarizations, while the real and imaginary parts in the co-doped PGQDs are

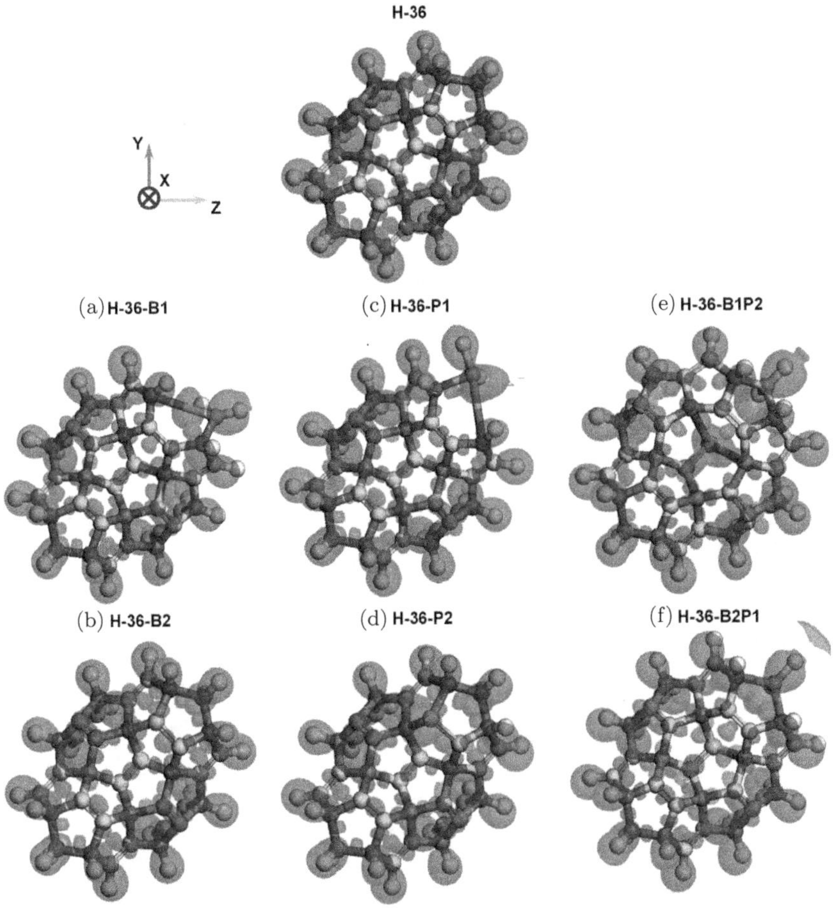

Fig. 13.7. The electron localization function of the doped PGQDs.

similar in the y- and z-directions. Specifically, we observe the real parts of the dielectric function with a similar form and insignificant differences in the values for H-36-B1 and H-36-B2. A similar result is obtained from the single-doped sample with P atoms. The maximum value of $\varepsilon_1(\omega)$ for them varies from 1.03 to 1.06 in the x- and y-directions, corresponding to an incident photon energy of about 3.5 eV. In addition, we also observe several peaks of ε_1, with lower values in the visible region for single-doped samples.

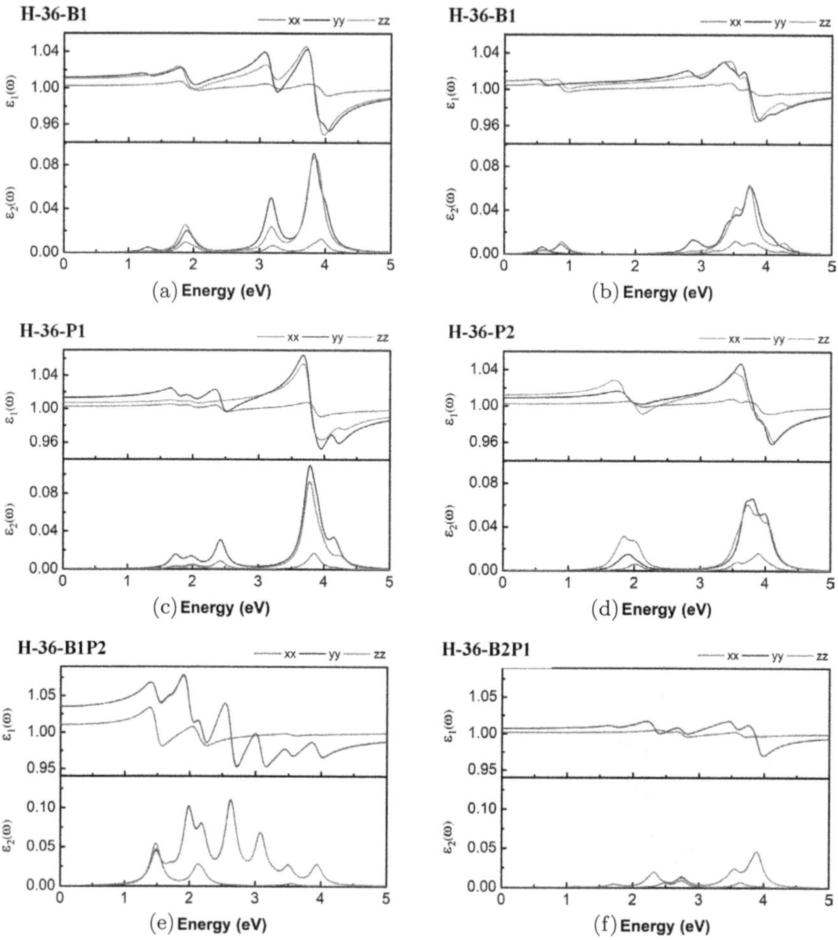

Fig. 13.8. The real and imaginary parts of the dielectric function: (a) H-36-B1, (b) H-36-B2, (c) H-36-P1, (d) H-36-P2, (e) H-36-B1P2, and (f) H-36-B2P1.

For co-doped samples (H-36-B1P2 and H-36-B2P1), more peaks of ε_1 appear in the incident photon energy range of 1.5–4 eV. In Fig. 13.8, the imaginary parts of the dielectric function $\varepsilon_2(\omega)$ of doped PGQDs are also presented. This result will recognize the optical characteristics similar to the absorption spectrum in Fig. 13.9.

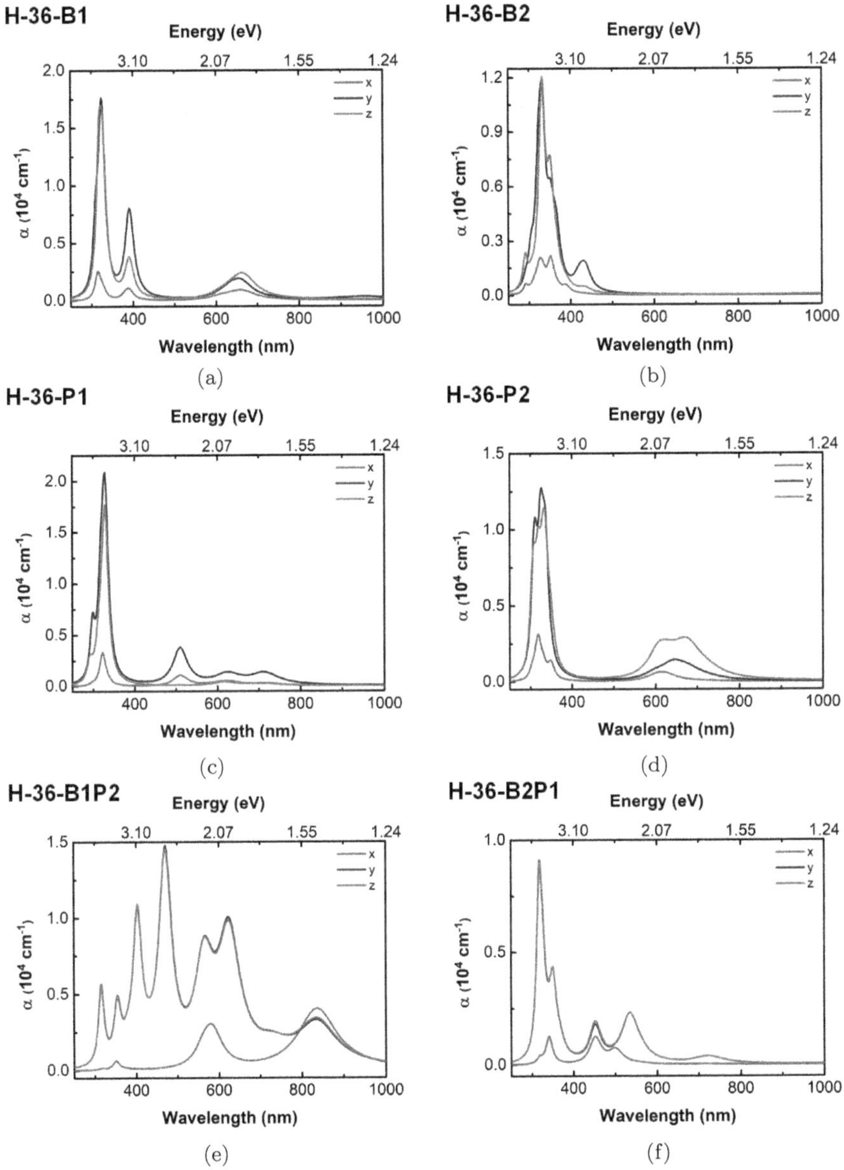

Fig. 13.9. The optical absorption spectrum of (a) H-36-B1, (b) H-36-B2, (c) H-36-P1, (d) H-36-P2, (e) H-36-B1P2, and (f) H-36-B2P1.

Explicitly, absorption peaks are observed in the ultraviolet and visible regions. For single-doped PGQDs, the adsorption peaks ranged from 330 to 650 nm, depending on the site and doping element. For the co-doped samples, the absorption peaks extend from the ultraviolet to the near-infrared region (310–850 nm).

The obtained results have shown the influence of the doping element and the doping site on the electronic and optical properties of PGQDs. The narrowing of the band structure due to doping makes the transition from the valence to the conduction band of electrons more sensitive, which can in turn increase the possible transitions, thereby increasing the number of observed spectral peaks. In addition, changes in the electronic properties of PGQD samples containing the impurities B, P, or BP also cause a shift in the peak of the spectrum to the visible region from the ultraviolet region of the pure H-36 sample. The obtained results are similar to those obtained from the investigation of the optical properties of GQDs through phosphorization or boronization. The absorption peaks of GQDs through boronization appear at an energy value of 400 nm. In comparison, phosphorization produces new absorption peaks at a more extended wavelength range, from 500 to 800 nm.[23,24]

13.4. Effect of Passivation on Electronic and Optical Properties of PGQDs

The edge shape of the H-36 structure is similar to that of the zigzag graphene nanoribbon. So, this structure is denoted by ZZ-36.

ZZ-36 is passivated by Si, P, O, and F, which are denoted by Si-ZZ-36, P-ZZ-36, O-ZZ-6, and F-ZZ-36, respectively. The parameters for optimization, electronic characterization, and absorption spectra are similar to those mentioned in Section 13.1. However, the cutoff energy varies from 750 to 1000 eV, depending on the passivation atom.

Figure 13.10 presents the Si-ZZ-36, P-ZZ-36, O-ZZ-36, and F-ZZ-36 structures before and after optimization. The stability of the structures is determined using the binding energy E_B:

$$E_B = \frac{(E_{\text{total}} - n_C E_C - n_H E_X)}{n_C + n_X}, \qquad (13.6)$$

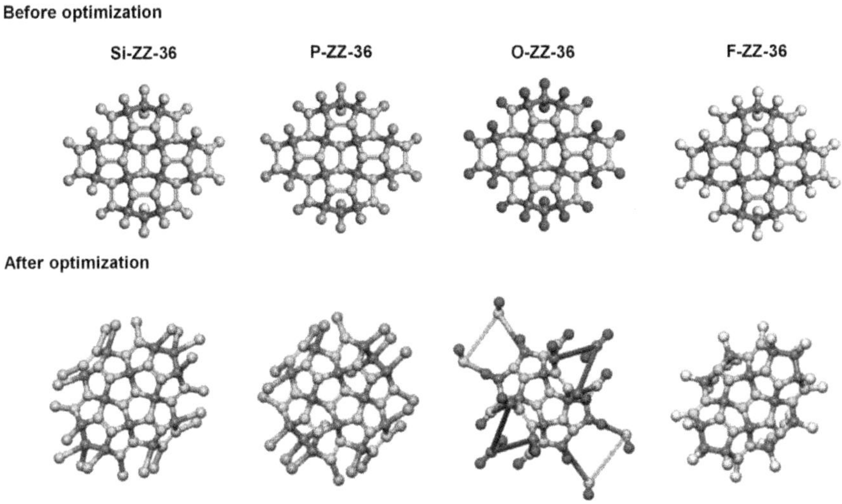

Fig. 13.10. Configurations of PGQDs passivated by Si, P, O, and F atoms before and after optimization: The gray- and green-colored balls represent sp^3- and sp^2-hybridized carbon atoms, respectively. The balls in yellow, violet, red, and sky blue colors represent the Si, P, O, and F atoms, respectively.

with the total energy for one unit cell being E_{total}. The isolated energies of one C and one X atom are E_C and E_X, respectively. The number of C and X atoms is denoted by n_C and n_X, respectively.

The binding energy values of Si-ZZ-36, P-ZZ-36, O-ZZ-36, and F-ZZ-36 are -5.34, -5.17, -5.96, and $-5.24\,\text{eV}$, respectively. In the above order, the structure passivated by O atoms has the lowest binding energy, and the structure passivated by P atoms has the highest binding energy. Similar to the quantum dot structures with different sizes, the average bond lengths of the four structures, Si-ZZ-36, P-ZZ-36, O-ZZ-36, and F-ZZ-36, before and after optimization are presented in Table 13.2. After optimization, the change in the average bond lengths between the sp^3-hybridized carbon atoms with Si, P, and F atoms (sp^3–edge) is significant. Especially for the quantum dot passivated with the Si and P atoms, this value increases by about 80% after optimization. In contrast, the average bond lengths between sp^3-hybridized C and O atoms show little difference. This can be explained by the fact that the atomic radii of Si and P are larger than those of O and F. The hybridization form also affects the bond length,

Table 13.2. The average bond lengths of Si-ZZ-36, P-ZZ-36, O-ZZ-36, and F-ZZ-36.

		Before optimization		
	Si-ZZ-36	**P-ZZ-36**	**O-ZZ-36**	**F-ZZ-36**
sp^2–sp^2	1.631	1.631	1.631	1.631
sp^2–sp^3	1.536	1.536	1.536	1.536
sp^3–sp^3	1.536	1.536	1.536	1.536
sp^2–edge	1.140	1.140	1.140	1.140
sp^3–edge	1.140	1.140	1.140	1.140
		After optimization		
	Si-ZZ-36	**P-ZZ-36**	**O-ZZ-36**	**F-ZZ-36**
sp^2–sp^2	1.391	1.426	1.914	1.346
sp^2–sp^3	1.511	1.533	2.072	1.525
sp^3–sp^3	1.526	1.568	7.176	1.594
sp^2–edge	1.098	1.828	1.141	1.331
sp^3–edge	2.095	1.947	1.181	1.383

leading to the difference in the average bond lengths between the sp^2-hybridized C atom with the P atom and other cases. In particular, the effect of the O atom on the average bond lengths of sp^2–sp^2, sp^2–sp^3, and sp^3–sp^3 is huge.

It can be seen that the change in bandgap depends on the passivation atom. Notably, in the P-ZZ-36 structure (Fig. 13.11), there are two subbands very close to the Fermi level, making the bandgap equal to 0.075 eV, which has metallic properties. Since the P atom belongs to the VA group with five electrons in the outermost orbit, after bonding with neighboring atoms (C atoms), there is still one free electron, so the passivation of the PGQD with the P atom corresponds to the formation of an n-type semiconductor from a pure semiconductor. However, because the proportion of P atoms in the P-ZZ-36 structure is huge, the quantum dot passivated with P shows metallic properties. This is similar to the results obtained with P doping or P passivation in the PG and PGNRs.[11,25] Other structures (Si-ZZ-36, O-ZZ-36, and F-ZZ-36) have semiconductor properties, with their bandgaps ranging from 0.7 to 3.3 eV. The difference in subbands shifting in the samples can be elucidated by edge orbital

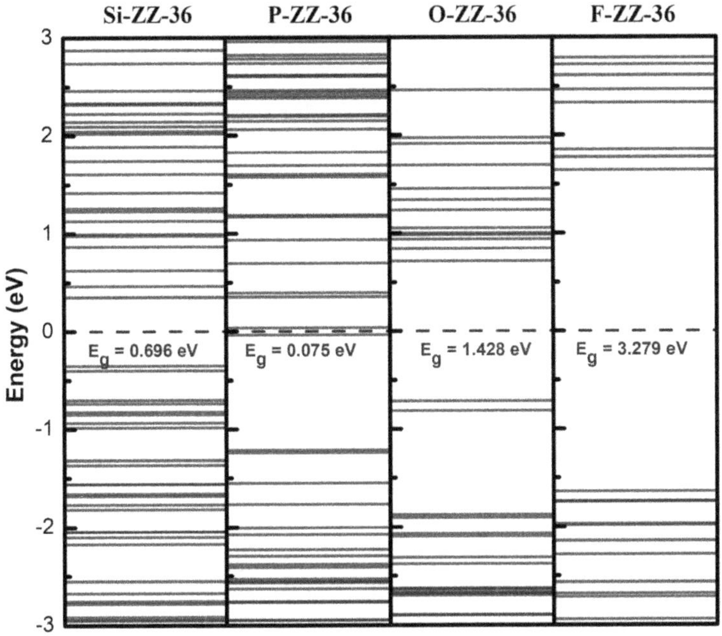

Fig. 13.11. Band structures of Si-ZZ-36, P-ZZ-36, O-ZZ-36, and F-ZZ-36.

hybridization and charge transfer between PGQDs and edge atoms, like in GQDs.[26] This result is further clarified by the DOS and PDOS depicted in Fig. 13.12. The PDOS plot shows the contribution of C and edge atoms. Specifically, for Si-ZZ-36, P-ZZ-36, and O-ZZ-36, there are significant contributions of Si, P, and O atoms around the Fermi level, leading to a dramatic reduction in the bandgap compared to H-36. Meanwhile, for F-ZZ-36, the contribution of the F atom around the Fermi level compared to that of the H atom in H-36 has not changed significantly. Therefore, the bandgap of these two structures is not much different.

The absorption spectra of Si-ZZ-36, O-ZZ-36, P-ZZ-36, and F-ZZ-36 are shown in Fig. 13.13. In general, the anisotropy of the absorption spectrum is acquired in all of the above structures. In addition, the absorption peaks of the studied structures correspond to wavelengths in the visible or near-infrared region. Except for the F-ZZ-36 structure, the absorption peak intensity in the Oy direction

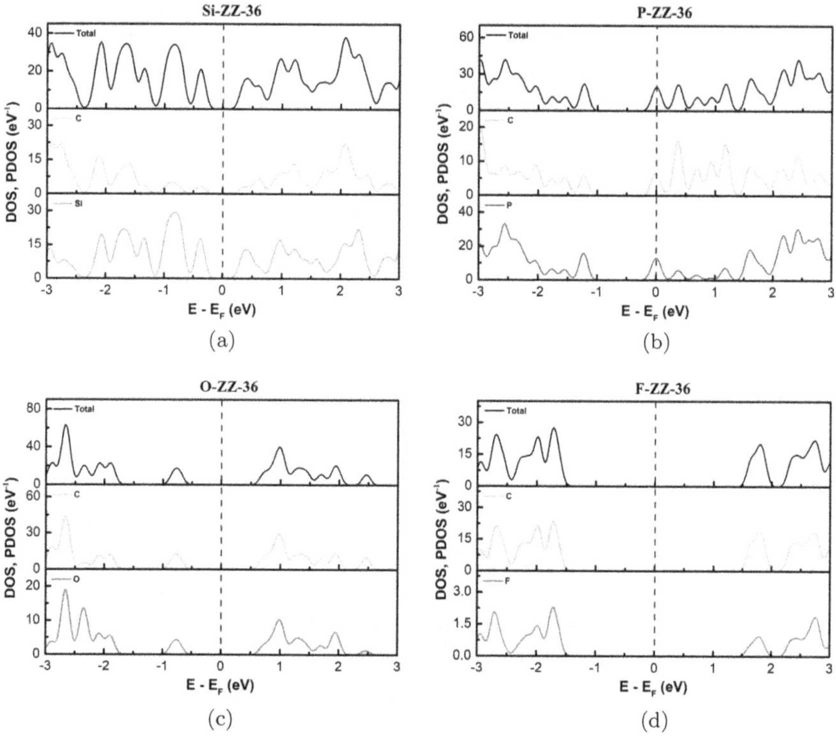

Fig. 13.12. Density of state (DOS) and partial density of state (PSOS): (a) Si-ZZ-36, (b) P-ZZ-36, (c) O-ZZ-36, and (d) F-ZZ-36.

is superior to the absorption peak intensity in the Ox and Oz directions. Specifically, for the Si-ZZ-36 structure, the intensity of the absorption peak in the Oy direction at 1500 nm is about eight times higher than that in the Ox and Oz directions over the wavelength range of 800–900 nm. Meanwhile, for P-ZZ-36, the spectral peak with the highest intensity, corresponding to 1000 nm, was observed in the Oy direction. There are also some additional spectral peaks at wavelengths corresponding to the visible or near-infrared regions in this direction. In particular, the PGQD is passivated by O atoms; the spectral peaks with the most incredible intensity in all three directions are in the visible light region, with wavelengths of 700, 450, and 435 nm in the Ox, Oy, and Oz directions, respectively. Compared with the H-ZZ-36 structure, substituting the H atom with the F atom

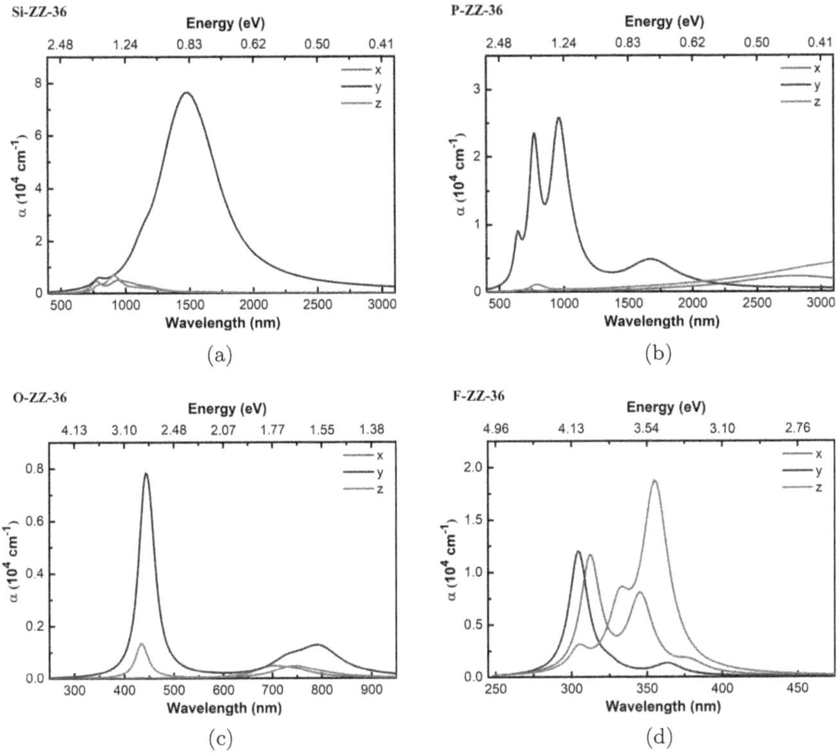

Fig. 13.13. Optical absorption spectrum of dielectric function: (a) Si-ZZ-36, (b) P-ZZ-36, (c) O-ZZ-36, and (d) F-ZZ-36.

at the edge causes a minor spectral peak shift. The obtained spectral peaks are shifted to the higher-energy region, but these peaks are still within the ultraviolet range.

The above results show a close relationship between the electronic and optical properties of PGQDs. Obviously, the changes in the band structure and bandgap of H-ZZ-36 and F-ZZ-36 are not much. This leads to similarities in the obtained absorption spectra. The narrowing of the bandgap of F-ZZ-36 compared with H-ZZ-36 leads to a shift in the absorption peak in the lower-energy region. The bandgap is very narrow in the structure with a P edge atom, and some spectral peaks appear in the red-light region or in its vicinity. This result is similar to that reported by several studies.[27,28] For the O-ZZ-36 structure, the semiconductor characteristics are most clearly demonstrated by the bandgap of 1.428 eV. The band structure shows that, in the valence band, the first subband lies in the 0.7–0.8 eV range, and the next is about 2 eV (in the neighborhood of the Fermi level). The two spectral peaks observed on Oy and Oz correspond to the transition from these two energy levels to the conduction band.

13.5. Conclusion

In summary, we have applied DFT calculations to systematically investigate the effect of chemical modification on the structural stability and optoelectronic properties of PGQDs. We found that larger PGQDs are thermodynamically more stable than the smaller ones. For the same system sizes, in the order of increasing absolute values of formation energies, we can arrange the passivated PGQDs as follows: P-ZZ-36, F-ZZ-36, Si-ZZ-36, and O-ZZ-36. For single-doped PGQDs, there appears to be a subband that coincides with the Fermi level. This result implies a transition from semiconductor to metal, compared to the pristine structure. In contrast, B and P co-doped PGQDs exhibit semiconductor properties. Specifically, their bandgaps are 1.50 and 1.14 eV, respectively. Tailoring the optoelectronic properties of semiconductor QDs is essential to designing functionalized nanoscale devices.

References

[1] Li, X.; Rui, M.; Song, J.; Shen, Z.; Zeng, H. Carbon and graphene quantum dots for optoelectronic and energy devices: A review. *Advanced Functional Materials* **2015**, *25* (31), 4929–4947.

[2] Li, M.; Chen, T.; Gooding, J. J.; Liu, J. Review of carbon and graphene quantum dots for sensing. ACS Sens. **2019**, *4* (7), 1732–1748.

[3] Prabhu, S. A.; Kavithayeni, V.; Suganthy, R.; Geetha, K. Graphene quantum dots synthesis and energy application: A review. *Carbon Letters* **2021**, *31* (1) 1–12.

[4] Espinosa-Ortega, T.; Luk'yanchuk, I. A.; Rubo, Y. G. Magnetic properties of graphene quantum dots. *Physical Review B* **2013**, *87* (20), 205434.

[5] Liu, H.; Qin, G.; Lin, Y.; Hu, M. Disparate strain dependent thermal conductivity of two-dimensional penta-structures. *Nano Letters* **2016**, *16* (6), 3831–3842.

[6] Krishnan, R.; Su, W. S.; Chen, H. T. A new carbon allotrope: Penta-graphene as a metal-free catalyst for CO oxidation. *Carbon* **2017**, *114*, 465–472.

[7] Nazir, M. A.; Hassan, A.; Shen, Y.; Wang, Q. Research progress on penta-graphene and its related materials: Properties and applications. *Nano Today* **2022**, *44*, 101501.

[8] Li, X.; Zhang, S.; Wang, F. Q.; Guo, Y.; Liu, J.; Wang, Q. Tuning the electronic and mechanical properties of penta-graphene via hydrogenation and fluorination. *Physical Chemistry Chemical Physics* **2016**, *18* (21), 14191–14197.

[9] Quijano-Briones, J. J.; Fernandez-Escamilla, H. N.; Tlahuice-Flores, A. Doped penta-graphene and hydrogenation of its related structures: A structural and electronic DFT-D study. *Physical Chemistry Chemical Physics* **2016**, *18*, 15505–15509.

[10] Tien, N. T.; Thao, P. T. B.; Phuc, V. T.; Ahuja, R. Electronic and transport features of sawtooth penta-graphene nanoribbons via substitutional doping. *Physica E: Low-dimensional Systems and Nanostructures* **2019**, *114*, 113572.

[11] Tien, N. T.; Thao, P. T. B.; Phuc, V. T.; Ahuja, R. Influence of edge termination on the electronic and transport properties of sawtooth penta-graphene nanoribbons. *Journal of Physics and Chemistry of Solids* **2020**, *146*, 109528.

[12] Dang, M. T.; Thao, P. T. B.; Thao, T. T. N.; Tien, N. T. First-principles study of electronic and optical properties of small edge-functionalized penta-graphene quantum dots. *AIP Advances* **2022**, *12* (6), 065008.

[13] Li, P.; Zhang, J.; Zhu, C.; Shen, W.; Hu, C.; Fu, W.; Yan, L.; Zhou, L.; Zheng, L.; Lei, H.; Liu, Z. Penta-PdPSe: A new 2D pentagonal material with highly in-plane optical, electronic, and optoelectronic anisotropy. *Advanced Materials* **2021**, *33* (35), 2102541.

[14] Clark, S.J.; Segall, M. D.; Pickard, C. J.; Hasnip, P. J.; Probert, M. I.; Refson, K.; Payne, M. C. First-principles methods using CASTEP. *Zeitschrift für Kristallographie – Crystalline Materials* **2005**, *220* (5–6), 567–570.

[15] Perdew, J. P.; Burke, K.; Ernzerhof, M. Generalized gradient approximation made simple. *Physical Review Letters* **1996**, *77* (18), 3865.

[16] Schlipf, M.; Gygi, F.; Optimization algorithm for the generation of ONCV pseudopotentials. *Computer Physics Communications* **2015**, *196*, 36–44.

[17] A. Dal Corso, A.M. Conte, Spin-orbit coupling with ultrasoft pseudopotentials: Application to au and Pt. *Physical Review B* *71* (11) (2005) 115106.

[18] Monshi, M. M.; Aghaei, S. M.; Calizo, I. Band gap opening and optical absorption enhancement in graphene using ZnO nanocluster. *Physics Letters A* **2018**, *382* (17), 1171–1175.

[19] Derbenyova, N. V.; Konakov, A. A.; Shvetsov, A. E.; Burdov, V. A. Electronic structure and absorption spectra of silicon nanocrystals with a halogen (Br, Cl) coating. *JETP Letters* **2017**, *106* (4), 247–251.

[20] Cao, L.; Li, X.; Li, Y.; Zhou, G. Electrical properties and spintronic application of carbon phosphide nanoribbons with edge functionalization. *Journal of Materials Chemistry C* **2020**, *8* (27), 9313–9321.

[21] Shahrokhi, M. Tuning the band gap and optical spectra of monolayer penta-graphene under in-plane biaxial strains. *Optik* **2017**, *136*, 205–214.

[22] Tien, N. T.; Thao, P. T. B.; Thuan, L. V. P.; Chuong, D. H. First-principles study of electronic and optical properties of defective saw-tooth penta-graphene nanoribbons. *Computational Materials Science* **2022**, *203*, 111065.

[23] Feng, J.; Guo, Q.; Liu, H.; Chen, D.; Tian, Z.; Xia, F.; Ma, S.; Yu, L.; Dong, L. Theoretical insights into tunable optical and electronic properties of graphene quantum dots through phosphorization. *Carbon* **2019**, *155*, 491–498.

[24] Feng, J.; Dong, H.; Pang, B.; Chen, Y.; Yu, L.; Dong, L. Tuning the electronic and optical properties of graphene quantum dots by selective boronization. *Journal of Materials Chemistry C* **2019**, *7* (2), 237–246.

[25] Dai, X. S.; Shen, T.; Feng, Y.; Liu, H. C. Structure, electronic and optical properties of Al, Si, P doped penta-graphene: A first-principles study. *Physica B: Condensed Matter* **2019**, *574*, 411660.

[26] Li, Y.; Shu, H.; Niu, X.; Wang, J. Electronic and optical properties of edge-functionalized graphene quantum dots and the underlying mechanism. *The Journal of Physical Chemistry C* **2015**, *119* (44), 24950–24957.

[27] Feng, J.; Dong, H.; Pang, B.; Chen, Y.; Yu, L.; Dong, L. Tuning the electronic and optical properties of graphene quantum dots by selective boronization. *Journal of Materials Chemistry C* **2019**, *7* (2), 237–246.

[28] Feng, J.; Guo, Q.; Liu, H.; Chen, D.; Tian, Z.; Xia, F.; Dong, L. Theoretical insights into tunable optical and electronic properties of graphene quantum dots through phosphorization. *Carbon* **2019**, *155*, 491–498.

CHAPTER 14

GRAPHENE QUANTUM DOTS: POSSIBLE STRUCTURE, APPLICATION, AND EFFECT OF OXYGEN-CONTAINING FUNCTIONAL GROUP

Nguyen Thi Ngoc Anh

Vinh Long University of Technology and Education, Vinh Long Province, Vietnam

Abstract

Graphene quantum dot (GQD), a zero-dimensional material with size less than 10 nm, is a newly developed member of the graphene family. GQDs have become highly promising materials in various applications, including catalysis, bioimaging, energy conversion, and storage, owing to their unique physicochemical properties, such as excellent photoluminescence, biocompatibility, low toxicity, easy fabrication, and edge effect. Additionally, oxygen-containing functional groups such as −COOH, −COC-, −OH, −CHO, and $-OCH_3$ play a crucial role in the structure of GQDs. Oxygen is located in different edges and the basal plane of GQDs. The unique architecture of oxygen inside the structure of GQDs and the outstanding properties and performance of GQDs provide a powerful impetus to use GQDs as a promising material with applicability in various fields. In this chapter, we describe the possible structure, application, and critical role of the oxygen-containing functional group on the essential properties and application of GQDs. Finally, we provide a brief outlook to point out the issues that need to settle for further development.

Keywords: GQDs, functional group, oxygen, properties.

14.1. Introduction

Carbon-based nanomaterials have attracted the attention of the scientific community. Geim and Novoslov were awarded the 2010 Nobel prize for their research on graphene. Tian *et al.*[1] summarized the entire development of carbon since the 19th century to date (Fig. 14.1). During the last century, there have been many efforts to evaluate the contribution of carbon materials, including graphene, fullerene, carbon nanotubes (CNTs), graphite, graphene oxide (GO), reduced graphene oxide (rGO), carbon dots (CDs), graphene quantum dots (GQDs), doped GQDs, graphene oxide quantum dots (GOQDs), and reduced graphene oxide quantum dots (rGOQDs).[1] Graphene, fullerene, CNT, GO, CDs, and GQDs are all important resource materials for various important applications.

Fig. 14.1. History of carbon material development.[1]

GQDs are zero-dimensional carbon-based nanomaterials, which have recently emerged as excellent fluorescent nanomaterials for environmental and biological applications because of their unique properties, such as excellent photoluminescence, tunable bandgap, good stability, low toxicity, and easy preparation. Moreover, they show outstanding properties, compared with other carbon materials, such as the capacity to disperse and solubilize in different solvents. Figure 14.2 shows the increase in the number of yearly publications related to the numerous, interesting properties of GQDs.[2] The important morphology and structure of GQDs are based on their method of fabrication.

Actually, most of the structure of GQDs contains three main elements: C, H, and O. The ratio of these three elements differ on the basis of the chemical, the method and the precursor used in the fabrication. The GQDs also have two types of edges, armchair and zigzag,[3] which affect their shape, and electronic and optical properties (Fig. 14.3).[4–6]

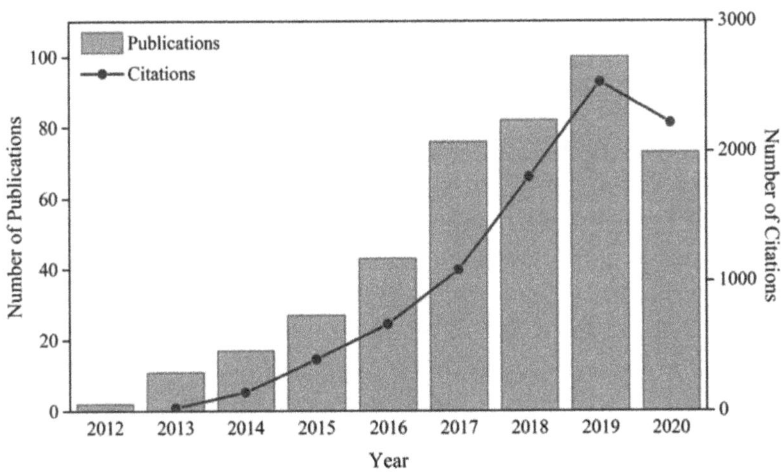

Fig. 14.2. Number of scientific publications and citations per year related to GQD composites. Data obtained from Web of Science accessed on September 2020 using the words "GQDs composites".[2]

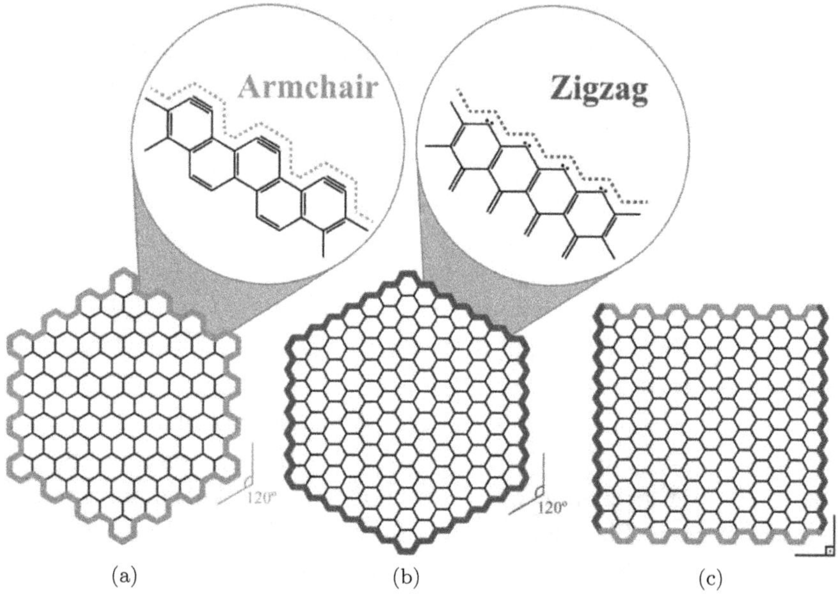

Fig. 14.3. The shapes of GQDs depending on the type of edge: (a) armchair, (b) zigzag, and (c) hybrid armchair–zigzag GQDs. The amplified images in (a) and (b) show the structural details of the armchair and zigzag edges, respectively.[3]

14.2. Doped GQDs

It has been shown that GQDs have some limitations as a result of the low quantum yield due to their zero bandgap. Doping with heteroatom will endow GQDs tuning their properties, including photoluminescence (PL) and bandgap. Among the heteroatom, nitrogen (N) or sulfur (S) is an attractive dopant due to the tuning of the physical and chemical properties of GQDs. There are many studies that have successfully used N atom-doped GQDs. Compared with other dopant elements, N is a strongly coupled interface and imparts a relatively high positive-charge density to the adjacent C atoms in GQDs.[7–14] To the best of our knowledge, there are only limited works that have reported using sulfur-doped GQDs. The main reason to let sulfur hardly combine inside the framework of GQDs is that S has an electronegativity of 2.58, larger than C, 2.55, and S atom is much larger than C. Compared with other heteroatoms, sulfur (S) atom is markedly larger than carbon (C) atom. Thus, difficulty arises from the chemical doping of S atom into the framework of GQDs. Li *et al.* prepared S-GQDs through a hydrothermal approach with fructose and sulfuric acid as source materials and proved that incorporation of ~1% of S atom into QDs could effectively modify the electronic structure of S-GQDs by introducing S-related energy levels between π and π^* of C.[15] However, S-doped GQDs can significantly enhance the fluorescence quantum yield of GQDs. In contrast, S has attractive interaction with some elements, which can create more active sites.[16–22]

In addition, heteroatom doping may also provide a solution because the GQDs are likely to process tailored or new photoluminescence properties with a heterogeneously hybridized carbon network. Taking advantage of N and S atom doping in carbon lattice, hetero-co-doping GQDs also have a great contribution to enhancing fluorescence quantum yield and improving the efficiency of different applications (Fig. 14.4).[23] Plenty of works have successfully fabricated nitrogen, sulfur co-doped graphene quantum dots (N,S-GQDs) using various techniques to improve the efficiency of different applications (Table 14.1).[24–30] Other chemicals and heteroatom dopants, such as B, P, F, Cl, Na, Nd, Tm, Er atoms, have also been investigated and can amplify the photoluminescence and emission energy behavior.[31–36]

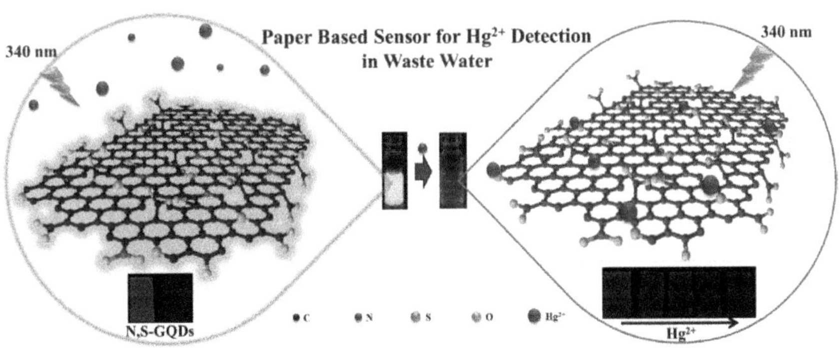

Fig. 14.4. Schematic diagram of Hg^{2+} detection using turn-off fluorescence property of N,S-GQDs and paper-based sensor images in real wastewater samples.[23]

Table 14.1. List of reports to fabricate doping GQDs and their applications.

Precursor	Doping	Method	Percentage doping	FLQY	Application	References
1,3,6-trinitropyrene 3-Mercaptopropionic acid		Hydrothermal	3.5	9.2	Ag^{+} detection	17
1,3,6-trinitropyrene Na_2S		Hydrothermal 200°C 10 h	1.3	11.6	Pb^{2+} detection	19
Graphite Sodium *p*-toluenesulfonate		Electrolysis 3 h	4.25	—	Fe^{3+} detection	21
GO $ZnSO_4.7\ H_2O$	Sulfur	Hydrothermal	5.95	—	Fluorescence	22
GQDs Sulfur powder		Hydrothermal 180°C 12 h	19.53	—	Fluorescence	18
Citric acid NaHS		Hydrothermal 170°C 24 h	1.92	—	Hydrogen regeneration	20
Durian Platinum catalyst		Hydrothermal 80–140°C	3.2	79	Bioimage	37
GO NH_3		Hydrothermal 500°C	2.9	—	Water splitting	7
Lemon juice		Pyrolysis 200°C 20 min	2.63	—	2,4,6-trinitrophenol detection	14
Citric acid Amonia		Hydrothermal 100°C 1 h	—	—	Nitroaniline detection	9
Citric acid Urea	Nitrogen	Autoclave 160°C 4 h	13.9	—	Oxygen reduction	11
Carbon fiber Acrylonitrile		Hydrothermal 2880°C	—	—	Solar cell	13
Citric acid Amonia		Hydrothermal 200°C 3 h	—	—	Hg^{2+} and Trypsin detection	10

(*Continued*)

Table 14.1. (*Continued*)

Precursor	Doping	Method	Percentage doping	FLQY	Application	References
Orange juice		Pyrolysis 200°C 40 min			TNP detection	8
Citric acid Ethylenediamine		Hydrothermal 140–190°C, 5–8 h	14.5	75.2	Photoluminescence	12
Citric acid L-cysteine		Pyrolysis 200°C 5 min	3.91%N 4.7%N	74.5	Fe^{3+} detection	24
GO Thiourea		Hydrothermal 200°C 15 h	—	—	TNP detection	25
Citric acid L-cysteine		Pyrolysis 240°C 12 min	3.31%N 3.42%S	67	Cyanine detection	26
GO Amonia	N,S-co-doped	Hydrothermal 180°C 12 h	—	18.6	Photoluminescence	27
Polythyophene HNO_3		Hydrothermal 150–180°C 1 h	3.7%N 8.6%S	2.35	Fe^{3+} detection	28
Glusosamine-HCl Thiourea		Hydrothermal 40 min 450 W	—	22	Photoelectro- luminescence	29
Citric acid Thiourea		Hydrothermal 160°C 4 h	—	71	Photocatalytic	30
Citric acid Thiourea		Hydrothermal 160°C 4 h	8%N 1.5%S	40	Hg^{2+} detection	23
Graphite Sodium Phytate	P	Electrochemical	7	—	Radical scavenging	32
BGO HNO_3	B	Hydrothermal 300°C 24 h	3.45	—	Glucose sensing	31
GQDs HCl, NH_4OH, BH_3O_3, KOH	Chemical and dopants	Hydrothermal 85°C 10 min	2%Na 4%K 1–2%B	—	Tuning emission	34

14.3. GQD-based Gold Nanocomposite

Nanocomposite is defined as materials including at least one constituent less than 100 nm. Composite materials are now promising materials in nanotechnology that help overcome the limitation of monoclinic materials in various applications. Particularly, GQD-based gold nanocomposites are potential materials due to their broad range of applications (Table 14.2). Until now, there are not many studies that successfully fabricated GQD-based gold nanocomposite. Zhu *et al.* prepared Au-PDA-GQD for the detection of enzymes in H_2O_2.[38] In addition, Hai *et al.* successfully fabricated r-GQD-coated AuNPs for highly sensitive and selective detection of cysteine[39] Ting *et al.* found that GQDs functionalized gold nanoparticles to detect metal ions.[40] Qin *et al.* reported novel core–shell-like Au_x/C_6/GQDs by using the hydrothermal method.[41] Moreover, Ju *et al.* successfully fabricated N-GQDs-AuNPs by using a simple method without adding any reductant agent.[42] Anh *et al.* prepared gold@S-GQD nanocomposite by simply mixing S-GQDs with $AuCl^-$ at room temperature,[43] thus reporting an open new strategy to fabricate nanocomposite by using carbon source (Fig. 14.5).

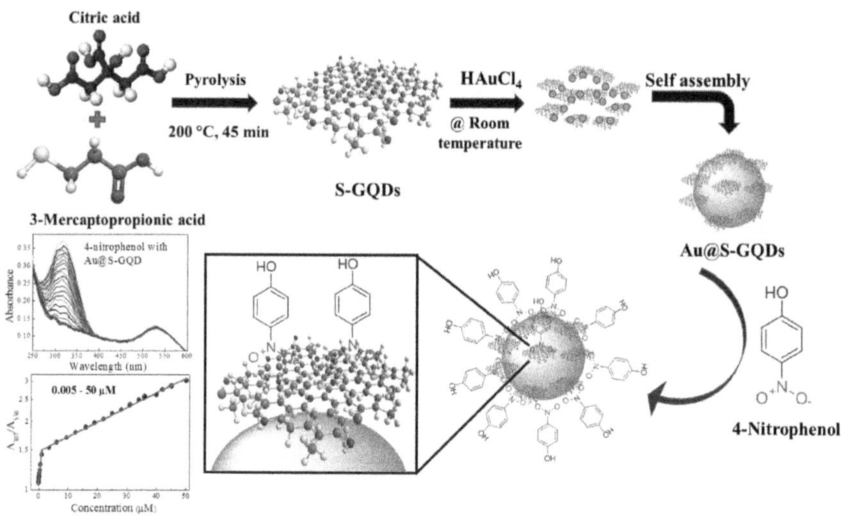

Fig. 14.5. Schematic diagram of optical detection of 4-nitrophenol using Au@S-GQD nanocomposites as the sensing probe.[43]

Table 14.2. Fabrication and applications of GQDs-AuNPs nanocomposite.

Precursor	Nanocomcosite	Reduction agent	Time and Temperature	Application	Size and shape	References
$AuCl^-$	N-GQDs/ AuNPs	N-GQDs	100°C 30 min	H_2O_2 detection	5.2 nm, spherical	42
Cysteamine $NaBH_3$ EDC, $AuCl^-$	GQDs-AuNPs	$NaBH_3$	dark, RT 1.5 h	Hg^{2+} Cu^{2+} detection	50 nm, Spherical	40
ITO AuNPs Dopamine GQDs	Au-DPA-GQD	–	RT 1 h	H_2O_2 detection	35 nm, spherical	38
Sodium citrate $AuCl^-$ G_6	$Au_x/G_6/GQD$	Sodium citrate	Heating 1.5 h	SERS	32–107 nm, Spherical	41
$AuCl^-$ r-GQDs	AuNPs@r-GQDs	r-GQDs	Boiling, 40 min	Cysteine detection	18–21 nm, Spherical	39
S-GQDs $AuCl^-$	Au@S-GQDs	S-GQDs	Room Temperature	4-Nitrophenol detection	5–17 nm, Spherical	43

14.4. Synthesis

Successful methods have been used to fabricate GQDs with tunable sizes and shapes. These can be divided into two methods: top-down and bottom-up approaches. Top-down nano-cutting methods involve neither physical nor chemical disruption of bulk carbon materials. Conversely, bottom-up methods mainly involve the carbonization of organic carbon sources from small molecules.

Multiple bulk carbon resources were used as raw materials, including GO,[44,45] graphene,[46] coal,[47] graphite,[48] and multi-walled CNTs (MWCNTs),[49] to synthesize GQDs (Fig. 14.6). The most common chemicals used as scissors, including HNO_3 or H_2SO_4/HNO_3 mixture, have been utilized to directly cut graphene sheet. The limitations of GQDs using the top-down technique include low QY and non-homogeneous size.[3] Top-down methods normally included multiple steps via solvothermal and hydrothermal oxidation. In this method, GQDs are used to obtain oxygen-containing functional groups, such as hydroxy, epoxy, carbonyl, and ether groups.[5]

Bottom-up methods have been used to synthesize homogenous GQDs using small organic molecules, such as citric acid, glucose, and amino acid. The bonding contained in the organic carbon sources such as –OH, –COOH, –C=O can dehydrate due to the

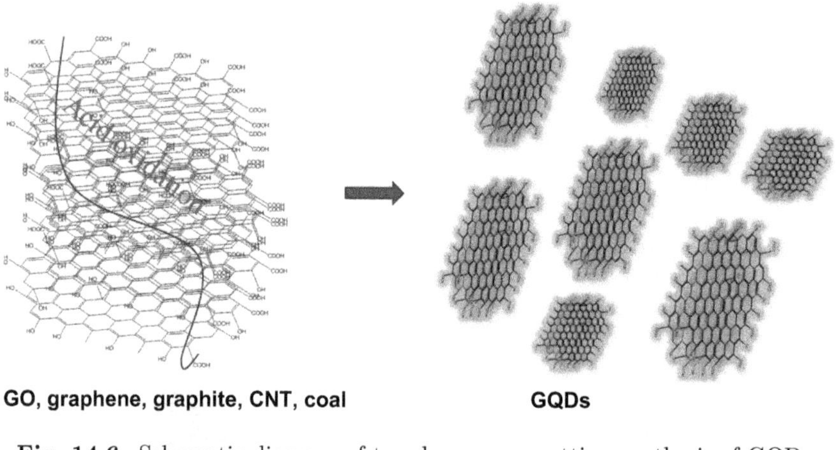

Fig. 14.6. Schematic diagram of top-down nano-cutting synthesis of GQDs.

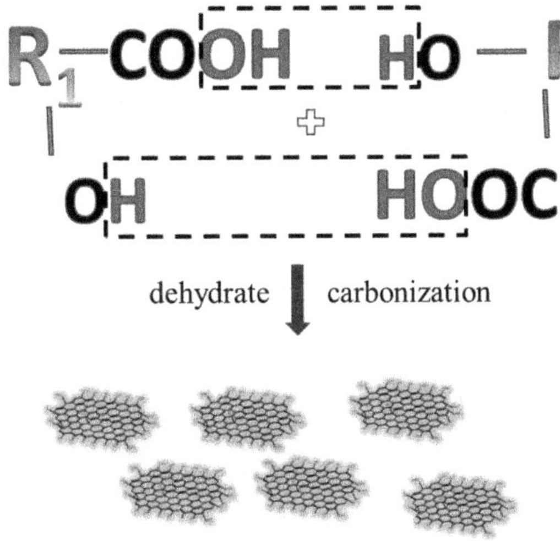

Fig. 14.7. Scheme of bottom-up synthesis GQDs from organic molecules.

elevated temperature (Fig. 14.7). The homogeneous size and surface modification of GQDs can be easily controlled by carefully designing precursors. Several methods have been developed to synthesize GQDs using bottom-up routes, including solution chemistry and carbonization.[3,50,51] One of the limitations of bottom-up methods is the low crystallization of GQDs. Until now, there have been several reports demonstrating the successful fabrication of GQDs with high crystallization using green carbon source and bottom-up technique. There is still a big challenge to investigate the intrinsic structure with different levels of crystallization of GQDs using bottom-up methods related to their application.

14.5. Effect of Oxygen-Containing Functional Group on the Properties and Applications

The presence of oxygen functional groups plays an important role in the structure and properties of GQDs. Fuhrer *et al.* demonstrated the outstanding properties of GQDs, such as lower toxicity, good biocompatibility, excellent physical properties, chemical properties, and optical properties.[52] One of the most important properties of GQDs is optical property, especially fluorescence property, in terms of mechanism explanation based on fluorescence. The intrinsic properties, including the size and shape of GQDs, are also important in influencing the photoluminescence of GQDs. Typically, GQDs show strong optical absorption in the UV region from 230 to 320 nm. In the aromatic structure of C–C bonds, a maximum peak at 230 nm is attributed to the π–π^* transition, while a shoulder peak at 300 nm is typically ascribed to the n–π^* transition of C=O bonds. The fluorescence (FL) properties of GQDs and the emission spectra of GQDs are roughly symmetrical on the excitation wavelength scale related to different sizes and surface groups.[23,46,53–56]

One of the important contributions of oxygen to the edges, interfaces, and surface of GQDs is that they show green PL emissions.[57,58] Besides, for the sensor field, oxygen-containing functional groups allow GQDs to attach to the targets and also improve their stability. Interestingly, GQDs with the armchair edge will have the bigger bandgap resulting in a blue shift in the PL emission.[3,59] Pan *et al.*[60] used a hydrothermal approach to successfully synthesize GQDs with less epoxy group and carbonyl group from graphene sheets resulting in the blue luminescence of GQDs. Another study shows that the oxygen-containing functional groups on the carbon lattice of S-GQDs will obtain strong PL properties and can decorate TiO_2 for catalytic application.[61] Li *et al.* aimed to achieve this with the oxygen-rich structure of N-GQDs that allows the blue shift in the PL emission of N-GQDs (Fig. 14.8).[62] The XPS and FTIR are important tools used to identify the composition and functional group in the framework of GQDs.

Additionally, Feng *et al.*[63] used density functional theory (DFT) to determine the oxygen-containing groups on the edges of GQDs, including –OH, –COOH, OCH_3, –CHO, and –COC, resulting in the different bandgaps corresponding to 2.34, 2.32, 2.31, 2.30, 2.27, and 2.15 eV. The results show that based on the HOMO–LUMO gap, (–COC–) plays an essential role in the electronic structure compared to other oxygen-containing functional groups (Fig. 14.9).[63] Quite a lot of research have successfully synthesized GQDs with a desirable functional group of carboxyl (–COOH), epoxy (–O–), and carbonyl

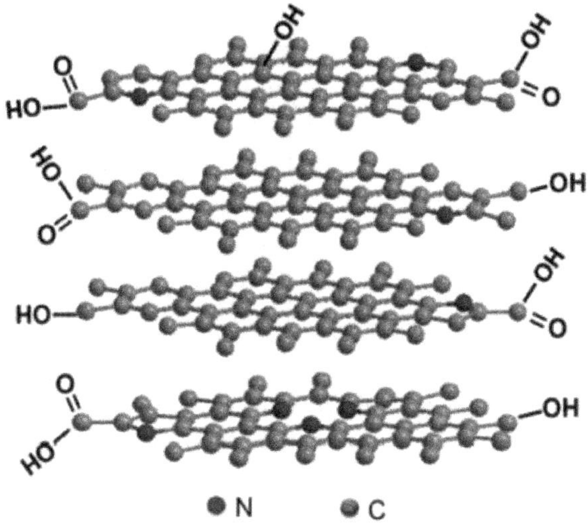

Fig. 14.8. Possible structure of oxygen-rich N-GQDs.[62]

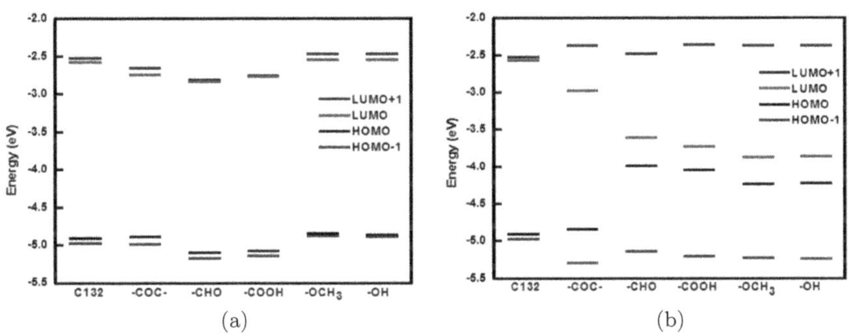

Fig. 14.9. Energy diagrams of GQDs with different oxygen-containing groups: (a) edge-functionalized GQDs and (b) surface-functionalized GQDs. The predicted energy levels corresponding to HOMO-1, HOMO, LUMO and LUMO+1 are shown for all GQDs.[63]

(–(C=O)–), which is able to tune the fluorescence of GQDs.[64–66] Based on DFT results, oxygen functional groups reveal that the epoxy group plays an important role in the distribution of the optical mechanism of GQDs than others. These results provide a new design for further desired applications.[63]

In most situations, the surface groups of GQDs result in the different colors of GQDs, from blue to green, rather than affecting the sizes of GQDs. Fluorescence quantum yield (FLQY) is defined as the number of emitted photons. Normally, GQDs possessed low FLQY in previous reports.[51, 67–69] Bear in mind that the fluorescence properties of GQDs are the most important in sensing applications and belonging to the bandgap. The mixture of sp^2 and sp^3carbon atom after dehydrating results in the bandgap. The electronic energy transitions lead to the formation of fluorescence. The fluorescence of GQDs can be tuned from visible light to near-infrared (IR) based on the varying ratios of sp^2/sp^3.[70] After surface modification or passivation, the FLQY can be dramatically increased. The enhanced FL properties are attributed to the strong FL centers on the surface, the synergy between the carbon core and the chemical groups, or solely to the presence of fluorophores.[71] The carbon network in the GQD structure arranged with oxygen-containing groups (–COOH, –OH, –C–O–C–, C=O) at their edge results in strong FL.

On the other hand, some researchers have found a special structure with more oxygen-containing groups in the functional groups of GQDs like graphene oxide quantum dots (GOQDs), which contain thin sheets of GO and a size less than 10 nm.[72] GOQDs can use for various applications in many fields: bioimage, drug delivery, and tissue engineering.[72] In 2020, Ebrahim *et al.* successfully fabricated GOQDs by using citric acid and ascorbic acid and claimed that oxygen-containing functional groups of GQDs have the potential to inhibit the fibrillation process under *in vitro* conditions. The results have demonstrated that oxygen-containing functional groups of GOQDs and rGOQDs have the capacity to inhibit fibrillation and to promote protein assembly. What is more, GOQDs and rGOQDs also can change the secondary structure of bovine insulin and is not toxic to SHSY5Y cells at a concentration of 10 μM.[73]

This approach lays the foundation for research to continue to find more evidence to apply GOQDs in biosensing, bioimage, and even drug delivery (Fig. 14.10).

Significantly, Liu *et al.*[74] reported that graphite could be used to synthesize GO and GOQDs through the modified Humer method with high quantum yields and applied for the detection of Eu^{3+} and Tb^{3+} due to –COOH and –OH groups in the structure of GOQDs (Fig. 14.11). Moreover, oxygen contributes to the high luminescence yields of GOQDs (about 8.8%).

Interestingly, Park *et al.*[75] reported GQDs-S/CB composites for the Li–S battery for the first time. The result shows that GQDs-S/CB with rich oxygen-containing functional groups can enhance the charge transfer resulting in excellent cycling and rate performance in the physicochemical properties of GQDs (Fig. 14.12).

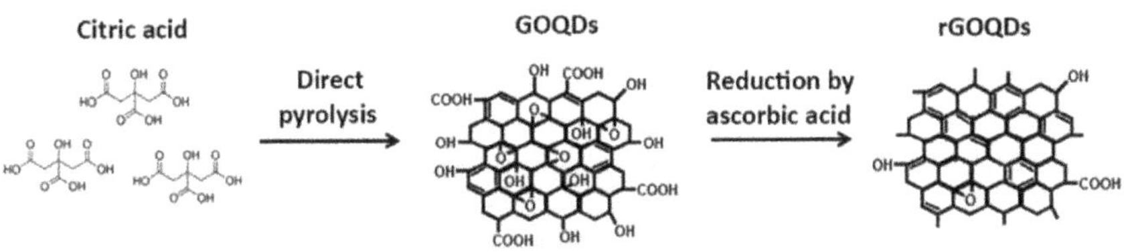

Fig. 14.10. Schematic presentation of GOQDs synthesis and reduction.[73]

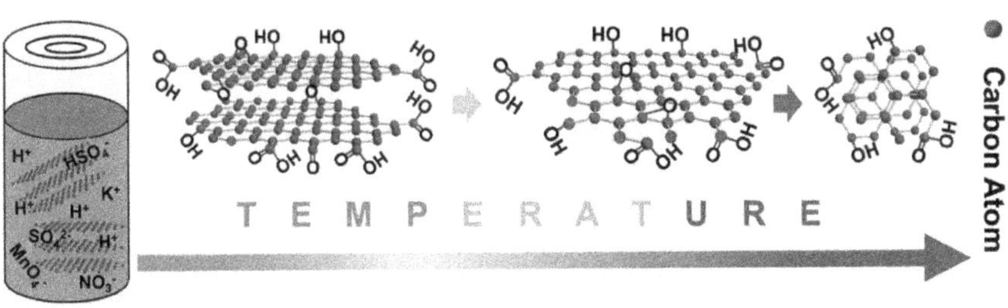

Fig. 14.11. Schematic representation of the solvothermal reaction route, which involves exfoliating graphite to GO and further cutting GO layers to GOQDs at given experimental conditions.[74]

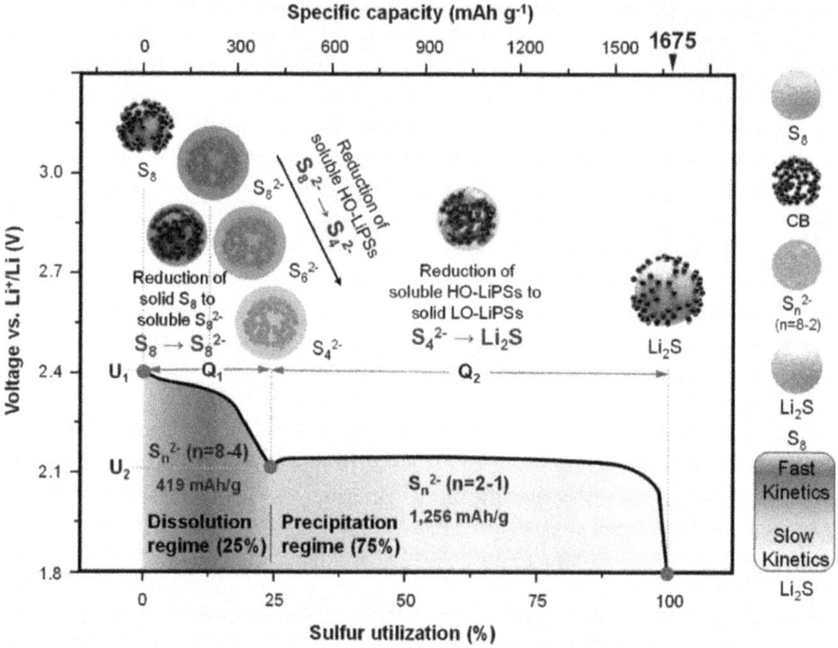

Fig. 14.12. Electrochemical properties of S/CB and GQDs-S/CB electrodes: (a) schematic illustration of the discharge profile of a conventional Li–S battery.[75]

14.6. Conclusion and Outlook

In summary, as a new member of carbon materials, GQDs have attracted great attention from scientists due to their outstanding properties. Various methods have been developed to determine the performance of GQDs, doped GQDs, and GQD composites. Up to now, limited studies have revealed the mechanism and contribution of these elements in the structure of GQDs. Only a few results show that the oxygen functional group induces the PL of GQDs and some applications. Additionally, researchers still struggle to synthesize GQDs with a single layer and high quantum yield. Future work needs to focus on controlling the GQD structure for specific applications and enhancing their properties in different fields. Moreover, there is still high demand for research into the contribution, distribution, and binding of these elements in the structure of GQDs.

References

[1] Tian, P.; Tang, L.; Teng, K.; Lau, S., Graphene quantum dots from chemistry to applications. *Materials Today Chemistry* **2018**, *10*, 221–258.

[2] Facure, M. H.; Schneider, R.; Mercante, L. A.; Correa, D. S., A review on graphene quantum dots and their nanocomposites: From laboratory synthesis towards agricultural and environmental applications. *Environmental Science: Nano* **2020**, *7* (12), 3710–3734.

[3] Zhu, S.; Song, Y.; Zhao, X.; Shao, J.; Zhang, J.; Yang, B., The photoluminescence mechanism in carbon dots (graphene quantum dots, carbon nanodots, and polymer dots): Current state and future perspective. *Nano Research* **2015**, *8* (2), 355–381.

[4] Kundu, S.; Pillai, V. K., Synthesis and characterization of graphene quantum dots. *Physical Sciences Reviews* **2020**, *5* (4).

[5] Li, L.; Wu, G.; Yang, G.; Peng, J.; Zhao, J.; Zhu, J.-J., Focusing on luminescent graphene quantum dots: Current status and future perspectives. *Nanoscale* **2013**, *5* (10), 4015–4039.

[6] WooáLee, M.; SangáSuh, J., Characteristics of graphene quantum dots determined by edge structures: Three kinds of dots fabricated using thermal plasma jet. *RSC Advances* **2015**, *5* (83), 67669–67675.

[7] Yeh, T. F.; Teng, C. Y.; Chen, S. J.; Teng, H., Nitrogen-doped graphene oxide quantum dots as photocatalysts for overall water-splitting under visible light Illumination. *Advanced Materials* **2014**, *26* (20), 3297–3303.

[8] Kaur, M.; Mehta, S. K.; Kansal, S. K., A fluorescent probe based on nitrogen doped graphene quantum dots for turn off sensing of explosive and detrimental water pollutant, TNP in aqueous medium. *Spectrochimica Acta Part A: Molecular and Biomolecular Spectroscopy* **2017**, *180*, 37–43.

[9] Chen, S.; Chen, X.; Xia, T.; Ma, Q., A novel electrochemiluminescence sensor for the detection of nitroaniline based on the nitrogen-doped graphene quantum dots. *Biosensors and Bioelectronics* **2016**, *85*, 903–908.

[10] Su, D.; Wang, M.; Liu, Q.; Qu, Z.; Su, X., A novel fluorescence strategy for mercury ions and trypsin activity assay based on nitrogen-doped graphene quantum dots. *New Journal of Chemistry* **2018**.

[11] Fan, M.; Zhu, C.; Yang, J.; Sun, D., Facile self-assembly N-doped graphene quantum dots/graphene for oxygen reduction reaction. *Electrochimica Acta* **2016**, *216*, 102–109.

[12] Gu, J.; Zhang, X.; Pang, A.; Yang, J., Facile synthesis and photoluminescence characteristics of blue-emitting nitrogen-doped graphene quantum dots. *Nanotechnology* **2016**, *27* (16), 165704.

[13] Moon, B. J.; Jang, D.; Yi, Y.; Lee, H.; Kim, S. J.; Oh, Y.; Lee, S. H.; Park, M.; Lee, S.; Bae, S., Multi-functional nitrogen self-doped graphene quantum dots for boosting the photovoltaic performance of BHJ solar cells. *Nano Energy* **2017**, *34*, 36–46.

[14] Kaur, M.; Mehta, S. K.; Kansal, S. K., Nitrogen doped graphene quantum dots: Efficient fluorescent chemosensor for the selective and sensitive detection of 2, 4, 6-trinitrophenol. *Sensors and Actuators B: Chemical* **2017**, *245*, 938–945.

[15] Li, X.; Lau, S. P.; Tang, L.; Ji, R.; Yang, P., Sulphur doping: A facile approach to tune the electronic structure and optical properties of graphene quantum dots. *Nanoscale* **2014**, *6* (10), 5323–5328.

[16] Wang, G.; Guo, Q.; Chen, D.; Liu, Z.; Zheng, X.; Xu, A.; Yang, S.; Ding, G., Facile and highly effective synthesis of controllable lattice sulfur-doped graphene quantum dots via hydrothermal treatment of durian. *ACS Applied Materials & Interfaces* **2018**.

[17] Bian, S.; Shen, C.; Qian, Y.; Liu, J.; Xi, F.; Dong, X., Facile synthesis of sulfur-doped graphene quantum dots as fluorescent sensing probes for Ag^{+} ions detection. *Sensors and Actuators B: Chemical* **2017**, *242*, 231–237.

[18] Luo, Y.; Li, M.; Sun, L.; Xu, Y.; Hu, G.; Tang, T.; Wen, J.; Li, X.; Zhang, J.; Wang, L., High fluorescent sulfur regulating graphene quantum dots with tunable photoluminescence properties. *Journal of Colloid and Interface Science* **2018**.

[19] Bian, S.; Shen, C.; Hua, H.; Zhou, L.; Zhu, H.; Xi, F.; Liu, J.; Dong, X., One-pot synthesis of sulfur-doped graphene quantum dots as a novel fluorescent probe for highly selective and sensitive detection of lead (II). *Rsc Advances* **2016**, *6* (74), 69977–69983.

[20] Gliniak, J.; Lin, J. H.; Chen, Y. T.; Li, C. R.; Jokar, E.; Chang, C. H.; Peng, C. S.; Lin, J. N.; Lien, W. H.; Tsai, H. M., Sulfur-doped graphene oxide quantum dots as photocatalysts for hydrogen generation in the aqueous phase. *ChemSusChem* **2017**, *10* (16), 3260–3267.

[21] Li, S.; Li, Y.; Cao, J.; Zhu, J.; Fan, L.; Li, X., Sulfur-doped graphene quantum dots as a novel fluorescent probe for highly selective and sensitive detection of Fe^{3+}. *Analytical Chemistry* **2014**, *86* (20), 10201–10207.

[22] Kharangarh, P. R.; Umapathy, S.; Singh, G., Thermal effect of sulfur doping for luminescent graphene quantum dots. *ECS Journal of Solid State Science and Technology* **2018**, *7* (3), M29–M34.

[23] Anh, N. T. N.; Chowdhury, A. D.; Doong, R.-a., Highly sensitive and selective detection of mercury ions using N,S-codoped graphene quantum dots and its paper strip based sensing application in wastewater. *Sensors and Actuators B: Chemical* **2017**, *252*, 1169–1178.

[24] Xia, C.; Hai, X.; Chen, X.-W.; Wang, J.-H., Simultaneously fabrication of free and solidified N,S-doped graphene quantum dots via a facile solvent-free synthesis route for fluorescent detection. *Talanta* **2017**, *168*, 269–278.

[25] Mondal, T. K.; Dinda, D.; Saha, S. K., Nitrogen, sulphur co-doped graphene quantum dot: An excellent sensor for nitroexplosives. *Sensors and Actuators B: Chemical* **2018**, *257*, 586–593.

[26] Chen, C.; Zhao, D.; Hu, T.; Sun, J.; Yang, X., Highly fluorescent nitrogen and sulfur co-doped graphene quantum dots for an inner filter effect-based cyanide sensor. *Sensors and Actuators B: Chemical* **2017**, *241*, 779–788.

[27] Zhang, B.-X.; Gao, H.; Li, X.-L., Synthesis and optical properties of nitrogen and sulfur co-doped graphene quantum dots. *New Journal of Chemistry* **2014**, *38* (9), 4615–4621.

[28] Xu, H.; Zhou, S.; Xiao, L.; Yuan, Q.; Gan, W., Time-efficient syntheses of nitrogen and sulfur co-doped graphene quantum dots with tunable luminescence and their sensing applications. *RSC Advances* **2016**, *6* (43), 36554–36560.

[29] Hasan, M. T.; Gonzalez-Rodriguez, R.; Ryan, C.; Faerber, N.; Coffer, J. L.; Naumov, A. V., Photo-and electroluminescence from nitrogen-doped and nitrogen–sulfur codoped graphene quantum dots. *Advanced Functional Materials* **2018**, 1804337.

[30] Qu, D.; Zheng, M.; Du, P.; Zhou, Y.; Zhang, L.; Li, D.; Tan, H.; Zhao, Z.; Xie, Z.; Sun, Z., Highly luminescent S, N co-doped graphene quantum dots with broad visible absorption bands for visible light photocatalysts. *Nanoscale* **2013**, *5* (24), 12272–12277.

[31] Zhang, L.; Zhang, Z.-Y.; Liang, R.-P.; Li, Y.-H.; Qiu, J.-D., Boron-doped graphene quantum dots for selective glucose sensing based on the "abnormal" aggregation-induced photoluminescence enhancement. *Analytical Chemistry* **2014**, *86* (9), 4423–4430.

[32] Li, Y.; Li, S.; Wang, Y.; Wang, J.; Liu, H.; Liu, X.; Wang, L.; Liu, X.; Xue, W.; Ma, N., Electrochemical synthesis of phosphorus-doped graphene quantum dots for free radical scavenging. *Physical Chemistry Chemical Physics* **2017**, *19* (18), 11631–11638.

[33] Kundu, S.; Yadav, R. M.; Narayanan, T.; Shelke, M. V.; Vajtai, R.; Ajayan, P. M.; Pillai, V. K., Synthesis of N, F and S co-doped graphene quantum dots. *Nanoscale* **2015**, *7* (27), 11515–11519.

[34] Eriksson, M. O.; Schmidt, S.; Asghar, M.; Lin, P.-C.; Holtz, P. O.; Syväjärvi, M.; Yazdi, G. R., Tuning the emission energy of chemically doped graphene quantum dots. *Nanomaterials* **2016**, *6* (11), 198.
[35] Hasan, M. T.; Gonzalez-Rodriguez, R.; Lin, C. W.; Campbell, E.; Vasireddy, S.; Tsedev, U.; Belcher, A. M.; Naumov, A. V. J. A. O. M., Rare-earth metal ions doped graphene quantum dots for near-IR in vitro/in vivo/ex vivo imaging applications. **2020**, *8* (21), 2000897.
[36] Huynh, T. V.; Anh, N. T. N.; Darmanto, W.; Doong, R.-A., Erbium-doped graphene quantum dots with up-and down-conversion luminescence for effective detection of ferric ions in water and human serum. *Sensors and Actuators B: Chemical* **2021**, *328*, 129056.
[37] Wang, G.; Guo, Q.; Chen, D.; Liu, Z.; Zheng, X.; Xu, A.; Yang, S.; Ding, G., Facile and highly effective synthesis of controllable lattice sulfur-doped graphene quantum dots via hydrothermal treatment of durian. *ACS Applied Materials & Interfaces* **2018**, *10* (6), 5750–5759.
[38] Zhu, Y.; Lu, S.; Manohari, A. G.; Dong, X.; Chen, F.; Xu, W.; Shi, Z.; Xu, C., Polydopamine interconnected graphene quantum dots and gold nanoparticles for enzymeless H_2O_2 detection. *Journal of Electroanalytical Chemistry* **2017**, *796*, 75–81.
[39] Hai, X.; Lin, X.; Chen, X.; Wang, J., Highly selective and sensitive detection of cysteine with a graphene quantum dots-gold nanoparticles based core-shell nanosensor. *Sensors and Actuators B: Chemical* **2018**, *257*, 228–236.
[40] Ting, S. L.; Ee, S. J.; Ananthanarayanan, A.; Leong, K. C.; Chen, P., Graphene quantum dots functionalized gold nanoparticles for sensitive electrochemical detection of heavy metal ions. *Electrochimica Acta* **2015**, *172*, 7–11.
[41] Qin, L.; Liu, J.; Kang, S.-Z.; Li, G.; Li, X., The strong dependence of the bi-functionalities of core–shell-like gold-based nanocomposites on the size of gold nanoparticles. *Journal of Materials Chemistry C* **2017**, *5* (44), 11411–11415.
[42] Ju, J.; Chen, W., In situ growth of surfactant-free gold nanoparticles on nitrogen-doped graphene quantum dots for electrochemical detection of hydrogen peroxide in biological environments. *Analytical Chemistry* **2015**, *87* (3), 1903–1910.
[43] Anh, N. T. N.; Doong, R.-a., One-step synthesis of size-tunable gold@ sulfur-doped graphene quantum dot nanocomposites for highly selective and sensitive detection of nanomolar 4-nitrophenol in aqueous solutions with complex matrix. *ACS Applied Nano Materials* **2018**, *1* (5), 2153–2163.

[44] Nair, R. V.; Thomas, R. T.; Sankar, V.; Muhammad, H.; Dong, M.; Pillai, S., Rapid, acid-free synthesis of high-quality graphene quantum dots for aggregation induced sensing of metal ions and bioimaging. *ACS Omega* **2017**, *2* (11), 8051–8061.

[45] Kellici, S.; Acord, J.; Power, N. P.; Morgan, D. J.; Coppo, P.; Heil, T.; Saha, B., Rapid synthesis of graphene quantum dots using a continuous hydrothermal flow synthesis approach. *RSC Advances* **2017**, *7* (24), 14716–14720.

[46] Ananthanarayanan, A.; Wang, X.; Routh, P.; Sana, B.; Lim, S.; Kim, D. H.; Lim, K. H.; Li, J.; Chen, P., Facile synthesis of graphene quantum dots from 3D graphene and their application for Fe^{3+} sensing. *Advanced Functional Materials* **2014**, *24* (20), 3021–3026.

[47] Ye, R.; Xiang, C.; Lin, J.; Peng, Z.; Huang, K.; Yan, Z.; Cook, N. P.; Samuel, E. L.; Hwang, C.-C.; Ruan, G., Coal as an abundant source of graphene quantum dots. *Nature Communications* **2013**, *4*, 2943.

[48] Zhu, H.; Liu, A.; Xu, Y.; Shan, F.; Li, A.; Wang, J.; Yang, W.; Barrow, C.; Liu, J., Graphene quantum dots directly generated from graphite via magnetron sputtering and the application in thin-film transistors. *Carbon* **2015**, *88*, 225–232.

[49] Kushwaha, N.; Mittal, J.; Pandey, S.; Kumar, R., High temperature acidic oxidation of multiwalled carbon nanotubes and synthesis of graphene quantum dots. *International Journal of Nano Dimension* **2018**, *9* (2), 191–197.

[50] Yan, X.; Cui, X.; Li, B.; Li, L.-s., Large, solution-processable graphene quantum dots as light absorbers for photovoltaics. *Nano Letters* **2010**, *10* (5), 1869–1873.

[51] Liu, R.; Wu, D.; Feng, X.; Müllen, K., Bottom-up fabrication of photoluminescent graphene quantum dots with uniform morphology. *Journal of the American Chemical Society* **2011**, *133* (39), 15221–15223.

[52] Fuhrer, M. S.; Lau, C. N.; MacDonald, A. H., Graphene: Materially better carbon. *MRS Bulletin* **2010**, *35* (4), 289–295.

[53] Lu, J.; Yan, M.; Ge, L.; Ge, S.; Wang, S.; Yan, J.; Yu, J., Electrochemiluminescence of blue-luminescent graphene quantum dots and its application in ultrasensitive aptasensor for adenosine triphosphate detection. *Biosensors and Bioelectronics* **2013**, *47*, 271–277.

[54] Sun, H.; Wu, L.; Wei, W.; Qu, X., Recent advances in graphene quantum dots for sensing. *Materials Today* **2013**, *16* (11), 433–442.

[55] Wang, F.; Gu, Z.; Lei, W.; Wang, W.; Xia, X.; Hao, Q., Graphene quantum dots as a fluorescent sensing platform for highly efficient

detection of copper (II) ions. *Sensors and Actuators B: Chemical* **2014**, *190*, 516–522.

[56] Dutta Chowdhury, A.; Doong, R.-a., Highly sensitive and selective detection of nanomolar ferric ions using dopamine functionalized graphene quantum dots. *ACS Applied Materials & Interfaces* **2016**, *8* (32), 21002–21010.

[57] Wang, L.; Zhu, S.-J.; Wang, H.-Y.; Qu, S.-N.; Zhang, Y.-L.; Zhang, J.-H.; Chen, Q.-D.; Xu, H.-L.; Han, W.; Yang, B., Common origin of green luminescence in carbon nanodots and graphene quantum dots. *ACS Nano* **2014**, *8* (3), 2541–2547.

[58] Zhu, S.; Wang, L.; Li, B.; Song, Y.; Zhao, X.; Zhang, G.; Zhang, S.; Lu, S.; Zhang, J.; Wang, H., Investigation of photoluminescence mechanism of graphene quantum dots and evaluation of their assembly into polymer dots. *Carbon* **2014**, *77*, 462–472.

[59] Haque, E.; Kim, J.; Malgras, V.; Reddy, K. R.; Ward, A. C.; You, J.; Bando, Y.; Hossain, M. S. A.; Yamauchi, Y., Recent advances in graphene quantum dots: Synthesis, properties, and applications. *Small Methods* **2018**, *2* (10), 1800050.

[60] Pan, D.; Zhang, J.; Li, Z.; Wu, M., Hydrothermal route for cutting graphene sheets into blue-luminescent graphene quantum dots. *Advanced Materials* **2010**, *22* (6), 734–738.

[61] Luo, Y.; Li, M.; Hu, G.; Tang, T.; Wen, J.; Li, X.; Wang, L., Enhanced photocatalytic activity of sulfur-doped graphene quantum dots decorated with TiO_2 nanocomposites. *Materials Research Bulletin* **2018**, *97*, 428–435.

[62] Li, Y.; Zhao, Y.; Cheng, H.; Hu, Y.; Shi, G.; Dai, L.; Qu, L., Nitrogen-doped graphene quantum dots with oxygen-rich functional groups. *Journal of the American Chemical Society* **2012**, *134* (1), 15–18.

[63] Feng, J.; Dong, H.; Yu, L.; Dong, L., The optical and electronic properties of graphene quantum dots with oxygen-containing groups: A density functional theory study. *Journal of Materials Chemistry C* **2017**, *5* (24), 5984–5993.

[64] Hamilton, I. P.; Li, B.; Yan, X.; Li, L.-s., Alignment of colloidal graphene quantum dots on polar surfaces. *Nano Letters* **2011**, *11* (4), 1524–1529.

[65] Zhu, S.; Zhang, J.; Tang, S.; Qiao, C.; Wang, L.; Wang, H.; Liu, X.; Li, B.; Li, Y.; Yu, W., Surface chemistry routes to modulate the photoluminescence of graphene quantum dots: From fluorescence mechanism to up-conversion bioimaging applications. *Advanced Functional Materials* **2012**, *22* (22), 4732–4740.

[66] Shen, J.; Zhu, Y.; Chen, C.; Yang, X.; Li, C., Facile preparation and upconversion luminescence of graphene quantum dots. *Chemical Communications* **2011**, *47* (9), 2580–2582.

[67] Mørk, J., Quantum dots for optoelectronics. *Applied Physics Letters* **2007**, *90*, 33508.

[68] Mijatovic, D.; Eijkel, J. C.; van den Berg, A., Technologies for nanofluidic systems: Top-down vs. bottom-up — a review. *Lab on a Chip* **2005**, *5* (5), 492–500.

[69] Kaur, R.; Kim, K.-H.; Deep, A., A convenient electrolytic assembly of graphene-MOF composite thin film and its photoanodic application. *Applied Surface Science* **2017**, *396*, 1303–1309.

[70] Hai, X.; Feng, J.; Chen, X.; Wang, J., Tuning the optical properties of graphene quantum dots for biosensing and bioimaging. *Journal of Materials Chemistry B* **2018**, *6* (20), 3219–3234.

[71] Wang, X.; Sun, G.; Routh, P.; Kim, D.-H.; Huang, W.; Chen, P., Heteroatom-doped graphene materials: Syntheses, properties and applications. *Chemical Society Reviews* **2014**, *43* (20), 7067–7098.

[72] Dong, Y.; Shao, J.; Chen, C.; Li, H.; Wang, R.; Chi, Y.; Lin, X.; Chen, G., Blue luminescent graphene quantum dots and graphene oxide prepared by tuning the carbonization degree of citric acid. *Carbon* **2012**, *50* (12), 4738–4743.

[73] Rostampour Ghareghozloo, E.; Mahdavimehr, M.; Meratan, A. A.; Nikfarjam, N.; Ghasemi, A.; Katebi, B.; Nemat-Gorgani, M., Role of surface oxygen-containing functional groups of graphene oxide quantum dots on amyloid fibrillation of two model proteins. *PloS One* **2020**, *15* (12), e0244296.

[74] Liu, B.; Xie, J.; Ma, H.; Zhang, X.; Pan, Y.; Lv, J.; Ge, H.; Ren, N.; Su, H.; Xie, X., From graphite to graphene oxide and graphene oxide quantum dots. *Small* **2017**, *13* (18), 1601001.

[75] Park, J.; Moon, J.; Kim, C.; Kang, J. H.; Lim, E.; Park, J.; Lee, K. J.; Yu, S.-H.; Seo, J.-H.; Lee, J., Graphene quantum dots: Structural integrity and oxygen functional groups for high sulfur/sulfide utilization in lithium sulfur batteries. *NPG Asia Materials* **2016**, *8* (5), e272–e272.

CHAPTER 15

BONDING, INTERACTION, AND IMPACT OF HYDROGEN ON 2D SiC MATERIALS

Nguyen Minh Phi and Tran Thi Thu Hanh

Laboratory of Computational Physics, Faculty of Applied Science, Ho Chi Minh City University of Technology (HCMUT), VNU-HCM, Ho Chi Minh City, Vietnam

Abstract

A series of density functional theory calculations was performed to understand the bonding and interaction of hydrogen adsorption on two-dimensional (2D) silicon carbide (SiC). The converged energy results pointed out that the H atom can sufficiently bond to 2D SiC at the top sites (atop Si and C), of which the most stable adsorption site is T_{Si}. The vibrational properties, along with the zero-point energy, were incorporated into the energy calculations to further understand the phonon effect of the adsorbed H. Most of the 2D SiC structure deformations caused by the H atoms were found at the adsorbent atom along the vertical axis. For the first time, five SiC defect formations, including the quadrilateral-octagon linear defect (8-4), the silicon interstitial defect, the divacancy (4-10-4) defect, the divacancy (8-4-4-8) defect, and the divacancy (4-8-8-4) defect, were investigated. The linear defect (8-4) has the lowest formation energy and is most likely to form. This study could be an interesting step toward future applications of hydrogen energy.

Keywords: Density functional theory calculation, two dimensional, silicon carbide, hydrogen adsorption.

15.1. Introduction

Since 2004, after the discovery of graphene, two-dimensional (2D) thin hexagonal forms of materials have increasingly attracted attention.[1–6] Among the Group IV elements, besides the special element C, there are other elements such as Si, Ge, Sn, and Pb. Khan *et al.* show that for Si, it is very complicated to construct a stable structure of nanotubes or graphene-like sheets.[7] However, the silicene (a mixture of Si) on the silver substrates was synthesized in the studies of Aufray *et al.* (2010) and Vogt *et al.* (2012).[8,9] Besides, using the experimental method of the high-temperature thermochemical Si powder's reactions, a quasi-2D SiC flake was created.[10]

Previous theoretical studies have shown that the 2D SiC has a planar structure like graphene but with a longer bond distance (1.77–1.79 Å) and a larger band gap (2.5–2.6 eV).[11,12] With the characteristic of large-bandgap materials, SiC materials are predicted to have great applications for nanodevices under operating conditions of high temperature and high frequency.[16] Recently, many theoretical research groups have continued to focus on studying the configurations and characteristics of 2D SiC models using MD and DFT simulation methods.[3,4,13–15]

Especially in the most recent research paper of our group, we have carefully verified the 2D SiC model with different interaction potentials by using the MD simulation method.[4] Based on the obtained accuracy of the 2D SiC model, we continue to conduct a series of DFT calculations to determine the hydrogen storage properties of 2D SiC in this study. The study has been carefully performed with convergence control and considering the zero-point vibrational effect.

In Section 15.3.1, the adsorption sites and the adsorption energy were precisely investigated, in consideration of the zero-point vibrational energy (ZPE) effect. Several hydrogenated 2D SiC configurations were simulated to understand the most favorable adsorption site for hydrogen atoms. The structural deformations of 2D SiC caused by the adsorbed hydrogen atom are also noted and investigated. Moreover, in the 2D SiC obtained from previous studies, defects have also appeared, but they have not been studied in detail.[3,4,6] Therefore, for the first time in this study, five different types of defects are investigated and discussed in Section 15.3.2. They are the quadrilateral-octagon linear defect (8-4), the silicon interstitial defect, the divacancy (4-10-4) defect, the divacancy (8-4-4-8) defect, and the divacancy (4-8-8-4) defect.

15.2. Calculation

15.2.1. *SIESTA simulation*

We use the density functional theory method implemented in the SIESTA package, which is a combination of a self-consistent loop, a norm-conserving pseudopotential, and a plane-wave basis set, to calculate the ground-state energies and optimized structures of all the simulation models.[17–19] For the exchange and correlation energy, we use the generalized gradient approximation (GGA) in the form of a revised Perdew–Burke–Ernzerhof (revPBE) functional.[20,21] We also employ the double-zeta polarized basis sets while using an energy cutoff of 200 Ry for all the simulations. The structural relaxations are performed under the 2D periodic boundary condition while using a vacuum gap of 40 Å in the z-axis to neglect most of the interaction between the periodic lateral layers. The self-consistent loop is stopped when the maximum stress component is less than 0.02 eV/Å. In the previous DFT calculations, these selected parameters have provided reasonable accuracy.[2,22–25]

Besides, for long-range interactions of adsorbed hydrogen atoms, it is important for the bond lengths and energies of the hydrogen molecules to be calculated correctly. Therefore, the extended basic set is included to calculate the adsorbed hydrogen atoms. An energy shift of 60 meV and a split norm of 0.53 for the second zeta were used for the H atoms. The obtained H_2 bond length, the binding energy, and the zero-point energy are 0.754 Å, 4.525 eV, and 0.269 eV, respectively, which are in good agreement with the experiment.[30,31]

At first, a periodic 7×7 supercell SiC monolayer, consisting of 49 silicon and 49 carbon atoms (see Fig. 15.1), is cut out of the obtained MD simulated structure (see Ref. [4]). We decided to choose a 7×7 supercell because a previous study on graphene showed that

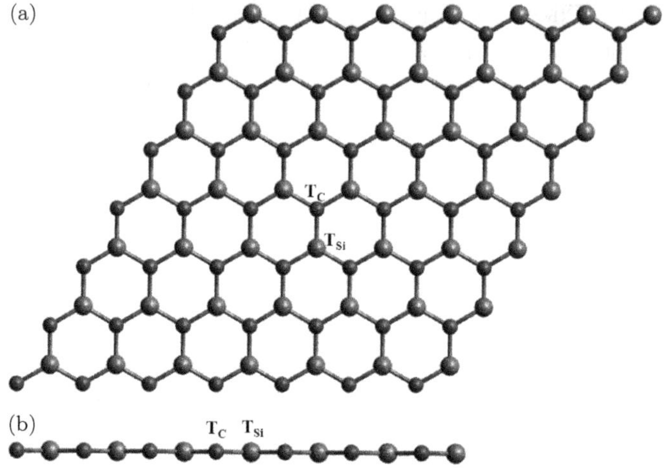

Fig. 15.1. The (a) top view and (b) non-buckling side view of the two-dimensional silicon carbide model that includes a total of 98 atoms, where T_C and T_{Si} sites are atop C (gray balls) and Si (blue balls) atoms, respectively.

the size effect on hydrogen adsorption was negligible when the size was larger than a 4×4 cell.[29] The Si–C bond length is 1.85 Å. This bond length is much larger than in other studies (1.77–1.79 Å),[3,26] but it is closer to 1.89 Å found in bulk SiC.[27,28] And this value is within the range defined in Ref. [6] when embedding Si atoms in a supercell of graphene and relaxing its atomic structure via DFT. The model has no buckling, as expected from the sp^2 hybridization type.[3,26] The system is optimized until the residual forces are smaller than 0.01 eV $Å^{-1}$.

To test the adsorption site, a hydrogen atom is put atop the SiC sheet at the possible sites. After that, we check the converged data regarding the increasing *k*-point grid from $(1 \times 1 \times 1)$ to $(13 \times 13 \times 1)$ Monkhorst–Pack (MP) scheme. Zero-point vibrational energy (ZPE) is also considered for better results.

15.2.2. *Zero-point energy and defect calculations*

Because the hydrogen vibrations around the equilibrium position significantly affect the interaction energy of the adsorption system, the accuracy calculation is taken into account for the hydrogen adsorption energy.[24,25,32,33] Additionally, in this study, the vibrational energy of the adsorbed H atoms on the SiC surface, which is called the hydrogen ZPE, was calculated using the following equation:

$$\mathrm{ZPE} = \frac{h}{4\pi}\sqrt{\frac{k(m_1+m_2)}{m_1 m_2}},$$

where m_1 is the H mass, m_2 is the Si/C mass, h is the Planck's constant, and k is the force constant. The force constant k can be calculated by using the minimum square calculation method when the H atoms are displaced around the obtained equilibrium sites in both the normal and parallel directions. This ZPE calculation method has been presented in detail in a previous study by our group.[25]

To study the 2D SiC defects, a model of 98 atoms (49 Si atoms and 49 C atoms) is taken into account. The following five defects are found: the quadrilateral-octagon linear defect (8-4), the silicon interstitial defect, the divacancy (4-8-8-4) defect, the divacancy (8-4-4-8) defect, and the divacancy (4-10-4) defect. The defect formation energy is found by subtracting the total energy of the non-defective model from the total energy of the defective model:

$$E_{\text{form}} = E_{\text{non-defect}} - E_{\text{defect}},$$

where E_{form} is the defect formation energy, $E_{\text{non-defect}}$ is the total energy of the non-defective model, and E_{defect} is the total energy of the defective model.

15.3. Results and Discussions

15.3.1. *Possible hydrogen adsorption sites*

To test the possible adsorption sites, the hydrogen atoms are added to a relaxed SiC sheet at the following possible sites: Top C (T_C), Top Si (T_{Si}), Center (C), and Bridge (B). After optimizing the system with the $(3 \times 3 \times 1)$ k-point, it is shown that only the top sites (T_C and T_{Si}) are the stable sites (see Fig. 15.2). Similar results on the

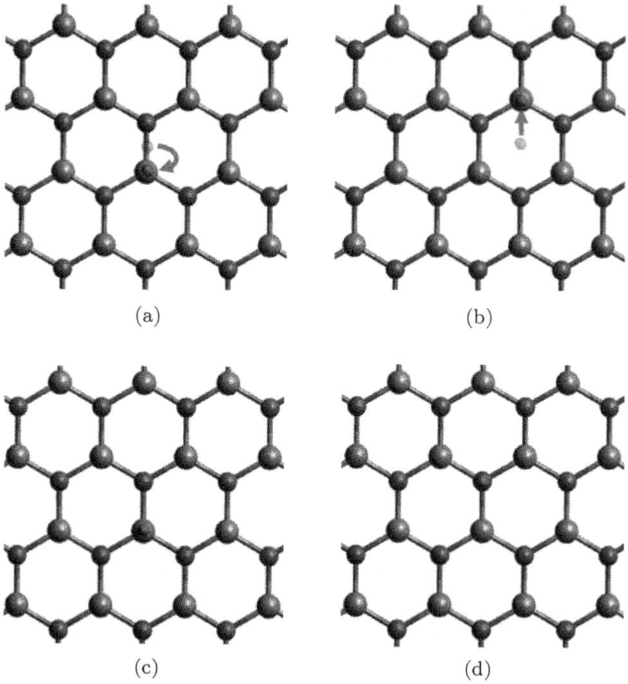

Fig. 15.2. The H atoms (red balls) on (a) the bridge and (b) the center sites are not stable and shift to the T_{Si} site after optimization, while the (c) T_S and (d) T_C sites are stable adsorption sites.

stable adsorption sites of hydrogen on the surfaces of 2D materials, such as graphene and germane, have also been published.[2,29] Next, to be sure of the model's convergence results on the k-point value, the adsorption energies of T_C and T_{Si} systems are calculated with different MP grids, from $(1\times1\times1)$ to $(13\times13\times1)$ (see Table 15.1 and Fig. 15.3). It is shown that the adsorption energy does not change when the MP grid of $(3 \times 3 \times 1)$ is used. Therefore, we decide to use the k-point of $(3 \times 3 \times 1)$ MP grid for all the remaining systems studied in the following. This is a rather low number of k-points but has been used before in larger-cell 2D SiC DFT calculations.[7]

To calculate the hydrogen adsorption energy, the following equation is used:

$$E_{\text{ads}} = E(n_H) - E(0) - \frac{n_H}{2} E_{H_2},$$

Table 15.1. The adsorption energy of the H atom atop SiC monolayer surface on the T_C and T_{Si} sites with different k-points (eV).

k-points	Top C	Top Si
$(1 \times 1 \times 1)$ MP	−0.346426	−0.715646
$(2 \times 2 \times 1)$ MP	−0.37178	−0.711654
$(3 \times 3 \times 1)$ MP	−0.373746	−0.711698
$(4 \times 4 \times 1)$ MP	−0.374103	−0.711379
$(5 \times 5 \times 1)$ MP	−0.373733	−0.711111
$(6 \times 6 \times 1)$ MP	−0.373024	−0.711994
$(7 \times 7 \times 1)$ MP	−0.372715	−0.711514
$(8 \times 8 \times 1)$ MP	−0.372744	−0.711974
$(9 \times 9 \times 1)$ MP	−0.373913	−0.711033
$(10 \times 10 \times 1)$ MP	−0.37377	−0.71156
$(11 \times 11 \times 1)$ MP	−0.37377	−0.71156
$(12 \times 12 \times 1)$ MP	−0.37377	−0.71156
$(13 \times 13 \times 1)$ MP	−0.37407	−0.71187

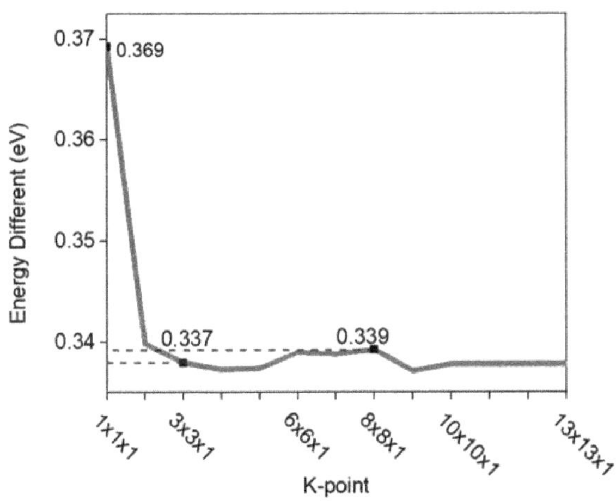

Fig. 15.3. k-point dependence of the adsorption energy difference ($E_{ads}(T_c) - E_{ads}(T_{Si})$) shows a convergence value at the $(3 \times 3 \times 1)$ k-point.

where E_{ads} is the hydrogen adsorption energy, $E(n_H)$ is the total energy of the H/SiC system with n_H hydrogen atoms, $E(0)$ is the total energy of bare 2D SiC, and E_{H_2} is the total energy of the isolated H_2 molecule.

Table 15.2. The bond length of the adsorbent atom (the Si and C atoms) with the H atom on top of itself ($d_{-\mathrm{H}}$, Å), the distance of the adsorbent atom with their neighbors ($d_{\mathrm{Si-C}}$, Å), the vertical displacement of the adsorbent atom compared with their relaxed positions (h, Å), and the hydrogen adsorption energies (E_{ads}, eV).

	Fixed 2D SiC				Non-fixed 2D SiC			
Site	$d_{-\mathrm{H}}$	$d_{\mathrm{Si-C}}$	h	E_{ads}	d_{H}	$d_{\mathrm{Si-C}}$	h	E_{ads}
T_{C}	1.149	1.865	0	−0.374	1.117	1.975	0.460	−1.073
T_{Si}	1.580	1.865	0	−0.712	1.525	1.929	0.369	−1.244

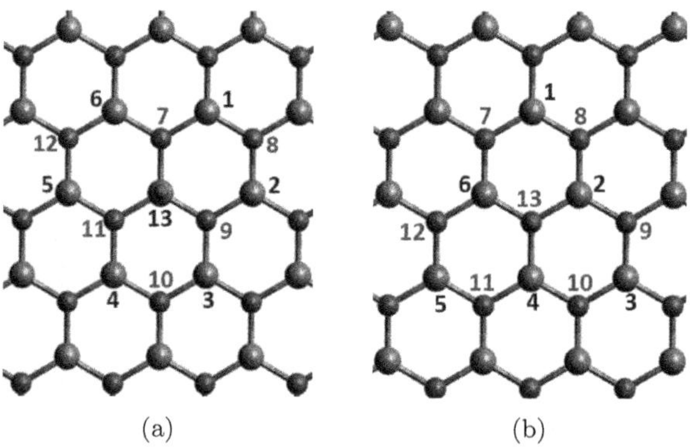

Fig. 15.4. There are no distortions observable in the 2D SiC supercell when viewed from the top. The H atoms are marked in red color in both adsorption configurations: (a) T_{Si} and (b) T_{C}. The neighborhoods of the adsorbent atom are numbered to correspond with Table 15.3.

The adsorption energies of hydrogen on T_{Si} and T_{C} sites with the optimized Si/C–H bond lengths are shown in Table 15.2. The distortion around the adsorption position is displayed in Figs. 15.4 and 15.5. The distortions are not observable when viewed from the top, which shows there is little displacement in the in-plane axes. Table 15.3 shows the displacement of Si and C atoms when a hydrogen atom is adsorbed on the SiC surface. The numbering in Table 15.3 corresponds to that in Fig. 15.4. At the adsorption site, the carbon atom is attracted up along the vertical axis more strongly than the Si atoms ($z_{\mathrm{C}} = 0.460$ Å, $z_{\mathrm{Si}} = 0.369$ Å). The first

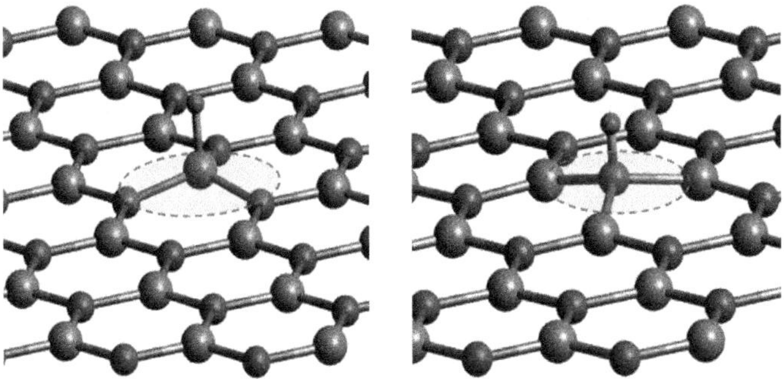

(a) (b)

Fig. 15.5. The out-of-plane distortion can be observed at the adsorbent atom (highlighted) in both adsorption configurations: (a) T_{Si} and (b) T_C.

Table 15.3. The displacement of non-fixed 2D SiC atoms with a hydrogen adsorption on top sites compared with the optimized model. The first-order neighborhoods of the adsorbent atom have been highlighted.

	Top C				Top Si			
	Atom	***x*-axis**	***y*-axis**	***z*-axis**	**Atom**	***x*-axis**	***y*-axis**	***z*-axis**
1	Si	0.016	0.017	0.011	Si	0.014	0.011	0.014
2	**Si**	**0.087**	**0.021**	**0.066**	Si	0.006	0.018	0.014
3	Si	0.025	0.014	0.013	Si	0.011	0.001	0.015
4	**Si**	**0.004**	**0.072**	**0.058**	Si	0.017	0.006	0.014
5	Si	0.001	0.008	0.014	Si	0.013	0.016	0.014
6	**Si**	**0.037**	**0.050**	**0.061**	Si	0.006	0.004	0.014
7	C	0.012	0.014	0.003	**C**	**0.002**	**0.029**	**0.004**
8	C	0.020	0.012	0.006	C	0.008	0.003	0.011
9	C	0.027	0.005	0.010	**C**	**0.017**	**0.020**	**0.006**
10	C	0.020	0.014	0.004	C	0.003	0.002	0.011
11	C	0.008	0.009	0.007	**C**	**0.034**	**0.009**	**0.005**
12	C	0.004	0.001	0.009	C	0.004	0.004	0.011
13	**C**	**0.017**	**0.000**	**0.460**	**Si**	**0.005**	**0.001**	**0.369**

neighboring Si atoms are also attracted upward to about 0.06 Å, forming local deformations around the adsorbed C atom at the H adsorption site (Fig. 15.5(b)). These local hill-forming properties were also noted in previous studies of germanene and graphene.[2,29] However, the displacements of the first neighboring C atoms around the adsorbed Si atom were not clearly observed (Fig. 15.5(a)). This means that the Si atoms are more stable than the C atoms in the 2D SiC system, and this is in agreement with the adsorption energies shown in Table 15.2.

To study the hydrogen vibrational effect, we consider the ZPE shown in Table 15.4. Similar to previous studies on hydrogen adsorption,[2,24] strong fluctuations of hydrogen are observed along the z-axis, while fluctuations along the x and y axes are very small (Fig. 15.6). The vibrational ZPE of the hydrogen atom on the SiC surface (~5 meV, Table 15.4) is less than half of the ZPE value on

Table 15.4. Force constants (k, eV/Å^2), vibrational frequencies (f, cm^{-1}), the ZPE (meV), and the total adsorption energy (E_{ads}, eV) of the H adsorption on the optimized fixed 2D SiC with consideration of ZPE.

	Top C			**Top Si**		
	x-axis	y-axis	z-axis	x-axis	y-axis	z-axis
k, eV/Å^2	0.0357	0.135	1.454	0.0771	0.0756	0.681
f, cm^{-1}	8.483	16.526	54.165	12.474	12.353	37.063
ZPE (meV)	0.526	1.02	3.36	0.773	0.766	2.30
Total ZPE (meV)		4.91			3.84	
E_{ads} (eV)		−0.3787			−0.7155	

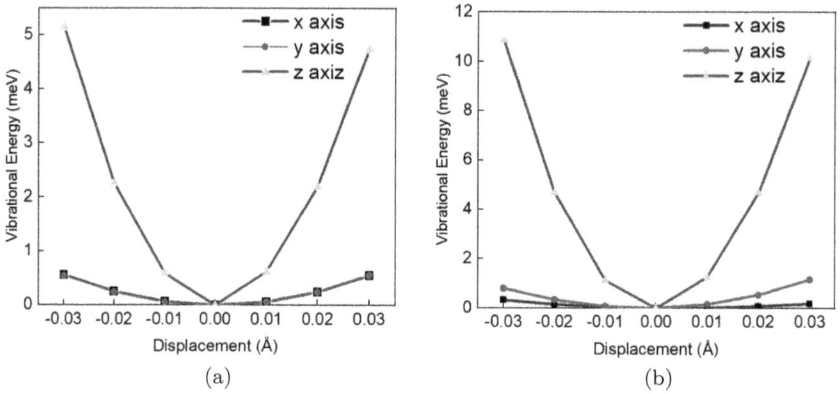

Fig. 15.6. The vibrational energies of the adsorbed hydrogen at the (a) T_{Si} and (b) T_C sites have good correspondence with the harmonic approximation.

the Ge surface (~10 meV).[2] If we compare the ZPE of H on the 3D Pt surface (~160 meV)[24] with that on the 2D surfaces, we see that the vibrational effect of the hydrogen atom on the 2D surface is very small. The conclusion of the vibrational zero-point energy's effect on the 2D SiC system, which has not been reported to date, can provide the needed information for further research on the interaction of hydrogen on the 2D SiC surface. The value of adsorption energy, taking zero-point energy into account (see Table 15.4), again shows that the adsorbed hydrogen at the Si site is much more stable than the hydrogen at the C site.

15.3.2. *Structural defects of 2D silicon carbide*

From the results of the 2D SiC model obtained by the MD method in Ref. [4], five defects are found. They are the quadrilateral-octagon linear defect (8-4 defect), the silicon interstitial defect, and three other defects that appear in conjunction with the linear 8-4 defect, namely the divacancy (4-8-8-4) defect, the divacancy (8-4-4-8) defect, and the divacancy (4-10-4) defect (see Fig. 15.7).

The previous studies have shown that the non-magnetic ground state is more commonly observed, and the dangling bonds are rarely found in the system.[2,34] Similar results were also obtained in our study. Table 15.5 shows that the lowest formation energy is found for the quadrilateral-octagon linear defect. This means that, in the 2D SiC system, this 8-4 defect is the most stable and the most commonly formed defect. This is consistent with the study on defects of the germanene system, which showed that the pentagon-heptagon linear defect is the most commonly formed defect.[2] However, our defectivity result is contrary to the previous studies of Padilha,[34] which reported that the single vacancy defect is the most stable defect. The difference between our results on defect formation and Padilha's results may be attributed to the buckled structure of germanene, which is not observed in SiC. Besides, the larger value of the lattice constant and the smaller value of the in-plane nearest-neighbor distance also affect the formation of defects. There are three other variations in the 8-4 defect, which are the divacancy (4-8-8-4) defect, the divacancy

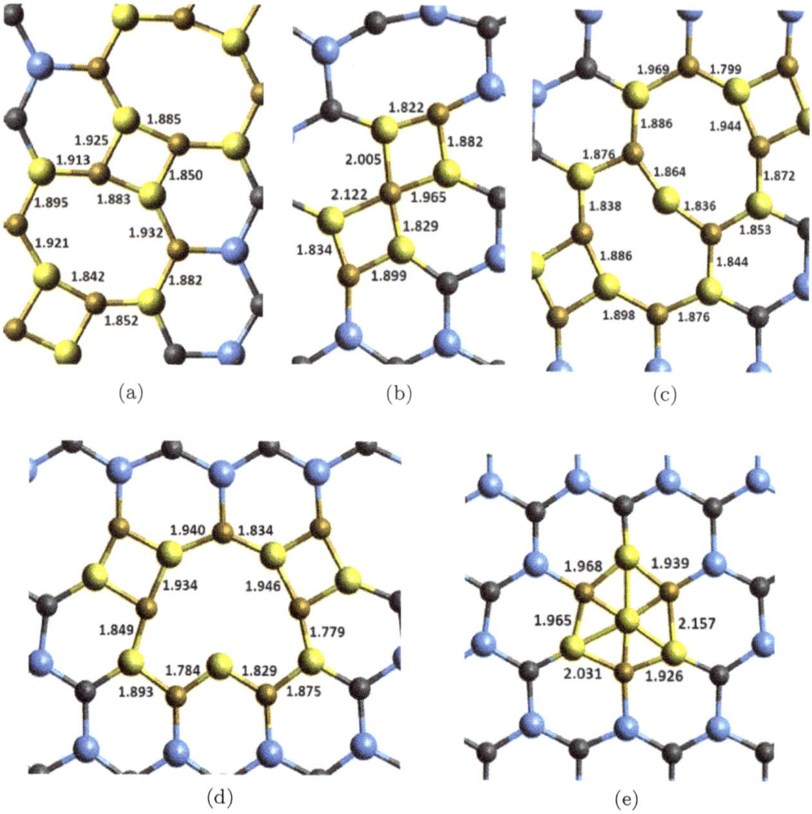

Fig. 15.7. The structural defects (yellow highlight) in 2D silicon carbide: (a) the linear (8-4) defect; (b) the divacancy (4-8-8-4) defect; (c) the divacancy (8-4-4-8) defect; (d) the divacancy (4-10-4) defect; and (e) the silicon interstitial defect.

Table 15.5. The formation energy of the defects in 2D SiC (eV).

Structural defect	Formation energy
Linear (8-4) defect	7.60537
Silicon interstitial defect	10.0246
Divacancy (4-10-4)	12.7898
Divacancy (8-4-4-8)	8.46357
Divacancy (4-8-8-4)	9.51503

Fig. 15.8. Overview of the 2D SiC model relaxed at 300 K using the MD simulation[4] with a system of 6,240 Si and C atoms.

(8-4-4-8) defect, and the divacancy (4-10-4) defect. All of them have higher formation energies than the original linear defect, while the divacancy (4-10-4) defect has the highest formation energy compared to all the other defects, proving that this defect is the most unlikely to form. We also noted that most of the Si–C bonding in all defects fluctuates between 1.80 and 2.00 Å and that there is no buckling defect. In addition to the quadrilateral-octagon linear defects, we have another interstitial defect (Fig. 15.7(e)). This defect is caused by the Si atom appearing in the middle of the hexagonal ring, causing the honeycomb structure to be divided into six triangular structures. The bonding of the Si interstitial atom with the hexagonal ring is about ~2.20 Å (Si–Si) and ~1.80 Å (Si–C). This defect formation energy is higher than the linear 8-4 defect formation energy by a value of 2.419 eV. Furthermore, the Stone–Wales defect, which has been found in previous 2D defect studies, is not found in this model. Figure 15.8 depicts the SiC monolayer surface structure, with the strains highlighted in brown.

15.4. Conclusions

We calculated the hydrogen adsorption energy of pristine 2D SiC using the DFT approach. Only two adsorption sites have been found stable, both on the top of the adsorbent atom (the C and Si atoms). The highest hydrogen adsorption energy in consideration of the zero-point energy was found in the T_{Si} configuration. We conclude that the most favorable adsorption site is the T_{Si} site. This agrees well with the vibrational energy calculation, where the hydrogen atom in the T_{Si} configuration has the lowest vibrational energy. We observe that most of the hydrogen vibrational energy comes from the z-axis vibration in all the configurations. The local deformations of 2D SiC caused by H adsorption are also shown. In the T_c configuration, the H atom caused more distortion compared to T_{Si}. The distortion of the adsorbent atom is greatest along the vertical axis for both configurations. Finally, we present five defects that could be found in 2D SiC and calculate their formation energies. Those defects are the linear (8-4) defect, the silicon interstitial defect, the divacancy (4-8-8-4) defect, the divacancy (8-4-4-8) defect, and the divacancy (4-10-4) defect. The linear (8-4) defect is shown to have the lowest formation energy and is the most expected to form. The results of this study can be an important step toward the future of hydrogen energy applications, with the defects showing many more possibilities for experimentation.

References

[1] Geim, A. K.; Novoselov, K. S. The rise of graphene. *Nat. Mater.* **2007**, *6*, 183–191.

[2] TTTHanh, NMPhi, NVHoa, Hydrogen adsorption on two-dimensional germanene and its structural defects: an ab initio investigation, *Phys. Chem. Chem. Phys.* **2020**, *22*, 7210–7217.

[3] Sun, L.; *et al.* Electronic structures of SiC nanoribbons, *The Journal of Chemical Physics* **2008**, *129*, 174114.

[4] DTN Tranh; *et al.* Modeling glassy SiC nanoribbon by rapidly cooling from the liquid: An af rmation of appropriate potentials, *Physica B* **2021**, *608*, 412746.

[5] NHGiang; *et al.* Formation of graphene on BN substrate by vapor deposition method and size effects on its structure, *Physica B: Condensed Matter* **2018**, *534*, 26–33.

[6] Susi, T.; *et al.* Computational insights and the observation of SiC nanograin assembly: towards 2D silicon carbide, *Scientific Reports* **2017**, *volume 7* (Article number), 4399.

[7] Khan, F.; Broughton, J.; *Phys. Rev. B* **1991**, *43*, 11754.

[8] Aufray, B.; *et al.* Graphene-like silicon nanoribbons on Ag(110): A possible formation of silicene. *Applied Physics Letters* **2010**, *96*, 183102.

[9] Vogt, P.; *et al.* Silicene: Compelling experimental evidence for graphenelike two-dimensional silicon. *Phys. Rev. Lett.* **2012**, *108*, 155501.

[10] Lin, S.; *et al.* Quasi-two-dimensional sic and sic2: Interaction of silicon and carbon at atomic thin lattice plane. *e Journal of Physical Chemistry C 119*, 19772.

[11] Hsueh, H. C.; Guo, G. Y.; Louie, S. G. Excitonic effects in the optical properties of a SiC sheet and nanotubes. *Phys. Rev. B* **2011**, 84, 085404.

[12] Lin, X.; *et al.* Ab initio study of electronic and optical behavior of two-dimensional silicon carbide. *J. Mater. Chem. C* **2013** *1*, 2131.

[13] Zhao, M.; Xia, Y.; Li, F.; Zhang, R. Q.; Lee, S.-T.; *Phys. Rev. B* **2005**, *71*, 085312.

[14] Gali, A.; *Phys. Rev. B* **2006**, *73*, 245415.

[15] Baumeier, B.; Krüger, P.; Pollmann, J.; *Phys. Rev. B* **2007**, *76*, 085407.

[16] Feng, Z. C.; Mascarenhas, A. J.; Choyke, W. J.; Powell, J. A.; *J. Appl. Phys.* **1988**, *64*, 3176.

[17] Kohn, W.; Sham, L. J.; Self-consistent equations including exchange and correlation effects, *Phys. Rev.* **1965**, *140*, A1133.

[18] Ordejón, P.; Artacho, E.; Soler, J. M.; Self-consistent order- N density-functional calculations for very large systems, *Phys. Rev. B: Condens. Matter Mater. Phys.* **1996**, *53*, R10441.

[19] Soler, J. M.; Artacho, E.; Gale, J. D.; Garcia, A.; Junquera, J.; Ordejon, P.; The SIESTA method for ab initio order-N materials simulation, *J. Phys.: Condens. Matter* **2002**, *14*, 2745.

[20] Perdew, J. P.; Burke, K.; Ernzerhof, M.; Generalized gradient approximation made simple, *Phys. Rev. Lett.* **1996**, *77*, 3865–3868.

[21] Zhang, Y.; Yang, W.; Comment on "Generalized gradient approximation made simple, *Phys. Rev. Lett.* **1998**, *80*, 890.

[22] Giang, N. H.; Tran, T. T. H.; Hoang, V. V.; Structural and thermodynamic properties of two-dimensional confined germanene: a molecular dynamics and DFT study, *Mater. Res. Express* **2019**, *6*, 086411.

[23] Hanh, T. T. T.; Takimoto, Y.; Sugino, O.; First-principles thermodynamic description of hydrogen electroadsorption on the Pt (111) surface, *Surf. Sci.* **2014**, *625*, 104.

[24] Hanh, T. T. T.; Hang, N. T. T.; A DFT study of hydrogen electroadsorption on the missing row Pt (1 1 0)-(1 2) surface, *Comput. Mater. Sci.* **2017**, *13*, 295.

[25] Hanh, T. T. T.; Hoa, N. V.; Zero-point vibration of the adsorbed hydrogen on the Pt(110) surface, *Adsorption* **2020**, *25*, 1–7.

[26] Chabi, S.; Guler, Z.; Brearley, A. J.; Benavidez, A. D.; Luk, T. S.; *Nanomaterials* **2021**, *11*, 1799.

[27] Susi, T.; Skákalová, V.; Mittelberger, A.; Kotrusz, P.; Hulman, M.; Pennycook, T. J.; Mangler, C.; Kotakoski, J.; Meyer, J. C.; Computational Insights and the Observation of SiC Nanograin Assembly: Towards 2D Silicon Carbide. *Sci. Rep.* **2017**, *7*, 1–13.

[28] Freeman, C. L.; Claeyssens, F.; Allan, N. L.; Harding, J. H.; Graphitic Nanofilms as Precursors to Wurtzite Films: Theory. *Phys. Rev. Lett.* **2006**, *96*, 066102.

[29] Ivanovskaya, V. V.; Zobelli, A.; Teillet-Billy, D.; Rougeau, N.; Sidis, V.; Briddon, P. R.; Hydrogen adsorption on graphene: a first principles study, *Eur. Phys. J. B* **2010**, *76*, 481–486.

[30] Roger, L. D.; Harry, B. G.; Chemical Structure and Bonding, University Science Books, **1989**, p. 199.43.

[31] Kunimatsu, K.; Uchida, H.; Osawa, M.; Watanabe, M.; In situ infrared spectroscopic and electrochemical study of hydrogen electro-oxidation on Pt electrode in sulfuric acid, *J. Electroanal. Chem.* **2006**, *587*, 299.

[32] Tomonari, M.; Sugino, O.; DFT calculation of vibrational frequency of hydrogen atoms on Pt electrodes: Analysis of the electric field dependence of the Pt–H stretching frequency, *Chem. Phys. Lett.* **2007**, *437*, 170.
[33] Hanh, T. T. T.; The nature of the hydrogen interaction on the unreconstructed platinum (110) surface: ab-initio study, *Phys. Scr.* **2021** *96* 025707.
[34] Padilha, J. E.; Pontes, R. B.; Electronic and transport properties of structural defects in monolayer germanene: an ab initio investigation, *Solid State Commun.* **2016**, *225*, 38–43.

CHAPTER 16

STRUCTURAL, ELECTRONIC, AND ELECTRON TRANSPORT PROPERTIES OF CHEMICALLY MODIFIED PENTAGONAL SiC_2 NANORIBBONS

Nguyen Thanh Tien*, Pham Thi Bich Thao*,
Tran Yen Mi* and Ming-Fa Lin†

*College of Natural Science, Can Tho University, Vietnam
†Department of Physics, National Cheng Kung University, Taiwan

Abstract

Two-dimensional monolayer pentagonal silicon dicarbide ($pSiC_2$) and its one-dimensional nanoribbon derivative ($pSiC_2NR$) possess novel electronic properties, possibly leading to many potential applications. In this chapter, we have systematically investigated the structural, electronic, and transport properties of pSiC2NRs using chemical function and strain engineering. The energy gaps of the $pSiC_2NRs$ (ZZ-ribbon, ZA-ribbon, AA-ribbon, and SS-ribbon) are created mainly owing to the competition in the edge structures, finite-size confinements, and asymmetry of chemical bonds in the tetrahedral lattice or chemical modification. By applying uniaxial tensile or compressive strain, it is possible to modulate the physical properties of $pSiC_2NRs$, which subsequently change the transport behavior of the carriers. Interestingly, the pentagon network of SS-$pSiC_2NRs$ is still maintained, but the bond length along the strained direction undergoes a large change. The electronic band structure and bandgap are strongly affected by the uniaxial compressive strain. The evolution of bandgap versus strain is linear. The I–V characteristic of SS-$pSiC_2NR$ seems to be more sensitive to compressive strain than stretch strain. The ultrahigh and strain-modulated carrier mobility in monolayer penta-SiC_2 may lead to many novel applications in high-performance electronic devices. Furthermore, the unusual properties of the $pSiC_2$ nanoribbons render them with great potential for applications in optoelectronic devices, especially in photovoltaics.

Keywords: Pentagonal SiC_2 nanorribons, Structural, electronic, and electronic transports.

16.1. Introduction

Recently, various two-dimensional (2D) nanomaterials with novel pentagonal structures[1–4] have been predicted. Penta-graphene (PG) is the earliest proposed pentagonal 2D structure by Zhang *et al.* in 2015. PG is made of pentagons characterized by a three-layer thick structure and resembling Cairo pentagonal tilling with a central sp^3 carbon sublattice and two perpendicular, symmetrically non-equivalent sp^2 carbon sublattices located on the top and bottom layers.[1] PG showed a semiconducting feature with an intrinsic quasi-direct bandgap of 3.25 eV and an atypical negative Poisson's ratio. Later, penta-silicene (PS), penta-CN_2 (pCN_2), penta-CB_2 (pCB_2), penta-SiC_2 ($pSiC_2$), penta-SiN_2 ($pSiN_2$), and so on have been proposed. Liu *et al.*[5] found that $pSiC_2$ and $pSiN_2$ reveal different strains depending on thermal conductivity and the strong tunability of thermal conductivity, which may inspire fruitful research on other derivatives of pentagonal structures to explore potential materials for next-generation digital devices. Very recently, the structural stability of the 2D pentagonal $PdSe_2$ and PdPSe materials has been analyzed.[6,7] Additionally, by a combination of crystal-chemical design and high-pressure synthesis, Bykov *et al.*[8] have discovered the compound $pNiN_2$, which possesses atomic-thick layers featuring pentagonal Cairo tessellations. The material is recoverable at ambient conditions and is a potential precursor for a pentagonal 2D material. It is predicted that there may be a wide scope for the study of various 2D materials of future technological interest. In 2021, research on the stability of the penta-carbides family, a novel family of 2D Group-IV materials with C2 dimers, was published.[9–11] This work reveals that the unitary (pC: PG), binary (pXC_2, X = Ge, Si, or Sn), and ternary penta-carbides have excellent energetic, dynamical, thermal, and mechanical stability with remarkable properties. Penta-SiC_2 belongs to the binary penta-carbides. The quantum confinement of electrons in low-dimensional nanomaterials is expected to lead to fascinating physical properties distinct from their bulk materials.[12,13] One-dimensional (1D) nanoribbons etched or patterned from the above-mentioned 2D materials can offer even more tunability in electronic properties because of special quantum confinement and edge effects.[14–16] Similar to the penta-graphene nanoribbons (PGNRs), which are cut from PG sheets,[17] the $pSiC_2$ nanoribbons ($pSiC_2NRs$) are cut from $pSiC_2$ sheets.[18,19] We have explored the structural diversity and the electronic properties of $pSiC_2NRs$ under various edge shapes and widths in the first-principles theoretical insight. PGNRs have plentiful and unique physical features.[17,20,21] Therefore, it is of urgent importance to understand deeply $pSiC_2NRs$ to develop possible applications in emerging electronic devices. However, we know that nanoscale structures, including pentagonal structures, cannot be produced as perfectly as they are designed in theoretical studies.[22] Hence, some investigations have been conducted to understand how mechanical imperfections lead to changes in electronic properties. In other words, we should understand how to modulate their electronic properties by compressing or stretching them. Graphene nanoribbon has the capability to afford large deformations, making it an ideal material for flexible strain sensors.[23,24] The effects of uniaxial strain on field effect transistors based on the armchair graphene nanoribbon (AGNR) were studied in Ref. [25]. Fang *et al.*[26] studied the transport properties of the AGNRs under tensile strain and found that the negative differential resistance effect (NDR) is merely obtained for AGNR(3p-1). We also studied the electronic structures and transport properties of the sawtooth–sawtooth PGNR (SS-PGNR) under a sequence of uniaxial strains in Ref. [27]. Our calculations show that the curve representing the values of bandgap versus strain is in the form of an asymmetric parabola. This is contrary to the graphene nanoribbons (zigzag and armchair) whose bandgap versus strain curve is linear.[28] Furthermore, the current in SS-PGNRs under bias is sensitive to uniaxial strains. Nevertheless, the electronic and transport properties of $pSiC_2NRs$ under uniaxial strain have not been fully explored, even from a theoretical point of view.[29–31]

16.2. Methodology

From the 2D penta-SiC_2, penta-SiC_2 nanoribbons with four different edges, including zigzag–zigzag (ZZ-ribbon), zigzag–armchair (ZA-ribbon), armchair–armchair (AA-ribbon), and sawtooth–sawtooth (SS-ribbon), are generated (as shown in Fig. 16.1). The structures were optimized and their properties were studied using the Atomistix ToolKit (ATK) software package (version 2017.1)[32] based on the DFT method using the generalized gradient approximation (GGA)

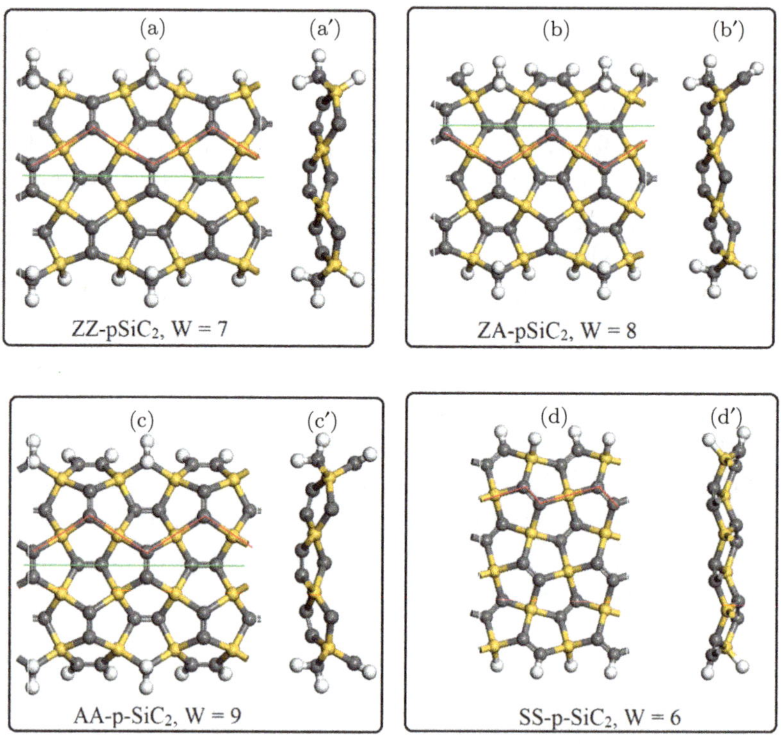

Fig. 16.1. Top and side views of zigzag–zigzag (ZZ-ribbon), zigzag–armchair (ZA-ribbon), armchair–armchair (AA-ribbon), and sawtooth–sawtooth (SS-ribbon). The widths (W) of ZZ-ribbon, ZA-ribbon, AA-ribbon, and SS-ribbon are 7, 8, 9, and 6, respectively.[19]

for the Perdew–Burker–Ernzerhof (PBE) exchange-correlation functional,[33] with the following conditions: a k-point of $1 \times 1 \times 5$ and a cutoff energy of 544 eV. To avoid interaction between periodic images of the structure, a 40 Å vacuum region was applied along a direction perpendicular to the layer of the studied samples. The convergence precision of energy for the maximum energy change, the maximum residual force on each atom, the maximum stress, and the maximum displacement were set at 2×10^{-5} Ha, 0.01 Ha/Å, 0.005 Å, and 10^{-4} GPa, respectively. In all calculations, the Fermi level is set to zero.

16.3. Results and Discussion

16.3.1. *Structural properties of the various edge ribbons*

To evaluate the stability of the four $pSiC_2$ nanoribbons, the formation energy is calculated using the following equation:

$$E_{\text{for}} = \frac{E_{\text{total}} - (n_{\text{Si}}E_{\text{Si}} + n_{\text{C}}E_{\text{C}} + n_{\text{H}}E_{\text{H}})}{n_{\text{Si}} + n_{\text{C}} + n_{\text{H}}}, \tag{16.1}$$

with the total energy for each ribbon being E_{total}. The isolated energies of each Si, C, and H atom are E_{Si}, E_{C}, and E_{H}, respectively, and the numbers of Si, C, and H atoms are denoted by n_{Si}, n_{C}, and n_{H}, respectively. The E_{for} values of all the structures are negative and vary from −5.84 to −7.01 eV, suggesting that these nanoribbons are stable in terms of thermodynamics.

Figure 16.2 depicts the dependence of the formation energy on the width of the ribbon as well as various edges. It can be clearly seen that the formation energies of all four ribbons decrease gradually as their widths increase, meaning that the $pSiC_2$ nanoribbons become more stable with increasing width. The obtained results also show that SS-ribbon is the most stable structure due to its lowest formation energy compared with the other types with similar widths. This result is similar to the structural characterization survey of the four types of PGNRs.[17]

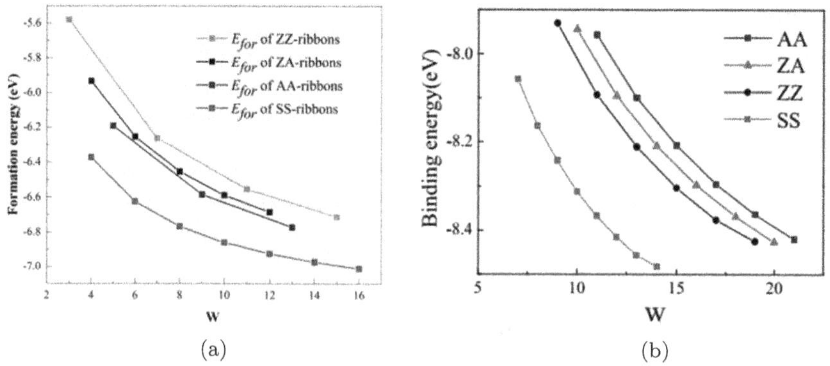

Fig. 16.2. The formation energy (binding energy) versus the width of the (a) $pSiC_2$ nanoribbons[19] and (b) penta-graphene nanoribbons.[17]

The numerical results also show that the C–C bond length is about 1.37 Å and the C–Si bond length is approximately 1.90 Å for the four studied $pSiC_2$ NRs. Buckling is a remarkable feature of the surveyed structures. Buckling decreases in the following order: ZA-$pSiC_2$ > AA-$pSiC_2$ > ZZ-$pSiC_2$ > SS-$pSiC_2$ (1.77 Å > 1.62 Å > 1.60 Å > 1.24 Å, respectively).

From the analysis of the structural properties, SS-$pSiC_2$ was evaluated as the most stable structure and exhibited the smallest buckling. Therefore, the studies of electronic structure and electron transport characteristics focus on this structure, similar to SS-PGNRs.

16.3.2. *Electronic properties of the various edge ribbons*

Figure 16.3 depicts the change in bandgap as a function of the structure width. Specifically, the bandgap decreases as the width increases. For a width of 2, the bandgap is about 2.8 eV, which abruptly decreases to 1.92 eV when the width is 4. Then, it gradually decreases to about 1.5 eV when the width is 12.

The reduction of the 1D confinement effects leads to the narrowing of the bandgap. As the width increases, the effect of the confinement becomes weaker and weaker, and the electronic properties

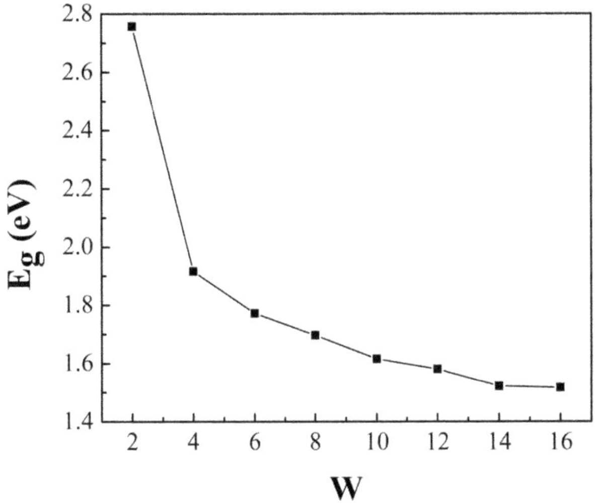

Fig. 16.3. The variation of bandgaps under the various widths of SS-pSiC_2-4.[19]

gradually correspond to those of the 2D materials. Therefore, for the widths of 12 and 14, the obtained bandgaps are approximately the same as the bandgap of the 2D pSiC_2 (1.5 eV).[34,35]

The band structures of SS-pSiC_2-4, SS-pSiC_2-8, SS-pSiC_2-12, and SS-pSiC_2-16 are shown in Fig. 16.4. All four surveyed structures have indirect bandgaps, with the valence band maximum (VMB) at the Γ point and the conduction band minimum (CBM) at the middle of the Γ and Z points. Notably, considering the subband in the valence band closest to the Fermi level, the distance between the VBM (at the Γ point) and the Z point is very small. This suggests that under a uniaxial strain, the VBM of this structure can shift. This could give some distinct electronic properties and widen the application of SS-pSiC_2. The density of states (DOS) and partial density of states (PDOS) of all systems examined were considered in Figs. 16.4 and 16.5. For all structures, the p-states of C and Si atoms contribute primarily to the DOS, while their s-states are negligible. With an increasing width, new electronic states of C and Si atoms appear in the energy zone near the Fermi level, leading to the reduction of the bandgap.

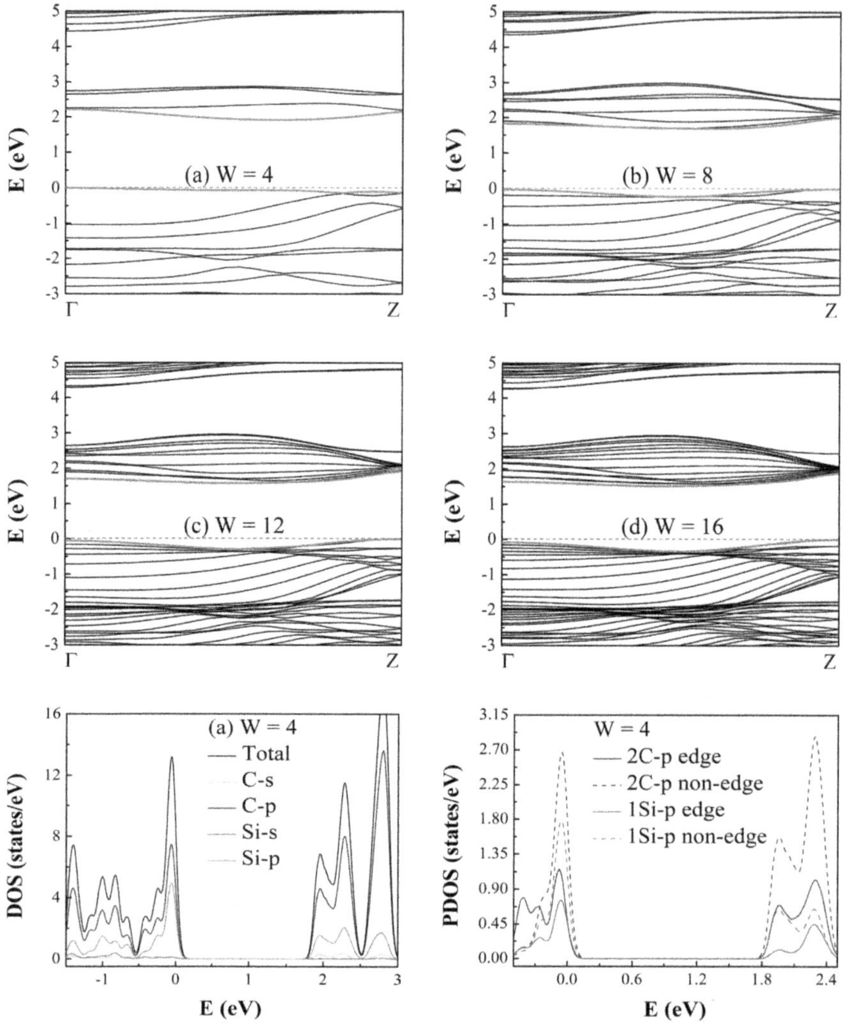

Fig. 16.4. Electronic band structures of SS-pSiC_2-4, SS-pSiC_2-8, SS-pSiC_2-12, and SS-pSiC_2-16.

16.3.3. *Structural properties of the uniaxial strain ribbons*

In this section, the electronic structures of SS-pSiC_2 nanoribbon with a width of 4 (SS-pSiC_2-4) under a sequence of uniaxial strains,

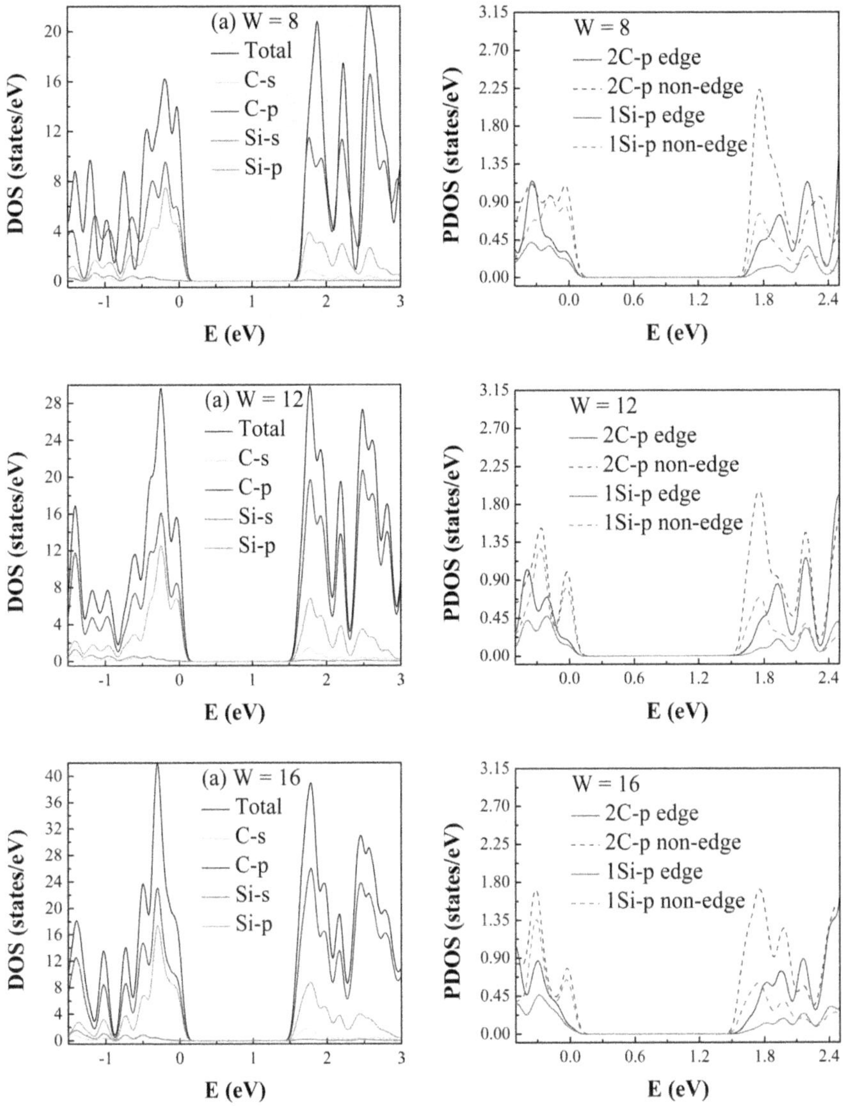

Fig. 16.5. Density of state of SS-pSiC_2-4, SS-pSiC_2-8, SS-pSiC_2-12, and SS-pSiC_2-16.

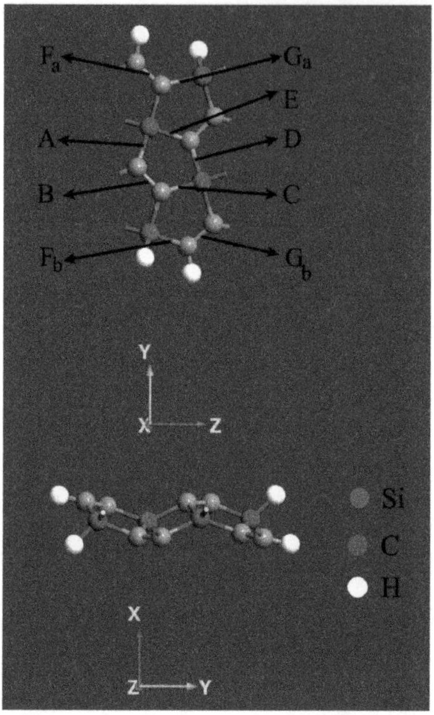

Fig. 16.6. Top and side views of a SS-pSiC_2-4 unit cell.

ranging from 10% compression to 10% stretching, are examined in detail.

The CASTEP package[36] using the plane-wave basis sets in the GGA with the PBE exchange-correlation functional[37] was used to investigate the structural relaxations and electronic structure calculations of SS-pSiC_2-4 under uniaxial strains (Fig. 16.6). The cutoff energy was set at 550 eV. A Monkhorst–Pack k-point grid of $1 \times 1 \times 13$ or $1 \times 1 \times 25$ is chosen for the structural optimization and the electronic calculations. The convergence energy precision and the maximum force on each atom were 10^{-6} eV and 0.01 eV Å^{-1}, respectively. To guarantee negligible interaction between periodic images, the separation between the layers is set to 15 Å (Ox and Oy directions).

The formula for uniaxial strain was defined as follows:

$$\varepsilon = \frac{c - c_0}{c_0}, \tag{16.2}$$

where c and c_0 are the lengths of a strained and an unstrained unit cell, respectively.

The total energy and cohesive energy were calculated to assess the energetic stability of the studied patterns:

$$E_{\text{coh}} = \frac{E_{\text{total}} - 4E_{\text{Si}} - 8E_{\text{C}} - 4E_{\text{H}}}{16}, \tag{16.3}$$

where E_{total} is the total energy of one unit cell and E_{Si}, E_{C}, and E_{H} denote the isolated energies of single Si, C, and H atoms, respectively.

As seen in Fig. 16.7, the total energy of the relaxed SS-pSiC_2-4 is lowest at -1789.64681 eV. This value increases slightly for strained SS-pSiC_2-4. Specifically, the effect of the tensile strain on the total energy is negligible, while the difference in total energy for SS-pSiC_2-4 under a compressive strain of 10% compared to SS-pSiC_2-4 is about 5 eV. The cohesive energy of those samples exhibits a similar

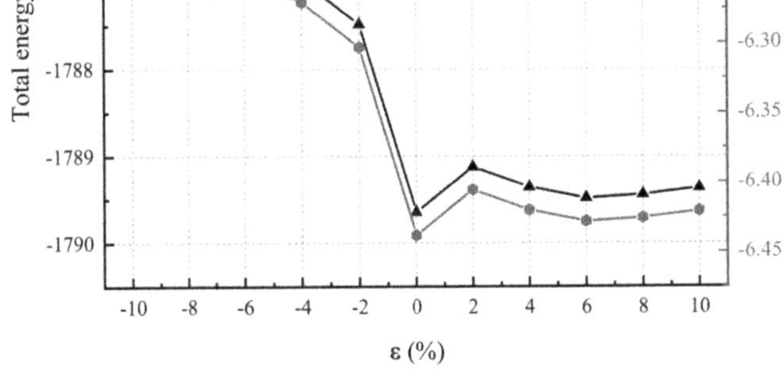

Fig. 16.7. The total energy and cohesive energy of relaxed and strained SS-pSiC2-4. The minus and plus signs refer to the compressive and tensile uniaxial strains, respectively.

tendency and falls between −6.15 and −6.45 eV. That means the considered structures are stable under uniaxial strains of up to 10% (for both the compressive and tensile strains). However, the greater the uniaxial strain, the less stable the structure, and the effect of compressive strain is greater than that of the tensile one.

To investigate the structural properties of the studied samples, the lengths of bonds A, B, C, D, E, F_a, G_a, F_b, and G_b (shown in Fig. 16.8) were determined. In the relaxed pattern, the length along the ribbons' direction of an optimized unit cell is $c_0 = 4.2$ Å and $d_{Si–C} = 1.79$ Å. The obtained results are consistent with previous studies.[18,19,35] In general, pentagonal networks in strained SS-pSiC_2-4 (from −10% to 10%) are maintained, although their bond lengths are slightly changed. The bond lengths decrease under a compressive strain and increase under a tensile strain. This is consistent with the deformation law of 2D materials and the results of previous studies. In particular, the bond lengths along the strained direction (C, E, G_a, and F_b) change significantly, while the other bond lengths vary slightly. The electronic properties of strained SS-pSiC_2-4 are predicted to be different from those of strained SS-PGNR-4 because greater variability in the bond length is observed.

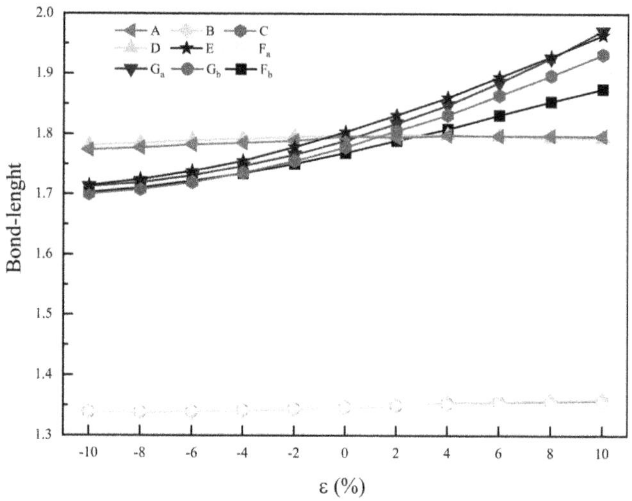

Fig. 16.8. The bond lengths (A, B, C, D, E, Fa, Ga, Fb, and Gb) of the strained SS-pSiC_2-4.

16.3.4. *Electronic properties of the uniaxial strain ribbons*

Band structures, total density of states (TDOS) and PDOS for the SS-pSiC_2-4 are represented in Fig. 16.9. Clearly, the SS-pSiC_2-4 is a semiconductor with an indirect bandgap of 1.87 eV due to a VMB appearing at the Γ point and a CBM appearing between the Γ and Z points. Considering v_1 (the upmost valence subbands), a small distance can be observed between the VBM (at the Γ point) and the Z point. In contrast, quite a large distance from the CBM to the Γ point or the Z point is observed at c_1. Therefore, it is difficult for the CBM to appear at point Γ or Z, while the VBM position can be swapped at point Γ or Z. 'The TDOS and PDOS analyses of SS-pSiC_2-4 show that the contribution of the C-2p state is the largest. Notably, a significant contribution of Si 3d orbitals (from 0.92 to 2 eV in the conduction band or from −1 to −2.5 eV in the valence band) results in a sp^2 d-like hybridization.

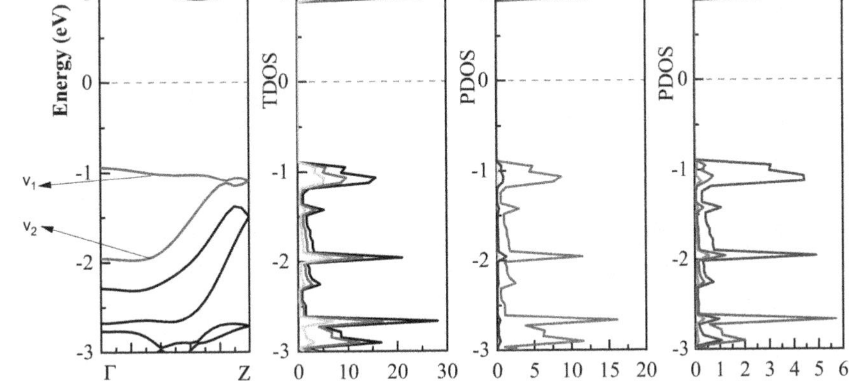

Fig. 16.9. Band structure and density of states for unstrained SS-pSiC_2-4.

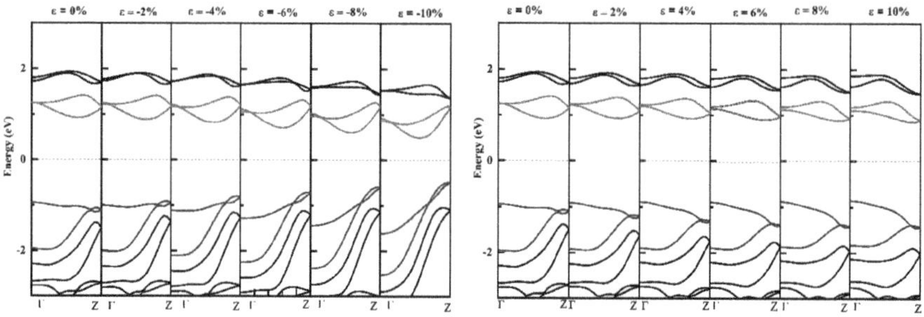

Fig. 16.10. Band structure and density of states of strained SS-pSiC_2-4 ($\varepsilon = -10\% : 10\%$).

Figure 16.10 shows the band structures of uniaxially strained SS-pSiC_2-4. These studied structures are still semiconductors with an indirect bandgap. However, under a uniaxial compressive strain, VBMs move from point Γ to point Z, while the CBMs in between Γ and Z are similar to the unstrained one. In contrast, under uniaxial compressive strain, the positions of VBMs and CBMs changed but were similar to the pristine structure (unstrained SS-pSiC_2-4). From the change in the shapes of v_1, v_2, c_1, and c_2 subbands in the investigated structures, it can be clearly seen that the compressive strain increases the dispersion curvatures of both valence and conduction bands around the Fermi level, while the influence of the tensile strain is not much.

The TDOS and PDOS under a series of uniaxial strains are plotted in Fig. 16.10. The contributions of the aforementioned states are similar to the TDOS and PDOS of the untrained SS-pSiC_2-4. The decrement in bandgap under both compressive and tensile strains increases the spread of the energy range of both valence and conduction bands when the strain value increases. However, the higher interaction between the occupied C-p and Si-d states in the conduction band is due to the reduction of the C–C bond length in SS-pSiC_2-4 under uniaxial strain. This trend leads to a lower overlapping of the atomic orbitals of the constituent atoms.

In all the uniaxially strained SS-pSiC_2-4, the bandgap (E_g) decreases compared to the unstrained one (in Fig. 16.11). The bandgap decreases rapidly in the compressive case. E_g peaks at 1.87 eV at 0% and falls at −10% to a value of 0.97 eV. In contrast,

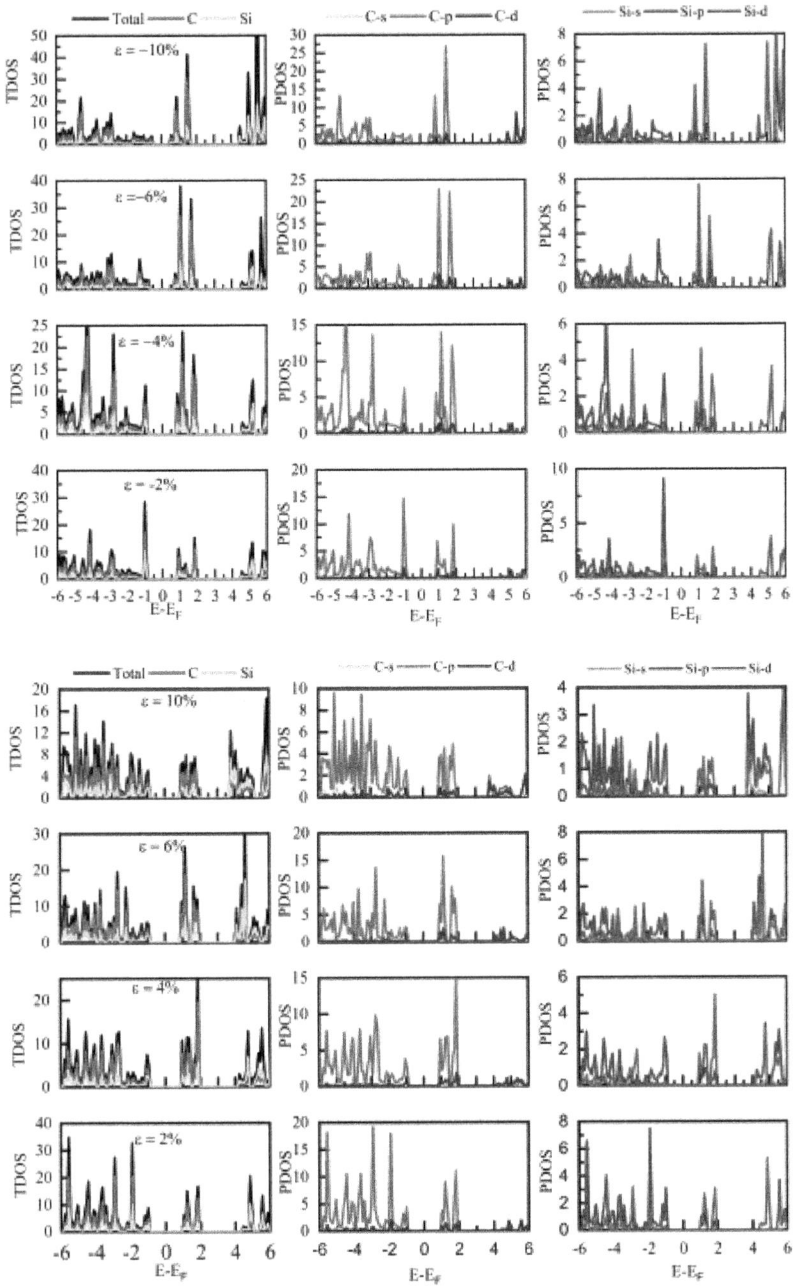

Fig. 16.11. The total and partial density of states of SS-pSiC_2-4 under uniaxial strain.

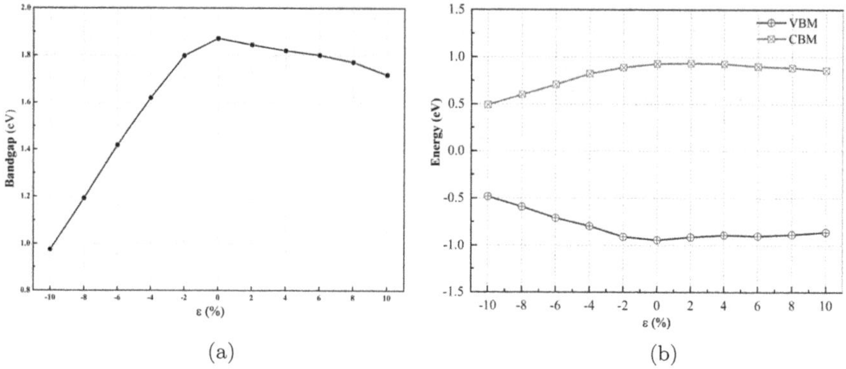

Fig. 16.12. (a) The bandgap of SS-pSiC_2-4 vs. strain. (b) The CBM/VBM of SS-pSiC_2-4 vs. strain.

E_g decreased slightly from 1.87 to 1.70 eV under a tensile strain of 0–10%. This reveals that the bandgap of SS-pSiC_2-4 is more sensitive to the compressive strain than the tensile one.

This result is similar to the uniaxial strain for 2D pSiC_2. The effect of the compressive uniaxial strain is superior, while the effect of the uniaxial compressive and uniaxial tensile strains on the bandgap is almost symmetric in 2D pSiC_2 (Fig. 16.12).[18] Meanwhile, the symmetry seems to be broken for the SS-pSiC_2-4 1D structure. This is an interesting distinction that can guide future research.

16.3.5. *Electron transport of the uniaxial strain ribbons*

The eight SS-pSiC_2-4 unit cells were used to construct model devices, including the central scattering region and two unit cells for each electrode, as shown in Fig. 16.14. Electron transport was calculated using the ATK software package[36] by the fully self-consistent DFT method with the GGA-PBE function and DZP basis sets[37] in combination with the NEGF method. BZ is sampled using the $1 \times 1 \times 21$ Monkhorst–Pack grid in the x-, y-, and z-directions, respectively. The z-axis is the direction of the nanoribbon length in the strained direction. The density mesh cutoff for the electrostatic potential is 680 eV.

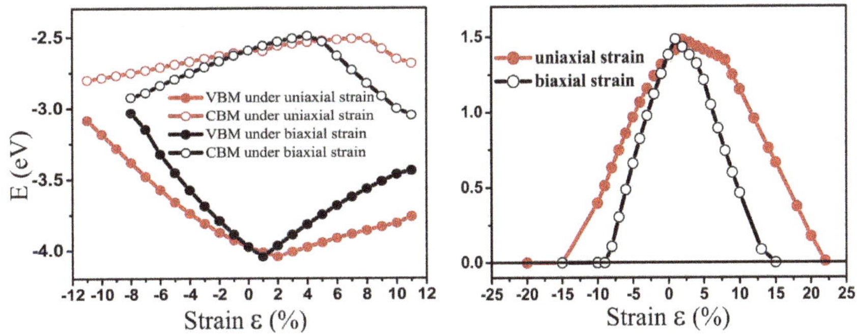

(a) (b)

Fig. 16.13. (a) The CBM/VBM with the applied uniaxial and biaxial strains. (b) The bandgap of penta-SiC_2 as a function uniaxial and biaxial strains.[18]

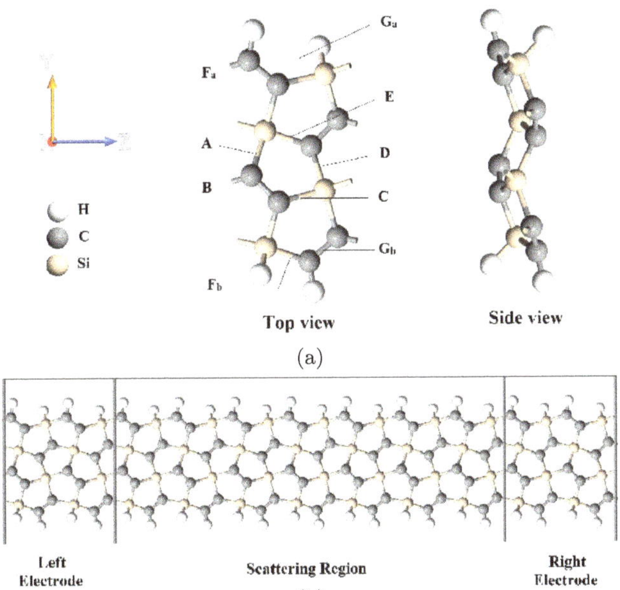

(b)

Fig. 16.14. Schematic illustration of the transport device: the central scattering region and the left and right electrodes.

The current was determined using the Landauer–Büttiker formula[38]:

$$I(V_b) = \frac{2e}{h} \int_{-\infty}^{\infty} [f(E - \mu_L) - f(E - \mu_R)] T(E, V_b) dE, \qquad (16.4)$$

where $T(E, V_b)$ is the transmission function at energy E and bias V_b[39]:

$$T(E, V_b) = Tr[\Gamma_L(E, V_b)G(E, V_b)\Gamma_R(E, V_b)G^\dagger(E, V_b)] \qquad (16.5)$$

and $f(E - \mu_L)$ and $f(E - \mu_R)$ are the Fermi distribution functions of the electrochemical potentials of the left and right leads.

$G^\dagger$ and G in Equation (16.5) are the retarded and advanced Green functions of the scattering region, while $\Gamma_{L/R}$ are the coupling functions of the left (L) and right (R) leads, respectively.

The correlation between the current and bias voltage of the compressive-strained and tensile-strained SS-pSiC_2-4 is shown in Fig. 16.15. For all the studied samples, the current intensity is almost

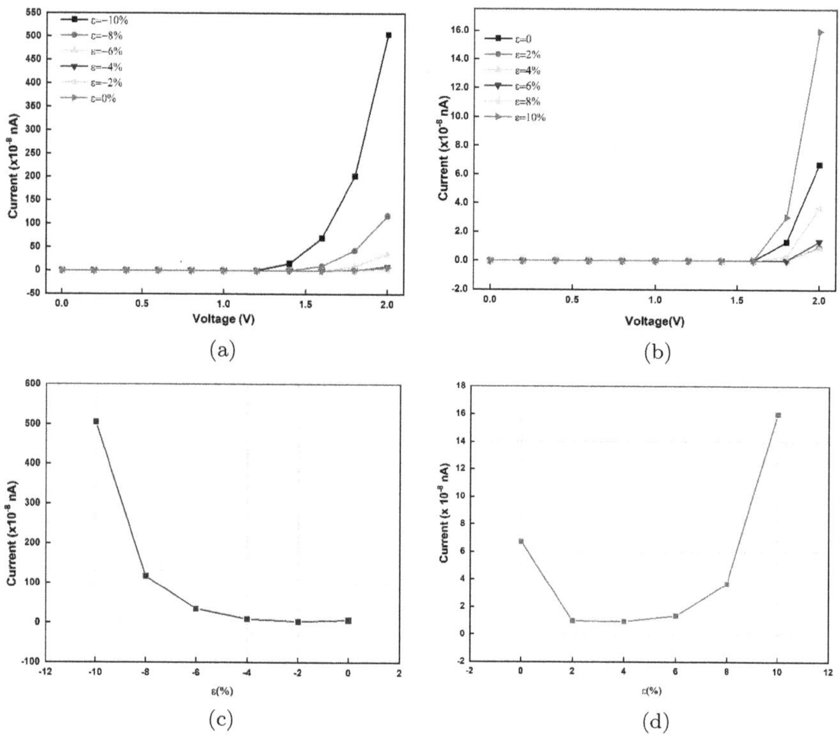

Fig. 16.15. The *I–V* curves of (a) compressive-strained SS-pSiC_2-4 and (b) tensile-strained SS-pSiC_2-4. The current at 2 V bias as a function of (c) compressive strain and (d) tensile strain.

unchanged until the bias voltage reaches a certain value. The currents of the two start to increase to this value up to 2 V and exhibit semiconductor *I–V* properties.

From Figs. 16.12 and 16.15, it can be clearly seen that there is an inverse relationship between the current strength and the bandgap. For compressive strain cases, an increase in the bandgap reduces the current, and vice versa.

The variations in the current at 2 V show a greater increase for the compressive-strained SS-pSiC_2-4 compared to that for the tensile-strained SS-pSiC_2-4. Especially, at -10%, the current increases by nearly 75 times compared to that at 0% and about 30 times compared to that at 10%.

To further understand the *I–V* curves of the studied systems, the transmission spectra as a function of the applied bias and electron energy were examined in Fig. 16.16. Similar to the current, at a bias of 2 V, the filled area bounded by the horizontal axis and the $T(E)$ curve in the bias window is proportional to ε for the compressive

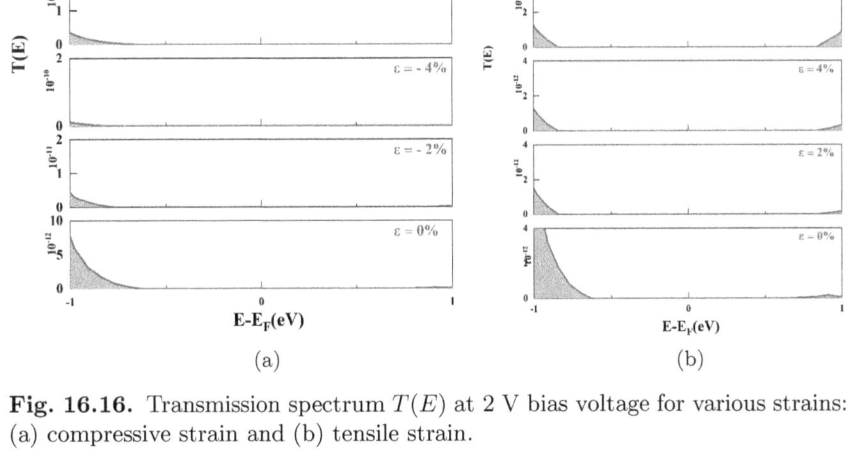

Fig. 16.16. Transmission spectrum $T(E)$ at 2 V bias voltage for various strains: (a) compressive strain and (b) tensile strain.

cases. Meanwhile, for the tensile cases, this area decreases with $\varepsilon = 0 : 4\%$ and increases with $\varepsilon = 4 : 10\%$. In addition, the order of magnitude of the increase in $T(E)$ corresponds to that of the current. Specifically, at $\varepsilon = -10\%$, $T(E)$ increases by three orders of magnitude compared to this value at $\varepsilon = 0\%$. After that, the $T(E)$ spectrum is very consistent with the $I(V)$ curve.

16.4. Potential applications

Similar to PG, $pSiC_2$ will be one of the potential candidates for device applications in heterojunctions, battery anodes, and gas sensing.

16.4.1. *Heterojunctions*

Penta-materials with diverse electronic and structural properties are predicted to be suitable candidates for electronic and multifunctional devices. In particular, the combination of a penta-material and another material can create new properties or improve their existing properties.[40–42] However, lattice mismatch is one of the biggest obstacles in heterostructures. Through theoretical calculations, it has been shown that a network deviation of less than 2% is optimal.[43] In addition, a number of theoretical studies have shown the feasibility of establishing heterostructures from penta-materials. In addition, edge banding with H or Cl atoms is more efficient for a bipolar spin filtering effect.[44] Research results on the metallization experience of penta-graphene under all chosen metal contacts, such as the PG–Au contact, can solve the problem of the lack of a low resistance metal contact in a real device.[45] These results open up potential applications for $pSiC_2$, with similar and more diverse properties than PG.

16.4.2. *Anode material*

By using first-principles simulations, the potential applications of PG as a promising anode material for Li/Na-ion batteries were studied by Xiao *et al.*[46] Their results show that a single Li/Na ion occurs at the top of the sp^2-hybridized C atoms of PG to increase its discharge rate and capacity. Compared with PG, $pSiC_2$ possesses a similar structure but even more diverse electronic properties because it is formed from two types of atoms. An anode material with outstanding performance is a promising research and development direction for $pSiC_2$.

16.4.3. *Gas sensing*

Sensors are playing an increasingly important role in health and the environment. Sensors are used to effectively detect and control toxic gases to ensure safety and protect the working and living environments. In medicine, sensors are a fast and efficient screening method. 2D materials can be used for gas sensing applications because of their high surface areas, effective carrier mobilities, and low electrical noise.[47] DFT-based calculations indicate good adsorption of gases or organic molecules onto the surface of some penta-materials.[48–50] These findings support the development of $pSiC_2$ for future sensor development.[51]

16.5. Conclusion

In conclusion, through first-principles calculations, the electronic properties of $pSiC_2$ nanoribbons (ZZ-ribbon, ZA-ribbon, AA-ribbon, and SS-ribbon) under various edge shapes are explored. Moreover, SS-$pSiC_2$ energy gaps with various widths and the energetic stability of SS-$pSiC_2$-4 under uniaxial strain of up to 10% are investigated in detail. The results show that the bandgap of SS-$pSiC_2$ decreases sharply as the width increases because of the significant effects of the 1D confinements on the feature, resulting in narrow widths. The electronic properties of SS-$pSiC_2$-4 are sensitive to uniaxial strain. In combination with NEGF formalism, the computed results show that the value of strain-dependent current differs significantly at bias potentials from 1.5 to 2 V. Consequently, SS-$pSiC_2$ is a potential candidate for electronic device development and can be an active part of strain sensors.

References

[1] Zhang, S.; Zhou, J.; Wang, Q.; Chen, X.; Kawazoe, Y.; Jena, P. Penta-graphene: A new carbon allotrope. *Proceedings of the National Academy of Sciences* **2015**, *112* (8), 2372–2377.

[2] Lan, Y.-S.; Chen, X.-R.; Hu, C.-E.; Cheng, Y.; Chen, Q.-F. Penta-pdx 2 (x = s, se, te) monolayers: Promising anisotropic thermoelectric materials. *Journal of Materials Chemistry A* **2019**, *7* (18), 11134–11142.

[3] Yagmurcukardes, M.; Sahin, H.; Kang, J.; Torun, E.; Peeters, F.; Senger, R. Pentagonal monolayer crystals of carbon, boron nitride, and silver azide. *Journal of Applied Physics* **2015**, *118* (10), 104303.

[4] Aierken, Y.; Leenaerts, O.; Peeters, F. M. A first-principles study of stable few-layer penta-silicene. *Physical Chemistry Chemical Physics* **2016**, *18* (27), 18486–18492.

[5] Liu, H.; Qin, G.; Lin, Y.; Hu, M. Disparate strain dependent thermal conductivity of two-dimensional penta-structures. *Nano Letters* **2016**, *16* (6), 3831–3842.

[6] Kuklin, A. V.; Ågren, H.; Avramov, P. V. Structural stability of single-layer PdSe2 with pentagonal puckered morphology and its nanotubes. *Physical Chemistry Chemical Physics* **2020**, *22* (16), 8289–8295.

[7] Li, P.; Zhang, J.; Zhu, C.; Shen, W.; Hu, C.; Fu, W.; Yan, L.; Zhou, L.; Zheng, L.; Lei, H.; *et al.* Penta-pdpse: A new 2d pentagonal material with highly in-plane optical, electronic, and optoelectronic anisotropy. *Advanced Materials* **2021**, *33* (35), 2102541.

[8] Bykov, M.; Bykova, E.; Ponomareva, A. V.; Tasnadi, F.; Chariton, S.; Prakapenka, V. B.; Glazyrin, K.; Smith, J. S.; Mahmood, M. F.; Abrikosov, I. A.; *et al.* Realization of an ideal cairo tessellation in nickel diazenide nin2: High-pressure route to pentagonal 2d materials. *ACS Nano* **2021**, *15* (8) 13539–13546.

[9] Kilic, M. E.; Lee, K.-R. Penta carbides: Two-dimensional group-iv semiconductors containing c2 dimers for nanoelectronics and photocatalytic water splitting. *Physical Review Materials* **2021**, *5* (6), 065404.

[10] Kilic, M. E.; Lee, K.-R. Four-penta-graphenes: Novel two-dimensional fenestrane-based auxetic nanocarbon allotropes for nanoelectronics and optoelectronics. *Carbon* **2022**.

[11] Cheng, Z.; Zhang, X.; Zhang, H.; Liu, H.; Yu, X.; Dai, X.; Liu, G.; Chen, G. Binary pentagonal auxetic materials for photocatalysis and energy storage with outstanding performances. *Nanoscale* **2022**.

[12] Quang, D. N.; Tuan, L.; Tien, N. T. Electron mobility in gaussian heavily doped zno surface quantum wells. *Physical Review B* **2008**, *77* (12), 125326.

[13] Thangadurai, T. D.; Manjubaashini, N.; Thomas, S.; Maria, H. J. Physics and chemistry of nanostructures. In *Nanostructured Materials*. Springer, 2020, pp. 47–53.

[14] Xia, Y.; Yang, P.; Sun, Y.; Wu, Y.; Mayers, B.; Gates, B.; Yin, Y.; Kim, F.; Yan, H. One-dimensional nanostructures: Synthesis, characterization, and applications. *Advanced Materials* **2003**, *15* (5), 353–389.

[15] Tien, N. T.; Phuc, V. T.; Ahuja, R. Tuning electronic transport properties of zigzag graphene nanoribbons with silicon doping and phosphorus passivation. *AIP Advances* **2018**, *8* (8), 085123.

[16] Tien, N. T.; Thao, P. T. B.; Phuc, V. T.; Ahuja, R. Influence of edge termination on the electronic and transport properties of sawtooth pentagraphene nanoribbons. *Journal of Physics and Chemistry of Solids* **2020**, *146*, 109528.

[17] Yuan, P.; Zhang, Z.; Fan, Z.; Qiu, M. Electronic structure and magnetic properties of penta-graphene nanoribbons. *Physical Chemistry Chemical Physics* **2017**, *19* (14), 9528–9536.

[18] Xu, Y.; Ning, Z.; Zhang, H.; Ni, G.; Shao, H.; Peng, B.; Zhang, X.; He, X.; Zhu, Y.; Zhu, H. Anisotropic ultrahigh hole mobility in two-dimensional penta-sic 2 by strain-engineering: Electronic structure and chemical bonding analysis. *RSC Advances* **2017**, *7* (72), 45705–45713.

[19] Mi, T. Y.; Khanh, N. D.; Ahuja, R.; Tien, N. T. Diverse structural and electronic properties of pentagonal sic2 nanoribbons: A first-principles study. *Materials Today Communications* **2021**, *26*, 102047.

[20] Tien, N. T.; Thao, P. T. B.; Phuc, V. T.; Ahuja, R. Electronic and transport features of sawtooth penta-graphene nanoribbons via substitutional doping. *Physica E: Low-dimensional Systems and Nanostructures* 2019, *114*, 113572.

[21] Tien, N. T.; Thao, P. T. B.; Chuong, D. H.; *et al.* First-principles study of electronic and optical properties of defective sawtooth penta-graphene nanoribbons. *Computational Materials Science* **2022**, *203*, 111065.

[22] Soldano, C.; Mahmood, A.; Dujardin, E. Production, properties and potential of graphene. *Carbon* **2010**, *48* (8), 2127–2150.

[23] Wang, Y.; Yang, R.; Shi, Z.; Zhang, L.; Shi, D.; Wang, E.; Zhang, G. Super-elastic graphene ripples for flexible strain sensors. *ACS Nano* **2011**, *5* (5), 3645–3650.

[24] Serry, M.; Sakr, M. A. Modeling and experimental characterization of flexible graphene composite strain sensors. In 2016 IEEE Sensors, IEEE, 2016, pp. 1–3.

[25] Kang, J.; He, Y.; Zhang, J.; Yu, X.; Guan, X.; Yu, Z. Modeling and simulation of uniaxial strain effects in armchair graphene nanoribbon tunneling field effect transistors. *Applied Physics Letters* **2010**, *96* (25), 252105.

[26] Fang, H.; Wang, R.-Z.; Chen, S.-Y.; Yan, M.; Song, X.-M.; Wang, B. Strain-induced negative differential resistance in armchair-edge graphene nanoribbons. *Applied Physics Letters* **2011**, *98* (8), 082108.

[27] On, V. V.; Thanh, L. N.; Tien, N. T. The electronic properties and electron transport of sawtooth penta-graphene nanoribbon under uniaxial strain: Ab-initio study. *Philosophical Magazine* **2020**, *100* (14), 1834–1848.

[28] Sun, L.; Li, Q.; Ren, H.; Su, H.; Shi, Q.; Yang, J. Strain effect on electronic structures of graphene nanoribbons: A first-principles study. *The Journal of Chemical Physics* **2008**, *129* (7), 074704.

[29] Abdi, M.; Bencherif, H.; Bendib, T. T.; Meddour, F.; Chahdi, M. Significant improvement of infrared graphene nanoribbon phototransistor performance: A quantum simulation study. *Sensors and Actuators A: Physical* **2021**, *317*, 112446.

[30] Jyoti, R.; Choudhary, B.; Singh, R.; Sharma, R. K.; *et al.* Enhanced sensitivity of graphene nanoribbon gas sensor for detection of oxides of nitrogen using boron and phosphorous co-doped system: A first principles study. *Sensors and Actuators A: Physical* **2021**, 112897.

[31] Tien, N. T.; Thao, D. N.; Thao, P. T. B.; Quang, D. N. Key scattering mechanisms limiting the lateral transport in a modulation-doped polar heterojunction. *Journal of Applied Physics* **2016**, *119* (21), 214304.

[32] Smidstrup, S.; Markussen, T.; Vancraeyveld, P.; Wellendorff, J.; Schneider, J.; Gunst, T.; Verstichel, B.; Stradi, D.; Khomyakov, P. A.; Vej-Hansen, U. G.; *et al.* Quantumatk: An integrated platform of electronic and atomic-scale *modelling tools. Journal of Physics: Condensed Matter* **2019**, *32* (1), 015901.

[33] Perdew, J. P.; Burke, K.; Ernzerhof, M. Generalized gradient approximation made simple. *Physical Review Letters* **1996**, *77* (18) 3865.

[34] Berdiyorov, G. R.; Madjet, M. E. A. First-principles study of electronic transport and optical properties of penta-graphene, penta-SiC_2 and penta-CN_2. *RSC Advances* **2016**, *6* (56), 50867–50873.

[35] Lin, S.; Zhang, S.; Li, X.; Xu, W.; Pi, X.; Liu, X.; Wang, F.; Wu, H.; Chen, H. Quasi-two-dimensional sic and sic2: Interaction of silicon and

carbon at atomic thin lattice plane. *The Journal of Physical Chemistry C* **2015**, *119* (34), 19772–19779.

[36] Clark, S. J.; Segall, M. D.; Pickard, C. J.; Hasnip, P. J.; Probert, M. I.; Refson, K.; Payne, M. C. First principles methods using castep. *Zeitschriftfür kristallographie-crystalline Materials* **2005**, *220* (5–6), 567–570.

[37] Perdew, J. P.; Burke, K.; Ernzerhof, M. Generalized gradient approximation made simple. *Physical Review Letters* **1996**, *77* (18), 3865.

[38] Taylor, J.; Guo, H.; Wang, J. Ab initio modeling of quantum transport properties of molecular electronic devices. *Physical Review B* **2001**, *63* (24), 245407.

[39] Landauer, R. Conductance as a consequence of incident flux. *IBM Journal of Research and Development* **1957**, *1*, 223.

[40] Datta, S. *Electronic Transport in Mesoscopic Systems.* Cambridge University Press, 1997.

[41] Choudhary, K.; Garrity, K. F.; Pilania, G.; Tavazza, F. Efficient computational design of 2D van der Waals Heterostructures: Band-alignment, lattice-mismatch, web-app generation and machine-learning. arXiv prep. arXiv:2004.03025, **2020**.

[42] Zhao, K.; Guo, Y.; Wang, Q. Contact properties of a vdW heterostructure composed of penta-graphene and penta-BN2 sheets. *Journal of Applied Physics* **2018**, *124*, 165103.

[43] Li, J.; Fan, X.; Wei, Y.; Chen, G. Penta-B_xN_y sheet: A density functional theory study of two-dimensional material. *Scientific Reports 2016*, *6*, 1–9.

[44] Guo, Y.; Wang, F.Q.; Wang, Q. An all-carbon vdW heterojunction composed of penta-graphene and graphene: Tuning the Schottky barrier by electrostatic gating or nitrogen doping. *Applied Physics Letters* **2017**, *111*, 073503.

[45] Yi, X.; Long, M.; Liu, A.; Zhang, S.; Li, M.; Xu, H. Study on electronic structures and transport properties of penta-graphene and penta-B2N4 heterojunctions. *Physica B: Condensed Matter* **2020**, *595*, 412362.

[46] Hassan, A.; Guo, Y.; Wang, Q.; Kawazoe, Y.; Jena, P. Interfacial properties of penta- graphene-metal contacts. *Journal of Applied Physics* **2019**, *125*, 065308.

[47] Xiao, B.; Li, Y.-C.; Yu, X.-F.; Cheng, J.-B. Penta-graphene: A promising anode material as the Li/Na-ion battery with both extremely high theoretical capacity and fast charge/discharge rate. *ACS Applied Materials* **2016**, *8*, 35342–35352.

[48] Meng, R.-S.; Cai, M.; Jiang, J.-K.; Liang, Q.-H.; Sun, X.; Yang, Q.; Tan, C.-J.; Chen, X.-P. First principles investigation of small molecules adsorption on antimonene. *IEEE Electron Device Letters* **2016**, *38*, 134–137.

[49] Cheng, M. Q.; Chen, Q.; Yang, K.; Huang, W. Q.; Hu, W. Y.; Huang, G. F. Penta-graphene as a potential gas sensor for NOx detection. *Nanoscale Research Letters* **2019**, *14* (1), 1–8.

[50] Gao, Y.; Liu, X.; Hu, W.; Yang, J. Tunable n-type and p-type doping of two-dimensional layered $PdSe_2$ via organic molecular adsorption. *Physical Chemistry Chemical Physics* **2020**, *22* (23), 12973–12979.

[51] Xia, X.; Guo, S.; Xu, L.; Guo, T.; Wu, Z.; Zhang, S. Sensing performance of SO_2, SO_3 and NO_2 gas molecules on 2D pentagonal PdSe2: A first-principle study. *IEEE Electron Device Letters* **2021**, *42* (4), 573–576.

CHAPTER 17

HYDROGEN ADSORPTION ONTO TWO-DIMENSIONAL GERMANENE AND ITS STRUCTURAL DEFECTS: AB INITIO INVESTIGATION

Tran Thi Thu Hanh[*,†], Nguyen Minh Phi[*,†] and Nguyen Van Hoa[*,†]

[*] *Computational Physics Lab, Ho Chi Minh City University of Technology (HCMUT), 268 Ly Thuong Kiet St., Dist. 10, Ho Chi Minh City, Vietnam*
[†] *Vietnam National University Ho Chi Minh City, Linh Trung Ward, Thu Duc District, Ho Chi Minh City, Vietnam*

Abstract

The hydrogen adsorption on pristine germanene is studied using *ab initio* calculations. By performing a converged density functional theory calculation, we find the nearly degenerated nature of hydrogen on the top sites (H_{T1} and H_{T2}), where H_{T1} is the most stable site. The adsorption of hydrogen atoms leads to local germanene structural changes. The localized surface curvature and the zero-point energy of hydrogen, which have not been studied to date for the 2D germanene, are investigated. We also demonstrate the properties of germanene defects via four obtained defects: Stone–Wales (55-77), divacancies (77-555-6 and 555-7), and pentagon-heptagon linear defect (5-7). The lowest formation energy of the pentagon-heptagon linear defect is shown for the first time in this study.

Keywords: 2D germanene, hydrogen adsorption, structural defects, density functional theory calculations.

17.1. Introduction

In recent years, the studies of two-dimensional (2D) materials have shown their excellent potential properties in nanotechnology applications.[1–3] Zhang's group showed the possibility of realizing the intrinsic quantum anomalous Hall (QAH) effect in the 2D Nb_2O_3 lattice,[3] and they also proposed a novel QAH insulator with unique states in a stanene film.[2] Among the 2D material researches, the investigations on graphene have been conducted extensively.[4–6] Because defect-free graphene is a gapless semiconductor, it is necessary to create an electronic bandgap to make graphene usable in nanoelectronics.[7,8] There have been many studies that have focused on solving this problem, such as those on strain, chemical functionalization, and the creation of defects.[9–12] Along with that, the studies on other 2D materials were also conducted,[1] especially on germanene.[13–22] This material brings out the possibility of overcoming the absence of an electronic bandgap in graphene. Germanene also has the same honeycomb structure as 2D graphene, but the atoms are buckled. The 2D buckling property has been attributed to pseudo Jahn–Teller distortion.[16] In addition, its electronic bandgap can be controlled by buckling under the application of an external electrical field.[17]

In 2009, Sahin *et al.* theoretically predicted the existence of germanene in its freestanding form.[18] The germanene sheets on Pt(111), Ag(111), and Al(111) surfaces were fabricated.[19–21] In 2014, the construction of 2D germanene on the Au(111) surface was demonstrated by Dávila *et al.*[22] Recently, many theoretical research groups have also begun studying the physical and chemical properties of this 2D material.[13,14,23] The buckling height of germanene (Δ) was found to be 0.69 Å, and the in-plane nearest-neighbor distance (d) was 2.34 Å when using the density functional theory (DFT) calculations.[24–26] However, Giang *et al.* demonstrated that the germanene buckling height was shorter ($\Delta \sim 0.66$ Å), while the distance d was longer ($d \sim 2.41$ Å) than previous results obtained using molecular dynamic (MD) and DFT studies.[14] This study has also shown that the germanene bandgap is 0.8 eV larger than in previous germanene studies.[14] This is an important and new feature that the authors have investigated. We, therefore, continue to study other properties of this obtained 2D germanene: its atomic adsorption and defects.

The interaction of the hydrogen atom (H) with 2D materials is a field of great interest currently owing to its involvement in several areas of fundamental science and technology, namely hydrogen storage. Experimental and theoretical studies on the adsorption of an H atom on graphite and graphene have been extensively conducted.[27–30] It has been experimentally shown that H atoms do not stick to a graphite surface at low temperatures (less than 30 K), but at higher temperatures, the adsorption of H atoms onto graphite or graphene was obtained.[31–33] Therefore, for the first time in this study, we investigate the adsorption of a single hydrogen atom on a 2D germanene. The convergence calculated data are checked, the zero-point energy (ZPE) of hydrogen is calculated, and the possible adsorbed sites of the hydrogen atom on the germanene and the localized germanene surface curvature are shown.

Besides, 2D germanene, which is obtainable through different techniques, can present structural defects, such as single vacancies, Stone–Wales defects, and divacancies.[13,14] So, it is important to investigate these defects too. In this work, four possible structural defects were considered: the Stone-Wales defect (55–77), the divacancy (77-555), the divacancy (555–7), and the pentagon-heptagon linear defect (5-7). We showed that the pentagon-heptagon linear germanene defect has the lowest formation energy, which means that this defect structure is the easiest to form in germanene.

17.2. Calculation Methods

17.2.1. *Computational method*

We used the DFT framework for all of the calculations.[34] The first-principles electronic structures and properties were obtained through a linear combination of atomic orbitals and a pseudopotential scheme implemented in SIESTA.[35,36] The generalized gradient approximation to the exchange-correlation functional based on revised Perdew–Burke–Ernzerhof was used.[37,38] The use of the dispersion-corrected functional has been shown to yield accurate results for hydrogen adsorption.[39,40] We adopted the double-zeta polarized basic set and a mesh cutoff of 200 Ry for the SIESTA calculation. The Fermi–Dirac function was employed with an electronic temperature of 300 K to carry out the Brillouin zone integrations. For germanene, an energy shift value of 200 meV, which determines the cutoff radius per angular momentum channel, was used. These chosen parameters provided reasonable accuracy in a previous DFT calculation for the 2D confined germanene.[14]

We used a supercell containing 60 atoms for pristine germanene (equivalent to (5×4) lateral unit cells, Fig. 17.1), which was cut from the MD calculation model.[14] The (5×4) cell was chosen because a previous study of graphene had shown that the effect of size on the adsorption of hydrogen is negligible when the size is larger than (4×4).[30] The buckling height of germanene ($\Delta \sim 0.66$ Å) and the in-plane nearest-neighbor distance ($d \sim 2.41$ Å) were used to construct

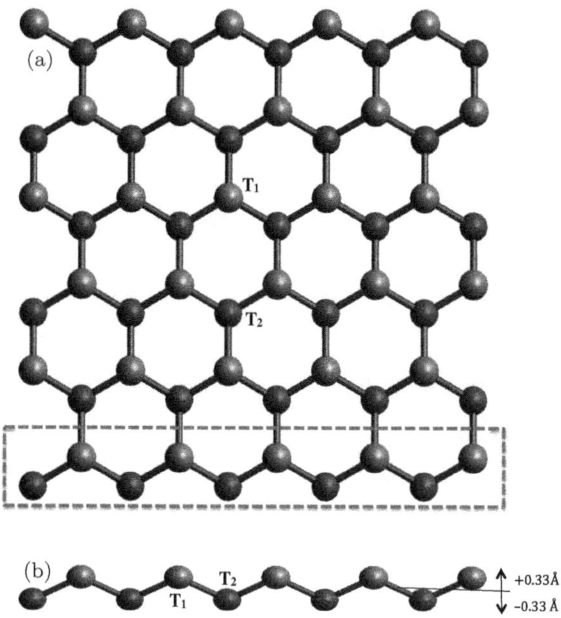

Fig. 17.1. (a) Top view and (b) side view of the cleaved part of the buckling germanene model, where T_1 and T_2 are Ge atoms with the vertical coordinate $z = +0.33$ and -0.33 Å, respectively.

the 2D Ge. The model was relaxed until the residual forces were smaller than 0.01 eV/Å.

However, for the adsorbed H atoms, a more extended basis was used to ensure that we obtain the correct bond length and accurate energy of the hydrogen molecule (H_2), which are important for long-range interactions.[41] The 60 meV value of the energy shift and the split norm of 0.53 for the second zeta were used for the H atoms. These standard computational parameters used in the SIESTA calculation provided reasonably accuracy in the calculations for a H_2 molecule. The obtained bond length of the hydrogen molecule was 0.754 Å, which is in good agreement with experimental data ~0.74 Å.[42] The binding and zero-point energies were 4.525 and 0.269 eV, respectively, which also agree well with the experimental data (4.530 and 0.270 eV, respectively).[43]

The calculation of H adsorption on germanene was done for the following sets of configurations. First, one H atom was adsorbed on the 2D germanene surface. A vacuum of 30.370 Å, by which the interaction energy between germanene monolayers can be reduced to 1 meV, was considered. The surface-irreducible Brillouin zone was sampled on the k-point mesh generated by the Monkhorst–Pack (MP) scheme of $(3 \times 3 \times 1)$ MP grid.[44] The total energy was obtained after relaxing all the H and Ge atoms. This calculation was done mainly to find the stability of the hydrogen binding sites. Second, we investigated the convergence property with respect to the k-point meshes. In the Brillouin zone integration, 1–221 special k-points were used to sample the $(1 \times 1 \times 1)$–$(21 \times 21 \times 1)$ MP grids, respectively.

17.2.2. *Zero-point energy calculation*

It has been shown that the vibrations of H around the possible adsorption sites also significantly affect the adsorption of the system.[45–48] Therefore, to calculate the H adsorption energy accurately, we continue calculating the vibrational energies of the adsorbed H atoms on the surface, which are called ZPEs. First, we displace the H atoms around the obtained equilibrium top positions both in the normal and parallel directions and calculate their energies using a harmonic approximation:

$$E = \frac{1}{2}kx^2,$$

where x (Å) is the displacement from the equilibrium position of the H atom and the coefficient k (eV/Å^2) is the force constant. Using the method of minimum square calculation, we find the force constant k. Then, using k, we continue with calculating the vibrational frequencies of the interaction of the H atoms with the germanene surface. The expression of the frequency has the following form:

$$\tilde{v} = \frac{1}{2\pi c}\sqrt{\frac{k(m_1 + m_2)}{m_1 m_2}},$$

where c is the speed of light, m_1 is the mass of hydrogen atom, and m_2 is the mass of platinum atom. Finally, ZPE is calculated through

$$\varepsilon = \frac{hv}{2} = \frac{hc}{2\lambda} = \frac{hc\tilde{v}}{2},$$

where h is Planck's constant.

17.2.3. *Defect calculations*

Defective germanene was also investigated in this work. We used 60 Ge atoms for each reconstructed model to investigate the formation energy. Four obtained structural defects were considered: the Stone–Wales defect (55-77), the divacancy (77-555), the divacancy (555-7), and the pentagon-heptagon linear defect (5-7). These defects presented a local reconstruction that can be identified using structural images. We showed that the pentagon-heptagon linear germanene defect has the lowest formation energy, which is in good agreement with the previously obtained defect structures.[14]

17.3. Results and Discussions

17.3.1. *Hydrogen adsorption on germanene*

We begin by showing the possible adsorption sites of H atoms on germanene. It was shown that, when we fix and unfix the relaxed

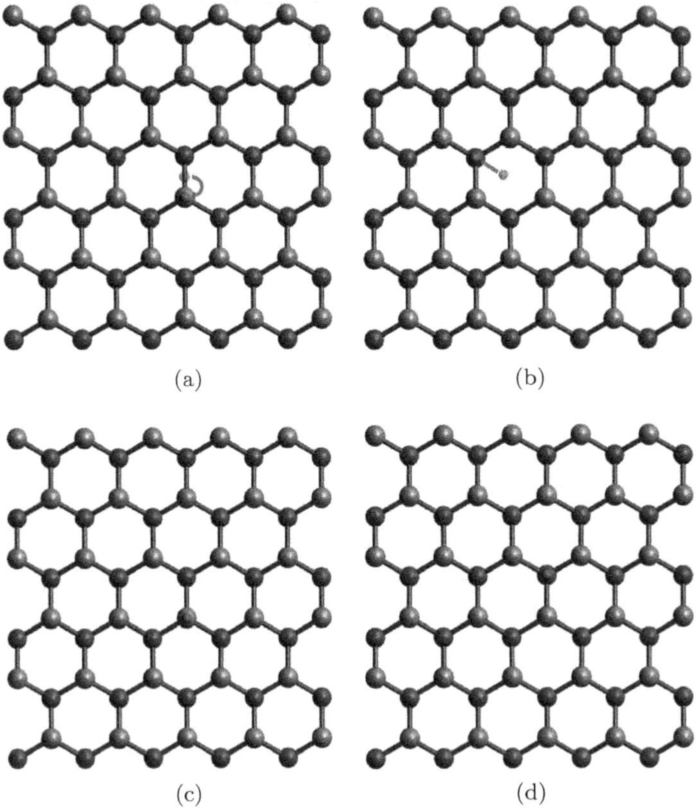

Fig. 17.2. After relaxing the system of the set H atoms (red points) on the (a) bridge, (b) center, (c) T_1, and (d) T_2 sites of germanene, the H atoms are adsorbed only on the T_1 and T_2 sites.

germanene monolayer by using the k-point of $(3 \times 3 \times 1)$ MP grid, the H atoms can be adsorbed only on the top sites (H_{T1} and H_{T2}, Fig. 17.2). This result is in agreement with that of a previous study of hydrogen adsorption on graphene.[30] Besides, the experimental studies of the hydrogen adsorption configurations on Ge(001) probed with the real-space scanning tunnel microscope method[49] and the adsorption of H on the Ge/Si(001) surface probed with the low-energy electron diffraction method[50] also show that the H atoms are adsorbed onto the Ge surface at room temperature. Then, we check the convergence of k-point by calculating the H_{T1} and H_{T2}

Table 17.1. The adsorption energy of the H atoms on the germanene surface on the T_1 and T_2 sites with different k-points.

k-point	T_1	T_2
$(3 \times 3 \times 1)$ MP	−1.833	−1.291
$(4 \times 4 \times 1)$ MP	−1.838	−1.288
$(5 \times 5 \times 1)$ MP	−1.839	−1.290
$(6 \times 6 \times 1)$ MP	−1.839	−1.292
$(7 \times 7 \times 1)$ MP	−1.839	−1.292
$(8 \times 8 \times 1)$ MP	−1.841	−1.293
$(9 \times 9 \times 1)$ MP	−1.839	−1.291
$(10 \times 10 \times 1)$ MP	−1.839	1.291
$(11 \times 11 \times 1)$ MP	−1.839	−1.291
$(12 \times 12 \times 1)$ MP	−1.839	−1.291
$(15 \times 15 \times 1)$ MP	−1.839	−1.291
$(18 \times 18 \times 1)$ MP	−1.839	−1.291
$(21 \times 21 \times 1)$ MP	−1.839	−1.291

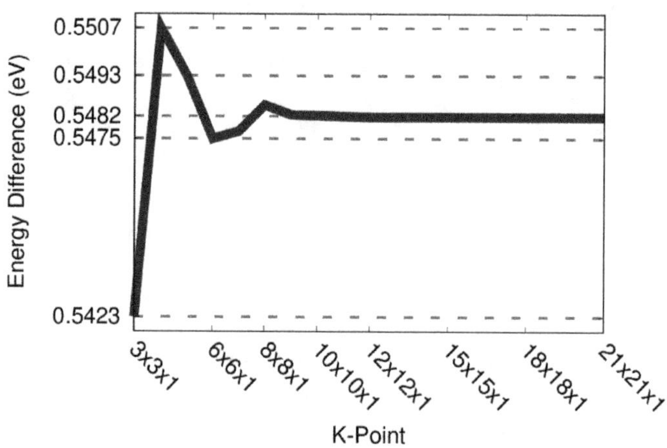

Fig. 17.3. k-point dependence of the adsorption energy difference $(E_{\text{tot}}(T_2) - E_{\text{tot}}(T_1))$.

adsorption energies when the MP grid is changed from $(1 \times 1 \times 1)$ to $(21 \times 21 \times 1)$ (see Table 17.1). Figure 17.3 plots the dependence of k-points, which shows that the result of the energy difference is unchanged when using the $(10 \times 10 \times 1)$ MP grid. Therefore, hereinafter, all the configurations are investigated with the k-point of $(10 \times 10 \times 1)$ MP grid.

The adsorption of the H atoms on the top sites T_1 and T_2 and their local distortion around the adsorbent Ge atom at the H chemisorption are shown in Fig. 17.4. Table 17.2 lists the optimized Ge–H bond lengths (d_{Ge-H}, Å), the Ge–Ge bond lengths of the adsorbent Ge atom with its neighbors ($d_{\text{Ge–Ge}}$, Å), the vertical displacement of the adsorbent Ge atom (h,Å),and the hydrogen adsorption energies (E_{ads}, eV) when one H atom is adsorbed onto the fixed and unfixed germanene surfaces. The hydrogen adsorption energy is calculated by

$$E_{\text{ads}} = E_{\text{tot}}(N_H) - E_{\text{tot}}(0) - \frac{N_H}{2} E_{H_2},$$

(a)

(b)

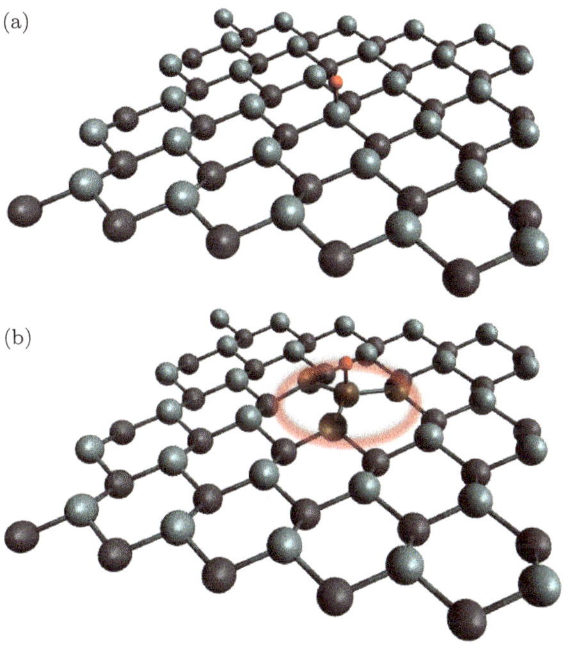

Fig. 17.4. The adsorption of H atoms (red balls) on the (a) T_1 and (b) T_2 sites at full (5×4) cell relaxation. The local distortions around the adsorbent Ge atom at the H chemisorption are observed at the T_2 site. The Ge atoms with an out-of-plane displacement upon H_{T2} atom chemisorption are indicated by the red circle (bronze-colored balls).

Table 17.2. The optimized Ge–H bond length ($d_{\text{Ge–H}}$, Å), the Ge–Ge bond length of the adsorbent Ge atom with its neighbors ($d_{\text{Ge–Ge}}$, Å), the vertical displacement of the adsorbent Ge atom (h, Å), and the hydrogen adsorption energies (E_{ads}, eV). The results from Ref. [27] are parenthesized.

	Fixed germanene				**Unfixed germanene**			
Site	$d_{\text{Ge–H}}$	$d_{\text{Ge–Ge}}$	h	E_{ads}	$d_{\text{Ge–H}}$	$d_{\text{Ge–Ge}}$	h	E_{ads}
T_1	1.58	2.49	0.00	−1.83	1.58	2.56	0.00	−1.89
T_2	1.67	2.49	0.00	−1.29	1.59	2.56	1.02 (0.50–0.70)	−1.86

where $E_{tot}(N_H)$ is the total energy of the germanene surface adsorbed with (N_H) H atoms and E_{H_2} is the total energy of the isolated H_2 molecule.

In the case of the unfixed germanene surface, when the H atom is adsorbed on the T_2 site, we find that the vertical displacement of the adsorbent Ge atom is about 1.02 Å, accompanied by an upward movement up to a maximum of 0.16 Å for the first neighboring atoms and 0.03 Å for the second neighboring atoms. This behavior results in a local hillock within the germanene layer, as in the results from a previous theoretical study of graphene: The vertical displacements are in the ranges of 0.5–0.7 Å for the adsorbent Ge atom, 0.2–0.3 Å for the first neighboring Ge atoms, and at 0.1 Å for the second neighboring Ge atoms.[30] It is interesting that when the H atom is adsorbed on the T_1 site of the unfixed germanene surface, the local hillock within the Ge layer is not observed. It means that the Ge atoms at the T_1 positions are more stable than the Ge atoms at the T_2 positions, which is consistent with the adsorption energy values shown in Table 17.1.

It has been shown that, to accurately calculate the hydrogen adsorption energy, the ZPEs of the adsorbed H atoms on the surface should be taken into account. Table 17.3 shows our vibrational frequency calculations and the ZPEs of the H_{T1} and H_{T2} atoms. We find that the vibration of the hydrogen atom along the z-axis is greatest for both the T_1 and T_2 positions, and along the x- and y-axes, the vibration is negligible (Fig. 17.5). Although the

Table 17.3. Force constants (k, eV/Å^2), vibrational frequencies (f, cm^{-1}), the ZPEs (eV), and the total adsorption energy (E_{ads}, eV) of H adsorption on the optimized fixed germanene.

	T_1			T_2		
	x	y	z	x	y	z
k(eV/Å^2)	0.679	0.692	13.656	0.861	0.428	9.521
f(cm^{-1})	26.907	27.158	120.621	30.295	21.174	100.720
ZPE (meV)	1.668	1.684	7.478	1.878	1.313	6.244
Total ZPE (eV)		0.011			0.009	
Total E_{ads} (eV)		−1.883			−1.849	

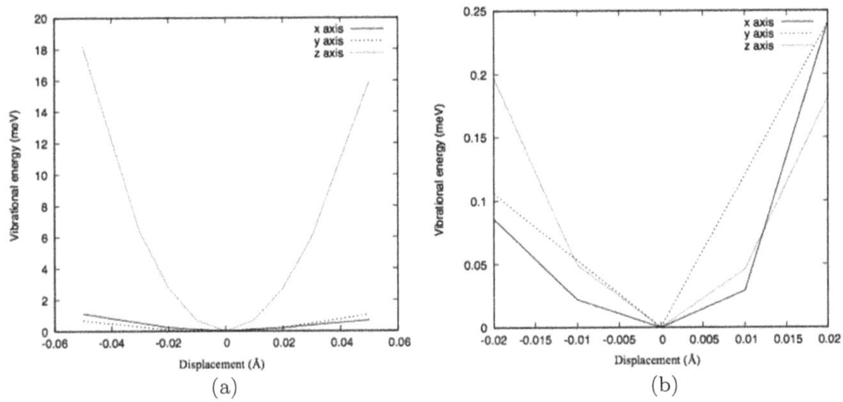

Fig. 17.5. The vibrational energy of the adsorbed hydrogen, which is displaced around the equilibrium position along the x-, y-, and z-axes.

vibrational ZPE of the H atom on the germanene surface (~10 meV) is much smaller than that on the Pt surface (~160 meV),[47] they have the same property: The hydrogen vibration is strongest along the z-axis. After adding the ZPE to the adsorption energy, the total adsorption energy of H_{T1} is found to be smaller than that of H_{T2}, which indicates that the H_{T1} site is a bit more stable than the H_{T2} site. This result once again confirms that T_1 is the most stable position in germanene.

The phonon spectra of the bare optimized germanene and the optimized unfixed germanene with adsorbed hydrogen are shown

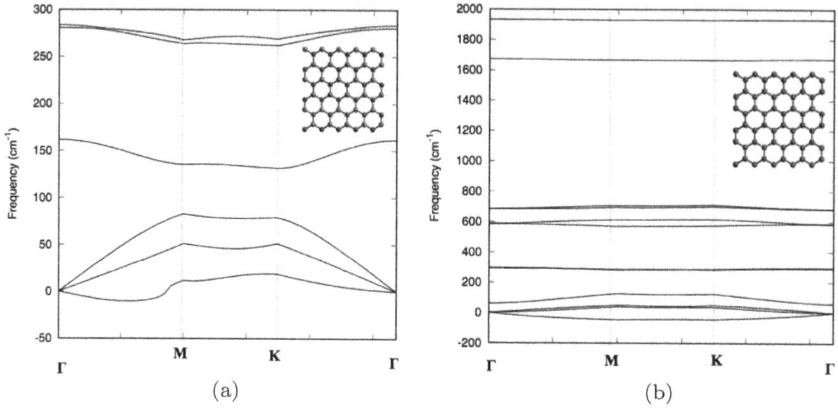

Fig. 17.6. The phonon spectrum of (a) bare germanene and (b) germanene with the adsorbed hydrogen on the T_1 and T_2 sites. The inset shows the optimized conformation of (a) bare germanene and (b) germanene with the adsorbed hydrogen (red balls).

in Fig. 17.6. The dispersion lines of the bare germanene are consistent with the previous theoretical study.[51] The crossings of the dispersion lines ZA and ZO (the acoustic and optical modes along the z-axis) are observed for the graphene system; however, they are not observed for the germanene system. This can be explained by the appearance of the buckling of the atoms in the germanene model. Additionally, for the germanene system with adsorbed hydrogen, the frequencies of ZA phonons show a quadratic dependence on the wave vectors. This is a characteristic feature of layered materials.[52] Besides, Fig. 17.6(b) shows that the vibrational frequency of the germanene system, which is much larger for the adsorbed hydrogen than for the bare germanene system.

Furthermore, to investigate the interaction of H atoms with the germanene surface, we let two adjacent hydrogen atoms adsorb onto the germanene surface at the positions T_1 and T_2 (see Fig. 17.7). The interaction energy of two adsorbed hydrogen atoms, H–H, on the germanene surface is calculated by

$$E_{\text{int}} = E_{\text{tot}}(N_H) - E_{\text{tot}}(0) - \frac{N_H}{2}E_{H_2} - E_{\text{ads}}(H_{\text{T1}}) - E_{\text{ads}}(H_{\text{T2}}),$$

(a) (b) (c)

Fig. 17.7. Two adsorbed hydrogen atoms (red balls) on (a) two T_1 sites, (b) two T_2 sites, and (c) T_1 and T_2 sites.

Table 17.4. The interaction energies (E_{int}, eV) of two hydrogen atoms on the germanene surface.

Interaction pair	$\mathbf{E_{int}}$
$H_{T1} - H_{T1}$	0.093
$H_{T2} - H_{T2}$	0.065
$H_{T1} - H_{T2}$	−0.473

where $E_{ads}(H_{T1})$ and $E_{ads}(H_{T2})$ are the adsorption energies of H_{T1} and H_{T2}, respectively. Table 17.4 lists the interaction energies of two H atoms. From this energy result, we find that two adjacent hydrogen atoms on the same type of site (T_1 or T_2) interacted via a repulsive force (positive energy), while two adjacent hydrogen atoms on different-type sites (T_1 and T_2) interacted via an attractive force (negative energy). These hydrogen interaction results for 2D germanene, which have not been reported to date, could provide the needed information for further research on the interaction of hydrogen on a 2D germanene surface.

17.3.2. *Germanene structural defects*

From the results of the germanene structure formation at room temperature in Ref. [14], we selected the four most commonly

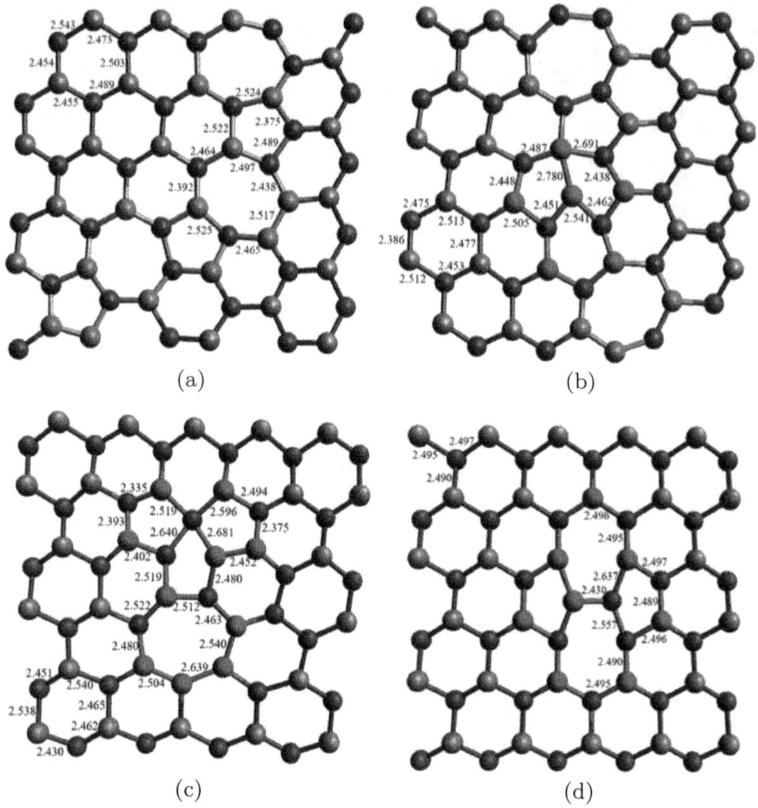

Fig. 17.8. Structural defects in germanene: (a) pentagon-heptagon linear defect (5-7), (b) divacancy (77-555-6), (c) divacancy (555-7), and (d) Stone–Wales defect (55-77). The numbers represent the Ge–Ge bond lengths in Å.

formed defects, which are shown in Fig. 17.8. In Fig. 17.8(a), the pentagon-heptagon linear defect (5-7) can be observed when moving one neighboring Ge atom. It has been shown that there are no dangling bonds in the system, and the non-magnetic ground state is always more favorable.[13] From Table 17.5, we also observe that the pentagon-heptagon linear defect has the lowest formation energy. This means that this defect is the most stable defect, and it is also the most commonly formed defect in our germanene model (see Fig. 10 in Ref. [14]). This is in contrast to previous studies on graphene, where it was shown that the Stone–Wales defect has the lowest formation

Table 17.5. Formation energy of the defects in germanene (eV).

Structural defect	Formation energy
Stone–Wales (55-77)	2.64
divacancy (77-555-6)	1.84
divacancy (555-7)	1.29
linear defect (5-7-5-7)	0.96

energy.[53,54] However, in the study by Padilha *et al.* on germanene, they showed that the single-vacancy defect has the lowest formation energy when an atom is removed from the system.[13] The difference in the formation energies between graphene and germanene defects is due to the buckled structure of germanene, which increases the relaxation degree of freedom of the defect. Moreover, the formation energy difference between our study and Padilha's study may be due to the larger value of the lattice constant, the smaller value of the in-plane nearest-neighbor distance, and the buckling height (4.17, 2.41, and 0.66 Å from our model vs, 4.08, 2.46, and 0.7 Å from Ref. [13], respectively). Besides, the three pentagons and two heptagons reconstructions are presented in Fig. 17.8(b). Figure 17.8(c) presents the three pentagons and one heptagon reconstructions. These two defects have lower formation energies than the Stone–Wales defect (see Table 17.5). The Stone–Wales defect (see Fig. 17.8(d)) can be reconstructed when a pair of germanium atoms is rotated by 90° about the center of the Ge–Ge bond. This defect can be created during the growth process or upon application of low-energy electron irradiation, as shown by experiments.[13] In the Stone–Wales defect, the two heptagons and the two pentagons are formed from the four neighboring hexagons. The Ge–Ge bond in the Stone-Wales defect is stronger than in pristine germanene because of the 90° rotation of a dimer. From Fig. 17.8(d), we can see that the Ge–Ge bond length of the dimer decreases by approximately 1% (from 2.49 to 2.43 Å). The bond length reduction is consistent with that found in the previous germanene study by Padilha *et al.*[13]

17.4. Conclusions

The hydrogen adsorption onto the defect-free germanene and the germanene structural defects were studied using the DFT calculation. The most stable adsorption sites of the hydrogen atoms are the top sites: H_{T1} and H_{T2}. The localized germanene surface curvature and the hydrogen zero-point energy were shown. The vibration of the hydrogen atom along the z-axis is greatest for both the T_1 and T_2 positions, and along the x- and y-axes, the vibration is negligible. We find that the interaction between two adjacent hydrogen atoms on the same-type sites of germanene is repulsive, while the interaction between two adjacent hydrogen atoms on the different-type sites is attractive. Furthermore, the germanene defects were demonstrated via four defects: the Stone–Wales defect (55-77), the divacancies (77-555 and 555-7), and the pentagon-heptagon linear defect (5-7). The pentagon-heptagon linear defect is shown to exhibit the lowest formation energy for the first time in this study.

Acknowledgments

This research is funded by the Vietnam National University-Ho Chi Minh City (VNU-HCM) under grant number B2018-20-01.

This research was originally published in a journal: Phys.Chem. Chem.Phys., 2020, 22, 7210.

Open Issues

The effect of hydration can be calculated in a follow-up study to find more accurate results and reduce bias.

References

[1] Wang, Y. P.; Ji, W. X.; Zhang, C. W.; Li, P.; Zhang, S. F.; Wang, P. J.; Li, S. S.; Yan, S.S. Two-dimensional arsenene oxide: A realistic large-gap quantum spin Hall insulator. *Applied Physics Letters* **2017**, *110*, 213101.

[2] Zhang, M. H.; Zhang, C. W.; Wang, P. J.; Lim S. S. Prediction of high-temperature Chern insulator with half-metallic edge states in asymmetry-functionalized stanene. *Nanoscale* **2018**, *10*, 20226.

[3] Zhang, S. J.; Zhang, C. W.; Zhang, S. F.; Ji, W. X.; Li, P.; Wang, P. J.; Li, S. S.; Yan, S. S. Intrinsic Dirac half-metal and quantum anomalous Hall phase in a hexagonal metal-oxide lattice. *Physical Review B* **2017**, *96*, 205433.
[4] Novoselov, K. S.; Geim, A. K.; Morozov, S. V.; Jiang, D.; Zhang, Y.; Dubonos, S. V.; Grigorieva, I. V.; Firsov, A. A. Electric field effect in atomically thin carbon films. *Science* **2004**, *306*, 666.
[5] Geim, A. K.; Novoselov, K. S. The rise of graphene. *Nature Materials* **2007**, *6*, 183.
[6] Novoselov, K. S.; Morozov, S. V.; Mohinddin, T. M. G.; Ponomarenko, L. A.; Elias, D. C.; Yang, R.; Barbolina, I. I.; Blake, P.; Booth, T. J.; Jiang, D.; Giesbers, J.; Hill, E. W.; Geim, A. K. Electronic properties of graphene. *Physica Status Solidi B* **2007**, *244*, 4106.
[7] Novoselov, K. S.; Geim, A. K.; Morozov, S. V.; Jiang, D.; Katsnelson, M. I.; Grigorieva, I. V.; Dubonos, S. V.; Firsov, A. A. Two-dimensional gas of massless Dirac fermions in graphene. *Nature* **2005**, *438*, 197.
[8] Boukhvalov, D.; Katsnelson, M. Tuning the gap in bilayer graphene using chemical functionalization: Density functional calculations. *Physical Review B* **2008**, *78*, 085413.
[9] Schwierz, F. Graphene transistors. *Nature Nanotechnology* **2010**, *5*, 487.
[10] Hashimoto, A.; Suenaga, K.; Gloter, A.; Urita, K.; Iijima, S. Direct evidence for atomic defects in graphene layers. *Nature* **2004**, *430*, 870.
[11] Johari, P.; Shenoy, V. B. Modulating optical properties of graphene oxide: Role of prominent functional groups. *ACS Nano* **2011**, *5*, 7640.
[12] Singh, N.; Kaloni, T. P.; Schwingenschlšgl, U. A first-principles investigation of the optical spectra of oxidized graphene. *Applied Physics Letters* **2013**, *102*, 023101.
[13] Padilha, J. E.; Pontes, R. B. Electronic and transport properties of structural defects in monolayer germanene: An ab initio investigation. *Solid State Communications* **2016**, *225*, 38–43.
[14] Giang, N. H.; Tran, T. T. H.; Hoang, V. V. Structural and thermodynamic properties of two-dimensional confined germanene: A molecular dynamics and DFT study. *Materials Research Express* **2019**, *6*, 086411.
[15] Cahangirov, S.; Topsakal, M.; Akturk, E.; Sahin, H.; Ciraci, S. Two- and one-dimensional honeycomb structures of silicon and germanium. *PRL* **2009**, *102*, 236804.
[16] Jose, D.; Datta, A. Structures and chemical properties of silicene: Unlike graphene. *Accounts of Chemical Research* **2014**, *47*, 593.
[17] Drummond, N. D.; Zólyomi, V.; Falko, V. I. Electrically tunable band gap in silicene. *Physical Review B* **2012**, *85*, 075423.

[18] Sahin, H.; Cahangirov, H.; Topsakal, M.; Bekaroglu, E.; Akturk, E.; Senger, R. T.; Ciraci, S. Monolayer honeycomb structures of group-IV elements and III-V binary compounds: First-principles calculations. *Physical Review B* **2009**, *80*, 155453.

[19] Li, L.; Lu, S. Z.; Pan, J.; Qin, Z.; Wang, Y. Q.; Wang, Y.; Cao, G. Y.; Du, S.; Gao, H. J. Buckled germanene formation on Pt (111). *Advanced Materials* **2014**, *26*, 4820.

[20] Oughaddou, H.; Sawaya, S.; Goniakowski, J.; Aufray, B.; Le Lay, G.; Gay, J. M.; Trglia, G.; Bibrian, J. P.; Barrett, N.; Guillot, C.; Mayne, A.; Dujardin, G. Ge/Ag(111) semiconductor-on-metal growth: Formation of an surface alloy. *Physical Review B* **2000**, *62*, 16653.

[21] Derivaz, M.; Dentel, D.; Stephan, R.; Hanf, M.-C.; Mehdaoui, A.; Sonnet, P.; Pirri, C. Continuous germanene layer on Al (111). *Nano Letters* **2015**, *15*, 2510.

[22] Dávila, M. E.; Xian, L.; Cahangirov, S.; Rubio, A.; Le Lay, G. Germanene: A novel two-dimensional germanium allotrope akin to graphene and silicene. *New Journal of Physics* **2014**, *16*, 095002.

[23] Huang, L. F.; Gong, P. L.; Zeng, Z. Phonon properties, thermal expansion, and thermomechanics of silicene and germanene. *Physical Review B* **2015**, *91*, 205433.

[24] Matthes, L.; Pulci, O.; Bechstedt, F. Massive Dirac quasiparticles in the optical absorbance of graphene, silicene, germanene, and tinene. *Journal of Physics: Condensed Matter* **2013**, *25*, 395305–395312.

[25] Matthes, L.; Pulci, O.; Bechstedt, F. Universal infrared absorbance of two-dimensional honeycomb group-IV crystals. *Physical Review B* **2013**, *87*, 035438–9.

[26] Balendhran, S.; Walia, S.; Nili, H.; Sriram, H.; Bhaskaran, M. Elemental analogues of graphene: Silicene, germanene, stanene, and phosphorene. *Small* **2014**, *11*, 640–52.

[27] Elias, D. C.; Nair, R.; Mohiuddin, T.; Morozov, S.; Blake, P.; Halsall, M.; Ferrari, A.; Boukhvalov, D.; Katsnelson, M.; Geim, A. *Science* **2009**, *323*, 610.

[28] Ryu, S.; Han, M. Y.; Maultzsch, J.; Heinz, T. F.; Kim, P.; Steigerwald, M. L.; Brus, L. E. Reversible basal plane hydrogenation of graphene. *Nano Letters* **2008**, *8*, 4597.

[29] Jeloaica, L.; Sidis, V. DFT investigation of the adsorption of atomic hydrogen on a cluster-model graphite surface. *Chemical Physics Letters* **1999**, *300*, 157.

[30] Ivanovskaya, V. V.; Zobelli, A.; Teillet-Billy, D.; Rougeau, N.; Sidis, V.; Briddon, P.R. Hydrogen adsorption on graphene: A first principles study. *The European Physical Journal B* **2010**, *76*, 481–486.

[31] Vidali, G.; Roser, J.; Manico, G.; Pirronello, V.; Perets, H.; Biham, O. Formation of molecular hydrogen on analogues of interstellar dust grains: Experiments and modeling. *Journal of Physics: Conference Series* **2005**, *6*, 36.

[32] Baouche, S.; Gamborg, G.; Petrunin, V.; Luntz, A.; Baurichter, A.; Hornekær, L. High translational energy release in H_2 (D_2) associative desorption from H (D) chemisorbed on C(0001). *The Journal of Chemical Physics* **2006**, *125*, 084712.

[33] Thomas, C.; Angot, T.; Layet, J. Investigation of D (H) abstraction by means of high resolution electron energy loss spectroscopy. *Surface Science* **2008**, *602*, 2311.

[34] Kohn, W.; Sham, L. J. Self-consistent equations including exchange and correlation effects. *Physical Review Journals* **1965**, *140*, A1133.

[35] Ordejón, P.; Artacho, E.; Soler, J. M. Self-consistent order-N density-functional calculations for very large systems. *Physical Review B* **1996**, *53*, R10441.

[36] Soler, J. M.; Artacho, E.; Gale, J. D.; Garcia, A.; Junquera, J.; Ordejon, P. The SIESTA method for ab initio order-N materials simulation. *Journal of Physics: Condensed Matter* **2002**, *14*, 2745.

[37] Perdew, J. P.; Burke, K.; Ernzerhof, M. Generalized gradient approximation made simple. *Physical Review Letters* **1996**, *77*, 3865–3868.

[38] Zhang, Y.; Yang, W. Comment on "Generalized gradient approximation made simple." *Physical Review Letters* **1998**, *80*, 890.

[39] Andersson, M. P. Density functional theory with modified dispersion correction for metals applied to molecular adsorption on Pt (111). *Physical Chemistry Chemical Physics* **2016**, *18*, 19118–19122.

[40] Ilawe, N. V.; Zimmerman, J. A.; Wong, B. M. Breaking badly: DFT-D2 gives sizeable errors for tensile strengths in palladium-hydride solids. *Journal of Chemical Theory and Computation* **2015**, *11*, 5426–5435.

[41] Hanh, T. T. T.; Takimoto, Y.; Sugino, O. First-principles thermodynamic description of hydrogen electroadsorption on the Pt (111) surface. *Surface Science* **2014**, *625*, 104.

[42] Roger, L. D.; Harry, B. G. *Chemical Structure and Bonding.* University Science Books, 1989, p. 199.

[43] Kunimatsu, K.; Uchida, H.; Osawa, M.; Watanabe, M. In situ infrared spectroscopic and electrochemical study of hydrogen electro-oxidation on Pt electrode in sulfuric acid. *Journal of Electroanalytical Chemistry* **2006**, *587*, 299.

[44] Monkhorst, H. J.; Pack, D. Special points for Brillouin-zone integrations. *Physical Review B* **1976**, *13*, 5188.

[45] Tomonari, M.; Sugino, O. DFT calculation of vibrational frequency of hydrogen atoms on Pt electrodes: Analysis of the electric field dependence of the Pt–H stretching frequency. *Chemical Physics Letters* **2007**, *437*, 170.

[46] Kunimatsu, K.; Senzaki, T.; Tsushima, M.; Osawa, M. A combined surface-enhanced infrared and electrochemical kinetics study of hydrogen adsorption and evolution on a Pt electrode. *Chemical Physics Letters* **2005**, *401*, 451.

[47] Hanh, T. T. T.; Hang, N. T. T. A DFT study of hydrogen electroadsorption on the missing row Pt (1 1 0)-(1 × 2) surface. *Journal Computational Materials Science* **2017**, *13*, 295.

[48] Hanh, T. T. T.; Hoa, N. V. Zero-point vibration of the adsorbed hydrogen on the Pt(110) surface. *Adsorption* **2020**, *25*, 1–7.

[49] Saedi, A.; Poelsema, B.; Zandvliet, H. J. W. Hydrogen adsorption configurations on Ge(001) probed with STM. *Physical Review B* **2009**, *79*, 153402.

[50] Fuse, T.; Ryu, J. T.; Fujino, T.; Inudzuka, K.; Katayama, M.; Oura, K. Adsorption of H on the Ge/Si(001) Surface as Studied by Time-of-Flight Elastic Recoil Detection Analysis and Coaxial Impact Collision Ion Scattering Spectroscopy. *Japanese Journal of Applied Physics* **1999**, *38*, 1359–1362.

[51] Ge, X. J.; Yao, K. L.; Lu, J. T. Comparative study of phonon spectrum and thermal expansion of graphene, silicene, germanene, and blue phosphorene. *Physical Review B* **2016**, *94*, 165433.

[52] Carrete, J.; Li, W.; Lindsay, L.; Broido, D. A.; Gallego, L. J.; Mingo, N. Physically founded phonon dispersions of few-layer materials and the case of borophene. *Materials Research Letters* **2016**, *4*, 204–211.

[53] Banhart, F.; Kotakoski, J.; Krasheninnikov, A. V. Structural defects in graphene. *ACS Nano* **2011**, *5*, 26–41.

[54] Saito, M.; Yamashita, K.; Oda, T. Magic numbers of graphene multivacancies. *Japanese Journal of Applied Physics* **2007**, *46*, L1185.

CHAPTER 18

POTENTIAL APPLICATIONS

Hsien-Ching Chung*, Shih-Yang Lin†, Nguyen Thi Dieu Hien†
and Ming-Fa Lin†

*RD Department, Super Double Power Technology Co., Ltd., Changhua City, Changhua County Taiwan
†Department of Physics, College of Natural Sciences, Can Tho University, Can Tho, Vietnam

Abstract

This chapter discusses the diverse applications of semiconductor materials in agriculture, 3D printing, biology, and electronic/optical devices. Semiconductors play a vital role in precision farming, 3D printing functional components, biosensors, and electronic/optical devices such as transistors, solar cells, LEDs, and lasers. These materials have revolutionized resource management, additive manufacturing, diagnostics, and advanced technology, enhancing various industries and improving quality of life.

Keywords: Semiconductor materials, Agriculture, 3D printing, Biosensors.

18.1. 3D Printing Principles and Applications

3D printing technology, also referred to as additive manufacturing, is one of the rapid prototyping techniques in which objects are made by adding layer upon layer of materials. The materials are added layer by layer at the millimeter scale. This is why 3D printing is also called an additive manufacturing process. Its applications are growing in many fields, including medical instrumentation, automobiles, construction, aircraft, engineering, manufacturing, entertainment, and education.

3D printing is an incredibly flexible technology, not only in the materials used but also in the objects that can be printed. It is an accurate, fast, and promising tool for the future of manufacturing, with a low-risk, low-cost, and fast method of producing prototypes that allow for testing the efficiency of new products and ramping up development without the need for expensive models or proprietary tools.

All 3D printing techniques are based on the same principle. A digital 3D model is sliced into many two-dimensional layers using computer-aided design (CAD) software. It is then converted to machine language for the 3D printer to execute, as shown in Fig. 18.1. The 3D-printed objects often require post-processing to achieve the optimal surface finish, depending on the type of printing technique and the size of the objects. It is quite different from traditional manufacturing processes, such as injection molding and CNC machining, which use various cutting tools to construct the desired structure from workpieces. In contrast, 3D printing requires no cutting tools. Objects are manufactured directly on the machine platform.

3D printing techniques can be classified in a variety of ways, one of which is by printing materials. 3D-printed materials are widely used and ever-growing, from plastics to metals, with even food and live tissues in development. The classification is shown in Fig. 18.2.

3D printing enables manufacturers to produce complex parts with speed and cost-effectiveness. Prototyping products using 3D printing technology is used in the research and development stage to fix design problems before mass production.

In the current paradigm of Industry 4.0, one of the key components is big cloud data. With cloud computing, the Internet of

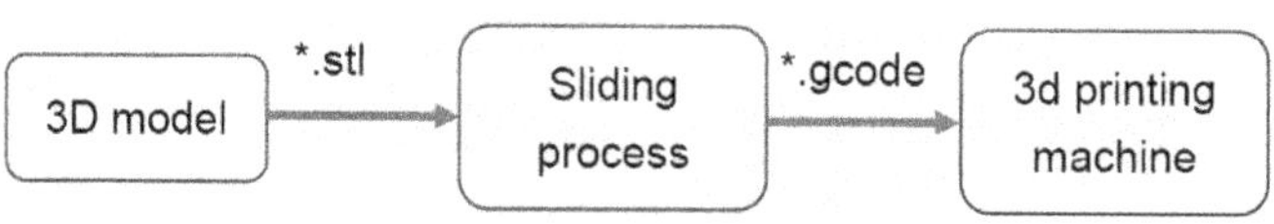

Fig. 18.1. 3D printing principle.

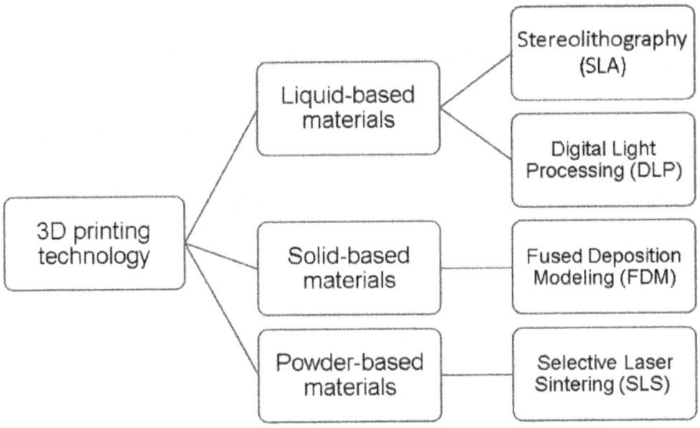

Fig. 18.2. 3D printing classification.

Things (IoT) and artificial intelligence (AI) have become available and are capable of processing big cloud data. This big data can be directly transferred to a 3D printer while prototyping without expensive molding and manufacturing tools. 3D printers play an important role in Industry 4.0 to create an evolution in the industry toward intelligent production.

18.2. Agriculture

Scientific research on nanomaterials applied to plant protection or fertilizer products has extensively increased since 2000. Most of the patents are released by the United States and Germany, and most of the scientific articles are published by Asian countries. Carbon-based nanomaterials contribute ~40% to this field.[1] The materials with sizes larger than 100 nm exhibit diverse forms, e.g., solid-doped particles, polymers, and oil-water-based structures. They often act as additives and active components. The majority of these applications are still under development, and a selection of promising approaches are presented in this section.

18.2.1. *Plant growth stimulators and fertilizers*

Many studies have reported a positive impact of carbon-based nanomaterials on plant growth, leading to research on carbon-based nanomaterial-containing fertilizers. The primary features of such fertilizers are based on amendments to mineral and organic fertilizers with carbon-based nanomaterials. In most cases, carbon-based nanomaterials act as fertilizer additions, with the aim of advancing plant nutrient availability, decreasing nutrient losses, and stimulating plant growth.

Zhai *et al.* have demonstrated that multi-wall carbon nanotubes (MWCNTs) can stimulate the growth of maize and inhibit the growth of soybean at the exposed doses. The cumulative transpiration of water in maize was almost twice as much as that in the maize control. The dry biomass of maize exposed to MWCNTs was larger than that of maize control.[2] Nair *et al.* have studied the effects of engineered carbon nanomaterials of various dimensionalities (e.g., graphene, carbon nanotubes (CNTs), and fullerene (C_{60})) on the germination of rice seeds. A pronounced increase in the rate of germination was identified for rice seeds in the presence of some of these carbon nanomaterials. Increased water content was identified in the carbon nanomaterial-treated seeds during germination compared to controls.[3] Zhang *et al.* have shown that graphene at a low concentration affected the germination of tomato seeds and seedling growth. Graphene-treated seeds germinated much more rapidly than control seeds. Analytical results showed that graphene penetrated seed husks. The penetration might break the husks to facilitate water uptake, leading to larger germination rates.[4]

18.2.2. *Nanoencapsulation and smart delivery systems*

Smart treatment delivery systems in plants, as a promising technology for the target delivery of agrochemicals, have many potential advantages. Nanoparticles tagged with agrochemicals or other substances could cut down the damage to other plant tissues and the amount of chemicals released into the environment.[5] Sarlak *et al.* have shown that fungicides encapsulated in MWCNTs functionalized with citric acid had a superior toxic influence on *Alternaria alternata* fungi compared to the non-encapsulated bulk pesticide.[6] Encapsulated agrochemicals have demonstrated improved stability, protection from degradation, a reduction in the amount of used agrochemicals, and increased efficiency of agrochemical usage.

The techniques of fertilizer broadcasting application in conventional agriculture are often associated with considerable losses of nutrients by leaching or evaporation. Controlled-release or slow-release fertilizer broadcasting techniques are developed for changing the nutrient supply according to the current demand of the plant, providing many advantages, such as avoiding overdoses, extending the function time of fertilizers, and counteracting losses by leaching. Zhang *et al.* have demonstrated a slow-release fertilizer developed by encapsulating KNO_3 pellets with graphene oxide (GO) films. After the reduced-GO (rGO) shell formed on KNO_3 pellets, the slow-release features of the fertilizer largely improved. The fertilizer release procedure was prolonged. Even for very soil-mobile nutrients (e.g., potassium nitrate), encapsulation by rGO considerably makes the procedure of fertilizer release longer. On the other hand, mass production of rGO encapsulation seems to be possible at a relatively low cost.[7]

18.2.3. *Antifungal and antibacterial agents*

Low-dimensional carbon-based nanomaterials[8–11] are promising materials for the development of fungicides owing to their antifungal properties. Wang *et al.* have evaluated the antifungal activity of six carbon-based nanomaterials, including single-walled carbon nanotubes (SWCNTs), MWCNTs, GO, rGO, C_{60}, and activated carbon, against two important plant pathogenic fungi (i.e., *Fusarium graminearum* (*F. graminearum*) and *Fusarium poae* (*F. poae*)). SWCNTs demonstrated the strongest antifungal activity, followed by MWCNTs, GO, and rGO. On the other hand, C_{60} and activated carbon exhibited no significant antifungal activity.[12]

Other studies have attributed the antimicrobial activity of GO sheets to their ability to cause damage to the microbial membrane, disrupt the membrane potential and electron transport,[13,14] and induce oxidative stress through reactive oxygen species.[15,16] The antibacterial properties of GO are also dependent on the size of the GO sheets, i.e., larger GO sheets can separate bacteria from their environment more effectively, exhibiting stronger antibacterial activity compared to small GO sheets.[17] On the other hand, the basal plane of GO sheets seems to have an important role in this antibacterial mechanism because the basal plane, masked by non-covalent protein adsorption and bovine serum albumin, leads to loss of antibacterial activity.[18] Lin *et al.* have compared the antibacterial activity of four types of graphene-based materials (graphite, graphite oxide, GO, and rGO). Under similar concentration and incubation conditions, GO dispersion exhibited the highest antibacterial activity, sequentially followed by rGO, graphite, and graphite oxide.[18] Li *et al.* have applied both Gram-positive *Staphylococcus aureus* (*S. aureus*) and Gram-negative *Escherichia coli* (*E. coli*) to study the antibacterial actions of large-area monolayer graphene sheets on a conductor, Cu, a semiconductor, Ge, and an insulator, SiO_2. The results demonstrated that the graphene sheets on Cu and Ge can surprisingly inhibit the growth of both bacteria.[19]

18.3. Biology

Graphene is formed by carbon atoms bonded by sp^2 hybridization arranged in a hexagonal lattice.[20–23] Up until now, it has been the simplest form of carbon and the thinnest material fabricated.[8,10,24–29] It can be synthesized as mono-, few-, or multi-layer graphite. X-ray diffraction and scanning tunneling microscopy (STM) techniques reveal the crystal structure of graphene as a closely packed honeycomb crystal structure.[30] It has stimulated enormous interest in various fields owing to its unique electrochemical properties, such as high thermal conductivity (3000–5000 W/mK),[31] high breakdown current density (10^8 A/cm^2),[32–34] optical transmittance for a wide range of wavelength (97.7%).[35,36] Furthermore, graphene has controllable permittivity and very high hydrophilicity[37] among its other distinctive properties.[38–41] The above-mentioned properties could enable the development of multifunctional biomedical devices.[42] The graphene family includes several members, such as monolayer graphene,[20,21] graphene oxide (GO),[43–47] reduced graphene oxide (rGO),[48–50] and multilayered graphene (MLG).[51,52]

GO has unique intrinsic physical and chemical properties, such as a large surface area,[53,54] an oxygen-containing functionality,[55] high conductivity,[53] and better biocompatibility.[56] The chemical properties of GO make it applicable to biosensing,[57] bioimaging,[58] and hypothermia capabilities.[59] GO exhibits featured characteristics as a biosensing platform owing to its capabilities for direct contact with biomolecules (i.e., making a heterogeneous chemical and electronic structure), the possibility to be processed in solution, and the ability to be modulated as an insulator, semiconductor, or semi-metal.[57] Kim *et al.* demonstrated that a hybrid gene carrier fabricated by conjugation of low-molecular-weight branched polyethylenimine (BPEI) with GO increased the effective molecular weight of BPEI, resulting in improved DNA binding, condensation, and transfection efficiency. Moreover, this hybrid material facilitated sensing and bioimaging owing to its tunable and intrinsic electrical and optical properties.[58]

On the other hand, pristine 2D graphene containing a single layer of carbon atoms arranged in a hexagonal lattice, shows superior properties to GO, e.g., surface area, diameter, thickness, stiffness, conductivity, as well as mechanical, electronic, thermal, biological, and optical properties.[44,60–62] These properties, especially biocompatibility and quick functionalization, make graphene applicable to biosensing,[63] bioimaging,[64] small molecular drug delivery,[65] cancer treatment,[66] and being an assuring platform in tissue engineering.[67] With the development of low-dimensional materials (e.g., Group IV,[68,69] Group III–V,[11] and transition metal dichalcogenides (TMDs),[70,71] these materials might play important roles in biological applications in the future.

18.4. Electronic Devices

Power semiconductor devices are applied in power supply[72,73] and power control applications.[74,75] They are characterized by high power and high speed, which help improve energy conversion efficiency. Over the years, silicon-based (Si-based) chip design architecture for power semiconductors has become mainstream.[76,77] The emergence of a third type of semiconductor materials, such as silicon carbide (SiC)[78,79] and gallium nitride (GaN),[80,81] has made the application of power semiconductor devices more diverse and efficient.

Power semiconductor devices are the core devices for power supply and power control applications, with functions such as reducing conducting resistance and improving power conversion efficiency. Among them, metal oxide semiconductor field effect transistors (MOSFETs)[81–83] and insulated gate bipolar transistors (IGBTs)[84,85] are widely used. Most importantly, both have their own advantages and disadvantages.

A MOSFET plays the role of power electronic control. According to the difference in conductivity and channel, it can be divided into *n*-type MOS (NMOS),[86,87] *p*-type MOS (PMOS),[88,89] and complementary MOS (CMOS).[90,91] In the field of high-power semiconductors, various structures of MOSFETs play different roles. The IGBT element possesses a composite structure: The input end is a MOSFET structure, and the output end is a bipolar structure. It has the characteristics of a low saturation voltage and a high switching speed, but its switching speed is lower than that of the MOSFET.

The main characteristics of traditional Si-based IGBTs are high voltage resistance and high current. They are mostly used in high-power, high-current power equipment or power infrastructure, such as railway power grids[92,93] and wind turbines.[94,95] The disadvantage is that it is relatively incapable of shrinking. MOSFETs have the characteristics of a small driving current and are mostly used in frequency conversion-oriented 3C equipment or consumer 3C products, such as mobile phone chargers and transformers for small household appliances. The disadvantage is that they cannot withstand excessive voltage and current.

In terms of technology development and application, MOSFET and IGBT have their own pain points to overcome. Due to their limited physical properties, Si materials have poor tolerance, poor conversion efficiency, and heat dissipation problems at high voltages, so they cannot fully meet the demand for new electronic power products.[96,97] Global warming is becoming more serious, and the energy policies of various countries are actively moving toward the goal of net zero carbon emissions.[98–100] In addition, in response to the needs of the times, people prefer short, small, light, thin, and easy-to-carry products, so how to shrink the products is also a major issue.

The rise of the third type of compound semiconductors, e.g., SiC and GaN, will help solve the difficulties encountered by traditional silicon-based components. The third type of compound semiconductor has the characteristics of high temperature resistance, high voltage resistance, fast action, etc., and can be widely used in high-power, high-frequency, and high-temperature electronic power systems, such as electric vehicles[101,102] and electric vehicle charging equipment,[103,104] large wind turbines,[105,106] solar panel inverters,[107] transformers,[108,109] and other applications.

The biggest advantage of SiC is its high temperature and high breakdown voltage tolerance,[110,111] while GaN has high stability and a melting point of up to 1700°.[112] Electrical conductivity and thermal conductivity are important factors mostly in transformers and chargers, such as laptops and tablets that require higher voltages, as well as mobile phone and watch charging products that require lower voltages, which can effectively shorten the charging time. The switching speed of GaN components is more than 10 times that of silicon-based devices.[113,114] Compared with Si, it is more suitable for high-frequency and high-efficiency electronic products, including 5G products. With the 2050 net-zero carbon emission goal approaching,[115,116] all countries are moving toward the electrification of fuel vehicles in terms of transportation policies and industry promotion, driving the overall electric vehicle industry. SiC and GaN can be used in the automotive industry at the same time, especially since SiC has advantages in the automotive field. The system applications of electric vehicles mainly include inverters, onboard battery chargers (OBCs), and direct current (DC) transformers.

References

[1] Gogos, A.; Knauer, K.; Bucheli, T. D. Nanomaterials in plant protection and fertilization: Current state, foreseen applications, and research priorities. *Journal of Agricultural and Food Chemistry* **2012**, *60*, 9781–9792.

[2] Zhai, G.; Gutowski, S. M.; Walters, K. S.; Yan, B.; Schnoor, J. L. Charge, size, and cellular selectivity for multiwall carbon nanotubes by maize and soybean. *Environmental Science & Technology* **2015**, *49*, 7380–7390.

[3] Nair, R.; Mohamed, M. S.; Gao, W.; Maekawa, T.; Yoshida, Y.; Ajayan, P. M.; Kumar, D. S. Effect of carbon nanomaterials on the germination and growth of rice plants. *Journal of Nanoscience and Nanotechnology* **2012**, *12*, 2212–2220.

[4] Zhang, M.; Gao, B.; Chen, J.; Li, Y. Effects of graphene on seed germination and seedling growth. *Journal of Nanoparticle Research* **2015**, *17*, 1–8.

[5] González-Melendi, P.; Fernández-Pacheco, R.; Coronado, M. J.; Corredor, E.; Testillano, P.; Risueño, M. C.; Marquina, C.; Ibarra, M. R.; Rubiales, D.; Pérez-de-Luque, A. Nanoparticles as smart treatment-delivery systems in plants: Assessment of different techniques of microscopy for their visualization in plant tissues. *Annals of Botany* **2008**, *101*, 187–195.

[6] Sarlak, N.; Taherifar, A.; Salehi, F. Synthesis of nanopesticides by encapsulating pesticide nanoparticles using functionalized carbon nanotubes and application of new nanocomposite for plant disease treatment. *Journal of Agricultural and Food Chemistry* **2014**, *62*, 4833–4838.

[7] Zhang, M.; Gao, B.; Chen, J.; Li, Y.; Creamer, A. E.; Chen, H. Slow-release fertilizer encapsulated by graphene oxide films. *Chemical Engineering Journal* **2014**, *255*, 107–113.

[8] Chung, H.-C.; Chang, C.-P.; Lin, C.-Y.; Lin, M.-F. Electronic and optical properties of graphene nanoribbons in external fields. *Physical Chemistry Chemical Physics* **2016**, *18*, 7573–7616.

[9] Chung, H.; Lee, M.; Chang, C.; Lin, M. Exploration of edge-dependent optical selection rules for graphene nanoribbons. *Optics Express* **2011**, *19*, 23350–23363.

[10] Chung, H.-C.; Lin, Y.-T.; Lin, S.-Y.; Ho, C.-H.; Chang, C.-P.; Lin, M.-F. Magnetoelectronic and optical properties of nonuniform graphene nanoribbons. *Carbon* **2016**, *109*, 883–895.

[11] Chung, H.; Chiu, C.; Lin, M. Spin-polarized magneto-electronic properties in buckled monolayer GaAs. *Scientific Reports* **2019**.

[12] Wang, X.; Liu, X.; Chen, J.; Han, H.; Yuan, Z. Evaluation and mechanism of antifungal effects of carbon nanomaterials in controlling plant fungal pathogen. *Carbon* **2014**, *68*, 798–806.

[13] Lin, N.; Dufresne, A. Surface chemistry, morphological analysis and properties of cellulose nanocrystals with gradiented sulfation degrees. *Nanoscale* **2014**, *6*, 5384–5393.

[14] Liu, S.; Hu, M.; Zeng, T. H.; Wu, R.; Jiang, R.; Wei, J.; Wang, L.; Kong, J.; Chen, Y. Lateral dimension-dependent antibacterial activity of graphene oxide sheets. *Langmuir* **2012**, *28*, 12364.

[15] áde Leon, A. On the antibacterial mechanism of graphene oxide (GO) Langmuir–Blodgett films. *Chemical Communications* **2015**, *51*, 2886–2889.

[16] Hui, L. Availability of the basal planes of GO determines whether it is antibacterial. *ACS Applied Materials & Interfaces* **2014**, *6*, 13183–13190.

[17] Hou, W.-C.; Chowdhury, I.; Goodwin Jr., D. G.; Henderson, W. M.; Fairbrother, D. H.; Bouchard, D.; Zepp, R. G. Photochemical transformation of graphene oxide in sunlight. *Environmental Science & Technology* **2015**, *49*, 3435–3443.

[18] Vecitis, C.; Zodrow, K.; Kang, S.; Elimelech, M. ACS Nano **2011**, *4*, 5471;(b) Liu, S.; Zeng, T. H.; Hofmann, M.; Burcombe, E.; Wei, J.; Jiang, R.; Kong, J.; Chen, Y. *ACS Nano 5*, 6971.

[19] Li, J.; Wang, G.; Zhu, H.; Zhang, M.; Zheng, X.; Di, Z.; Liu, X.; Wang, X. Antibacterial activity of large-area monolayer graphene film manipulated by charge transfer. *Scientific Reports* **2014**, *4*, 4359.

[20] Huang, J. Y.; Ding, F.; Yakobson, B. I.; Lu, P.; Qi, L.; Li, J. In situ observation of graphene sublimation and multi-layer edge reconstructions. *Proceedings of the National Academy of Sciences* **2009**, *106*, 10103–10108.

[21] Hong, Y.-L.; Liu, Z.; Wang, L.; Zhou, T.; Ma, W.; Xu, C.; Feng, S.; Chen, L.; Chen, M.-L.; Sun, D.-M. Chemical vapor deposition of layered two-dimensional MoSi2N4 materials *Science* **2020**, *369*, 670–674.

[22] Lin, C.-Y.; Yang, C.-H.; Chiu, C.-W.; Chung, H.-C.; Lin, S.-Y.; Lin, M.-F. *Rich Quasiparticle Properties of Low Dimensional Systems.* IOP Publishing, 2021.

[23] Nguyen, T. D. H.; Lin, S.-Y.; Chung, H.-C.; Tran, N. T. T.; Lin, M.-F. First-principles calculations for cathode, electrolyte and anode battery materials. *First-Principles Calculations for Cathode* 2021.

[24] Liang, L.; Wang, J.; Lin, W.; Sumpter, B. G.; Meunier, V.; Pan, M. Electronic bandgap and edge reconstruction in phosphorene materials. *Nano Letters* **2014**, *14*, 6400–6446.

[25] Geim, A.; Novoselov, K. The rise and rise of graphene. *Nature Nanotechnology* **2010**, *5*, 755.

[26] Chung, H.-C.; Yang, P.-H.; Li, T.-S.; Lin. M.-F. Effects of transverse electric fields on Landau subbands in bilayer zigzag graphene nanoribbons. *Philosophical Magazine* **2014**, *94*, 1859–1872.

[27] Chung, H.-C.; Su, W.-P.; Lin, M.-F. Electric-field-induced destruction of quasi-Landau levels in bilayer graphene nanoribbons. *Physical Chemistry Chemical Physics* **2013**, *15*, 868–875.

[28] Chung, H.-C.; Lee, M.-H.; Chang, C.-P.; Huang, Y.-C.; Lin, M.-F. Effects of transverse electric fields on quasi-landau levels in zigzag graphene nanoribbons. *Journal of the Physical Society of Japan* **2011**, *80*, 044602.

[29] Chung, H.; Huang, Y.; Lee, M.; Chang, C.; Lin, M. Quasi-landau levels in bilayer zigzag graphene nanoribbons. *Physica E: Low-dimensional Systems and Nanostructures* **2010**, *42*, 711–7114.

[30] Geim, A.; Novoselov, K. *Nature Materials* **2009**, *6* (3), 183.

[31] Dallimore, E. J.; Hertenstein, J. H.; Platt, M. B. Impact of cold-calling on student voluntary participation. *Journal of Management Education* **2013**, *37*, 305–341.

[32] Burg, G. W.; Prasad, N.; Fallahazad, B.; Valsaraj, A.; Kim, K.; Taniguchi, T.; Watanabe, K; Wang, Q.; Kim, M. J.; Register, L. F. Coherent interlayer tunneling and negative differential resistance with high current density in double bilayer graphene–WSe2 heterostructures. *Nano Letters* **2017**, *17*, 3919–3925.

[33] Lee, K.-J.; Chandrakasan, A. P.; Kong, J. Breakdown current density of CVD-grown multilayer graphene interconnects. *IEEE Electron Device Letters* **2011**, *32*, 557–559.

[34] Buh, G.-H.; Hwang, J.-H.; Jeon, E.-K.; So, H.-M.; Lee, J.-O.; Kong, K.-J.; Chang, H. On-chip electrical breakdown of metallic nanotubes for mass fabrication of carbon-nanotube-based electronic devices. *IEEE Transactions on Nanotechnology* **2008**, *7*, 624–627.

[35] Bae, S.; Kim, H.; Lee, Y.; Xu, X.; Park, J.-S.; Zheng, Y.; Balakrishnan, J.; Lei, T.; Ri Kim, H.; Song, Y. I. Roll-to-roll production of 30-inch graphene films for transparent electrodes *Nature Nanotechnology* **2010**, *5*, 574–578.

[36] Nair, R. R.; Blake, P.; Grigorenko, A. N.; Novoselov, K. S.; Booth, T. J.; Stauber, T.; Peres, N. M.; Geim, A. K. Fine structure constant defines visual transparency of graphene. *Science* **2008**, *320*, 1308.

[37] Doumeng, M.; Makhlouf, L.; Berthet, F.; Marsan, O.; Delbé, K.; Denape, J.; Chabert, F. A comparative study of the crystallinity of polyetheretherketone by using density, DSC, XRD, and Raman spectroscopy techniques. *Polymer Testing* **2021**, *93*, 106878.

[38] Coleman, J. N.; Lotya, M.; O'Neill, A.; Bergin, S. D.; King, P. J.; Khan, U.; Young, K.; Gaucher, A.; De, S.; Smith, R. J. Two-dimensional nanosheets produced by liquid exfoliation of layered materials. *Science* **2011**, *331*, 568–571.

[39] Novoselov, K. Nobel lecture: Graphene: Materials in the flatland. *Reviews of Modern Physics* **2011**, *83*, 837.
[40] Choi, W.; Lahiri, I.; Seelaboyina, R.; Kang, Y. *Solid State Material Science* **2010**, *35*, 52–71.
[41] Chen, J. Jang, C.; Xiao, S. D.; Ishigami, M.; Fuhrer, M. S. *Nature Nanotechnology* **2008**, *3*, 206.
[42] Novoselov, K.; Fal, V. Ko, L.; Colombo, P. R.; Gellet, M. G.; Schwab and Kim, K. *Nature* **2012**, *940*, 192–200.
[43] Zhu, Y.; Murali, S. W. I. Cai, X. Li, JW Suk, Potts, J. R.; Ruoff, R. S. *Advance Materials* **2010**, *22*, 3906–3924.
[44] Marcano, D. C.; Kosynkin, D. V.; Berlin, J. M.; Sinitskii, A.; Sun, Z.; Slesarev, A.; Alemany, L. B.; Lu; Tour, J. M. *Improved synthesis of graphene oxide. ACS Nano* **2010**, *4*, 4806.
[45] Marcano, D. C.; Kosynkin, D. V.; Berlin, J. M.; Sinitskii, A.; Sun, Z.; Slesarev, A. S.; Alemany, L. B.; Lu, W.; Tour, J. M. Improved synthesis of graphene oxide (Vol 4, p. 4806, 2010). *Acs Nano* **2018**, *12*, 2078.
[46] Gao, W. *Graphene Oxide*. Springer, **2015**, pp. 61–95.
[47] á Stankovich, S.; á Dikin, D.; á Piner, R.; á Kohlhaas, K.; áKleinhammes, A.; áJia, Y.; áWu, Y.; áNguyen, S. T.; áRuoff, R. S. *Carbon* **2007**, *45*, 1558.
[48] Eda, G.; Fanchini, G.; Chhowalla, M. *Nature Nanotechnology* **2008**, *3*, 270.
[49] Choi, W.; Chun, K.-Y.; Kim, J.; Han, C.-S. Ion transport through thermally reduced and mechanically stretched graphene oxide membrane. *Carbon* **2017**, *114*, 377–382.
[50] Compton, O. C.; Nguyen, S. T. Graphene oxide, highly reduced graphene oxide, and graphene: Versatile building blocks for carbon-based materials. *Small* **2010**, *6*, 711–723.
[51] Möller, G.; Moessner, R. Magnetic multipole analysis of kagome and artificial spin-ice dipolar arrays. *Physical Review B* **2009**, *80*, 140409.
[52] Shahil, K. M.; Balandin, A. A. Graphene–multilayer graphene nanocomposites as highly efficient thermal interface materials. *Nano Letters* **2012**, *12*, 861–867.
[53] Tien, H. N.; Hien, N. T. M.; Oh, E.-S.; Chung, J.; Kim, E. J.; Choi, W. M.; Kong, B.-S.; Hur, S. H. Synthesis of a highly conductive and large surface area graphene oxide hydrogel and its use in a supercapacitor. *Journal of Materials Chemistry A* **2013**, *1*, 208–211.
[54] Montes-Navajas, P.; Asenjo, N. G.; Santamaría, R.; Menendez, R.; Corma, A.; García, H. Surface area measurement of graphene oxide in aqueous solutions. *Langmuir* *29*, 13443–13448.

[55] Allahbakhsh, A.; Sharif, F.; Mazinani, S. The influence of oxygen-containing functional groups on the surface behavior and roughness characteristics of graphene oxide. *Nano* **2013**, *8*, 1350045.
[56] Jung, I.; Dikin, D. A.; Piner, R. D.; Ruoff, R. S. *Nano Letters* **2008**, *8*, 4283–4287.
[57] Morales-Narváez, E.; Merkoçi, A. Graphene oxide as an optical biosensing platform. *Advanced Materials* **2012**, *24*, 3298–3308.
[58] Kim, H.; Namgung, R.; Singha, K. Oh, I.-K.; Kim, W. J. Graphene oxide–polyethylenimine nanoconstruct as a gene delivery vector and bioimaging tool. *Bioconjugate Chemistry* **2011**, *22*, 2558–2567.
[59] Liu, Z.; Robinson, J. T.; Tabakman, S. M.; Yang, K.; Dai, H. Carbon materials for drug delivery & cancer therapy. *Materials Today* **2011**, *14*, 316–323.
[60] Sanchez, V. C.; Jachak, A.; Hurt, R. H.; Kane, A. B. Biological interactions of graphene-family nanomaterials: An interdisciplinary review. *Chemical Research in Toxicology* **2012**, *25*, 15–34.
[61] Jiang, Z.; Zhang, Y.; Stormer, H.; Kim, P. Quantum Hall states near the charge-neutral Dirac point in graphene. *Physical Review Letters* **2007**, *99*, 106802.
[62] Rao, C. N. R.; Sood, A. K.; Subrahmanyam, K. S.; Govindaraj, A. Graphene: The new two-dimensional nanomaterial. *Angewandte Chemie International Edition* **2009**, *48*, 7752–7577.
[63] Kuila, T.; Bose, S.; Khanra, P.; Mishra, A. K.; Kim, N. H.; Lee, J. H. *Biosensors and Bioelectronics* *26*, 4637–4648.
[64] Shen, J.; Zhu, Y.; Yang, X.; Li, C. Graphene quantum dots: emergent nanolights for bioimaging, sensors, catalysis and photovoltaic devices. *Chemical Ccommunications* **2012**, *48*, 3686–3699.
[65] Goenka, S.; Sant, V.; Sant, S. Graphene-based nanomaterials for drug delivery and tissue engineering. *Journal of Controlled Release* **2014**, *173*, 75–88.
[66] Cai, B.; Wang, S.; Huang, L.; Ning, Y.; Zhang, Z.; Zhang, G.-J. Ultrasensitive label-free detection of PNA–DNA hybridization by reduced graphene oxide field-effect transistor biosensor. *ACS Nano* **2014**, *8*, 2632–2638.
[67] Kulshrestha, S.; Khan, S.; Meena, R.; Singh, B. R.; Khan, A. U. A graphene/zinc oxide nanocomposite film protects dental implant surfaces against cariogenic *Streptococcus mutans*. *Biofouling* **2014**, 30, 1281–1294.
[68] Şahin, H.; Cahangirov, S.; Topsakal, M.; Bekaroglu, E.; Akturk, E.; Senger, R. T.; Ciraci, S. Monolayer honeycomb structures of group-IV elements and III-V binary compounds: First-principles calculations. *Physical Review B* **2009**, *80*, 155453.

[69] Li, W.-B.; Wang, Y.-M.; Chung, H.-C.; Lin, M.-F. 1D finite-width graphene nanoribbon systems: alkalization and hydrogenation. **2022**, *arXiv preprint arXiv:2206.11162*

[70] Yang, J.; Pickett, M.; Li, X.; Ohlberg, D.; Stewart, D.; Williams, R. *Nature Nanotechnology* **2008**.

[71] Xi, X.; Wang, Z.; Zhao, W.; Park, J.-H.; Law, K. T.; Berger, H.; Forró, L.; Shan, J.; Mak, K. F. Ising pairing in superconducting NbSe2 atomic layers. *Nature Physics* **2016**, *12*, 139–143.

[72] Hwang, S.-W.; Tao, H.; Kim, D.-H.; Cheng, H.; Song, J.-K.; Rill, E.; Brenckle, M. A.; Panilaitis, B.; Won, S. M.; Kim, Y.-S. A physically transient form of silicon electronics. *Science* **2012**, *337*, 1640–1644.

[73] Cook-Chennault, K. A.; Thambi, N.; Sastry, A. M. Powering MEMS portable devices — A review of non-regenerative and regenerative power supply systems with special emphasis on piezoelectric energy harvesting systems. *Smart Materials and Structures* **2008**, *17*, 043001.

[74] Fujita, H. A resonant gate-drive circuit with optically isolated control signal and power supply for fast-switching and high-voltage power semiconductor devices. *IEEE Transactions on Power Electronics* **2013**, *28*, 5423–5430.

[75] Iravani, M.; Maratukulam, D. Review of semiconductor-controlled (static) phase shifters for power systems applications. *IEEE Transactions on Power Systems* **1994**, *9*, 1833–1839.

[76] Disney, D.; Shen, Z. J. Review of silicon power semiconductor technologies for power supply on chip and power supply in package applications. *IEEE Transactions on Power Electronics* **2013**, *28*, 4168–4181.

[77] Zucker, O.; Long, J.; Smith, V.; Page, D.; Hower, P. Experimental demonstration of high-power fast-rise-time switching in silicon junction semiconductors. *Applied Physics Letters* **1976**, *29*, 261–263.

[78] Biela, J.; Schweizer, M.; Waffler, S.; Kolar, J. W. SiC versus Si — Evaluation of potentials for performance improvement of inverter and DC–DC converter systems by SiC power semiconductors. *IEEE Transactions on Industrial Electronics* **2010**, *58*, 2872–2882.

[79] Merkert, A.; Krone, T.; Mertens, A. Characterization and scalable modeling of power semiconductors for optimized design of traction inverters with Si-and SiC-devices. *IEEE Transactions on Power Electronics* **2013**, *29*, 2238–2245.

[80] Baliga, B. J. Trends in power semiconductor devices. *IEEE Transactions on electron Devices* **1996**, *43*, 1717–1731.

[81] Kordoš, P.; Heidelberger, G.; Bernát, J.; Fox, A.; Marso, M.; Lüth, H. High-power Si O2/Al Ga N/Ga N metal-oxide-semiconductor

heterostructure field-effect transistors. *Applied Physics Letters* **2005**, *87*, 143501.

[82] Frank, D. J.; Dennard, R. H.; Nowak, E.; Solomon, P. M.; Taur, Y.; Wong, H.-S. P. Device scaling limits of Si MOSFETs and their application dependencies. *Proceedings of the IEEE* **2001**, *89*, 259–288.

[83] Ren, Y.; Xu, M.; Zhou, J.; Lee, F. C. Analytical loss model of power MOSFET. *IEEE Transactions on Power Electronics* **2006**, *21*, 310–319.

[84] Xu, Y.; Ho, C. N. M.; Ghosh, A.; Muthumuni, D. An electrical transient model of IGBT-diode switching cell for power semiconductor loss estimation in electromagnetic transient simulation. *IEEE Transactions on Power Electronics* **2019**, *35*, 2979–2989.

[85] Morozumi, A.; Yamada, K.; Miyasaka, T.; Sumi, S.; Seki, Y. Reliability of power cycling for IGBT power semiconductor modules. *IEEE Transactions on Industry Applications* **2003**, *39*, 665–671.

[86] Ma, T.; Scoggan, G.; Leone, R. Comparison of interface-state generation by 25-keV electron beam irradiation in p-type and n-type MOS capacitors. *Applied Physics Letters* **1975**, *27*, 61–63.

[87] Pensl, G.; Beljakowa, S.; Frank, T.; Gao, K.; Speck, F.; Seyller, T.; Ley, L.; Ciobanu, F.; Afanas'ev, V.; Stesmans, A. Alternative techniques to reduce interface traps in n-type 4H-SiC MOS capacitors. *Physica Status Solidi (B)* **2008**, *245*, 1378–1389.

[88] Huang, W.; Khan, T.; Paul Chow, T. Comparison of MOS capacitors on n- and p-type GaN. *Journal of Electronic Materials* **2006**, *35*, 726–732.

[89] Fernandez, J.; Godignon, P.; Berberich, S.; Rebollo, J.; Brezeanu, G.; Millan, J. High frequency characteristics and modelling of p-type 6H-silicon carbide MOS structures. *Solid-State Electronics* **1996**, *39*, 1359–1364.

[90] Roy, K.; Mukhopadhyay, S.; Mahmoodi-Meimand, H. Leakage current mechanisms and leakage reduction techniques in deep-submicrometer CMOS circuits. *Proceedings of the IEEE* **2003**, *91*, 305–327.

[91] Guyot, A.; Abou-Samra, S. Low power CMOS digital design. In: *Proceedings of the Tenth International Conference on Microelectronics (Cat. No. 98EX186)*: IEEE), 1998, pp. IP6–I13.

[92] Perpina, X.; Serviere, J.-F.; Urresti-Ibañez, J.; Cortes, I.; Jorda, X.; Hidalgo, S.; Rebollo, J.; Mermet-Guyennet, M. Analysis of clamped inductive turnoff failure in railway traction IGBT power modules under overload conditions. *IEEE Transactions on Industrial Electronics* **2010**, *58*, 2706–2714.

[93] Busatto, G.; Fratelli, L.; Abbate, C.; Manzo, R.; Iannuzzo, F. Analysis and optimisation through innovative driving strategy of high power IGBT performances/EMI reduction trade-off for converter systems in railway applications. *Microelectronics Reliability* **2004**, *44*, 1443–1448.

[94] Senturk, O. S.; Helle, L.; Munk-Nielsen, S.; Rodriguez, P.; Teodorescu, R. Power capability investigation based on electrothermal models of press-pack IGBT three-level NPC and ANPC VSCs for multimegawatt wind turbines. *IEEE Transactions on Power Electronics* **2012**, *27*, 3195–3206.

[95] Senturk, O. S.; Helle, L.; Munk-Nielsen, S.; Rodriguez, P.; Teodorescu, R. Converter structure-based power loss and static thermal modeling of the press-pack IGBT three-level ANPC VSC applied to multi-MW wind turbines. *IEEE Transactions on Industry Applications* **2011**, *47*, 2505–2515.

[96] Shoji, M.; Omura, Y.; Tomizawa, M. Physical basis and limitation of universal mobility behavior in fully depleted silicon-on-insulator Si inversion layers. *Journal of Applied Physics* **1997**, *81*, 786–794.

[97] Stem, N.; Cid, M. Physical limitations for homogeneous and highly doped n-type emitter monocrystalline silicon solar cells. *Solid-State Electronics* **2004**, *48*, 197–205.

[98] Rogelj, J.; Schaeffer, M.; Meinshausen, M.; Knutti, R.; Alcamo, J.; Riahi, K.; Hare, W. Zero emission targets as long-term global goals for climate protection. *Environmental Research Letters* **2015**, *10*, 105007.

[99] Wigley, T. M. The relationship between net GHG emissions and radiative forcing with an application to article 4.1 of the Paris agreement. *Climatic Change* **2021**, *169*, 1–9.

[100] Tanaka, K.; O'Neill, B. C. The Paris agreement zero-emissions goal is not always consistent with the 1.5°C and 2°C temperature targets. *Nature Climate Change* **2018**, *8*, 319–324.

[101] Zhang, H.; Tolbert, L. M.; Ozpineci, B. Impact of SiC devices on hybrid electric and plug-in hybrid electric vehicles. *IEEE Transactions on Industry Applications* **2010**, *47*, 912–921.

[102] Wu, R.; Gonzalez, J. O.; Davletzhanova, Z.; Mawby, P. A.; Alatise, O. The potential of SiC cascode JFETs in electric vehicle traction inverters. *IEEE Transactions on Transportation Electrification* **2019**, *5*, 1349–1359.

[103] Nguyen, H. V.; Lee, D.-C.; Blaabjerg, F. A novel SiC-based multifunctional onboard battery charger for plug-in electric vehicles. *IEEE Transactions on Power Electronics* **2020**, *36*, 5635–5646.

[104] Yildirim, D.; Öztürk, S.; Çadirci, I.; Ermiş, M. All SiC PWM rectifier-based off-board ultrafast charger for heavy electric vehicles. *IET Power Electronics* **2020**, *13*, 483–494.
[105] Mubarok, F.; Armada, S.; Fagoaga, I.; Espallargas, N. Thermally sprayed SiC coatings for offshore wind turbine bearing applications. *Journal of Thermal Spray Technology* **2013**, *22*, 1303–1309.
[106] Gammeter, C.; Krismer, F.; Kolar, J. W. Comprehensive conceptualization, design, and experimental verification of a weight-optimized all-SiC 2 kV/700 V DAB for an airborne wind turbine. *IEEE Journal of Emerging and Selected Topics in Power Electronics* **2015**, *4*, 638–656.
[107] Barater, D.; Concari, C.; Buticchi, G.; Gurpinar, E.; De, D.; Castellazzi, A. Performance evaluation of a three-level ANPC photovoltaic grid-connected inverter with 650-V SiC devices and optimized PWM. *IEEE Transactions on Industry Applications* **2016**, *52*, 2475–2485.
[108] Dong, D.; Agamy, M.; Bebic, J. Z.; Chen, Q.; Mandrusiak, G. A modular SiC high-frequency solid-state transformer for medium-voltage applications: Design, implementation, and testing. *IEEE Journal of Emerging and Selected Topics in Power Electronics* **2019**, *7*, 768–778.
[109] Rothmund, D.; Guillod, T.; Bortis, D.; Kolar, J. W. 99% efficient 10 kV SiC-based 7 kV/400 V DC transformer for future data centers. *IEEE Journal of Emerging and Selected Topics in Power Electronics* **2018**, *7*, 753–767.
[110] Chang, S.-C.; Wang, S.-J.; Uang, K.-M.; Liou, B.-W. Design and fabrication of high breakdown voltage 4H-SiC Schottky barrier diodes with floating metal ring edge terminations. *Solid-state Electronics* **2005**, *49*, 437–444.
[111] Pelaz, L.; Orantes, J.; Vicente, J.; Bailon, L.; Barbolla, J. Avalanche breakdown of high-voltage pn junctions of SiC. *Microelectronics Journal* **1996**, *27*, 43–51.
[112] Harafuji, K.; Tsuchiya, T.; Kawamura, K. Molecular dynamics simulation for evaluating melting point of wurtzite-type GaN crystal. *Journal of Applied Physics* **2004**, *96*, 2501–2512.
[113] Brown, D. F.; Shinohara, K.; Corrion, A. L.; Chu, R.; Williams, A.; Wong, J. C.; Alvarado-Rodriguez, I.; Grabar, R.; Johnson, M.; Butler, C. M. High-speed, enhancement-mode GaN power switch with regrown n + GaN Ohmic contacts and staircase field plates. *IEEE Electron Device Letters* **2013**, *34*, 1118–1120.

[114] Augeau, P.; Bouysse, P.; Martin, A.; Nebus, J. M.; Quéré, R.; Lapierre, L.; Jardel, O.; Piotrowicz, S. A new GaN-based high-speed and high-power switching circuit for envelope-tracking modulators. *International Journal of Microwave and Wireless Technologies* **2014**, *6*, 13–21.

[115] Deutch, J. Is net zero carbon 2050 possible? *Joule* **2020**, *4*, 2237–2240.

[116] Oshiro, K.; Masui, T.; Kainuma, M. Transformation of Japan's energy system to attain net-zero emission by 2050. *Carbon Management* **2018**, *9*, 493–501.

CHAPTER 19

OPEN ISSUES AND NEAR-FUTURE FOCUSES

Nguyen Thanh Tien*, Vo Khuong Dien†, Nguyen Thi Han†, Nguyen Thi Dieu Hien† and Ming-Fa Lin†

**Department of Physics, College of Natural Sciences, Can Tho University, Can Tho, Vietnam*
†*Department of Physics, National Cheng Kung University, Tainan, Taiwan*

Abstract

The first-principles methods,[1,2] phenomenological models,[3,4] and experimental observations[5,6] present a number of merits and drawbacks under a grand quasi-particle framework. How to solve open issues by establishing their strong partnerships is one of the near-future research focuses. Obviously, the mainstream materials are outstanding candidates in developing more wide, broad and high viewpoints.

Keywords: Open issues, future works.

19.1. Emergent Materials

There is an urgent demand for the development of a number of atom- and molecule-enriched materials since they are expected to exhibit rich and unique quantum quasi-particle phenomena, e.g., the Coulomb-field excitations/decay rates,[4–7] the diversified magneto-optical selection rules,[8] and the quantum Hall transports in $B_xC_yN_z$ compounds,[9] composite bilayer heterojunctions,[10] binary III–VI[11]/III–V[12]/II–VI[13]/II–V[14]/IV–IV[15] semiconductor compounds, lithium oxides[16]/sulfides,[17] Group-IV nanotubes[18]/nanoribbons,[19] and their chemically modified partners.[20] How to overcome the numerous barriers, build reliable models, and overcome the experimental limitations have become the focuses of current and near-future scientific research. For example, the optical transitions, electronic excitations, and quasi-particle decay rates of an intermediate configuration of charging/discharging battery[21,22] and an adsorption-,[23] substitution-,[24] intercalation-, decoration-,[25] and heterojunction-enlarged[26] Moire superlattice with a high number of atoms/molecules cannot be evaluated through first principles because of too many decomposed wave functions. The occupied states with numerous strong energy dispersions below the Fermi level predicted by the VASP simulations are very difficult to examine using high-resolution ARPES measurements. Their strong cooperation might be the only way to reveal the measured three-dimensional (3D) band structures of the above-mentioned materials. Moreover, a reliable establishment, which covers the orbital-hybridization-enriched hopping integrals, on-site Coulomb potentials, and the spin-up and spin-down electron–electron many-body interactions, can greatly promote the enhancement of a grand quasi-particle framework with greater merits and lesser drawbacks, as clearly illustrated in a series of published books.[4–20] Most of the binary or ternary $B_xC_yN_z$ compounds have been successfully synthesized in the laboratory.[27,28] Interestingly, their crystal phases clearly show the 3D diamond-, graphite-, two-dimensional (2D) graphene-, graphene-nanoribbon-, carbon-nanotube-, and zero-dimensional (0D) carbon-disk-like crystals. Furthermore, there exist the metallic, semi-metallic, and narrow-/middle-/wide-gap semiconducting quasi-particle behavior, in which the rich band properties are greatly diversified by the chemical bondings of multi-/single-orbital hybridizations, spin configurations, dimensionalities, stacking configurations, layer numbers, open/closed edges, and periodic/open boundary conditions. Obviously, such factors play critical roles in other quasi-particle phenomena, such as the dimension-crossover behavior of layered ternary $B_xC_yN_z$ compounds with band/optical gaps, Coulomb-field excitations/de-excitations, quantized Landau levels/Landau subbands [Refs/Refs], optical/magneto-optical transitions, and Hall transport effects. Most of the wide-gap semiconductors belong to the third-generation ones with special functionalities. However, buckled honeycomb structures have not been found in experimental and theoretical studies. Systematic investigations are available for developing a grander quasi-particle framework.

A bilayer composite, an extension of bilayer graphene[29,30] or double-walled carbon nanotubes (CNTs),[31,32] is a specific hetero-junction. In general, there exist very weak interlayer atomic interactions, as shown by van der Waals ones between the neighboring graphitic sheets. Such materials mainly arise from direct bilayer superposition, e.g., the unusual IV–IV[33] and IV–BN.[34] Using the delicate VASP simulations, graphene–silicene/–germanene/–tinene/–plumbene bilayer composites[10] are predicted to exhibit diversified quasi-particle phenomena in terms of crystal symmetries, low-energy band structures, spatial charge densities, and van Hove singularities. A giant Moire superlattice is formed under a very poor commensurate case through the buckled honeycomb lattices. It is very sensitive to the intralayer and interlayer orbital hybridizations, in which the latter are much smaller than the former, while they play critical roles in determining the essential quasi-particle properties. The linear and isotropic Dirac cone valence and conduction subbands of monolayer graphene clearly show the strong modifications, e.g., two pairs of low-lying energy dispersions with simultaneous C–Si/–Ge/–Sn/–Pb dominances, and the greatly enhanced asymmetry of hole and electron spectra about the sensitive Fermi level. The low but observable charge density variations are identified along the perpendicular direction. The merged van Hove singularities can reveal the dominating orbital hybridizations, covering the sp^2, $2p_z$-$3p_z$, and sp^3 bondings in the C–C, C–Si, and Si–Si bonds. On the other hand, the spin-orbital couplings[35] might be non-negligible for Ge-/Sn-/Pb-related composites. Too many active orbitals will induce super-high simulation barriers. If the first-principles method cannot work inr such cases, the development of empirical formulas becomes very important for the full investigation of the many-body effects, e.g., the Coulomb-field excitations/de-excitations[36] and the optical transitions in the presence of stable excitons.[37–39]

The current theoretical studies only reveal the well-characterized carbon–carbon interactions for the sp^2-dominated graphene-associated systems. The position-dependent C–Si, C–Ge, C–Sn, and C–Pb atomic interactions need to be thoroughly explored in current and near-future scientific research. The greatly enlarged unit cells are very useful in verifying the magnetic-flux quantum (a periodic Aharonov–Bohm effect),[40,41] as examined by the graphene–BN composite.[42]

Group-IV materials comprise carbon (C), silicon (Si), germanium (Ge), tin (Sn), lead (Pb), and flerovium (FI), which mainly have four active orbitals (2s, $2p_x$, $2p_y$, and $2p_z$). In general, natural and man-made nanostructures might come into existence in many forms, i.e., 3D, 2D, 1D, and 0D chemical compounds. Among them, quantum dots (QDs) are artificial crystal phases that present diverse and rich chemical, physical, and material characteristics. Many significant factors can determine the essential properties of QDs, including their range of size, shape, composition, and structures, leading to manufacturing applications and technology. The variety of uses of QDs in catalysis, electronics, energy storage, biomedical technology, and imaging processing requires high-quality QD products with suitable physical and chemical parameters. QDs have clearly illustrated high-symmetry chemical environments and presented diverse quasi-particle phenomena. They are an achievement of basic and applied sciences under the strong cooperation among first-principles simulations,[3–43] phenomenological models,[3–44] and experimental observations.[45,46] The systematic studies include those on the optimal geometries, electronic/magneto-electronic properties, Coulomb-field excitations or Coulomb decays, optical/magneto-optical excitations, quantum Hall transports, and phonon spectra. In addition to pure QDs, chemical modifications play a critical role in the controllable feature properties of the nanostructural systems. For example, OH, COOH, F, and H adsorptions on few-layer Group-IV QDs might be a powerful technique to enhance the essential properties and broaden their applications. The delicate STM,[47,48] RHEED,[49] and TEM[45–50] examinations based on the DFT are urgently required to identify the position-dependent chemical bonds, which are very useful in capturing their rich pictures of physics and chemistry.

19.2. Time-Dependent LDA

In the 1960s, Hohenberg, Kohn, and Sham introduced the DFT, which is based on the concept of using the electron charge density to replace the N electron many-body systems in the ground-state electronic structure calculations.[51] As a result of how the electrons are treated in the calculations, traditional DFT is not able to treat time-dependent problems and thus the excited electronic states. Later, the time-dependent density functional theory (TD-DF) was introduced by Runge and Gross in 1984,[52] because of which the understanding of electronic excited states became possible through their theory of the linear response. This method was widely used in the calculation of optical absorption spectra for mainstream materials in the 1990s owing mainly to its better accuracy and ease of use. To date, the applications of TD-DFT have grown diversely.[53–56]

Following a large number of studies performed in cooperation by theoretical and experimental scientists, the strengths and weaknesses of the TD-DFT emerged.[57–61] TD-DFT has been found to work best for one-electron excitations, low-energy regions, little or almost vanished charge transfers, and non-delocalized systems. The problems of TD-DFT could be related to the exchange-correlation energy, exchange-correlation potential, and the frequency-dependent part of the many-body perturbation theory. The first one arises from exchange-correlation energy, typically due to the geometries related to bonds forming or breaking in the ground state. The other problem comes from exchange-correlation potential occurring in the Franck–Condon region near the ground-state equilibrium geometry. Some advances have been achieved by numerous scientists in the most recent years; however, greater functionals are still needed for extending the applications of TD-DFT. A common use of the

TD-DFT for photochemistry also needs to be introduced. The studies of TD-RDMFT indicate that TD-DFT functionals may also act as variables of the occupation number and the time-dependent phase of the orbitals.

19.3. Semiconductor Compounds

According to their electronic properties, materials could be classified into five types (Fig.19.1): (i) p-/n-type metallic materials, which have a conduction/valence band across the Fermi level and thus are capable of producing free holes or electrons at low temperatures; (ii) semi-metals, in which the top of the conduction and bottom of the valence bands are overlapped and therefore making possible rich concentrations of electrons and holes surviving simultaneously; (iii) zero-gap semiconductors, in which the occupied and unoccupied states touch each other at the Fermi level and thus are without free carriers; (iv) semiconductors, in which the occupied and unoccupied

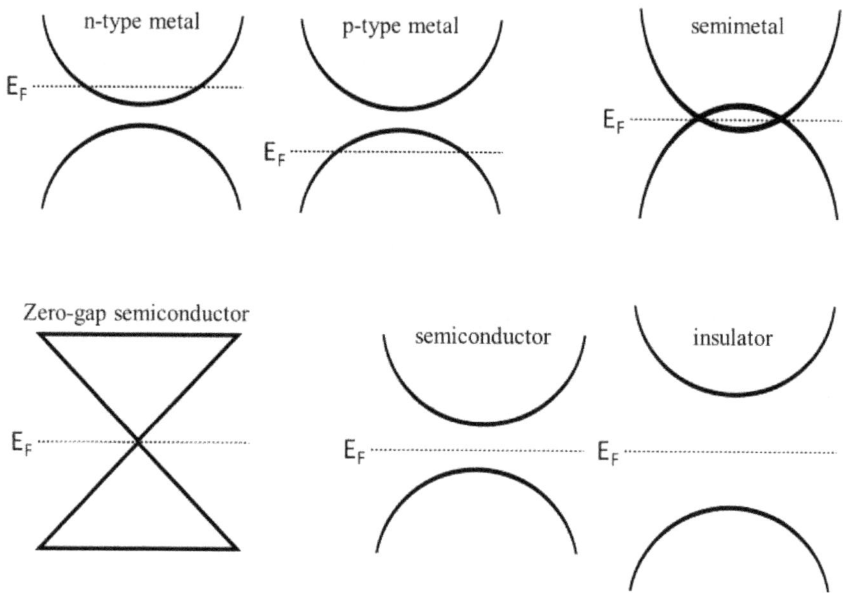

Fig. 19.1. The electronic band structures of n-/p-type metal, semi-metal, zero-gap semiconductor, semiconductor, and insulator.

states are separated by a gap; and finally, (v) large-gap semiconductors or insulators, which are materials that consist of a very large bandgap (> 4eV). Among them, semiconductor materials have attracted great attention in the past several decades owing to their exceptional physical properties and potential applications. Indeed, semiconductor compounds clearly reveal diverse physical, chemical, and material phenomena, such as good engineering in crystal phases, orbital mixings in spatial charge density distributions, bandgaps, optical gaps, bound states of stable excitons, prominent absorption peaks, and van Hove singularities across the Fermi levels.

Initially, the primary material in semiconductor products was germanium. Later, silicon emerged with superior characteristics, e.g., a sizable electronic bandgap and cheaper and more stable than germanium. Such kinds of materials have become the main substances for the semiconductor industry until now. Silicon, germanium, and other semiconductors which are made from only one element are so-called simple semiconductor compounds. Numerous studies have been done to look beyond simple semiconductors toward new kinds of materials to satisfy the increasing technological demands. Among them, compound semiconductors, which are made of at least two elements, have been receiving considerable attention due to their physical property values greatly exceeding those of silicon-based semiconductors.

In general, semiconductor compounds consist of rich composites of Group-II, -III, -V, -V, and -VI elements. The four active orbitals of the outermost s, p_x, p_y and p_z take part in diverse covalent bondings and thus greatly diversify the composite quasi-particles, as systematically shown for III–VI semiconductor compounds. In general, semiconductor compounds could survive in 3D, 2D, 1D, and 0D forms due to the presence of sp3–sp3 hybridizations. Delicate first-principles calculations and experimental examinations are needed to check whether the diamond structure, other cubic symmetries, or graphite-like layered structures can survive in the 3D crystals of such compounds. For example, binary semiconductor compounds of gallium, oxygen, and selenium systems are examined to present the non-diamond cubic symmetry. layered graphitic stackings, and buckled 2D honeycomb lattices. There also exits the weak but very significant interlayer orbital hybridizations in determining the 3D stacking configurations and dimensionality crossing behaviors. More interestingly, the modulation strategy could be easily achieved through obvious changes in the different atoms, relative concentrations (ternary/quaternary/pentanary compounds), stacking configurations, and layer numbers. In addition, the chemical modifications and physical perturbations, as discussed earlier, are able to create more composite quasi-particle phenomena and thus expand their framework.

19.4. Inter-Metallic Compounds

The ternary intermetallic compounds of $RE_2T_3X_5$, where RE = rare-earth elements, T = transition metals, and X = Ga, Si, Ge, or Sn, present rich features and unique fundamental material, chemical, and physical properties. Examples include the anomalous magnetoresistances, the diverse stable geometric phases, the meta-magnetic transitions or the antiferromagnetic/ferromagnetic spin configurations, the unusual superconductivity behaviors in the transition-temperature range of $T_c \sim 1-10$ K, the oscillatory valence and conduction energy spectra along the high-symmetry points with significant band overlaps, and the charge density waves.[62–70] The well-developed first-principles calculations are capable of clearly exhibiting the quite complicated essential properties due to the large number of atoms present in a primitive unit cell of such materials, the critical multi-orbital hybridizations of the different chemical/metal bonds, and the RE/T atom-induced spin configurations.[69–71]

Among the different kinds of inter-metallic compounds, the $U_2Co_3Si_5$ type[72–75] includes the meta-stable compounds of the tetragonal $Sc_2Fe_3Si_5$ type[74–77] and the monoclinic $Lu_2Co_3Si_5$ type, as well as very few cases in the orthorhombic $Y_2Rh_3Sn_5$ and $Yb_2Pt_3Sn_5$ types.[77–80] Specifically, their geometric structures possess a large unit cell with many atoms, where the various kinds of chemical bondings can provide non-uniform chemical environments, resulting in high numerical barriers to identifying the concise multi-/single-orbital hybridizations in various chemical/metallic bonds and the spin-dependent ferromagnetism or antiferromagnetism through the density functional theory. The well-developed theoretical framework, which is presented in the current book, would be useful for fully exploring the diversified geometric, electronic, and magnetic properties of the intermetallic $RE_2T_3X_5$ compounds.

For all the ternary intermetallic compounds, the critical mechanisms could be thoroughly examined from their optimal lattice geometrics, atom- and spin-dominated energy bands, spatial charge bonding schemes, spatial spin densities, magnetic moments, and atom-, orbital- and spin-projected van Hove singularities, in the current and future investigations. Interestingly, the specific chemical bonds and orbital hybridizations will be proposed to determine the valence and conduction band overlaps across the Fermi level. Specifically, the seven/five/three/one categories of 4f/3d/3p/3s orbitals in rare-earth/transition-metal/Si/Si atoms might pose high barriers to numerical calculations under the projections of Bloch wave functions. Moreover, it might be almost impossible to characterize the low-energy electronic structure using the tight-binding model with a high-dimensional Hamiltonian matrix.

19.5. Ion Transports

It is well-known that secondary batteries have become increasingly important for industrial products, particularly electronic devices and electric vehicles. The combination of the cathode, anode, and electrolyte materials might contribute to enhancing the efficiency of ion-based batteries.[81–84] During the charging and discharging processes, the existence of intermediate crystal structures and the ionic or electric states require systematic research. The crucial mechanism of charge flows will determine the ion transports and electrons moving into and out of electric circuits. For instance, in lithium-ion batteries, the cathode is typically made from chemical compounds, with recent examples being lithium-cobalt oxide ($LiCoO_2$)[85] and lithium iron phosphate ($LiFePO_4$),[86,87] while the anode component is generally made from carbon (graphite), and the electrolyte varies from liquid, solid, and polymer materials. In the charging process, lithium ions migrate from the cathode to graphite (anode) through the electrolyte. Due to the chemical interactions, lithium ions will tend to combine with other ions and form neutral atoms. It means that at the negative electrode, ion transport might play a minor role in forming chemical and physical properties. On the other hand, building a comprehensive theoretical framework to understand the vital mechanism of ion transport might become a crucial problem in basic and applied sciences. To date, physical numerical analyses and X-ray measurements are available to identify the crystal structures. However, they are not sufficient to investigate the whole main characteristics because their concerns are about the measurements of higher lattice structures in a few stable or metastable configurations. The dramatic transformations of various intermediate structures based on ion migrations might become a fascinating topic to classify. The basic principles of quantum mechanics show the importance of combining diverse physical pictures to fully investigate condensed-matter systems. For example, the Heisenberg picture and Kubo formulas of perturbation theory can be used to understand many-body scattering. Significant scatterings between charged ions might determine the diverse charge distributions, which play a critical role in evaluating the ion transports of various secondary batteries. These issues might be interesting topics for near-future studies, which require many endeavors from theoretical and experimental researchers.

19.6. Solar Cells

To date, solar cell systems are useful to study in basic sciences of properties and applications.[88–93] Indeed, their environment friendliness, high power conversion efficiency,[94,95] and special process of converting the energy of light into electrical energy have been investigated successfully.[96–99] In addition to the above merits of solar cells, they have some drawbacks, such as seasonal energy availability, a high upfront cost, installation requirements, and location sensitivity.[100]

Among the different kinds of solar cells, perovskite solar cells are one that consists of an absorber layer sandwiched between the electron- and hole-transport layers.[101,102] This core component efficiently absorbs the external electromagnetic (EM) waves or photons and leads to a number of electron–hole pairs due to vertical optical transitions. Apparently, the featured optical properties are strongly dependent on the main characteristics of the electronic energy spectra and wave functions, such as the threshold absorption frequency and its relation with bandgap, the prominent structures connected with the joint van Hove singularities, and the hole–electron pair (excitonic) interactions.[103]

How to establish a relationship among the featured chemical and physical properties is worthy of systematic investigation. Our frameworks provide similarly reliable verifications of different properties that require high-precision experimental measurements. For example, the X-ray diffraction (XRD) peaks[104,105] and the angle-resolved photoemission spectroscopy (ARPES)[106,107] measurements of the occupied energy spectrum can examine the crystal symmetry and the energy gap in the band structure, respectively. The scanning tunneling spectroscopy (STS) can confirm the van Hove singularities in the density of states (DOS). The reflectance, absorption, transmittance, and photoluminescence spectra could be tested using optical spectroscopies.[107,108]

Until now, the metal halide perovskites[109,110] with the general formula of ABX_3 (A = cation, B = metal cation, and X = halide anion) are interesting candidates and are expected to exhibit outstanding properties.[111–113] Such perovskite solar cell systems should be considered in the current development of the quasi-particle framework,[114,115] such as $CsGeX_3$ (X = Cl, Br, or I), $BiFeO_3$, and $BaTiO_3$.[116–122]

19.7. Hydrogen Energy

Rapid developments around the world have led to an increase in energy demand. For decades, fossil fuels were considered the most common source of energy. However, they are a non-renewable source of energy, and their burning also causes many problems for our planet, such as the greenhouse effect, acid rain, ozone depletion, and pollution of the atmosphere. The above disadvantages of fossil fuels have promoted the introduction of many other types of alternative energy sources, such as wind,[123–125] solar,[126–128] and hydrogen.[129–131] Among them, hydrogen energy systems are considered an effective solution to replace the current fossil fuel system because their combustion does not produce harmful gases and causes little or no environmental pollution. Besides, they are also a renewable energy source, which will create a permanent energy system that helps reduce replacement costs after a long time of use.

There are many different hydrogen energy production processes, such as water vapor conversion of natural gas, catalytic decomposition of natural gas, partial oxidation of heavy oil, coal gasification, and steam gasification of coal. These processes have all reached commercially exploitable levels. Besides, many processes are still in the research and development stage. Specifically, at present, most of the world's hydrogen energy is still obtained from the main natural gases, namely methane, which predominates. Hydrogen energy production is a catalyzed methane conversion. In addition, water electrolysis is also one of the most interesting hydrogen production processes today. The three main technologies in this electrolysis process include alkali, polymer film, and ceramic oxide electrolysis. Theoretically, photoelectrochemistry is seen as a viable method to directly convert solar energy into hydrogen as a storage source. In the fields of high-temperature thermochemistry and photochemistry, the use of solar quantum and thermal photons is a huge step forward in technology.

Compared with other types of energy, hydrogen is of interest because of its significant advantages. First, hydrogen is the easiest fuel to transport compared to other fuels, such as gasoline or alcohol. Second, unlike chemical fuels that are only converted into useful energy through combustion, hydrogen energy is converted by many different processes, which makes hydrogen the most versatile energy source. In addition, the efficiency of this type of energy is higher than that of fossil fuels (about 39%). In other words, this is the most energy-saving form of energy. In terms of safety, hydrogen is very safe because it limits the possibility of fire or explosion, like with gasoline and oil.

Besides the benefits of hydrogen energy, to develop a hydrogen economy, scientists need to make constant efforts to improve hydrogen storage technology. Currently, the most commonly used hydrogen storage methods are gaseous high pressure or liquid; however, these methods are no longer suitable for future use. Therefore, we must focus on the research and development of more advanced methods, such as the use of metal–organic framework hydrides and carbon materials. The studies performed on metal hydrides to improve their gaseous hydrogen storage properties and electron chemistry (in terms of kinetics, cyclic stability, pressure, and thermal response) depend on the composition and properties of the alloys.

Since its discovery in 2004,[132] graphene has been considered a good candidate for hydrogen storage applications because of its specific surface area. Hydrogen can interact with graphene surfaces by physical or chemical absorption. The physically absorbed hydrogen exhibits very fast kinetics due to the London force[133] but is thermodynamically not very stable because the binding energy with the graphene surface is small. On the other hand, chemically absorbed hydrogen can be stored very efficiently due to its strong interactions with graphene layers. In this case, we can design efficient storage materials, where the bottleneck is the reversibility of the adsorption process. In addition, using the scanning tunneling microscopy (STM)[134] technique has further extended the study of the adsorption and release of hydrogen atoms on monolayer graphene as a function of the curvature of the atomic layer. These measurements confirm that by controlling the local curvature of the graphene layers, we can create a promising material for hydrogen storage applications.

19.8. Group-Iv Nanotubes and Nanoribbons

Single-walled nanotubes comprising carbon, silicon, and germanium with non-uniform curved surfaces are formed under strong competition between atomic interactions and mechanical strains.

Their geometric structures exhibit various diversities, such as chiral and achiral systems. The achiral systems can be further categorized into two types: armchair nanotubes and zigzag nanotubes. There are obvious differences in the appearances of the two, which are observable from their top and side views. Both nanotubes show various electronic properties, including semi-metallic and semiconducting characteristics. Each optimal structure possesses a spiral profile after self-consistent relaxations. The total energy exhibits declining and fluctuating behaviors with the increase in chiral vectors and the size of tubs because of the decreased curvature variations. Armchair nanotubes exhibit lower total energy values per atom compared with zigzag nanotubes.

Armchair and zigzag systems have the significant orbital hybridizations related to the 2D layer materials, leading to unusual low-lying energy dispersions and bandgaps. Armchair CNTs show gapless behavior due to their symmetric geometry. Other kinds of armchair nanotubes, such as silicon and germanium, exhibit a semiconducting behavior. The gap values scale as $1/r^2$ with the increase in radii. The appearance of an energy gap in silicon and germanium nanotubes is due to the buckling structures. These buckling systems cause the tubular hollow structures to become unsymmetrical and further appear in the bandgap near the fermi level $E_F = 0$. Owing to the small size of silicon and germanium nanotubes (Si(5,5) and Ge(5,5)), the gap values do not follow the scaling law of $1/r^2$ because of the large curvature of the small-sized nanotubes.

Zigzag CNTs can be either metals or semiconductors, depending on the chiral vectors (m, n). If $m + n = 3q$, where q is an integer is, it is a metal; if $m + n \neq 3q$, it is a semiconductor. Other zigzag silicon and germanium nanotubes do not have a regular appearance like the zigzag carbon ones.

Tubular structures are one of the most fascinating materials. Since CNTs were first created experimentally in 1991, they have aroused great interest due to their specific electronic properties. Other tubular structures, such as silicon and germanium nanotubes, have also subsequently been fabricated successfully a few years later. A wide array of their features have been investigated, such as their mechanical, electrical, and optical properties and thermodynamics. However, there are still some interesting topics which can be further discussed. Adatoms on tubular structures can be classified into two types: adsorbed and substituted. The former focus on the interaction of adatoms and tubular structure atoms. The latter, combined with other atomic structures, can influence the covalent interaction and orbital hybridization. The application of electric field to tubular nanostructures is another interesting topic for investigation. The intensity of the electric field could induce the electrical properties of tubular nanostructures and further expand their applicability.

References

[1] Nielsen, O.; Martin, R. M. First-principles calculation of stress. *Physical Review Letters* **1983**, *50*, 697.

[2] Alfe, D.; Gillan, M. J. First-principles calculation of transport coefficients. *Physical Review Letters* **1998**, *81*, 5161.

[3] Foulkes, W. M. C.; Haydock, R. Tight-binding models and density-functional theory. *Physical Review B* **1989**, *39*, 12520.

[4] Lin, C.-Y.; Do, T.-N.; Huang, Y.-K.; Lin, M.-F. Optical properties of graphene in magnetic and electric fields. 2017.

[5] Wong, Y.-Y.; Chang, E. Y.; Yang, T.-H.; Chang, J.-R.; Ku, J.-T.; Hudait, M. K.; *et al.* The roles of threading dislocations on electrical properties of AlGaN/GaN heterostructure grown by MBE. *Journal of the Electrochemical Society.* **2010**, *157*, H746.

[6] Liu, C.-W.; Dai, J.-J.; Wu, S.-K.; Diep, N.-Q.; Huynh, S.-H.; Mai, T.-T.; *et al.* Substrate-induced strain in 2D layered GaSe materials grown by molecular beam epitaxy. *Scientific Reports.* **2020**, *10*, 1–8.

[7] Lin, C.-Y.; Ho, C.-H.; Wu, J.-Y.; Do, T.-N.; Shih, P.-H.; Lin, S.-Y.; *et al. Diverse Quantization Phenomena in Layered Materials.* CRC Press, 2019.

[8] Ho, Y.-H.; Chiu, Y.-H.; Lin, D.-H.; Chang, C.-P.; Lin, M.-F. Magneto-optical selection rules in bilayer Bernal graphene. *ACS Nano* **2010**, *4*, 1465–1472.

[9] Mieno, M.; Satoh, T. Study of BCN compounds prepared by the chemical vapor deposition with dimethylamineborane. *Journal of Materials Science* **2001**, *36*, 3925–3931.

[10] Lin, S.-Y.; Liu, H.-Y.; Nguyen, D. K.; Tran, N. T. T.; Pham, H. D.; Chang, S.-L.; *et al. Silicene-Based Layered Materials.* IOP Publishing Ltd. 2020.

[11] Xu, K.; Yin, L.; Huang, Y.; Shifa, T. A.; Chu, J.; Wang, F.; *et al.* Synthesis, properties and applications of 2d layered m iii x vi (m = ga, in; x = s, se, te) materials. *Nanoscale* **2016**, *8*, 16802–16818.

[12] Rideout, V. A review of the theory and technology for ohmic contacts to group III–V compound semiconductors. *Solid-State Electronics* **1975**, *18*, 541–550.

[13] Rajeshwar, K. Electrosynthesized thin films of group II–VI compound semiconductors, alloys and superstructures. *Advanced Materials* **1992**, *4*, 23–29.

[14] Sieberer, M.; Redinger, J.; Khmelevskyi, S.; Mohn, P. Ferromagnetism in tetrahedrally coordinated compounds of I/II-V elements: Ab initio calculations. *Physical Review B* **2006**, *73*, 024404.

[15] Corkill, J. L.; Cohen, M. L. Band gaps in some group-IV materials: A theoretical analysis. *Physical Review B* **1993**, *47*, 10304.

[16] Seriani, N. Ab initio thermodynamics of lithium oxides: From bulk phases to nanoparticles. *Nanotechnology* **2009**, *20*, 445703.

[17] Manthiram, A.; Fu, Y.; Su, Y.-S. Challenges and prospects of lithium–sulfur batteries. *Accounts of Chemical Research* **2013**, *46*, 1125–1134.

[18] Ghosh, S.; Banerjee, A. S.; Suryanarayana, P. Symmetry-adapted real-space density functional theory for cylindrical geometries: Application to large group-IV nanotubes. *Physical Review B* **2019**, *100*, 125143.

[19] Lu, Y.; Feng, Y. Band-gap engineering with hybrid graphane–graphene nanoribbons. The *Journal of Physical Chemistry C* **2009**, *113*, 20841–20844.

[20] Tran, N. T. T.; Lin, S.-Y.; Lin, C.-Y.; Lin, M.-F. *Geometric and Electronic Properties of Graphene-related Systems: Chemical Bonding Schemes*. CRC Press, 2017.

[21] Nikiforidis, G.; Van de Sanden, M.; Tsampas, M. N. High and intermediate temperature sodium–sulfur batteries for energy storage: Development, challenges and perspectives. *RSC Advances* **2019**, *9*, 5649–5673.

[22] Manthiram, A.; Chung, S. H.; Zu, C. Lithium–sulfur batteries: Progress and prospects. *Advanced Materials.* **2015**, *27*, 1980–2006.

[23] Nguyen, D. K.; Tran, N. T. T.; Chiu, Y.-H.; Lin, M.-F. Concentration-diversified magnetic and electronic properties of halogen-adsorbed silicene. *Scientific Reports* **2019**, *9*, 1–15.

[24] Pham, H. D.; Nguyen, T. D. H.; Vo, K. D.; Huynh, T. M. D.; Lin, M.-F. Rich essential properties of boron, carbon, and nitrogen substituted germanenes. *Applied Physics Express* **2020**, *13*, 085502.

[25] Li, W.-B.; Lin, S.-Y.; Tran, N. T. T.; Lin, M.-F.; Lin, K.-I. Essential geometric and electronic properties in stage-n graphite alkali-metal-intercalation compounds. *RSC Advances* **2020**, *10*, 23573–23581.

[26] Raja, M.; Marnadu, R.; Balaji, M.; Ravikumar, K.; Krishna, V. G.; Kumar, M.; *et al.* Fabrication and characterization of novel Ga-doped WO3 films and n-Ga@ WO3/p-Si junction diode for optoelectronic device applications. *Inorganic Chemistry Communications* **2022**, *139*, 109291.

[27] Terrones, M.; Benito, A.; Manteca-Diego, C.; Hsu, W.; Osman, O.; Hare, J.; *et al.* Pyrolytically grown BxCyNz nanomaterials: Nanofibres and nanotubes. *Chemical Physics Letters* **1996**, *257*, 576–582.

[28] Blase, X. Properties of composite BxCyNz nanotubes and related heterojunctions. *Computational Materials Science* **2000**, *17*, 107–114.

[29] Wong, J.-H.; Wu, B.-R.; Lin, M.-F.Strain effect on the electronic properties of single layer and bilayer graphene. *The Journal of Physical Chemistry C* **2012**, *116*, 8271–8277.

[30] Tran, N. T. T.; Lin, S.-Y.; Glukhova, O. E.; Lin, M.-F. Configuration-induced rich electronic properties of bilayer graphene. *The Journal of Physical Chemistry C* **2015**, *119*, 10623–10630.

[31] Sing Li, T.; Hsuan Lee, C.; Fa Lin, M. Transport properties of double-walled carbon nanotubes in a transverse electric field. *Journal of the Physical Society of Japan* **2007**, *76*, 104706.

[32] Shyu, F.-L.; Lin, M.-F. π-electronic excitations in multiwalled carbon nanotubes. *Journal of the Physical Society of Japan* **1999**, *68*, 3806–3809.

[33] Hu, Z.; Ding, Y.; Hu, X.; Zhou, W.; Yu, X.; Zhang, S. Recent progress in 2D group IV–IV monochalcogenides: Synthesis, properties and applications. *Nanotechnology* **2019**, *30*, 252001.

[34] Sutter, P.; Cortes, R.; Lahiri, J.; Sutter, E. Interface formation in monolayer graphene-boron nitride heterostructures. *Nano Letters* **2012**, *12*, 4869–4874.

[35] Fang, C.; Chen, Y.; Kee, H.-Y.; Fu, L. Topological nodal line semimetals with and without spin-orbital coupling. *Physical Review B* **2015**, *92*, 081201.

[36] Lin, C.-Y.; Wu, J.-Y.; Chiu, C.-W.; Lin, M.-F. *Coulomb Excitations and Decays in Graphene-related Systems.* CRC Press, 2019.

[37] Han, N. T.; Dien, V. K.; Lin, M.-F. Excitonic effects in the optical spectra of Li2SiO3 compound. *Scientific Reports* **2021**, *11*, 1–10.

[38] Dien, V. K.; Pham, H. D.; Tran, N. T. T.; Han, N. T.; Huynh, T. M. D.; Nguyen, T. D. H.; *et al.* Orbital-hybridization-created optical excitations in Li2GeO3. *Scientific Reports* **2021**, *11*, 1–10.

[39] Dien, V. K.; Han, N. T.; Su, W.-P.; Lin, M.-F. Spin-dependent optical excitations in LiFeO2. *ACS Omega* **2021**, *6*, 25664–25671.

[40] Timp, G.; Chang, A.; Cunningham, J.; Chang, T.; Mankiewich, P.; Behringer, R.; *et al.* Observation of the Aharonov-Bohm effect for ω c $\tau > 1$. *Physical Review Letters* **1987**, *58*, 2814.

[41] Huefner, M.; Molitor, F.; Jacobsen, A.; Pioda, A.; Stampfer, C.; Ensslin, K.; *et al.* The Aharonov–Bohm effect in a side-gated graphene ring. *New Journal of Physics* **2010**, *12*, 043054.

[42] Ramasubramaniam, A.; Naveh, D.; Towe, E. Tunable band gaps in bilayer graphene– BN heterostructures. *Nano Letters* **2011**, *11*, 1070–1075.

[43] Nguyen, T. D. H.; Lin, S.-Y.; Chung, H.-C.; Tran, N. T. T.; Lin, M.-F. First-principles calculations for cathode, electrolyte and anode battery materials.

[44] Nielsen, E.; Rahman, R.; Muller, R. P. A many-electron tight binding method for the analysis of quantum dot systems. *Journal of Applied Physics* **2012**, *112*, 114304.

[45] Bacon, M.; Bradley, S. J.; Nann, T. Graphene quantum dots. *Particle & Particle Systems Characterization* **2014**, *31*, 415–428.

[46] Kortshagen, U.; Anthony, R.; Gresback, R.; Holman, Z.; Ligman, R.; Liu, C.-Y.; *et al.* Plasma synthesis of group IV quantum dots for luminescence and photovoltaic applications. *Pure and Applied Chemistry* **2008**, *80*, 1901–1908.

[47] Zhang, Y.; Yan, L.; Xie, S.; Pang, S.; Gao, H.-J. Formation of ordered Ge quantum dots on the Si (111)-(7× 7) surface. *Applied Physics Letters* **2001**, *79*, 3317–3319.

[48] Moutanabbir, O. Group-IV epitaxial quantum dots: Growth subtleties unveiled through stable isotope nanoengineering. *Science of Advanced Materials* **2011**, *3*, 312–321.

[49] Akinwande, D.; Brennan, C. J.; Bunch, J. S.; Egberts, P.; Felts, J. R.; Gao, H.; *et al.* A review on mechanics and mechanical properties of 2D materials — Graphene and beyond. *Extreme Mechanics Letters* **2017**, *13*, 42–77.

[50] Campbell, E.; Hasan, M. T.; Gonzalez Rodriguez, R.; Akkaraju, G. R.; Naumov, A. V. Doped graphene quantum dots for intracellular multicolor imaging and cancer detection. *ACS Biomaterials Science & Engineering* **2019**, *5*, 4671–4682.

[51] Kohn, W.; Sham, L. J. Self-consistent equations including exchange and correlation effects. *Physical Review* **1965**, *140*, A1133.

[52] Runge, E.; Gross, E. K. Density-functional theory for time-dependent systems. *Physical Review Letters* **1984**, *52*, 997.

[53] Maitra, N. T. Perspective: Fundamental aspects of time-dependent density functional theory. *The Journal of Chemical Physics* **2016**, *144*, 220901.

[54] Pemmaraju, C. Simulation of attosecond transient soft x-ray absorption in solids using generalized Kohn–Sham real-time time-dependent density functional theory. *New Journal of Physics* **2020**, *22*, 083063.

[55] Hong, X.; Hu, M.; Wang, F. The plasmon excitations in small lithium clusters: A time-dependent density functional theory study. *Chemical Physics Letters* **2020**, *745*, 137235.

[56] Whitfield, J. D.; Yung, M.; Tempel, D. G.; Boixo, S.; Aspuru-Guzik, A. Computational complexity of time-dependent density functional theory. *New Journal of Physics* **2014**, *16*, 083035.

[57] Lai, L.; Lu, J.; Wang, L.; Luo, G.; Zhou, J.; Qin, R.; *et al.* Magnetism in carbon nanoscrolls: Quasi-half-metals and half-metals in pristine hydrocarbons. *Nano Research* **2009**, *2*, 844–850.

[58] Wachsmuth, P.; Hambach, R.; Kinyanjui, M. K.; Guzzo, M.; Benner, G.; Kaiser, U. High-energy collective electronic excitations in free-standing single-layer graphene. *Physical Review B* **2013**, *88*, 075433.

[59] Zhang, Y.; Chen, S.; Chen, G. First-principles time-dependent quantum transport theory. *Physical Review B* **2013**, *87*, 085110.

[60] Gulans, A.; Kontur, S.; Meisenbichler, C.; Nabok, D.; Pavone, P.; Rigamonti, S.; *et al.* Exciting: A full-potential all-electron package implementing density-functional theory and many-body perturbation theory. *Journal of Physics: Condensed Matter* **2014**, *26*, 363202.

[61] Botti, S.; Schindlmayr, A.; Del Sole, R.; Reining, L. Time-dependent density-functional theory for extended systems. *Reports on Progress in Physics* **2007**, *70*, 357.

[62] Kojima, T.; Kameoka, S.; Tsai, A.-P. Screening of ternary intermetallic catalysts is possible using metallurgical synthesis: Demonstration on Heusler alloys. *KONA Powder and Particle Journal* **2021**, 2021008.

[63] Cheng, K.; Sun, J.; Xu, H.; Wang, J.; Zhan, C.; Ghomashchi, R.; *et al.* Diffusion growth of ϕ ternary intermetallic compound in the Mg-Al-Zn alloy system: In-situ observation and modeling. *Journal of Materials Science & Technology* **2021**, *60*, 222–229.

[64] Lin, X.; Zhang, W.; Mao, Z.; Jian, X.; Wu, P. Exploring the structural, mechanical, thermodynamic, and electronic properties of

(Ni0. 66, Zn0. 33) 3Sn4 ternary intermetallic compounds by the first-principles study. *Journal of Materials Research* **2020**, *35*, 263–271.

[65] Maitra, T.; Gupta, S. Intermetallic compound formation in Fe–Al–Si ternary system: Part II. *Materials Characterization* **2002**, *49*, 293–311.

[66] Leonard, B. M.; Bhuvanesh, N. S.; Schaak, R. E. Low-temperature polyol synthesis of AuCuSn2 and AuNiSn2: Using solution chemistry to access ternary intermetallic compounds as nanocrystals. *Journal of the American Chemical Society* **2005**, *127*, 7326–7327.

[67] Öğüt, S.; Rabe, K. M. Band gap and stability in the ternary intermetallic compounds NiSnM (M= Ti, Zr, Hf): A first-principles study. *Physical Review B* **1995**, *51*, 10443.

[68] Fürtauer, S.; Flandorfer, H. Formation enthalpies of various ternary intermetallic compounds in the Cu-Li-Sn system. *Journal of Alloys and Compounds* **2019**, *781*, 1110–1118.

[69] Zhou, L.-H.; Li, Z.; Wang, S.-S.; Hu, R.-M.; Wang, S.-H.; Qin, Z.-W.; *et al.* Calculation of phase equilibria in Al-Fe-Mn ternary system involving three new ternary intermetallic compounds. *Advances in Manufacturing* **2018**, *6*, 247–257.

[70] Evans, M. J.; Wu, Y.; Kranak, V. F.; Newman, N.; Reller, A.; Garcia-Garcia, F. J.; *et al.* Structural properties and superconductivity in the ternary intermetallic compounds M A B (M= Ca, Sr, Ba; A= Al, Ga, In; B= Si, Ge, Sn). *Physical Review B* **2009**, *80*, 064514.

[71] Dshemuchadse, J.; Steurer, W. More statistics on intermetallic compounds–ternary phases. *Acta Crystallographica Section A: Foundations and Advances* **2015**, *71,* 335–345.

[72] Morozkin, A.; Seropegin, Y. D. Sm–Ru–Ge system at 1070 K. *Journal of Alloys and Compounds* **2004**, *365*, 168–172.

[73] Freccero, R.; Choi, S.; Solokha, P.; De Negri, S.; Takeuchi, T.; Hirai, S.; *et al.* Synthesis, crystal structure and physical properties of Yb2Pd3Ge5. *Journal of Alloys and Compounds* **2019**, *783*, 601–607.

[74] Kuo, C.; Hsu, C.; Tseng, C.; Chen, W.; Lin, S.; Liu, W.; *et al.* Charge density wave like behavior with magnetic ordering in orthorhombic S m 2 R u 3 G e 5. *Physical Review B* **2020**, *101*, 155140.

[75] Chabot, B.; Parthe, E. Ce2Co3Si5 and R2Ni3Si5 (R≡ Ce, Dy, Y) with the orthorhombic U2Co3Si5-type structure and the structural relationship with the tetragonaL Sc2Fe3Si5-type structure. *Journal of the Less Common Metals* **1984**, *97*, 285–290.

[76] Chabot, B.; Parthé, E. Dy2Co3Si5, Lu2Co3Si5, Y2Co3Si5 and Sc2Co3Si5 with a monoclinic structural deformation variant of the orthorhombic U2Co3Si5 structure type. *Journal of the Less Common Metals* **1985**, *106*, 53–59.

[77] Bugaris, D. E.; Malliakas, C. D.; Han, F.; Calta, N. P.; Sturza, M.; Krogstad, M. J.; *et al.* Charge density wave in the new polymorphs of RE 2Ru3Ge5 (RE= Pr, Sm, Dy). *Journal of the American Chemical Society* **2017**, *139*, 4130–4143.

[78] Bugaris, D. E.; Malliakas, C. D.; Bud'ko, S. L.; Calta, N. P.; Chung, D. Y.; Kanatzidis, M. G. Flux crystal growth of the RE 2Ru3Ge5 (RE= La, Ce, Nd, Gd, Tb) series and their magnetic and metamagnetic transitions. *Inorganic Chemistry* **2017**, *56*, 14584–14595.

[79] Galadzhun, Y. V.; Hoffmann, R.-D.; Pöttgen, R.; Adam, M. Complex three-dimensional [Au3In5] polyanions in Ln2Au3In5 (Ln= Ce, Pr, Nd, Sm). *Journal of Solid State Chemistry* **1999**, *148*, 425–432.

[80] Thomas, E. L.; Kim, M. S.; Sokolov, D.; Bennett, M. C.; Aronson, M.; Chan, J. Y. Structure, and magnetic and transport behavior of twinned Ce2Rh3 (Pb, Bi) 5. *Journal of Solid State Chemistry* **2007**, *180*, 2356–2362.

[81] Nitta, N.; Wu, F.; Lee, J. T.; Yushin, G. Li-ion battery materials: Present and future. *Materials Today* **2015**, *18*, 252–264.

[82] Aurbach, D.; Suresh, G. S.; Levi, E.; Mitelman, A.; Mizrahi, O.; Chusid, O.; *et al.* Progress in rechargeable magnesium battery technology. *Advanced Materials* **2007**, *19*4260–4267.

[83] Ambroz, F.; Macdonald, T. J.; Nann, T. Trends in aluminium-based intercalation batteries. *Advanced Energy Materials* **2017**, *7*, 1602093.

[84] Slater, M. D.; Kim, D.; Lee, E.; Johnson, C. S. Sodium-ion batteries. *Advanced Functional Materials* **2013**, *23*, 947–958.

[85] Donders, M.; Arnoldbik, W.; Knoops, H.; Kessels, W.; Notten, P. Atomic layer deposition of LiCoO2 thin-film electrodes for all-solid-state Li-ion micro-batteries. *Journal of The Electrochemical Society* **2013**, *160*, A3066.

[86] Ouyang, X.; Lei, M.; Shi, S.; Luo, C.; Liu, D.; Jiang, D.; *et al.* First-principles studies on surface electronic structure and stability of LiFePO4. *Journal of Alloys and Compounds* **2009**, *476*, 462–465.

[87] Wang, L.; He, X.; Sun, W.; Wang, J.; Li, Y.; Fan, S. Crystal orientation tuning of LiFePO4 nanoplates for high rate lithium battery cathode materials. *Nano Letters* **2012**, *12*, 5632–5636.

[88] Khare, S.; Dell'Amico, M.; Knight, C.; McGarry, S. Selection of materials for high temperature latent heat energy storage. Solar *Energy Materials and Solar Cells* **2012**, *107*, 20–27.

[89] Solomin, E.; Kaixuan, G.; Yanming, J.; Minghan, D. Off-shore PV and wind based energy generation. In *2019 International Conference on Industrial Engineering, Applications and Manufacturing (ICIEAM)*, 2019, pp. 1–6.

[90] Birnbaum, J.; Eck, M.; Fichtner, M.; Hirsch, T.; Lehmann, D.; Zimmermann, G. A direct steam generation solar power plant with integrated thermal storage. **2010**.

[91] Sharma, S.; Jain, K. K.; Sharma, A. Solar cells: In research and applications — A review. *Materials Sciences and Applications* **2015**, *6*, 1145.

[92] Hou, W.; Xiao, Y.; Han, G.; Lin, J.-Y. The applications of polymers in solar cells: A review. *Polymers* **2019**, *11*, 143.

[93] Day, J.; Senthilarasu, S.; Mallick, T. K. Improving spectral modification for applications in solar cells: A review. *Renewable Energy* **2019**, *132*, 186–205.

[94] Zhao, J.; Li, Y.; Yang, G.; Jiang, K.; Lin, H.; Ade, H.; *et al.* Efficient organic solar cells processed from hydrocarbon solvents. *Nature Energy* **2016**, *1*, 1–7.

[95] Hao, F.; Stoumpos, C. C.; Cao, D. H.; Chang, R. P.; Kanatzidis, M. G. Lead-free solid-state organic–inorganic halide perovskite solar cells. *Nature Photonics* **2014**, *8*, 489–494.

[96] Bagher, A. M.; Vahid, M. M. A.; Mohsen, M. Types of solar cells and application. *American Journal of optics and Photonics* **2015**, *3*, 94–113.

[97] Pearson, G. Conversion of solar to electrical energy. *American Journal of Physics* **1957**, *25*, 591–598.

[98] Würfel, P. Photovoltaic principles and organic solar cells. *CHIMIA International Journal for Chemistry* **2007**, *61*, 770–774.

[99] Fahrenbruch, A.; Bube, R. *Fundamentals of Solar Cells: Photovoltaic Solar Energy Conversion*. Elsevier, 2012.

[100] Ranabhat, K.; Patrikeev, L.; Antal'evna-Revina, A.; Andrianov, K.; Lapshinsky, V.; Sofronova, E. An introduction to solar cell technology. *Journal of Applied Engineering Science* **2016**, *14*, 481–491.

[101] Arora, N.; Dar, M. I.; Hinderhofer, A.; Pellet, N.; Schreiber, F.; Zakeeruddin, S. M.; *et al.* Perovskite solar cells with CuSCN hole extraction layers yield stabilized efficiencies greater than 20%. *Science* **2017**, *358*, 768–771.

[102] Muduli, S. K.; Varrla, E.; Kulkarni, S. A.; Han, G.; Thirumal, K.; Lev, O.; *et al.* 2D black phosphorous nanosheets as a hole transporting material in perovskite solar cells. *Journal of Power Sources* **2017**, *371*, 156–161.

[103] Albrecht, S.; Reining, L.; Del Sole, R.; Onida, G. Ab initio calculation of excitonic effects in the optical spectra of semiconductors. *Physical Review Letters* **1998**, *80*, 4510.

[104] Warren, B. E. *X-ray Diffraction*. Courier Corporation, 1990.

[105] Whittig, L.; Allardice, W. X-ray diffraction techniques. *Methods of Soil Analysis: Part 1 Physical and Mineralogical Methods* **1986**, *5*, 331–362.

[106] Palczewski, A. D. *Angle-resolved Photoemission Spectroscopy (ARPES) Studies of Cuprate Superconductors.* Iowa State University, 2010.

[107] Lv, B.; Qian, T.; Ding, H. Angle-resolved photoemission spectroscopy and its application to topological materials. *Nature Reviews Physics* **2019**, *1*, 609–626.

[108] French, R. H.; Müllejans, H.; Jones, D. J. Optical properties of aluminum oxide: Determined from vacuum ultraviolet and electron energy-loss spectroscopies. *Journal of the American Ceramic Society* **1998**, *81*, 2549–2557.

[109] Zhang, W.; Eperon, G. E.; Snaith, H. J. Metal halide perovskites for energy applications. *Nature Energy* **2016**, *1*, 1–8.

[110] Baranowski, M.; Plochocka, P. Excitons in metal-halide perovskites. *Advanced Energy Materials* **2020**, *10*, 1903659.

[111] Han, N. T.; Dien, V. K.; Lin, M.-F. Excitonic effects in the optical spectra of Li 2 SiO 3 compound. *Scientific Reports* **2021**, *11*, 1–10.

[112] Han, N. T.; Dien, V. K.; Tran, N. T. T.; Nguyen, D. K.; Su, W.-P.; Lin, M.-F. First-principles studies of electronic properties in lithium metasilicate (Li 2 SiO 3). *RSC Advances* **2020**, *10*, 24721–24729.

[113] Dien, V. K.; Pham, H. D.; Tran, N. T. T.; Han, N. T.; Huynh, T. M. D.; Nguyen, T. D. H.; *et al.* Orbital-hybridization-created optical excitations in Li 2 GeO 3. *Scientific Reports* **2021**, *11*, 1–10.

[114] Van Neck, D.; Verdonck, S.; Bonny, G.; Ayers P.; Waroquier, M. Quasiparticle properties in a density-functional framework. *Physical Review A* **2006**, *74*, 042501.

[115] Bäppler, S. A.; Plasser, F.; Wormit, M.; Dreuw, A. Exciton analysis of many-body wave functions: Bridging the gap between the quasi-particle and molecular orbital pictures. *Physical Review A* **2014**, *90*, 052521.

[116] Rahaman, M. Z.; Hossain, A. M. A. Effect of metal doping on the visible light absorption, electronic structure and mechanical properties of non-toxic metal halide CsGeCl 3. *RSC Advances* **2018**, *8*, 33010–33018.

[117] Schwarz, U.; Wagner, F.; Syassen, K.; Hillebrecht, H. Effect of pressure on the optical-absorption edges of CsGeBr 3 and CsGeCl 3. *Physical Review B* **1996**, *53*, 12545.

[118] Wang, Y.; Zhang, T.; Kan, M.; Zhao, Y. Bifunctional stabilization of all-inorganic α-CsPbI3 perovskite for 17% efficiency photovoltaics. *Journal of the American Chemical Society* **2018**, *140*, 12345–12348.

[119] Palewicz, A.; Sosnowska, I.; Przenioslo, R.; Hewat, A. BiFeO3 crystal structure at low temperatures. *Acta Physica Polonica-Series A General Physics* **2010**, *117*, 296.

[120] Acosta, M.; Novak, N.; Rojas, V.; Patel, S.; Vaish, R.; Koruza, J.; *et al.* BaTiO3-based piezoelectrics: Fundamentals, current status, and perspectives. *Applied Physics Reviews* **2017**, *4*, 041305.

[121] Choi, K. J.; Biegalski, M.; Li, Y.; Sharan, A.; Schubert, J.; Uecker, R.; *et al.* Enhancement of ferroelectricity in strained BaTiO3 thin films. *Science* **2004**, *306*, 1005–1009.

[122] Buttner, R.; Maslen, E. Structural parameters and electron difference density in BaTiO3. *Acta Crystallographica Section B: Structural Science* **1992**, *48*, 764–769.

[123] Şahin, A. D. Progress and recent trends in wind energy. *Progress in Energy and Combustion Science* **2004**, *30*, 501–543.

[124] Walker, J. F.; Jenkins, N. Wind energy technology. **1997**.

[125] Nelson, V. *Wind Energy: Renewable Energy and the Environment.* CRC press, 2009.

[126] Kabir, E.; Kumar, P.; Kumar, S.; Adelodun, A. A.; Kim, K.-H. Solar energy: Potential and future prospects. *Renewable and Sustainable Energy Reviews* **2018**, *82*, 894–900.

[127] Granqvist, C. G. Solar energy materials. *Advanced Materials* **2003**, *15*, 1789–1803.

[128] Gong, J.; Li, C.; Wasielewski, M. R. Advances in solar energy conversion. *Chemical Society Reviews* **2019**, *48*, 1862–1864.

[129] Edwards, P. P.; Kuznetsov, V.; David, W. I. Hydrogen energy. *Philosophical Transactions of the Royal Society A: Mathematical, Physical and Engineering Sciences* **2007**, *365*, 1043–1056.

[130] Rand, D. A. J.; Dell, R. M. d Hydrogen energy: Challenges and prospects. **2007**.

[131] Rosen, M. A.; Koohi-Fayegh, S. The prospects for hydrogen as an energy carrier: An overview of hydrogen energy and hydrogen energy systems. *Energy, Ecology and Environment* **2016**, *1*, 10–29.

[132] Mikhailov, S. *Physics and Applications of Graphene: Experiments.* BoD–Books on Demand, 2011.

[133] Pitzer, K. S. London force contributions to bond energies. *The Journal of Chemical Physics* **1955**, *23*, 1735–1735.

[134] Goler, S.; Coletti, C.; Tozzini, V.; Piazza, V.; Mashoff, T.; Beltram, F.; *et al.* Influence of graphene curvature on hydrogen adsorption: Toward hydrogen storage devices. *The Journal of Physical Chemistry C* **2013**, *117*, 11506–11513.

CHAPTER 20

CONCLUDING REMARKS

Nguyen Thanh Tien*, Vo Khuong Dien†, Nguyen Thi Han†,
Nguyen Thi Dieu Hien† and Ming-Fa Lin†

**Department of Physics, College of Natural Sciences, Can Tho University, Can Tho, Vietnam*
†Department of Physics, National Cheng Kung University, Tainan, Taiwan

The current book is successful in expanding a grand quantum quasi-particle framework though delicate first-principles calculations and analyses of the chemical modifications of mainstream materials. The strong partnerships between phenomenological models and experimental observations are established in detail. All significant orbitals from the outermost (s, p_x, p_y, p_z), five types of 3d/4d/5d/6d and seven types of 4f/5f are thoroughly identified to dominate the multi-/single-orbital hybridizations and ferromagnetic/antiferromagnetic spin configurations under chemical adsorptions, intercalations, substitutions, decorations, and heterojunctions. The concise pictures of physics and chemistry directly reflect the featured quasi-particle properties: the position-dependent crystal symmetries, the atom- and spin-dominated band structures, the spatial charge/spin-density distributions, the atom- and spin-decomposed magnetic moments, the atom-, orbital-, and spin-decomposed densities of states, and the prominent absorption peaks associated with the photon–orbital couplings. The drastic changes in orbital probability distributions (the dramatic transformations between the spherical and highly non-uniform charge densities) and the merged van Hove singularities are very useful in verifying the critical mechanisms in chemical bonds. A number of potential applications and open issues are discussed in detail. The basic and applied sciences require close combinations, especially for functional chips with many semiconducting components. Similar studies could be generalized to other mainstream materials, e.g., binary semiconductor compounds and ternary lithium oxides/sulfides, solar cells, and hydrogen energies. The time-dependent intermediate crystal phases of cathode/electrolyte/anode materials in ion-based batteries need to be thoroughly investigated under the first-principles method and molecular dynamics. Theoretical models are urgently needed to explain stationary ion transports. Their near-future researches can greatly promote the quasi-particle development in a series of published books and be helpful for the establishment of an international quantum quasi-particle center.

The delicate first-principles calculations have successfully identified the concise multi-orbital hybridizations and spin configurations due to the 14 kinds of active orbitals. All the active bondings are as follows: (1) (2s, $2p_x$, $2p_y$) – (2s, $2p_x$, $2p_y$) and $2p_z$–$2p_z$ for a planar carbon-honeycomb crystal, (2) (3s, $3p_x$, $3p_y$, $3p_z$) – (3s, $3p_x$, $3p_y$, $3p_z$) for a buckled silicene/3D silicon, (3) $2p_z$–$2p_z$ interlayer for the weak but significant van der Waals interaction, (4) $2p_z$ – (d_{z^2}, $d_{x^2-y^2}$, d_{yz}, d_{zx}, d_{xy}) for the interlayer carbon–transition-metal interaction, (5) (d_{z^2}, $d_{x^2-y^2}$, d_{yz}, d_{zx}, d_{xy}) – (d_{z^2}, $d_{x^2-y^2}$, d_{yz}, d_{zx}, d_{xy}) for the intercalant–intercalant interaction, (6) $2p_z$ – (f_{x^3}, f_{xz^2}, f_{yz^2}, f_{xyz}, $f_{z(x^2-y^2)}$, $f_{x(x^2-y^2)}$, $f_{x(x^2-3y^2)}$) for the interlayer carbon–rare-earth-metal interaction, (7) (f_{x^3}, f_{xz^2}, f_{yz^2}, f_{xyz}, $f_{z(x^2-y^2)}$, $f_{x(x^2-y^2)}$, $f_{x(x^2-3y^2)}$) – (f_{x^3}, f_{xz^2}, f_{yz^2}, f_{xyz}, $f_{z(x^2-y^2)}$, $f_{x(x^2-y^2)}$, $f_{x(x^2-3y^2)}$) for two rare-earth-metal atomic interaction, (8) (2s, $2p_x$, $2p_y$, $2p_z$) – (d_{z^2}, $d_{x^2-y^2}$, d_{yz}, d_{zx}, d_{xy}) for the 2D dominating/3D carbon–transition-metal bonds, and (9) (2s, $2p_x$, $2p_y$, $2p_z$) – (f_{x^3}, f_{xz^2}, f_{yz^2}, f_{xyz}, $f_{z(x^2-y^2)}$, $f_{x(x^2-y^2)}$, $f_{x(x^2-3y^2)}$) for the 2D dominating/3D carbon–rare-earth orbital hybridization. The critical mechanisms, which are achieved under a unified quasi-particle framework, are responsible for the featured energy spectra, the orbital- and spin-enriched wave functions, the unusual van Hove singularities, the unique optical transitions with/without excitonic effects for the semiconducting/metallic cases, and the diverse quantum phonons. The low-lying valence and conduction energy subbands are useful in establishing the tight-binding/generalized tight-binding model. The orbital-hybridization-dependent hopping integrals and on-site ionization energies and the Hubbard-like spin-induced electron–electron Coulomb interactions play critical roles in the Hamiltonian. The first-principles predictions require high-resolution experimental examinations, such as X-ray/STM/TEM/RLLED, ARPES, STS, and reflectance/absorption/transmission/photoluminescence spectroscopies, respectively, corresponding to crystal structures, occupied energy spectra, energy-dependent densities of states, and optical properties. The close relations among the first-principles method, phenomenological models, and experimental observations are successfully established in this book.

Based on detailed calculation results, we conclude that the magnetic and electronic properties of graphene are enriched and diversified after Mn and Cr adsorption. After optimization, both Mn and Cr prefer to be adsorbed at the hollow sites rather than at the top or bridge sites. There exists a strong and complex hybridization between the orbitals of the transition metals and graphene, which can be clearly seen in the graph of the density of states and the charge and charge density difference. Besides, the band structure shows that the adsorption of Mn and Cr can change graphene from behaving as a semiconductor with zero bandgap to a metallic one. In addition, the separation of spin states in the band structure and density of states and the positive magnetic moment and spin distribution indicate that the adsorbed graphene has a ferromagnetic configuration.

The rare-earth metal La/Ce-adatom adsorptions on monolayer graphene described in Chapter 4 have clearly indicated the diverse modification phenomena through an enlarged quasi-particle framework. The composite quasi-particles, which are very complicated in the delicate VASP simulations, are thoroughly examined in the unique adsorption structures (the hollow-site positions without the buckled carbon honeycomb), the semiconductor–metal transitions under the stronger electron affinity of guest adatoms, the blueshifts of the Fermi level (the distinct Fermi surface), the spin-degenerate/spin-split hole and electron energy spectra with C and La/Ce dominances within the specific ranges, the vanishing/finite magnetic moments with the C-sp^3- and La-d^5/Ce-f^7-decomposed contributions (the zero/net spin polarizations), the highly non-uniform and anisotropic charge density distributions in the presence of many active orbitals of C-sp^3 and La-d^5/Ce-f^7 orbitals, the spin-up and spin-down spatial configurations under the Ce adsorption cases, the atom-, orbital-, and spin-enriched van Hove singularities (the merged structures associated with the orbital mixings through the same spin-up/spin-down configurations), and the optical transitions of the dielectric function, energy loss function, reflectance, and absorption coefficient (the special orbital hybridizations for each prominent transition from the initial and final states). These featured properties are consistent with one another, making them very useful in identifying the concise pictures of charge- and spin-dependent interactions. The well-defined Hubbard model and the generalized tight-binding model need to be urgently developed in near-future scientific researches for further verifications and developments of the critical mechanisms. The current chemical adsorptions are totally different from the alkalizations, hydrogenations, oxidizations, and halogenations since only La- and Ce-related graphene adsorptions, respectively, present the almost perpendicular bondings of $2p_z$–$2p_z$ and ($2s$, $2p_x$, $2p_y$) – ($2s$, $2p_x$, $2p_y$) (the almost perpendicular π and σ bondings in the absence of honeycomb buckling), $2p_z$ – (f_{xyz}, f_{yz^2}, f_{z^3}, f_{xz^2}, $f_{z(x^2-3y^2)}$, $f_{x(x^2-3y^2)}$, $f_{y(y^2-3x^2)}$) and (f_{xyz}, f_{yz^2}, f_{z^3}, f_{xz^2}, $f_{z(x^2-3y^2)}$, $f_{x(x^2-3y^2)}$, $f_{y(y^2-3x^2)}$) through the C–C and C–La/C–Ce bonds. More interestingly, the orbital- and spin-enriched composite quasi-particles are thoroughly illustrated by the seven kinds of f^7 orbitals and non-magnetic/ferromagnetic spin configurations in the La-/Ce-based materials. Similar studies are urgently needed in the current and near-future scientific researches, e.g., the extremely observable ranges of the magnetic moments and the blueshift Fermi level for all rare-earth and transition metals under various chemical modifications. The above-mentioned phenomena could be greatly diversified through chemical adsorptions, substitutions, intercalations/de-intercalations (charging/discharging processes in ion-based batteries), decorations on open edges, heterojunctions, and compositions.

Through the quasi-particles framework implemented in VASP calculations, the structural, electronic, and optical properties of the 4d transition metal-intercalated graphite were investigated, as discussed in Chapter 5. The structural features indicated the modification of Zr/Nb intercalation, resulting in non-uniform environments in the system and charge density. The electronic properties of GICs were also adjusted based on the effect of the intercalants, indicating metallic behavior, and the strong dominance of Zr and Nb in the conduction band and the energy band near the Fermi level. Importantly, the induced magnetism was present in the system, demonstrating the main contribution of these transition-metal atoms to magnetic behavior. The optical properties, as shown in absorption, reflectivity, and refraction, also displayed the effect of intercalants in the system compared to pristine graphite. In addition, the strong hybridization of s-, p- orbitals and s-, d- orbitals of C and Zr/Nb atoms, respectively, was significantly exhibited, corresponding to the hybridized C–C and C–Zr/C–Nb bondings. The results portray the similarities and distinctions with alkali intercalation into graphite, indicating the unique and interesting properties of transition-metal atoms in intercalated compounds.

The delicate VASP simulations and analyses are very useful in exploring the intercalation effects of graphite rare-earth metal compounds, especially for the magnetic seven-5f orbitals. A wider/broader/deeper quasi-particle framework is clearly revealed in Chapter 6 through the close superposition of complicated orbital hybridizations and Pa-/U-initiated ferromagnetic spin-density distributions. The concise mechanisms of chemistry and physics are thoroughly examined from the unified results: the stage-n periodic structure with the planar carbon-honeycomb lattices and intercalant layers, the featured band structures and wave functions due to the significant intercalant and zone-folding effects (many strong energy dispersions and more critical points in the energy–wave-vector space), the ferromagnetic metals in the presence of multi-Fermi surfaces, the sensitive atom dominances at different energy ranges, the subband non-anticrossings of π- and σ-electronic states, the intralayer σ/π and Pa–Pa/U–U bondings and the interlayer C–P/C–U and van der Walls interactions, the atom- and orbital-decomposed magnetic moment, the intercalant-based spatial spin-density distributions, the atom-, orbital-, and spin-projected density of states, and the spin-dependent dielectric functions and energy loss spectra. More interestingly, the very complicated multi-orbital hybridizations include (I) C–Pa/C–U about (2s, $2p_x$, $2p_y$, $2p_z$) $-$ ($5f_{x^3}$, $5f_{xz^2}$, $5f_{yz^2}$, $5f_{xyz}$, $5f_{z(x^2-y^2)}$, $5f_{x(x^2-y^2)}$, $5f_{x(x^2-3y^2)}$), (2) Pa–Pa/U–U for ($5f_{x^3}$, $5f_{xz^2}$, $5f_{yz^2}$, $5f_{xyz}$, $5f_{z(x^2-y^2)}$, $5f_{x(x^2-y^2)}$, $5f_{x(x^2-3y^2)}$) $-$ ($5f_{x^3}$, $5f_{xz^2}$, $5f_{yz^2}$, $5f_{xyz}$, $5f_{z(x^2-y^2)}$, $5f_{x(x^2-y^2)}$, $5f_{x(x^2-3y^2)}$), (3) the intralayer C–C (2s, $2p_x$, $2p_y$) $-$ (2s, $2p_x$, $2p_y$) and $2p_z$–$2p_z$, and (4) the interlayer 2pz–2pz hybridizations. All orbital hybridizations, which are fully supported by the spin-density distributions, are rather helpful for developing the tight-binding/generalized tight-binding model, in which the orbital-dependent hopping integrals and on-site ionization energies and the spin-related electron–electron Coulomb interactions (the Heisenberg-like ones) are required as the reliable parameters. The first-principles predictions on the optimal stage-n layer structures, the ferromagnetic valence and conduction subbands and spin-split multi-Fermi surfaces, and the spin-dependent dielectric functions/energy loss function/reflectance spectra/absorption coefficients in the absence of excitonic effects could be verified, respectively, using the X-ray diffractions/RLLED, the spin-polarized ARPES, and the spin-polarized optical reflectance/absorption/transmission spectroscopy. Similar studies could be generalized for the intercalation effects of layered materials, e.g., the binary/ternary semiconductor compounds of GaS/GaSe/GaTe and InS/InSe/InTe. These scientific studies can greatly enhance the composite quasi-particles, especially for effective optical excitations associated with the significant atom/orbital/spin dependences of the initial and final subband states. More interestingly, the Coulomb-field perturbations need to be thoroughly investigated in the current work, in which the time-dependent LDA is partially supported by the phenomenological models. The many-body electron–electron Coulomb interactions could be fully understood from the diverse (momentum, frequency)-excitation phase diagrams or the featured electron–hole pair excitations and plasmon modes.

Chapter 7 thoroughly investigated thc crucial properties of substitution systems. The four outer orbital hybridizations (2s, $2p_x$, $2p_y$, $2p_z$) of carbon and the six orbitals (6s, $5d_{x^2-y^2}$, $5d_{xy}$, $5d_{yz}$, $5d_{xz}$, $5d_{z^2}$) of transition metals play substantial roles in determining the rich properties of modification systems. The optimal stability structures show fascinating buckling structures, mainly originating from the complicated chemical bonds of transition metals with carbon and among the carbon atoms. Tungsten (W), osmium (Os), and platinum (Pt) substitution systems reveal semiconducting behavior, whose bandgaps are in the range of 0.22–0.40 eV, depending on the atoms and (4×4) or (5×5) supercells, while Iridium show a semi-metallic feature with a direct bandgap of 0.04 eV. The significant magnetic properties depict spin-split band structures, the spin configurations of the Heisenberg model, and diverse insensitive van Hove singularities with spin-up and spin-down configurations. Importantly, nine prominent peaks appear in the case of the W (1:31) case, resulting from the various transitions from the band-edge states and their strong combination with the joint densities of states.

The rare-earth element La/Gd-adatom substituted in monolayer graphene has clearly illustrated the diverse modification phenomena in its structural, electronic, and magnetic properties through an expanded quasi-particle framework, as further developed in a series of published books. The composite quasi-particles, which are very complicated in the delicate VASP simulations package, are thoroughly examined through a unique substitution in the carbon honeycomb of a graphene sheet. In addition, the spin degeneracy/spin split of hole and electron energy spectra with the C and La/Cd dominances within the specific ranges, the vanishing/finite magnetic moments of the C-sp^2- and La-d_5/Gd-f7-decomposed contributions (the zero/net spin polarizations), the highly non-uniform and anisotropic charge density distributions in the presence of many active orbitals of C-sp^2- and La-d$_5$-/Gd-f7-orbitals, the spin-up and spin-down partial configurations in the Gd-adsorption cases, the atom-, orbital-, and spin-enriched van Hove singularities, the merged structures associated with the orbital mixings through the same spin-up/spin-down configurations). These featured properties are consistent with one another, making them very useful in identifying the concise pictures of charge and spin-dependent interactions. The current chemical substitutions are totally different from the alkalizations, hydrogenations, oxidizations, and halogenations since only La- and Gd-related graphene substitutions, respectively, present the almost perpendicular bindings of 2p$_z$–2p$_z$ and (2s, 2p$_x$, 2p$_y$) − (2s, 2p$_x$, 2p$_y$) orbitals. Besides, 2p$_z$ − (4d$_{z^2}$, 4d$_{x^2-y^2}$, 4d$_{xy}$, 4d$_{yz}$, 4d$_{xz}$) − (4d$_{z^2}$, 4d$_{x^2-y^2}$, 4d$_{xy}$, 4d$_{yz}$, 4d$_{xz}$) and (4d$_{z^2}$, 4d$_{x^2-y^2}$, 4d$_{xy}$, 4d$_{yz}$, 4d$_{xz}$)/2p$_z$− (4f$_{x^3}$, 4f$_{xz^2}$, 4f$_{yz^2}$, 4f$_{xyz}$, 4f$_{z(x^2-y^2)}$, 4f$_{x(x^2-3y^2)}$, 4f$_{y(3x^2-y^2)}$) and (4f$_{x^3}$, 4f$_{xz^2}$, 4f$_{yz^2}$, 4f$_{xyz}$, 4f$_{z(x^2-y^2)}$, 4f$_{x(x^2-3y^2)}$, 4f$_{y(3x^2-y^2)}$) − (4f$_{x^3}$, 4f$_{xz^2}$, 4f$_{yz^2}$, 4f$_{xyz}$, 4f$_{z(x^2-y^2)}$, 4f$_{x(x^2-3y^2)}$, 4f$_{y(3x^2-y^2)}$) form through the interaction of C–C, C–La/C–Gd, and La–La/Gd–Gd bonds. On the other hand, the orbital- and spin-enriched composite quasi-particles are thoroughly illustrated by the five kinds of d^5 orbitals and seven kinds of f^7 orbitals and the non-magnetic/ferromagnetic spin configurations in the La/Gd-substituted graphene. As for experimental examinations, most of the current techniques are available for achieving simultaneous progress, such as high-precision measurement using STM/TEM, non-/spin-polarized ARPES, REELS, non-/spin-polarized STS, and optical transmission spectroscopy. Whether the phenomenological models are worthy of further development and need to be evaluated in detail because of the very complicated multi-hopping integrals, the on-site ionization potentials, and the spin-up and spin-down electron–electron Coulomb interactions.

The transition-metal atoms, which possess 5d orbitals, make great contributions to the compounds in low-lying energy, such as a number of many merged structures of the atom-, orbital-, and spin-projected van Hove singularities, the atom- and orbital-decomposed magnetic moments, the Ta-/W-induced spin arrangements, and the spin-enriched optical transitions in dielectric functions. The critical properties of the transition-metal-edged decorations arise mainly from the orbital hybridizations of host–guest, host–host, and guest–guest atoms. In addition, the bonding of C–C includes $2p_z$–$2p_z$. Moreover, according to the spatial spin distributions, the interaction of spin just appears between transition-metal atoms, irrespective of the system being an armchair or zigzag type. Based on the calculated results, the C-$2p_z$ and five 5d orbitals almost co-dominate the physical and chemical properties of the transition-metal-decorated graphene nanoribbons within the low-lying energy. In summary, the quasi-particle properties, including spin-split orbitals and optical properties, have been seldom studied in the past years. It is very difficult to fully and accurately investigate the transition-metal-based compounds due to the five complicated 5d orbitals: ($5d_{xy}$, $5d_{yz}$, $5d_{z^2}$, $5d_{xz}$, $5d_{x^2}$).

The specific heterojunctions, which frequently arise in composite compounds, are outstanding candidates for providing the diversified quasi-particle phenomena and thus developing a grander framework. For example, layered Group-IV systems as well as binary and ternary semiconductor compounds of III–VI/III–V/II–VI/II–V/IV–IV, have been successfully grown on various substrates. Furthermore, the composite cathode/electrolyte/anode materials clearly illustrate the stationary ion transport through real boundaries during the charging and discharging processes of ion-based batteries. Such mainstream materials, at least, present three different orbital-hybridization/spin-modulation regions because of the creation of significant host–guest chemical bonds within the complicated boundary. Generally, there exists a Moire superlattice with a number of active atoms, even in a commensurate crystal environment. This will lead to a giant barrier in the VASP simulations and delicate analyses. All calculated results, which are consistent with one another within an expanded framework, are available for achieving the critical mechanisms associated with charge- and spin-dependent interactions. The concise chemical bondings and the lattice-site-dependent spin-density distributions are mainly determined from the optimal crystal symmetries, the featured band structure and wave functions with strong atom and spin dependences, the atom- and orbital-projected magnetic moments, the spatial charge/spin-density modulation distribution along the top and side views, the merged structures of the atom-/orbital-/spin-decomposed van Hove singularities, and the orbital- and spin-enriched optical excitations in the presence/absence of excitonic effects. In Chapter 11, monolayer and bilayer graphene systems on Pt[111] are chosen for a model study. Their fundamental properties are expected to reveal dramatic transformations under the strong multi-orbital hybridizations and spin-created many-body interactions, e.g., the low-lying valence and conduction energy subbands due to Moire unit cells. It should be noted that a too thick substrate would become meaningless in the theoretical simulations since their electronic states cannot escape from the surface even under the high-resolution measurement of ARPES.

Chapter 11, which focused on the graphene–platinum composites, is very successful in expanding the quantum quasi-particle framework through unusual modulations of the spatial charge/spin-density distributions. The delicate first-principles simulations and phenomenological analyses can solve very complicated multi-orbital hybridizations. Such strategies are rather suitable for semiconductor–metal, metal–metal, semiconductor–semiconductor, and cathode/electrolyte/anode composite compounds. A number of unusual heterojunctions, which possess highly non-uniform chemical/low-symmetry physical environments, frequently come to exist in the successful sample syntheses on the various substrates and the combined materials of the commercial products. Their rich and unique quasi-particle phenomena are thoroughly explored using first-principles simulations, but not the phenomenological models. The concise physics/chemistry pictures are closely related to the featured quasi-particle behaviors, namely: (I) the highly active crystal environments through the intralayer/interlayer C–C and Pt–Pt bonds and C–Pt bonds at heterojunction, (II) the atom- and orbital-dominated electronic wave functions, (III) the unusual modulations of metallic bondings along the longitudinal and transverse directions, (V) many merged van Hove singularities from the multi-orbital hybridizations within five kinds of chemical bonds, and (VI) the prominent absorption structures in the imaginary-part dielectric functions associated with the specific orbitals about the initial and final states and the high-frequency plasmon peaks in the screened response functions. Specifically, the redshift of the Fermi level could be estimated from the minimum density of states close to or below E_F and the optical gap. The thorough analyses clearly present (1) (2s, $2p_x$, $2p_y$) – (2s, $2p_x$, $2p_y$) & $2p_z$–$2p_z$ perpendicular to each other for the C–C bonds on a planar carbon-honeycomb crystal, (2) the interlayer van der Waals interactions of $2p_z$–$2p_z$ mixings, (3) $2p_z$ – ($5d_{z^2}$, $5d_{x^2-y^2}$, $5d_{yz}$, $5d_{zx}$, $5d_{xy}$, 6s) for the carbon–transition-metal heterojunction, and ($5d_{z^2}$, $5d_{x^2-y^2}$, $5d_{yz}$, $5d_{zx}$, $5d_{xy}$, 6s) – ($5d_{z^2}$, $5d_{x^2-y^2}$, $5d_{yz}$, $5d_{zx}$, $5d_{xy}$, 6s) within the metallic substrate. More interestingly, the metallic bondings mainly come from the neighboring platinum atoms, reveal strong modulations near the two-subsystem boundary, and then exhibit a massive decline within the very strong covalent σ bondings of honeycomb lattices. The atom- and position-dependent spin-density distributions clearly illustrate the sensitive modulations of ferromagnetic configurations. The physical perturbations (detailed in Section 2.2; Refs) are expected to be very efficient in driving the quasi-particle response abilities. It should be noted that carbon and platinum atoms in the heterojunctions, respectively, display (2s, $2p_x$, $2p_y$, $2p_z$) and ($5d_{x^2-y^2}$, $5d_{yz}$, $5d_{zx}$, $5d_{xy}$, 6s). The delicate VASP studies are very reliable for analyzing the unusual hetero-junctions, including semiconductor–semiconductor/semiconductor–metal/metal–metal composites and the cathode/electrolyte/anode linked materials within the ion-based batteries. As to experimental observations, the high-resolution STM/TEM/RLEED/X-ray and optical spectroscopies can, respectively, verify the first-principles predictions on crystal structure, occupied quasi-particle energy spectra, van Hove singularities, and single-particle and collective excitations. However, the proper establishment of phenomenological models would become very difficult under the rich and unique orbital- and spin-dominated interactions, even with strong first-principles cooperation. This open issue would be presented in the near-future investigations. The well-characterized formulas of the atomic interactions are outstanding only for the C–C bonds, but not the C–Pt and Pt–Pt ones.

The fundamental properties of graphene are dramatically changed through the significant substrate effect, chemical bondings, zone-folding effects, enhanced bucklings, and stacking configurations. The sufficient substrate CeO_2 layers in the VASP calculations are utilized to simulate the stable geometric structures. These monolayer and bilayer compounds might exhibit similarities and differences in band structures, including newly merged critical points, the enhanced asymmetry of electron and hole states about the Fermi level, the absence of modified Dirac cone structures, the low-lying intersecting energy dispersions, the Ce-dominated weakly dispersion bands, the strongly anisotropic Fermi momentum states, and ferromagnetic spin configurations. The unique electronic and magnetic properties should be difficult to simulate through the tight-binding model due to the complicated orbital hopping integrals and atom-, orbital-, and spin-dependent on-site Coulomb potentials. The distinct substrate effects would greatly diversify the physical, chemical, or material phenomena in layered materials, e.g., the diverse phenomena among monolayer and few-layer silicene, germanene, or stanene on different substrate effects.

In Chapter 13, we have applied DFT calculations to systematically investigate the effect of the chemical modification on the structural stability and optoelectronic properties of penta-graphene quantum dots (PGQDs). We found that larger PGQDs are thermodynamically more stable than the smaller ones. For the same system sizes, in the order of increasing absolute values of formation energies, we can arrange the passivated PGQDs as follows: P-ZZ-36, F-ZZ-36, Si-ZZ-36, and O-ZZ-36. In single-doped PGQDs, a subband appears to coincide with the Fermi level. This result shows a transition from semiconductor to metal compared to the pristine structure. In contrast, B and P co-doped PGQDs exhibit semiconductor properties. Specifically, the bandgaps are 1.50 and 1.14 eV, respectively. Tailoring the optoelectronic properties of semiconductor QDs is essential to designing functionalized nanoscale devices.

Graphene quantum dots (GQDs), a zero-dimensional material with a size of less than 10 nm, is a newly developed member of the graphene family. GQDs have become highly promising materials in various applications, including catalysis, bio-imaging, energy conversion, and storage, owing to their unique physicochemical properties, such as excellent photoluminescence and biocompatibility, low toxicity, easy fabrication, and the edge effect. Additionally, oxygen-containing functional groups, such as -COOH, -COC-, -OH, -CHO, and -OCH_3, play a crucial role in the structure of GQDs. Oxygen is located at different edges and on the basal plane of GQDs. The unique architecture of oxygen inside the structure of GQDs and the outstanding properties and performance of GQDs provide a powerful impetus to use GQDs as a promising material with applicability in various fields. In Chapter 14, we describe the possible structure, application, and critical role of the oxygen-containing functional group on the essential properties and application of GQDs. We also provide a brief outlook to point out the issues that need to be resolved for further development.

In Chapter 15, we calculated the hydrogen adsorption energy of a pristine two-dimensional silicon carbide using the DFT approach. Only two adsorption sites have been found to be stable, both of which are on the top of the adsorbent atoms (C and Si atoms). The highest hydrogen adsorption energy in the consideration of zero-point energy was found in the TSi configuration. We conclude that the most favorable adsorption site is the TSi site. This agrees well with the vibrational energy calculation, where the hydrogen atom in the TSi configuration has the lowest vibrational energy. We observe that most of the hydrogen vibrational energy comes from the z-axis vibration in all the configurations. The local deformations of 2D SiC caused by H adsorption are also shown. In the Tc configuration, the H atom caused more distortion compared to TSi. The distortion of the adsorbent atom is greatest along the vertical axis for both configurations. Finally, we presented five defects that could be found in 2D silicon carbide and calculated their formation energies. Those defects are the linear (8-4) defect, the silicon interstitial defect, the divacancy (4-8-8-4) defect, the divacancy (8-4-4-8) defect, and the divacancy (4-10-4) defect. The linear (8-4) defect is shown to have the lowest formation energy and is the most expected to form. The results of this study can be an important basis for future hydrogen energy applications, with the defects showing many possibilities for further experiments to consider.

Through first-principles calculations, the electronic properties of p-SiC_2 nanoribbons (ZZ-ribbon, ZA-ribbon, AA-ribbon, and SS-ribbon) with various edge shapes are explored. Moreover, the SS-pSiC_2 energy gaps with various widths and the energetic stability of SS-pSiC_2-4 under a uniaxial strain of up to 10% are investigated in detail. The results show that the bandgap of SS-pSiC_2 decreases sharply as the width increases because the 1D confinements significantly affect the feature of the narrow widths. The electronic properties of SS-pSiC_2-4 are sensitive to uniaxial strain. In combination with the NEGF formalism, the computed results show that the value of the strain-effect current differs significantly at bias potentials from 1.5 to 2 V. Consequently, SS-pSiC_2 is a potential candidate for electronic devices development and an active part of strain sensors.

INDEX

www.ingramcontent.com/pod-product-compliance
Lightning Source LLC
Chambersburg PA
CBHW071247180926
51613CB00002B/158
* 9 7 8 9 8 1 1 2 6 7 9 3 2 *